Springer Theses

Recognizing Outstanding Ph.D. Research

Aims and Scope

The series "Springer Theses" brings together a selection of the very best Ph.D. theses from around the world and across the physical sciences. Each volume has been selected for its scientific excellence by an internationally renowned institute or department. For greater accessibility to non-specialists, the published versions include an extended introduction, as well as a foreword by the student's supervisor explaining the special relevance of the work for the field. The series as a whole provides a valuable resource both for newcomers to the research fields, and for other scientists seeking detailed background information on special questions. Finally, it serves as an accredited documentation of the valuable contributions made by today's younger generation of scientists.

How to Apply

Theses prepared at internationally leading universities and institutes may be nominated for publication by supervisors with the endorsement of the head of department. Each such department is entitled to nominate one thesis per year. Enquiries and requests for a nomination form should be sent, together with the thesis, to Ms. Nathalie Berg (see link on right).

Eligibility Criteria

To qualify for consideration, theses should fulfil the following criteria:

- They must be written in good English.
- The topic should fall within the confines of Chemistry, Physics, Earth Sciences, Engineering and related interdisciplinary fields such as Materials, Nanoscience, Chemical Engineering, Complex Systems and Biophysics.
- The work reported in the thesis must represent a significant scientific advance.
- A maximum of 30% of the text may be a verbatim reproduction from the author's previous publications.
- They must have been examined and passed during the 12 months prior to nomination.
- The theses should have a clearly defined structure including an introduction accessible to new PhD students and scientists not expert in the relevant field.
- After acceptance, theses should be supplemented by a "Supervisor's Foreword" outlining the wider scientific relevance of the work.

Indexed by zbMATH.

Jona Motta

Development of Machine Learning τ Trigger Algorithms and Search for Higgs Boson Pair Production

In the $bb\tau\tau$ Decay Channel with the CMS Detector at the LHC

Doctoral Thesis prepared at École Polytechnique, Palaiseau, France and accepted by Institut Polytechnique de Paris, Palaiseau, France

Springer

Author
Dr. Jona Motta
University of Zürich
Zürich, Switzerland

Supervisors
Dr. Alexandre Zabi
Director of Research, Laboratoire Leprince
Ringuet CNRS-IN2P3
École Polytechnique, Institut Polytechnique
de Paris
Palaiseau, France

Dr. Roberto Salerno
Director of Research, Laboratoire Leprince
Ringuet CNRS-IN2P3
École Polytechnique, Institut Polytechnique
de Paris
Palaiseau, France

ISSN 2190-5053 ISSN 2190-5061 (electronic)
Springer Theses
ISBN 978-3-031-96287-5 ISBN 978-3-031-96288-2 (eBook)
https://doi.org/10.1007/978-3-031-96288-2

This Springer imprint is published by the registered company Springer Nature Switzerland AG
The registered company address is: Gewerbestrasse 11, 6330 Cham, Switzerland

A Mamma e Papà

Tho' much is taken, much abides; and tho'
We are not now that strength which in old days
Moved earth and heaven, that which we are, we are;
One equal temper of heroic hearts,
Made weak by time and fate, but strong in will
To strive, to seek, to find, and not to yield.

— *Alfred Tennyson*

Supervisors' Foreword

Identifying the fundamental laws of Nature represents the core of modern particle physics research. A precise modeling of nature's fundamental interactions is the key to understanding the evolution of our Universe from its very beginning. Even the masses of elementary particles themselves are the result of their interactions with the Brout-Englert-Higgs field. Over the past 50 years, breakthrough discoveries have led to the establishment of the Standard Model of particle physics, which represents the most predictive and the most precise model ever built. The Standard Model elegantly regroups our understanding of the microscopic world, finding its inspiration from other branches of physics and its grounding through experimental developments. It is the ideal starting point to reveal Nature's deepest secrets. The discovery of the Higgs boson by the ATLAS and CMS experiments at CERN in 2012, was a major step forward leading to the evidence of the very first scalar field in the Universe. The characterisation of the Brout-Englert-Higgs field is among today's most fascinating scientific endeavours and constitutes the heart of Jona Motta's Ph.D. research. The experimental requirements of the studies pursued in High-Energy Physics research has been systematically pushing the limits of modern technologies. The tools for discoveries have constantly grown in complexity in many aspects, including computing, data handling and data processing techniques. The modelling of large background contributions and advanced signal extraction methods are relying today on the use of machine learning technology neatly described in this thesis.

The years following the discovery of the Higgs boson have allowed the scientific community to measure its couplings with elementary particles, and in particular with fermions. The pursuit of the characterisation of the so-called "Higgs sector" implies the measurement of the Brout-Englert-Higgs field parameters, and clearly represents our doorway to the unknown.

A central focus of Jona Motta's research has been the study of Higgs boson pair production (HH) in the $bb\tau\tau$ final state, with a particular emphasis on the Standard Model's non-resonant production. The search for non-resonant HH production is driven by the long-term goal of measuring the Higgs boson self-coupling—an essential parameter fundamentally linked to the Higgs potential responsible for the spontaneous breaking of electroweak symmetry via the Brout-Englert-Higgs mechanism.

Importantly, the precise shape of this potential may hold clues about the ultimate fate of our Universe, determining whether the current vacuum state is absolutely stable, metastable, or destined to decay on cosmological timescales. Measuring the self-coupling is therefore not only a test of the Standard Model, but also a window into the deep structure of the Universe itself. His contributions were both original and impactful: from a more robust, data-driven estimation of the dominant $t\bar{t}$ background, to an interpretation of the results within the framework of Effective Field Theory, thereby enhancing the theoretical scope of the study.

Another aspect of the research conducted in this thesis concerns a primordial step towards any data analysis performed at colliders: the selection of interesting collisions amongst the millions produced per second at CERN's Large Hadron Collider (LHC). The first stages of selection can only be performed by a dedicated hardware system referred to as the trigger system. Because of the very high volumes of data (in excess of Tb/s), the electronics used in particle physics detectors has traditionally been application-specific, but over the last decade advances in technology mean that flexible common processing platforms, including generic and flexible firmware/software frameworks to facilitate algorithm development and deployment, can now be used in many cases. The filtering requirements for such large datasets require microsecond latencies for sophisticated algorithms, steadily leading towards real time data analysis with hardware systems.

The starting point of the advanced low latency trigger algorithm developments described in this thesis were conducted in the context of the upgrade for the High-Luminosity LHC, where the instantaneous luminosity of the collider will be increased sevenfold, representing an even greater challenge for the trigger system. The huge volumes of data produced, which are highly structured, well understood and open, provide an ideal test bed for advanced machine learning algorithms. The design of dedicated trigger algorithms to target specific physics signatures can be achieved through the precise reconstruction of physics objects. The tau leptons are successfully reconstructed and identified by the TauMinator algorithm with unprecedented efficiency and resilience to background. The manuscript describes how this algorithm was designed and how its performance could be optimised. The techniques developed have been adapted to today's LHC running conditions, hence validating their principles and methodologies. These innovative developments clearly lead the way towards even more sophisticated approaches, where unsupervised machine learning algorithms could help us to expect the unknown at future collider experiments. It also has direct applications in the fast approaching epoch when our lives are going to be heavily influenced, and to some extent determined, by data and its interpretation by Artificial Intelligence algorithms. The ability to handle, process and draw conclusions from the vast amount of information, growing at an exponential rate, will create an ever-increasing challenge.

Jona Motta has conducted his Ph.D. thesis work within the CMS group of the École Polytechnique de Paris, Institut Polytechnique de Paris. Jona's strong motivation for particle physics research led him to achieve excellence in the field with an everlasting determination to obtain brilliant scientific results. He has shown the rare ability to conduct both complex data analysis and sophisticated technical developments. Jona has made outstanding individual contributions to major topics in modern high-energy physics. The design of a very challenging low latency tau lepton algorithm was only possible through innovative thinking, driven by Jona's impressive creativity and ingenuity. Jona's contributions to the Higgs boson pair production analysis were pivotal in pushing the sensitivity of the measurement beyond what was considered achievable just a few years ago. Both trigger algorithm developments, and modern data analysis techniques are described with great detail in this Ph.D. manuscript, which can be seen as a reference in the field of fundamental science.

December 2024

Dr. Alexandre Zabi
Director of Research, Laboratoire
Leprince Ringuet CNRS-IN2P3
École Polytechnique, Institut
Polytechnique de Paris
Palaiseau, France

Dr. Roberto Salerno
Director of Research, Laboratoire
Leprince Ringuet CNRS-IN2P3
École Polytechnique, Institut
Polytechnique de Paris
Palaiseau, France

Abstract

This Thesis presents the study of the Higgs boson pair (HH) production in the final state with a pair of quarks and a pair of τ leptons ($bb\tau\tau$), exploiting proton-proton collisions data collected at 13 TeV centre-of-mass energy with the CMS detector at the CERN Large Hadron Collider (LHC), corresponding to 138 fb^{-1} accumulated during the Run-2 data-taking period (2015–2018). The $bb\tau\tau$ decay channel gives a good trade-off between a sizable branching fraction (7.3%) and the purity of the τ selection, ensuring the good rejection of the background contributions. The study of production gives access to the measurement of the Higgs boson self-coupling (λ_{HHH}). In the context of the Standard Model (SM), this coupling is the only parameter governing the shape of the Higgs potential and it is precisely predicted by the theory; therefore, a measurement of λ_{HHH} is a test of the validity of the SM and allows us to shed light on the process of electroweak symmetry breaking. In the context of Beyond the SM (BSM) theories (with a particular interest in effective field theories), λ_{HHH} can assume values larger than that predicted by the SM, greatly enhancing the production cross section; the measurement of deviations from the SM prediction would open the road to yet another new era of physics. Upper limits on the SM signal are set at 95% Confidence Level (CL) to be 3.3 and 124 times the SM for $\sigma gg(\rightarrow)HH$ and $\sigma qq(\rightarrow)HH$, respectively. The results are also interpreted in the context of 20 different independent BSM scenarios for which 95% CL limits are set. The experimental context of this Thesis is the restart of LHC operations in 2022 for its Run-3, a new phase with collisions at an energy of 13.6 TeV and instantaneous luminosity of $2-2.6\times10^{34}\text{cm}^{-2}\text{s}^{-1}$. In Run-3, the hardware capabilities of the CMS Level-1 Trigger (L1T) are unchanged with respect to Run-2. This requires the development of bolder and more sophisticated approaches to optimise available algorithms to guarantee the success of the CMS physics program. Especially interesting is the optimisation of the L1T section that exploits calorimetric information. As part of this Thesis, a new machine learning method based on a neural network has been developed for the calibration applied in the L1T to calorimeter energy deposits; it exploits data for the calibration of single detector objects, and its promising performance is evaluated against the offline reconstruction of electrons and hadronic jets. The calorimetric information is then optimally used by the algorithm for the reconstruction and identification of hadronically decaying τ leptons (τ_h), whose optimisation for the Run-3 is performed in this Thesis employing a new, simple, and more informative approach; the same optimization scheme is also successfully employed for the e/γ algorithm. The performance of this approach is evaluated using data collected during Run-3. At the same time, the CMS collaboration is striving for its Phase-2 upgrade program, which is intended to match the ambitious High-Luminosity LHC (HL-LHC) physics program starting in 2029. The considerably increased volume of data collected by the HL-LHC will ensure the statistical power for the detailed study of λ_{HHH} and possibly its measurement; on the other hand, the larger instantaneous luminosity will require the full replacement of the L1T with hardware of increased capabilities based on state-of-the-art Field Programmable Gate Arrays (FPGAs) to efficiently collect data. To exploit the FPGA capabilities to the maximum, a new machine learning algorithm for the reconstruction, identification, and calibration of τ_h candidates in the L1T has been developed as part of this Thesis. This algorithm exploits convolutional neural networks implemented in FPGA firmware and ensures largely enhanced performance compared to standard approaches. All the technical advancement developed within this Thesis has one goal: improving the sensitivity of CMS analyses to the measurement of the Higgs boson self-coupling during the ongoing and future Runs of the LHC.

Preface

What are the basic constituents that compose the Universe, and what are the fundamental laws that regulate them?

This question constitutes the *red thread* that underlies the entire history of physics (from the ancient Greek φύσις meaning Nature). To answer this question, a long line of philosophers and physicists have devoted their lives to conceiving and testing theories that could give a rational explanation of the Nature that surrounded them (the extent of what can be considered rational is here deliberately left undefined, as nowadays we would consider irrational many theories originated in ancient times).

The first glimpses of such enquiry date back to the Vth Century BC when Empedocles was the foremost philosopher to theorise that Nature had an underpinning order based on the four fundamental elements of Fire, Air, Water, and Earth, which interacted through the powers of Love and Strife. This theory, which could be classified as a story rather than a theory in modern terms, already presents two essential components of contemporary physics: a set of *fundamental constituents* and *forces* regulating their interactions. In the same Century, two essential pillars of physics were conceived for the first time by Leucippus and Democritus: the concept of *atoms* (from the ancient Greek ἄτομος meaning uncut or indivisible) and of *void*, i.e. the empty space where atoms moved. It is worth noticing that the idea of atom at this stage is not the one we are acquainted with today but rather the simple concept of a fundamental building block of Nature, which is not further divisible. Therefore, from a purely conceptual standpoint, the basic notions that constitute our present description of the *infinitely small* scales through elementary particles and forces originated 2500 years ago.

The most significant turning point toward our current understanding of the Universe, and arguably in the history of physics itself, is represented by the remarkable set of theories developed and experimental observations performed at the beginning of the XXth Century. In this period, the two main theoretical pillars of particle physics were developed: quantum mechanics and special relativity. Combined, these two theories precisely describe the laws regulating the *infinitely small* scales (subatomic scales), which are not regulated by the deterministic laws of the macroscopic world but by probabilistic laws extensively corroborated by experiments. Concurrently, the theory of general relativity provides a mathematical description of the *infinitely large* scales (cosmological scales); several observations corroborate it, and experiments are at present verifying some of its longstanding predictions. These three pillars of modern physics ensure an unprecedented understanding of the Universe, with a predictive power spanning 40 orders of magnitude on the length scale and the potential ability to predict the state of the Universe at any point on the arrow of time (the adjective potential is employed to convey our current limited ability to make predictions for $t = 0$ and $t \to \infty$ effectively).

Quantum mechanics and special relativity paved the way for the theory currently representing our best understanding of the subatomic scales: the Standard Model (SM) of particle physics, formalised between the 1960s and the 1970s. The SM is constructed as a renormalisable quantum field theory with strict symmetry rules under gauge transformations. It accounts for all the fundamental forces of Nature, apart from gravity, and explains the existence and the categorisation of the so-called fundamental particles. The keystone of the SM is the Brout-Englert-Higgs (BEH) mechanism, which constitutes the simplest way to endow fundamental particles with mass while keeping the SM renormalisable and predictive. The BEH mechanism is responsible for the spontaneous Electro-Weak Symmetry Breaking (EWSB) process, through which the weak vector bosons acquire their mass, and allows for the definition of the Yukawa interaction, through which fermions gain their masses. From a historical perspective, the SM is the culmination of roughly 50 years of back-to-back progress in theory and experimental results that took place in the first half of the XXth Century. The list of contributors to this effort is far too long to be summarized in the scope of this Introduction. Still, to highlight the immense effort of human ingenuity required, it is sufficient to remember that Nobel prizes have been awarded to 87 physicists whose work has either led to the SM or corroborated it.

The SM has enjoyed remarkable success due to its predictive capability and extensive validation through numerous experimental observations. In the 1970s, not only was the SM able to explain the existence of all particles known at the time, but it could also predict the existence of yet undiscovered elementary particles observed in the ensuing 50 years. The first confirmation of the SM prediction was the observation of weak neutral currents at the Gargamelle experiment in 1973 and the subsequent discoveries of the $W^{\pm}$ and Z bosons at the CERN Super Proton Synchrotron (SPS) in 1983 by the UA1 and UA2 Collaborations. The $W^{\pm}$ and Z bosons were extensively characterised throughout the following two decades, with precise measurements at the CERN Large Electron-Positron Collider (LEP), at the Standford SLAC National Accelerator Laboratory, and at the FNAL Tevatron. The predictive power of the SM over six orders of magnitude on the energy scale was further confirmed in 1995 with the observation of the top quark at the FNAL Tevatron.

The ultimate triumph of the SM came on July 4th 2012, when the ATLAS and CMS Collaborations at the CERN Large Hadron Collider (LHC) announced the observation of a new particle with a mass of approximately 125 GeV. At the current level of precision and given the statistical power available, this particle is compatible with the SM Higgs boson (H) to a remarkable degree. The existence of a physical H boson amounts to an experimental corroboration of the BEH mechanism, which triggered the EWSB in the early Universe. Despite this outstanding milestone, the H boson still represents an important conundrum within the SM as it is the only known scalar boson, i.e. with spin zero; it is the sole particle that does not arise from a local gauge invariance property of the SM Lagrangian; it breaks the degeneracy between the three families of fermions with couplings spanning four orders of magnitude. Moreover, a central property of the H is its self-coupling (λ_{HHH}), which is proportional to its mass and whose experimental evidence is yet to be found.

The SM accurately describes all the measurements at colliders performed until now from the keV to TeV energy scales, thus spanning nine orders of magnitude. Notwithstanding these impressive achievements, the SM presents severe limitations and shortcomings. It does not include a quantum description of gravity; it predicts neutrinos to be massless particles, in contradiction with experimental observation; it falls short in describing the asymmetry between the matter and antimatter content of the Universe; it does not include the presence of dark matter and dark energy; the mass of the Higgs boson is a free parameter of the SM, not protected by any symmetry, making the magnitude of m_H sensitive to quantum fluctuations up to the Planck scale. Therefore, although it is extremely successful at describing particle phenomenology at collider experiments, the SM is still incomplete and should be regarded as the low-energy part of a more extended physics theory.

The Higgs boson is thought to play an essential role in extending the SM to a more complete theory owing to its intrinsic peculiarities compared to the other components of the SM. The exploration of the scalar sector of the SM currently represents the preferred avenue in the search for Beyond the SM (BSM) physics. Therefore, verifying the H compatibility with the SM prediction and precisely characterising its properties is of primary importance. While the H discovery and its preliminary study were conducted during the first operational run of the LHC (Run-1, 2010–2013), the second operational run (Run-2, 2015–2018) has been the testbench for its in-depth characterization. Given the state-of-the-art results, the particle discovered at 125 GeV shows a tremendous agreement with the SM Higgs boson prediction in the probed features (i.e. its production cross section via six different mechanisms, its decay branching ratio in seven distinct final states, and its spin-parity).

The λ_{HHH} coupling represents the final missing piece in the H characterization effort. Two specific aspects distinguish it from other couplings: it is not a free parameter of the theory, and it is the only parameter regulating the shape of the BEH potential. The former renders the measurement of λ_{HHH} a closure test of the SM; the latter has a deep relation with how EWSB happened in the early Universe and has broad implications in understanding the (meta)stability problem of the Universe. Moreover, many BSM theories predict large deviations of the λ_{HHH} value, which can be central to understanding the origin of the matter-antimatter asymmetry. Therefore, measuring the H self-coupling is one of the major endeavours undertaken by the particle physics community, as foreseen by recent updates of the European Strategy for Particle Physics.

The sole method to directly probe the H self-coupling is via the study of Higgs boson pair (HH) production, which involves λ_{HHH} at the leading-order approximation. This process has a small cross section, which is roughly 1500 times smaller than the single H production cross section, making the measurement of λ_{HHH} one of the most challenging analyses at the LHC. Nevertheless, this small cross section is highly sensitive to possible new physics at high energy scales, which could manifest itself in quantum loop corrections to the non-resonant HH production. Among the decays channels of HH, the one involving two bottom quarks and two τ leptons ($b\bar{b}\tau^+\tau^-$, or for simplicity bb$\tau\tau$) represents one of the best channels for HH searches. The final state benefits from a sizeable branching fraction of 7.3% and concurrently profits from the high selection purity of the τ leptons that keeps background contamination contained, making it one of the final states most sensitive to HH production.

In this context, I had the opportunity to work on the full Run-2 analysis of the HH $\rightarrow$ bb$\tau\tau$ decay channel to establish stringent limits on the HH production cross section and set constraints on the value of λ_{HHH}. I have had the chance to take part in most of the steps of the analysis, contributing to the evaluation of the event selection performance, the modelling of the backgrounds, and the optimisation of the discriminating variable used for the statistical interpretation. Moreover, I have been in charge of including in the analysis framework the necessary components needed for the interpretation of the results within an Effective Field Theory (EFT) Lagrangian formalism. The λ_{HHH} search presented in this Thesis is the second most sensitive analysis to the HH production in the gluon fusion production channel and the most sensitive to the vector boson fusion production mechanism. Moreover, it sets strict limits on the possible deviations of λ_{HHH} from the SM prediction and searches for deviations of other H couplings from their predicted value. The analysis results are presented in a paper, of which I am a co-editor, published in the Physics Letters B journal [Collaboration, Physics Letters B 842 (2023), p. 137531.].

Given the small cross section of HH production and many other SM phenomena, a high selection efficiency of final state particles in the detector is crucial. This need becomes particularly challenging in the busy environment of the LHC during its third operational run (Run-3, 2022–2025), where proton bunches collide at the centre of the CMS detector every 25 ns and up to 80 simultaneous interactions per bunch crossing can take place. At the core of the event selection process stands the Level-1 (L1) trigger, a hardware system that filters collision events to retain only those with potential interest for physics analysis. To perform this decision, the L1 trigger exploits a global but coarse view of the calorimeters and muon chambers of the CMS detector in the form of the so-called *Trigger Primitives* (TPs), which sophisticated algorithms employ to construct particle candidates. Specifically, in the case of the bb$\tau\tau$ analysis, it is fundamental to efficiently reconstruct and identify hadronically decaying τ leptons (τ_h), which account for roughly 65% of τ decays. This is a particularly demanding task as it requires to reject an abundant jet background, much larger than the expected signal.

In this context, I had the chance to contribute to the commissioning of the L1 trigger during the Run-3 data-taking restart in 2022 and 2023. I have been coordinating the effort for the optimisation of the L1 trigger algorithms reconstructing τ_h and e/γ

objects; I have developed a simple and yet more informative approach to optimising the algorithm's parameters, and I have introduced a new estimation of the isolation of the L1 candidates which is more resilient to the limited statistical power of the dataset used in the process. The results of this work and the performance of the τ_h and e/γ trigger algorithms have been published as two Detector Performance Summary (DPS) notes, of which I have been the main editor, for the 2022 and 2023 data-taking periods [Collaboration, CMS-DP-2023-008; Collaboration, CMS-DP-2023-055]. Moreover, I have been the leading developer of a new machine learning based technique for the calibration of the CMS calorimeter TPs. This algorithm is called Calibraton and is the first of its kind; it exploits data for calibrating single TPs based on offline reconstructed electrons and jets. This innovative technique, based on a neural network architecture, is still being advanced and improved, and it is presented for the first time in this Thesis. This work for the constant amelioration of the L1 trigger is fundamental toward the success of the ambitious CMS Run-3 physics program, in which a possible first evidence of HH production could be provided in the form of an exclusion of $\lambda_{HHH} = 0$ at a 3σ significance level.

To further probe the SM, with particular interest in the H self-coupling, the High-Luminosity LHC (HL-LHC) operations are scheduled to start in 2029. In this new phase of the accelerator, referred to as Phase-2, the HL-LHC will deliver instantaneous luminosity ranging between 5 and $7.5 \cdot 10^{34}\,\text{cm}^{-2}\,\text{s}^{-1}$ with a number of simultaneous collisions per bunch crossing of $\mathcal{O}(200)$. The HL-LHC is intended to be operational for ten years, during which a dataset almost 30 times larger than that of Run-2 will be collected. The unparalleled amount of data will open an extraordinary window into the subatomic nature of the Universe, providing even higher precision measurements of the SM and allowing for extensive searches of BSM physics. In the scalar sector of the SM, it will allow the observation of the H self-interaction and thus directly access the BEH potential for the first time. Moreover, it will permit measurements of the coupling of the H boson to second generation fermions and search for additional scalar bosons and rare phenomena. To reach these ambitious goals, the CMS Collaboration is planning an extensive upgrade program of its experimental apparatus by substituting vast portions of its hardware and software systems. Of particular interest are the substitution of the endcap calorimeter with the newly designed High Granularity Calorimeter (HGCAL) and the upgrade of the L1 trigger system featuring the extensive use of powerful state-of-the-art Field Programmable Gate Arrays (FPGAs) that will ensure the maintenance of the current discovery potential of the CMS detector and at the same time extend it into never before explored realms of physics.

In this context, I had the possibility to develop a novel τ_h triggering algorithm called TauMinator. In the harsh environment of the HL-LHC, the efficient reconstruction of τ_h will become an even more challenging task than it is during Run-2 and Run-3; the enhanced energy and luminosity of the accelerator will largely increase the jet background, which needs to be effectively mitigated to reach feasible L1 trigger rate values. To this end, I have single-handedly developed the TauMinator algorithm, which exploits the highly granular information available in the Phase-2 L1 trigger and, by means of a Convolutional Neural Network (CNN) architecture, addresses the τ_h triggering problem with an image recognition solution. The TauMinator algorithm is designed to be implemented in FPGA firmware; therefore, stringent requirements are enforced on the CNN architecture and the precision of the input variables. The algorithm has been successfully implemented in firmware, and the hardware-to-software comparison has been proven to be exact to the bit level. This algorithm is the first one to ensure a unified treatment of all available calorimeter TPs from the Phase-2 CMS detector; at the same time, it is the first calorimeter-based algorithm that performs the identification and the calibration of τh candidates via a machine learning technique. The algorithm guarantees substantial improvement over the standard triggering techniques, and it is presented in detail for the first time in this Thesis. The TauMinator algorithm has been published as an article in Proceedings of Science [Jona Motta, PoS EPS-HEP2023 (2024) 590], as well as a DPS note [CMS Collaboration, CMS-DP-2023-062];for both publications, I am the sole editor.]

This Thesis is structured as follows. Chapter 1 introduces the theoretical framework of the SM, focusing on its components, shortcomings, and extension with an EFT Lagrangian formalism; the theoretical motivations of a search for HH production at the LHC are presented alongside their experimental status prior to this Thesis. Chapter 2 familiarises the reader with the LHC accelerator complex and the CMS experimental apparatus; particular attention is reserved to the L1 trigger system and the reconstruction techniques relevant for the bb$\tau\tau$ search. The following Chapters present my first-hand contributions to the subject during the three years of my doctoral research. Chapter 3 examines the L1 τ_h algorithm, its optimisation toward the Run-3 data-taking and its maintenance throughout 2022 and 2023, focusing in particular on the new developments introduced as part of this Thesis; the discussion continues with the exposition of the Calibraton algorithm design and demonstrates its expected performance. Chapter 4moves the focus to the future L1 trigger system at the HL-LHC, presenting the development of the TauMinator algorithm; it covers the entire development chain, from conceptual design to firmware implementation, and highlights the performance gain it guarantees over a standard trigger approach. Chapter 5 introduces the bb$\tau\tau$ analysis, discussing the trigger strategy, the identification and isolation requirements imposed on physics objects, and the event selection and categorisation; the modelling of physics processes is also discussed in detail, followed by the evaluation of the systematic uncertainties affecting the search. Chapter 6 finally presents the statistical methods used in the analysis and the results of the search for HH production; the findings are interpreted within the SM and the EFT frameworks and compared to other HH searches performed during Run-2.

Geneva, Switzerland Dr. Jona Motta

Acknowledgments

At the culmination of my doctoral journey, I am compelled to look back and acknowledge those whose support and guidance have been instrumental in shaping this Thesis, my work, and me as a person. This manuscript represents, at the same time, the end of a journey and the start of a broader academic endeavour in the years to come; all that have contributed to its achievement deserve a place in it.

I want to thank the referees, Giovanni Calderini and Laurent Serin, and the other members of the Jury, Cristina Botta, Gudrun Heinrich, Anna Sfyrla, and Paris Sphicas, for accepting to review the manuscript and for the motivating and rewarding discussions that followed.

I must express my deepest gratitude to my supervisors, Alex and Roberto, for their unwavering support and guidance throughout my Ph.D. I am fortunate to have worked under your supervision, and I am deeply grateful for the impact you had on both my academic and personal growth. Thank you for believing in me from the beginning until the end, both in my best and worst days. You launched me in deep water since day one and compelled me to swim to find my own place in the big Collaboration we are part of. You have always identified the best opportunities for my work to be innovative and meaningful; you have never failed to support me and put me in the spotlight where I could thrive. But most of all, I am grateful because you passed down to me the passion and enthusiasm you both have for the wonderful job we have the privilege to do. All of this is what I think a good supervisor should do to extract the best out of a student.

I am particularly grateful to all the members of the CMS group at LLR who accompanied me in the adventure of these three years. You are brilliant minds and researchers who can always inspire fascinating discussions about the most diverse topics. A particular mention goes to Olivier; you have not been my supervisor, but you have been a fundamental mentor in building my expertise in the world of the Level-1 trigger. An additional appreciation goes to Shamik; I will always be in your debt for guiding my first steps in the awesome field of machine learning. Without you, gentlemen, this Thesis would have never been what it is now. I hope I will have the possibility to work with you again in the future.

I want to express my gratitude to the members of the Level-1 trigger team. It has been a privilege and an honour to work with all of you, not only for the development of new exciting machine learning techniques but also during the tiring shifts of data-taking. I want to thank Sioni and Gianluca for their constant aid; you have always answered my pedantic questions with the utmost clarity, and you made me discover the incredible world of firmware. A special thank you goes to Cristina for her continuous support and relentless backing of my work and efforts. You have been an extraordinary mentor and a true inspiration to me. I wish from the bottom of my heart that we will be working together in the future.

No good physics analysis has ever been done without a great team effort. So, I will be forever indebted to all the people who contributed to the analysis and made it possible. A particular tribute goes to Chiara, Francesco, and Davide, who taught me how to run an analysis and understand its steps, all the time enduring my countless mistakes with a smile or a laugh. I must also express my gratitude to Konstantin and Marcel for their tireless effort in teaching the youngsters how statistical analysis works and what its pitfalls are.

I want to thank Alex a second time for the opportunity he gave me to teach in the TREX. Sharing the teaching experience with you and the other *moniteurs* has been a challenging and rewarding experience. You have taught me a huge passion for teaching, and you made me understand that the work we do should always be directed toward the younger generations because their curiosity and unbiased view are what trigger the deepest considerations.

Thank you Matteo and Alessandro (or should I say Sandro), for your cheerful passion for physics; to me, you are a true example in how you do research, how you teach, and how you approach life. You are the colleagues (and also the friends) I look up to and who always push me to do and be better; you truly are brilliant physicists. Thank you Ana and Elena, for being there when my workaholic attitude brought me to some dark places; you were there to lend your hand and cheer me up, for this I am truly grateful. And Elena, working with you has been great; you taught me more than I taught you. Finally, thanks to all the *bbqtautau* group, it has been a pleasure to share with you the *cafétéria* times, the world-class *Magnan* food, the *ECLA pots* (but not the ECLA cinema), and the Friday nights at *As du Falafel*.

I must also express my gratitude to all the members of the administration at LLR for the infinite work you take care of behind the scenes. Without you, our work would not be possible.

Grazie ai miei amici di lunga data a Ciserano; ci siete sempre stati e ci sarete sempre, nonostante la distanza. Grazie a tutti i miei compagni di università che mi hanno spinto a essere uno studente migliore di quanto fossi.

Grazie alla mia seconda famiglia, il Corso Ghibli. Il tempo e lo spazio non diminuiranno mai il legame di fratellanza che ci unisce. Grazie Davide e Michele, siete due fratelli nel vero senso della parola; ci siete sempre stati e mi avete costantemente spinto a essere la versione migliore di me stesso (soprattutto nello sport). Grazie Lorenzo per essere un fulgido esempio di instancabile ricerca di miglioramento; la tua capacitá di gestire mille impegni e studi non smetterà mai di stupirmi e incitarmi. Grazie Fora per essere l'amico che mi scriveva anche quando io non ti scrivevo. Grazie Dani, Fabri, Gio, Oro, e Pippo, per la fantastica esperienza di aver corso insieme una maratona.

Dade, Mike, Fora, quando lo facciamo questo Iron-Man?

Finally, I wish to thank Theresa. Meeting you has been my biggest blessing; you have supported me every step of the way, and you have sustained me in all the good and the bad moments. You are one of the strongest people I know, and without your constant encouragement, this Thesis would not have been possible. In the past three years, you have made me grow more than anyone can imagine. No words can contain what I feel for you but know that I can't wait to share with you the years to come, whatever they will bring.

Infine, grazie alla mia Famiglia, per aver sempre suppportato le mie decisioni, nonostante siano sempre state atipiche e lontane da quello a cui eravate abituati. Grazie a mio fratello; nonostante le tante litigate, sei sempre stato al mio fianco pronto a fare una battuta per rallegrare tutti; sono fortunato ad averti nella mia vita e a poterti chiamare fratello. Finalmente, grazie ai miei genitori, per non aver mai smesso di credere in me (anche quando io stesso non ci credevo) e per non avermi mai posto limiti. Senza di voi non sarei mai diventato la persona che sono e non avrei mai raggiunto questo traguardo; spero un giorno di essere anche io un Genitore con la *G* maiuscola come lo siete voi. Questa Tesi è dedicata a voi, cui va il GRAZIE più grande ed importante!

Contents

Chapter 1
Higgs Boson Pair Production Theoretical Motivation

The Standard Model (SM) of particle physics [1, 2] represents the current best physical understanding of the subatomic world. The SM is constructed as a renormalizable quantum field theory with strict symmetry rules under gauge transformations. It accounts for all the fundamental forces of Nature, apart from gravity, and explains the existence and the categorization of the so-called fundamental particles.

The SM has been developed and refined in the second half of the XXth century via constant and back-to-back progress of theory and experimental results. During this time, the SM has been extensively corroborated, and multiple experiments in several different conditions precisely measured its predictions. The latest verification of the SM was the discovery of the Higgs boson by the ATLAS and CMS Collaborations at the CERN LHC, announced on July 4th 2012 [3–5].

Despite having huge and continued successes in providing experimental predictions, we know that the SM is not the ultimate theory as it cannot account for many experimental facts at different scales, from subatomic to cosmological and astrophysical observations. The presence of so-called physics beyond the SM (BSM) can be accounted for through different approaches, from introducing possible new fields to establishing new interactions within known particles. The Higgs field represents one important component in most approaches to BSM owing to its unique scalar nature. In this context, a deep understanding of the Higgs boson (H) and its properties represents the current most important objective of the high-energy physics community.

Among the Higgs boson properties, its self-interaction is of utmost importance as it can shed light on the nature of the Electro-Weak Symmetry Breaking (EWSB) process. The Higgs boson pair (HH) production represents the best channel to perform such measurements at the LHC experiments. Moreover, as HH production is one of the processes with the smallest predicted cross section, BSM physics can appear through very large deviations in experimental results.

This Chapter is organized into three main Sections. The discussion starts with the introduction of the SM, its properties, and the all-important Brout-Englert-Higgs

J. Motta, *Development of Machine Learning τ Trigger Algorithms and Search for Higgs Boson Pair Production*, Springer Theses,
https://doi.org/10.1007/978-3-031-96288-2_1

(BEH) mechanism. The second Section highlights the importance of HH production within the SM and possible BSM scenarios. The last Section covers the experimental approaches employed in the searches for HH production at the LHC.

1.1 The Standard Model of Particle Physics

The SM of particle physics brings forward the concept of *elementary particle*: an indivisible fundamental block that takes part in the processes of the Universe. Notwithstanding this simple and flawless definition, the notion of which particles are elementary is not well-defined in high-energy physics; it is a concept that evolves with time following the progress in the experimental techniques which, by constantly increasing the resolution power of the observations, can show how systems believed to be elementary are composed of smaller constituents. In the XXth century, the chain of *elementarization* has been the following: molecules $\rightarrow$ atoms $\rightarrow$ electrons and nuclei $\rightarrow$ electrons, protons, and neutrons $\rightarrow$ leptons, quarks, and force bosons $\rightarrow$ leptons, quarks, force bosons, and Higgs boson. The last step in this chain represents the current understanding of what elementary particles are. Nevertheless, the SM is not able to predict (and it is actually not designed to) if this is the last step in the chain, and even less if the chain will ever end.

The particle content of the SM is represented in Fig. 1.1, where elementary particles are organized according to their fundamental quantum numbers and their participation in interactions. The SM is composed of two main groups of elementary particles: *fermions* of half-integer spin and *bosons* of integer spin. While the former are the building blocks of matter (being it stable or unstable matter), the latter are the force carrier quanta, which are exchanged in the interactions described by the SM. It is worth noticing from the start that the SM does not describe the gravitational force, and its possible force carrier is the graviton; nevertheless, it is also important to highlight how, at the subatomic level, the strength of this interaction is negligible as it is 25 orders of magnitude lower than the weakest SM force. The SM particles present a surface symmetry that divides them into 12 bosons of $s = 1$ and 12 fermions of $s = 1/2$. The former encompass eight gluons g, which are the mediators of the strong force, the $W^{\pm}$ and Z, which are the carriers of the weak interaction, and the photon γ, which is the mediator of the electromagnetic force. The latter entail six quarks and six leptons. These particles are further divided into three *families*, each comprising two quarks with electric charge $Q = +2/3$ and $Q = -1/3$ and two leptons with charge $Q = -1$ and $Q = 0$. (Families are also referred to as *generations*, and particles pertaining to the same family are said to have the same *flavour*). The particle picture of the SM is completed by the Higgs boson, which is the only $s = 0$ boson, and it is involved in the EWSB mechanism through which the other SM bosons gain their mass.

Some remarks should be made about the particles just listed and their properties. There is currently no deep explanation for the number of fundamental leptons or their fundamental quantum numbers. It follows that there is no basic reason for their

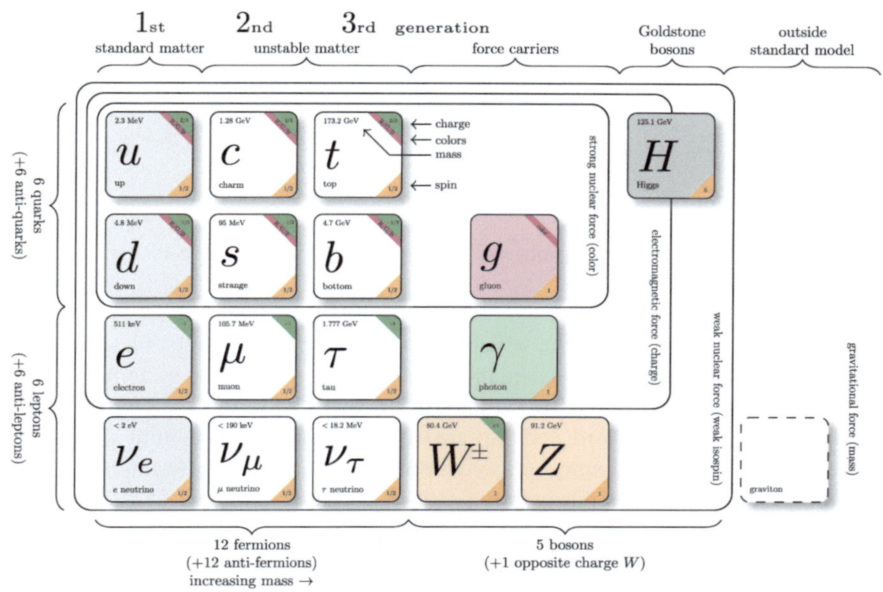

Fig. 1.1 Schematic illustration of the particle content of the SM. Although the SM is normally represented in a form that looks superficially like the periodic table of elements, the representation does not have any coherent two-dimensional logic. The graviton has not been observed yet and is therefore listed as being outside of the SM [6]

organization in three families. In contrast, the force carriers are $s = 1$ bosons (also referred to as *vector bosons*), and this property is well understood in the framework of gauge theories detailed below.

The mathematical framework employed to describe the content of the SM is that of quantum field theory, in which each particle can be understood as the excitation of a corresponding *field* defined in space-time. The fields are combined in the so-called *Lagrangian framework*, in which the Lagrangian equation encompasses the properties of the fields, and the symmetry qualities of the Lagrangian under specific transformation describe how the fields interact with each other. Interestingly enough, some properties in the SM are derived from the breaking of some symmetries.

From a poetic point of view, the SM is *A Tale of Symmetries and Fields* as described in the following.

1.1.1 Of Symmetries and Fields

The concept of symmetry is a fundamental one in physics; it follows from the assumption that a certain quantity is not measurable; thus, the equations of motion should not depend on it, and a conserved quantity exists. This important conservation law is

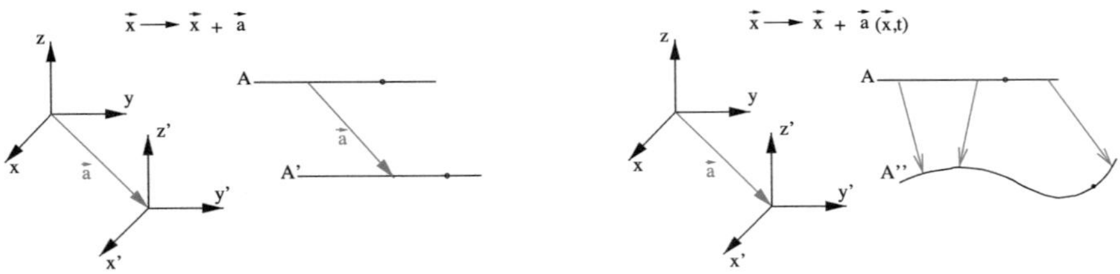

Fig. 1.2 Space translation by a constant vector $\vec{a}$ (left) and by a non-constant vector $\vec{a}(x)$ (right) [7]

epitomized in Noether's theorem, which states that if the Lagrangian of a system has a continuous symmetry, it exists a quantity that is conserved by the system, and vice versa. In classical mechanics, the symmetries of space and time are geometrical in the common sense of the word, and they are easy to visualise. In quantum mechanics, the concept of symmetry assumes a more abstract meaning. The most important type of symmetry in the present discussion is that of a *gauge symmetry*.

Figure 1.2 visually represents the concept of a local gauge transformation and its difference from a global transformation. The left panel shows that if A is the trajectory of a free particle in the (x, y, z) system, a transformation of the trajectory by a constant vector $\vec{\boldsymbol{a}}$ produces the image A$'$, which is also a possible trajectory of a free particle in the new (x', y', z') system. Therefore, the dynamics of free particles is invariant under space translations by a constant vector. In contrast, the right panel shows that if the transformation is generalized so that vector $\vec{\boldsymbol{a}}$ becomes a function of the position $\vec{\boldsymbol{x}}$, i.e. $\vec{\boldsymbol{a}}(\vec{\boldsymbol{x}})$, then the image A$''$ is no longer a possible trajectory of a free particle. This type of transformation is known as a local gauge transformation, and it is clear that the free particle dynamics is not invariant under such transformation.

From the purely geometrical consideration stated above arises the physical reflection: if trajectory A$''$ is not compatible with that of a free particle, then forces must have been at play. Can these forces be determined? The answer is yes. In order to estimate them, the approach is that of introducing interactions that restore the symmetry of the system under the local gauge transformation. This simple and visual introduction of how interactions can be described as arising from the requirement of symmetry properties under local gauge transformations is instrumental in understanding the mathematical nature of the SM. (The attentive reader will have realized that the example given above of a theory invariant under local translations turns out to be classical general relativity. Ironically, the gravitational force is the sole interaction not described by the SM).

The SM is a renormalizable quantum field theory whose Lagrangian is invariant under the linearly realized local gauge $\mathrm{SU(2)_L \otimes U(1)_Y \oplus SU(3)_C}$ symmetry. In particular, the $\mathrm{SU(3)_C}$ invariance results in the explanation of the strong force via the exchange of gluons as described in quantum chromodynamics. Moreover, the $\mathrm{SU(2)_L \otimes U(1)_Y}$ symmetry explains jointly the weak and electromagnetic forces, mediated by the $\mathrm{W^\pm}$ and Z bosons and the γ, respectively.

In the following, the fundamental fermionic fields will be introduced, followed by a discussion of the interactions described by the SM using the geometrical approach summarized above to derive them.

Fermions

Fermions are the building blocks of ordinary matter; they have spin $s = 1/2$, satisfy the Fermi-Dirac statistics, and abide by the Pauli exclusion principle. In the SM formulation, supported by the current state-of-the-art experimental observations, two types of fermions exist: *leptons* and *quarks*.

- **Leptons**
  Leptons are the first type of fermion; they interact only via weak and electromagnetic forces and are organized in three families. Each family presents a doublet with one lepton of charge $Q = -1$ and one of charge $Q = 0$. The three charged leptons are the electron (e), the muon (μ), and the tau (τ). These three particles behave in similar ways under the interactions described by the SM; nevertheless, an important difference sets them apart: their masses span four orders of magnitude. The electron has a mass of 511 keV, the muon of 105.7 MeV, and the τ of 1.8 GeV [8]. This outstanding difference has no fundamental explanation within the SM, and the attempts to understand it have been a very active field of theoretical physics in the past decades. Among the three charged leptons, only the electron is a stable particle, while the other two are unstable; the muon has a lifetime of 2.2 μs [8] and at the energies at which it is produced at the LHC can be considered as stable. In contrast, the τ lepton has a lifetime of $2.9 \cdot 10^{-13}$s [8] and can only be detected through its decay products.
  Associated with the three charged leptons are the three neutrinos of the same flavours: ν_e, ν_μ, and ν_τ. They are neutral particles that interact only via the weak force and, within the SM, are considered to be massless. Nonetheless, the measurement of neutrino oscillations [9] proves that neutrinos have non-zero masses, whose measurement attempts have yielded the upper limits $m_{\nu_e} < 2\,\text{eV}$, $m_{\nu_\mu} < 0.19\,\text{MeV}$, and $m_{\nu_\tau} < 18.2\,\text{MeV}$ [8]. It should be noted that the weak and mass eigenstates of neutrinos do not match and are thus subject to mixing, regulated by the Pontecorvo-Maki-Nakagawa-Sakata (PMNS) matrix [10].
  As leptons do not interact via the strong force, they are singlets of the $SU(3)_C$ transformation group.
- **Quarks**
  Quarks are the second type of fermions; they interact via all three forces of the SM and are the only fermions that carry a *colour* charge, to which Quantum Chromo-Dynamics (QCD) owes its name. Analogously to leptons, quarks are organized into three families, each presenting a doublet with one quark of charge $Q = +2/3$ and one of charge $Q = -1/3$. The first family of quarks is composed of the up (u) and down (d) quarks, with a mass of 2.2 MeV and positive charge, and 4.7 MeV and negative charge [8], respectively. They are the only stable quarks and together

compose ordinary baryonic matter. The second and third families follow the same charge scheme and are composed respectively of the charm (c) and strange (s) quarks of masses 1.3 GeV and 93 MeV, respectively, and the top (t) and bottom (b) quarks of masses 172.9 GeV and 4.18 GeV, respectively [8]. Quarks transform as triplets under the $SU(3)_C$ transformation group.
While all leptons can appear as free particles, the quarks are only detectable as bound states. This is a direct consequence of the *colour confinement* property of QCD, which is one of the deep unsolved problems in particle physics. Such bound states are generally referred to as *hadrons*, with the specific term of *mesons* for quark-antiquark states and *baryons* for tri-quarks states. QCD sets no limits on the number of quarks in a bound state. It should be noted that the t quark is the only one not found in hadrons as its lifetime of $0.5 \cdot 10^{-24}$ [8] is too short, so short that it decays before forming any bound state. A second important property of QCD is the so-called *asymptotic freedom*: as the momentum transfer in a given interaction increases, quarks asymptotically approach the state of free particles.
Quark flavour is conserved by electromagnetic and strong interactions, but it is not conserved by weak interactions. As a result of this property, like for neutrinos, the weak and mass eigenstates of quarks do not match and are thus subject to mixing, regulated by the Cabibbo - Kobayashi -Maskawa (CKM) matrix [11, 12].

In the following, leptons and quarks will be commonly denoted by the fermionic field ψ.

Strong Interaction

The strong interaction is formulated as arising from the symmetry properties of the SM Lagrangian under local gauge transformations of the non-abelian $SU(3)_C$ group, where the subscript C refers to the colour charge.

To understand how the strong force arises from the $SU(3)_C$ symmetry, it is instrumental to start from the Dirac Lagrangian of a free massive fermionic field ψ, which is not symmetric under such transformations. The Dirac Lagrangian can be written as:

$$\mathcal{L}_{\text{Dirac}} = \overline{\psi}(x)\left(i\partial\!\!\!/ - m\right)\psi(x) \tag{1.1}$$

where $\partial\!\!\!/ = \gamma^\mu \partial_\mu$ and γ are the Dirac matrices. Here and in the following, the Einstein notation for the summation over the indices is always implied. This Lagrangian is symmetric under the global SU(3) transformation:

$$\psi(x) \rightarrow e^{i g_s \theta^a \frac{\lambda_a}{2}} \psi(x), \tag{1.2}$$

where $\frac{\lambda^a}{2}$ represent the eight Gell-Mann matrices, corresponding to the generators of the SU(3) group and g_s is an arbitrary constant. Nevertheless, it is straightforward to notice that a local SU(3) transformation of the fermionic field ψ would give rise to terms $\sim \partial_\mu \theta^a(x)$. Therefore, the Dirac Lagrangian is not invariant under local gauge transformations.

To restore the invariance, Eq. 1.1 is modified to include a new field which transforms in such a way that the ensuing Lagrangian is symmetric under local SU(3) transformations. More precisely, the derivative $\partial\!\!\!/$ is promoted to the covariant derivative (D_μ) defined as:

$$D_\mu \equiv \partial_\mu + i g_s \frac{\lambda_a}{2} A_\mu^a, \tag{1.3}$$

where g_s is the coupling constant introduced above, and the vector bosons field A_μ^a is the strong force interaction field. With this minimal modification, the Lagrangian no longer describes a free Dirac field, and the requirement of invariance under local gauge transformation leads us to the *geometrical* introduction of interactions.

Given the properties of the SU(3) group, eight vector boson fields A_μ^a are associated with the eight generators of the group, and they are identified as the eight gluons which in the SM carry the strong force. To ensure that the modified Dirac Lagrangian is invariant under the desired transformations, the A_μ^a fields transform as:

$$A_\mu^a \rightarrow A_\mu^a + g_s \alpha_b(x) f^{abc} A_{\mu c} + \partial_\mu \alpha^a(x) \tag{1.4}$$

where the symbol f^{abc} encodes the non-abelianity of the $SU(3)_C$ group. The f^{abc} constants are the so-called *structure constants*, and they are defined based on the commutation rules of the group as:

$$i f^{abc} \frac{\lambda_c}{2} \equiv \left[\frac{\lambda^a}{2}, \frac{\lambda^b}{2} \right] \tag{1.5}$$

The final ingredient is represented by the kinetic term for the gluon field, which is generally expressed as:

$$-\frac{1}{4} F_{\mu\nu}^a F_a^{\mu\nu} \tag{1.6}$$

where the field strength tensor is defined as:

$$F_{\mu\nu}^a \equiv \partial_\mu A_\nu^a - \partial_\nu A_\mu^a + g_s f^{abc} A_{\mu b} A_{\nu c} \tag{1.7}$$

Following this approach, the obtained QCD Lagrangian is invariant under local $SU(3)_C$ gauge transformations and can be written as:

$$\mathcal{L}_{\text{QCD}} = \overline{\psi}(i\partial\!\!\!/ - m)\psi - g_s \overline{\psi} \gamma^\mu \frac{\lambda_a}{2} \psi A_\mu^a - \frac{1}{4} F_a^{\mu\nu} F_{\mu\nu}^a, \tag{1.8}$$

In this new formulation, the first term is analogous to the one originally present in Eq. 1.1 and represents the free-field propagation of the quark; the second encodes the interactions between the quarks and the gluons; the third term encompasses the propagation of the gluons. It is important to notice how the last term of the equation is linear in terms of the gluon field, thus introducing tri-linear and quadri-linear terms in the Lagrangian, corresponding to the self-interactions of gluons. Therefore, the propagation of the gluons is never that of a free field, and very complex kinematics arise in their propagation. The requirement of the invariance of under local $SU(3)_C$ gauge transformations also enforces a final property: gluons must be massless, as the introduction of a mass term of the form $m^2 A^a_\mu A^\mu_a$ would break the symmetry.

In this framework, the gluons are mathematically represented by the adjoint representation of the group (**8**) and differ by the two colour charges they carry. Quarks and anti-quark are instead defined in the simplest non-trivial irreducible representations of the $SU(3)_C$ group (**3** and $\bar{\mathbf{3}}$), which explains why quarks carry one colour charge. Finally, the g_s coupling constant is generally used to define the strong coupling constant $\alpha_s = g_s/4\pi$.

Electroweak Interaction

The Electro-Weak (EW) interaction arises from the unification of the electromagnetic force with the weak force. In the SM, it arises from the invariance of the Lagrangian under transformations of the non-abelian $SU(2)_L \otimes U(1)_Y$ group, where the subscripts L and Y refer to the preferred left-handedness of the interactions and the weak hypercharge, respectively. In the following, the electromagnetic and weak components are represented separately before being unified.

The electromagnetic compartment of the EW interaction is represented by Quantum Electro-Dynamics (QED), which can be derived using the same approach used above for the strong force. The Dirac Lagrangian in Eq. 1.1 is invariant under a global U(1) symmetry, but it is not under a local $U(1)_{EM}$ gauge symmetry. To restore the invariance, a similar generalization of the derivative term is done by the definition of the covariant derivative:

$$D_\mu \equiv \partial_\mu - ieA_\mu \tag{1.9}$$

where e is the coupling constant identified as the unit electric charge, while A_μ is the electromagnetic force interaction field, i.e. the photon. Given the $U(1)_{EM}$ invariance requirement, the gauge vector boson field transforms as $A_\mu(x) \to A_\mu(x) - \frac{1}{e}\partial_\mu\theta(x)$ and its field strength tensor is defined as $F^{\mu\nu} = \partial^\mu A^\nu - \partial^\nu A^\mu$. In contrast to the strong force, the electromagnetic field strength tensor is not linear in the field as the $U(1)_{EM}$ group is abelian; therefore, no interaction of the photon with itself is predicted in the SM. Moreover, as for the gluons, the photon is predicted to be massless to preserve the symmetry of the QED Lagrangian, which can be written as follows:

$$\mathcal{L}_{QED} = \overline{\psi}(i\not{D} - m)\psi - \frac{1}{4}F^{\mu\nu}F_{\mu\nu}. \tag{1.10}$$

The weak compartment of the EW theory can be understood as originating from the invariance of the Lagrangian under a local non-abelian $SU(2)_L$ gauge transformation. The procedure is the same as the one employed above, and the weak covariant derivative is defined to be:

$$D_\mu \equiv \partial_\mu + igT_i W^i_\mu \tag{1.11}$$

where g is an arbitrary coupling constant and the vector bosons field W^i_μ is the weak gauge interaction field. The three matrices denoted as $T_i = \frac{\sigma_i}{2}$ are the generators of the $SU(2)_L$ group, which are proportional to the Pauli matrices. Analogously to what was done above, the field strength tensor associated with W^i_μ can be defined as:

$$W^i_{\mu\nu} \equiv \partial_\mu W^i_\nu - \partial_\nu W^i_\mu + g\epsilon^{ijk} W_{\mu j} W_{\nu k}, \tag{1.12}$$

where the Levi-Civita tensor ϵ^{ijk} encodes the non-abelianity of the $SU(2)_L$ group and is defined by its commutation rules:

$$i\epsilon^{ijk}\sigma_k \equiv \left[\frac{\sigma^i}{2}, \frac{\sigma^j}{2}\right] \tag{1.13}$$

Besides the definitions just made, the integration of the weak force in the SM presents an additional difficulty compared to the strong and electromagnetic interactions: it has to account for the experimental observation of parity violation. This is achieved by introducing the concept of *chirality*, which distinguishes between left- and right-handed fermions, and by introducing different interactions for fermions of different chirality. From a mathematical standpoint, this is achieved with the definition of the fifth gamma matrix $i\gamma^5 \equiv -\gamma^0\gamma^1\gamma^2\gamma^3$ and the chirality projectors:

$$P_L = \frac{1-\gamma^5}{2} \quad , \quad P_R = \frac{1+\gamma^5}{2} \tag{1.14}$$

In the limit of a massless particle, the chirality corresponds to the helicity, defined as the normalized projection of the spin vector onto the spatial momentum vector. Fermionic fields can then be decomposed into their left and right chirality states. This approach allows us to define the representation of each fermionic field of the SM as one left chirality doublet (Ψ) and two right chirality singlets (ψ_R, ψ'_R):

$$\Psi_L = P_L\begin{pmatrix}\psi \\ \psi'\end{pmatrix} = \begin{pmatrix}\psi_L \\ \psi'_L\end{pmatrix} \tag{1.15}$$

$$\psi_R = P_R\psi \tag{1.16}$$

$$\psi'_R = P_R\psi' \tag{1.17}$$

In this notation, the ψ field represents the neutrinos and the up-type quarks, while the ψ' field represents the charged leptons and the down-type quarks. The weak Lagrangian can then be written as follows:

$$\mathcal{L}_{\text{weak}} = i\overline{\Psi}_L \not{D}\Psi_L + i\overline{\psi}_R \not{D}\psi_R + i\overline{\psi}'_R \not{D}\psi'_R - \frac{1}{4}W^i_{\mu\nu}W^{\mu\nu}_i \tag{1.18}$$

It can be appreciated how the weak force now acts in different ways on the left- and right-handed components of the fermionic field. The quantum number associated with the $SU(2)_L$ group is the weak isospin (I) whose third component (I_3) regulates how the fields transform. The right chirality fields have $I_3 = 0$ and transform as singlets of the $SU(2)_L$ group; the left chirality fields have $I_3 = \binom{1/2}{-1/2}$ and transform as doublets of the $SU(2)_L$ group. This effectively enforces the requirement of having a parity-breaking theory: the weak force acts only on the left-chirality doublets with the interaction term $-g\overline{\Psi}_L\gamma^\mu T_i \Psi_L W^i_\mu$.

Besides the discussion just made, the weak interaction presents one final problem: the three fields W^i_μ do not correspond to the physical fields of the weak bosons observed experimentally, which are denoted as $W^\pm_\mu$ and Z_μ. To solve this mismatch, we need to merge the weak and QED Lagrangians to build one single EW theory. To do so, the QED Lagrangian can be modified and replaced by a Lagrangian density that is invariant under the $U(1)_Y$ symmetry, where the subscript Y refers to the weak hypercharge. This symmetry is analogous to the one used to derive the QED Lagrangian, but they should not be confused as different charges are associated with them; one is the physical electrical charge, and the other is the unphysical weak hypercharge. The Lagrangian associated with the $U(1)_Y$ is

$$\mathcal{L}_Y = i\overline{\psi}\gamma^\mu D_\mu \psi - \frac{1}{4}B_{\mu\nu}B^{\mu\nu} \tag{1.19}$$

where we defined the covariant derivative and the field strength tensor to be:

$$D_\mu \equiv \partial_\mu - ig'\frac{Y}{2}B^i_\mu, \qquad B_{\mu\nu} = \partial_\mu B_\nu - \partial_\nu B_\mu \tag{1.20}$$

where g' is an arbitrary coupling constant and the gauge field B_μ is unphysical. The Lagrangian just defined results in an interaction that acts indistinctly on the left- and right-handed chirality fields through the gauge field B_μ. The quantum number associated with the $U(1)_Y$ symmetry is the weak hypercharge Y, which is strictly related to the physical electric charge by the Gell-Mann-Nishijima formula $Q = I_3 + \frac{Y}{2}$.

The unification of QED and the weak interaction is done by combining the Lagrangian densities defined in Eqs. 1.18 and 1.19, to obtain a single Lagrangian density which is invariant under local gauge transformation of the $SU(2)_L \otimes U(1)_Y$. In so doing, we obtain the following EW Lagrangian:

$$\mathcal{L}_{EW} = \overline{\Psi}_L(i\not{D})\Psi_L + \overline{\psi}_R(i\not{D})\psi_R + \overline{\psi}'_R(i\not{D})\psi'_R - \frac{1}{4}W_i^{\mu\nu}W^i_{\mu\nu} - \frac{1}{4}B^{\mu\nu}B_{\mu\nu} \tag{1.21}$$

where the covariant derivative that is introduced to ensure the overall gauge invariance is:

$$D_\mu \equiv \partial_\mu + igT_i W^i_\mu - ig'\frac{Y}{2}B_\mu \tag{1.22}$$

One interesting outcome of this unification is that the Lagrangian in Eq. 1.21 can be rewritten to highlight four distinct components:

$$\mathcal{L}_{EW} = \mathcal{L}_{F\text{-kin.}} + \mathcal{L}_{CC} + \mathcal{L}_{NC} + \mathcal{L}_{G\text{-kin.}} \tag{1.23}$$

where we define the Fermionic kinetic term $\mathcal{L}_{\text{F-kin.}}$, the Charged Current (CC) term $\mathcal{L}_{\text{CC}}$, the Neutral Current (NC) term $\mathcal{L}_{\text{NC}}$, and the gauge kinetic term $\mathcal{L}_{\text{G-kin.}}$ as:

$$\mathcal{L}_{\text{F-kin.}} = \overline{\Psi}_{\text{L}}(i\partial\!\!\!/)\Psi_{\text{L}} + \overline{\psi}_{\text{R}}(i\partial\!\!\!/)\psi_{\text{R}} + \overline{\psi}'_{\text{R}}(i\partial\!\!\!/)\psi'_{\text{R}} \tag{1.24}$$

$$\begin{aligned}\mathcal{L}_{\text{CC}} &= \frac{g}{\sqrt{2}} W^+_\mu \overline{\Psi}_{\text{L}}\gamma^\mu\sigma^+\Psi_{\text{L}} + \frac{g}{\sqrt{2}} W^-_\mu \overline{\Psi}_{\text{L}}\gamma^\mu\sigma^-\Psi_{\text{L}} = \\ &= \frac{g}{\sqrt{2}} W^+ \left(\overline{\psi}_{\text{L}}\gamma^\mu\psi'_{\text{L}}\right) + \frac{g}{\sqrt{2}} W^- \left(\overline{\psi}'_{\text{L}}\gamma^\mu\psi_{\text{L}}\right)\end{aligned} \tag{1.25}$$

$$\begin{aligned}\mathcal{L}_{\text{NC}} &= \frac{g}{\sqrt{2}} W^3_\mu \left[\overline{\psi}_{\text{L}}\gamma^\mu\psi_{\text{L}} - \overline{\psi}'_{\text{L}}\gamma^\mu\psi'_{\text{L}}\right] + \\ &+ \frac{g'\text{Y}}{\sqrt{2}} B_\mu \left[\overline{\psi}_{\text{L}}\gamma^\mu\psi_{\text{L}} + \overline{\psi}'_{\text{L}}\gamma^\mu\psi'_{\text{L}} + \overline{\psi}_{\text{R}}\gamma^\mu\psi_{\text{R}} + \overline{\psi}'_{\text{R}}\gamma^\mu\psi'_{\text{R}}\right]\end{aligned} \tag{1.26}$$

$$\mathcal{L}_{\text{G-kin.}} = -\frac{1}{4} B_{\mu\nu}B^{\mu\nu} - \frac{1}{4} W^i_{\mu\nu}W^{\mu\nu}_i \tag{1.27}$$

where we denoted $\sigma^\pm = \left(\sigma^1 \pm i\sigma^2\right)/\sqrt{2}$ and more importantly, we now introduced the physical fields associated with the weak bosons. At the same time, we retained the physical field of the photon, which is now understood as a linear combination of the other unphysical gauge fields. The physical fields are defined as follows:

$$W^\pm_\mu = \frac{1}{\sqrt{2}}\left(W^1_\mu \mp W^2_\mu\right) \tag{1.28}$$

$$\begin{pmatrix} A_\mu \\ Z_\mu \end{pmatrix} = \begin{pmatrix} \cos\theta_W & \sin\theta_W \\ -\sin\theta_W & \cos\theta_W \end{pmatrix} \begin{pmatrix} B_\mu \\ W^3_\mu \end{pmatrix} \tag{1.29}$$

where θ_W is the Weinberg angle. The full extent of the unification is grasped when the Gell-Mann-Nishijima formula is rewritten in terms of the weak coupling constants g and g', the electric charge e and the Weinberg mixing angle to yield:

$$e = g' \sin\theta_W = g\cos\theta_W \tag{1.30}$$

The separation of the EW Lagrangian into the four components above allows us to identify the two different types of weak interaction to which the fermions are subjects, either involving the $W^\pm_\mu$ field or the Z_μ and A_μ fields. The former corresponds to the charged current interaction, it involves only left-handed chirality fields, maximally violates parity, and can be flavour-changing. The latter are the neutral current interactions, they involve left- and right-handed chirality fields, do not violate parity, and cannot be flavour changing as they are highly suppressed by the GIM mechanism [13]. It should also be noted that, as $\text{SU}(2)_\text{L} \otimes \text{U}(1)_\text{Y}$ is a non-abelian group, a large array of multiboson vertices are predicted (i.e. ZWW, γWW, ZZWW, $\gamma\gamma$WW, γZWW, WWWW).

The last important remark to be made is that in the formulation just made, the physical $W^\pm_\mu$ and Z_μ fields are massless, as the introduction of a gauge mass term would break the theory invariance. Moreover, direct fermion mass terms are not allowed either as they are not invariant under $SU(2)_L \otimes U(1)_Y$ transformations which affect differently the left- and right-chirality fields, while the mass term could be decomposed as $m\overline{\psi}\psi = m(\overline{\psi}_R\psi_L + \overline{\psi}_L\psi_R)$.

The Unified Picture

The geometrical approach to deriving the mathematical description of the fundamental forces has proven extremely powerful. The strong force is understood as originating from the $SU(3)_C$ non-abelian group, while the unified weak and electromagnetic forces stem from the $SU(2)_L \otimes U(1)_Y$ non-abelian group. The two compartments can be juxtaposed to obtain a formulation of the SM of extraordinary beauty and elegance under the group $SU(2)_L \otimes U(1)_Y \oplus SU(3)_C$.

In this notation, fermions of left-handed chirality are identified as members of doublets of the $SU(2)_L$ group. In contrast, fermions of right chirality transform as singlets of the $SU(2)_L$ group. Moreover, fermions are identified by three specific quantum numbers: the charge Q, the weak hypercharge Y, and the weak isospin I_3. A summary of the fermionic fields is given in Table 1.1 under their $SU(2)_L$ representation.

Notwithstanding the immense predictive power of the theory as it has been explained thus far, one important point still misses: the existence of masses. Experimental evidence shows that the fermions and the weak bosons are all bestowed with mass; nevertheless, the theory described above only predicts massless particles. An elegant solution to this problem is the introduction of the Brout-Englert-Higgs mechanism, which is the basis for the spontaneous breaking of the electroweak symmetry through which the weak gauge bosons gain mass. This mechanism can then be extended to give mass to the fermions through the Yukawa interaction.

Table 1.1 Fermionic fields under their $SU(2)_L$ representation. The L and R subscripts denote the left and right chiralities, respectively. Lepton and quark fields are separately listed. The weak isospin I_3, weak hypercharge Y, and electric charge Q are reported alongside the transformation property under the $SU(3)_C$ group

Type	1st gen.	2nd gen.	3rd gen.	I_3	Y	Q	$SU(3)_C$
Leptons	$\begin{pmatrix} \nu_{eL} \\ e_L \end{pmatrix}$	$\begin{pmatrix} \nu_{\mu L} \\ \mu_L \end{pmatrix}$	$\begin{pmatrix} \nu_{\tau L} \\ \tau_L \end{pmatrix}$	$\begin{pmatrix} 1/2 \\ -1/2 \end{pmatrix}$	-1	$\begin{pmatrix} 0 \\ -1 \end{pmatrix}$	Singlet
	e_R	μ_R	τ_R	0	-2	1	
	$\nu_{e,R}$	$\nu_{\mu,R}$	$\nu_{\tau,R}$	0	0	0	
Quarks	$\begin{pmatrix} u_L \\ d_L \end{pmatrix}$	$\begin{pmatrix} c_L \\ s_L \end{pmatrix}$	$\begin{pmatrix} t_L \\ b_L \end{pmatrix}$	$\begin{pmatrix} 1/2 \\ -1/2 \end{pmatrix}$	1/3	$\begin{pmatrix} 2/3 \\ -1/3 \end{pmatrix}$	Triplet
	u_R	c_R	t_R	0	4/3	2/3	
	d_R	s_R	b_R	0	$-2/3$	$-1/3$	

1.1.2 *The Brout-Englert-Higgs Mechanism*

The BEH mechanism was proposed independently by Brout and Englert [14] and by Higgs [15] as a solution to generate the gauge boson masses and explain the presence of fermion masses. The BEH mechanism is said to induce the EWSB, and it is based on the concept of Spontaneous Symmetry Breaking (SSB), i.e. the lowest-lying energy state of a theory can spontaneously violate the symmetries of the theory itself. SSB is a widely known process in physics, finding proof from the classical theory of continuum (bent rod problem [7]) to magnetism (Ising spontaneous magnetization model [16]).

In the SM, to account for EWSB via the BEH mechanism, we have to introduce a new complex doublet to the theory, the BEH field doublet defined as:

$$\phi = \begin{pmatrix} \phi^+ \\ \phi^0 \end{pmatrix} = \frac{1}{\sqrt{2}} \begin{pmatrix} \phi^1 + i\phi^2 \\ \phi^3 + i\phi^4 \end{pmatrix} \tag{1.31}$$

To satisfy space isotropy and homogeneity, it must be a scalar and its vacuum expectation value (v.e.v. or v) a constant. Moreover, since the introduction of the BEH doublet must break the original local $\mathrm{SU(2)_L} \otimes \mathrm{U(1)_Y}$ symmetry while preserving the $\mathrm{U(1)_{EM}}$ one, it must have weak hypercharge $\mathrm{Y} = 1$. All of these requirements result in a scalar omni-pervasive field of electromagnetic charge $Q = 0$, whose quantum excitations manifest themselves as Higgs bosons. The Lagrangian that introduces this field in the SM is:

$$\mathcal{L}_{\mathrm{BEH}} = \left(D_\mu \phi\right)^\dagger \left(D^\mu \phi\right) - V(\phi^\dagger \phi) \tag{1.32}$$

where the covariant derivative is that introduced in Eq. 1.22 and the potential $V(\phi^\dagger \phi)$ can be explicitely written as:

$$V(\phi^\dagger \phi) \equiv -\mu^2 \phi^\dagger \phi + \lambda \left(\phi^\dagger \phi\right)^2 \tag{1.33}$$

where both μ^2 and λ are strictly positive. The functional form of such potential is depicted in Fig. 1.3, where the typical so-called *Mexican hat* shape can be appreciated. The potential presents an unstable local maximum for $\phi = 0$ and a continuum of stable ground states satisfying:

$$\left|\phi^\dagger \phi\right| = \frac{\mu^2}{2\lambda} = \frac{v^2}{2} \tag{1.34}$$

The choice of ground state among the continuum defined above is what spontaneously breaks the local $\mathrm{SU(2)_L} \otimes \mathrm{U(1)_Y}$ symmetry while preserving the $\mathrm{U(1)_{EM}}$ symmetry as it is parallel to the ϕ^0 component of the doublet. The small perturbations expansion around the minimum can be written as:

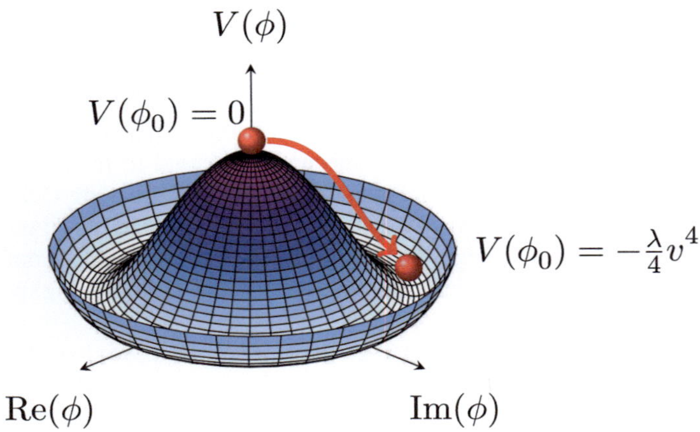

Fig. 1.3 Schematic illustration of spontaneous symmetry breaking via the BEH mechanism. The represented surface has the functional form of the Higgs potential (mexican hat), while the red ball illustrates the process of passing from an unstable local maximum to the continuum of ground states where v is referred to as vacuum expectation value

$$\phi(x) = \frac{1}{\sqrt{2}} e^{\frac{i}{v}\sigma_j\theta^j(x)} \begin{pmatrix} 0 \\ v + H(x) \end{pmatrix} \tag{1.35}$$

where $\theta^j(x)$ with $j = 1, 2, 3$ correspond to three unphysical massless scalar bosons originating as a consequence of Goldstone's theorem [17], which states that the spontaneous breaking of a continuous symmetry produces massless bosons in equal number to the broken generators of the symmetry, while $H(x)$ is a massive physical field. Being unphysical, the massless Goldstone bosons need to be reabsorbed by the theory. This is achieved by applying a local $SU(2)_L$ transformation of the BEH field doublet, referred to as the *unitary gauge*, under which the Lagrangian is invariant. The unitary gauge is parametrized as follows:

$$\phi(x) \to e^{-\frac{i}{v}\sigma_j\theta^j(x)}\phi(x) = \frac{1}{\sqrt{2}} \begin{pmatrix} 0 \\ v + H(x) \end{pmatrix} \tag{1.36}$$

The BEH Lagrangian can then be rewritten explicitly, yielding:

$$\begin{aligned} \mathcal{L}_{BEH} &= \frac{1}{2}\partial^\mu H \partial_\mu H - \frac{1}{2}\left(2\lambda v^2\right) H^2 \\ &+ \left[\left(\frac{gv}{2}\right)^2 W^{\mu +} W^-_\mu + \frac{\left(g^2 + g'^2\right) v^2}{8} Z^\mu Z_\mu\right] \left(1 + \frac{H}{v}\right)^2 \\ &+ \lambda v H^3 + \frac{1}{4}\lambda H^4 - \frac{1}{4}\lambda v^4 \end{aligned} \tag{1.37}$$

where three mass terms are now explicit:

$$m_{\mathrm{H}}^2 = 2\lambda v^2 = 2\mu^2 \tag{1.38}$$

$$m_{\mathrm{Z}}^2 = \frac{(g^2 + g'^2)v^2}{4} \tag{1.39}$$

$$m_{\mathrm{W}}^2 = \frac{g^2 v^2}{4} = (m_{\mathrm{Z}} \cos\theta_W)^2 \tag{1.40}$$

The introduction of the BHE field introduced unphysical Goldstone bosons which have been translated through the unitary gauge into additional degrees of freedom of the $W^{\pm}$ and Z fields, which can now assume longitudinal polarization. This effectively amounts to bestowing mass upon the $\mathrm{W}^{\pm}$ and Z bosons associated with the respective fields. The remaining massive field corresponds to a new physical massive particle: the Higgs boson whose mass m_{H} is a free parameter of the theory and is directly dependent on the v.e.v.

Equation 1.37 is written on three lines to separate different contributions to the theory. The first line represents the kinematics of the H doublet field and its associated mass. The second line defines the interaction of the weak bosons with the BEH field and their newly gained masses. The predicted vertices are HWW, HZZ, HHWW, and HHZZ all with couplings proportional to the masses of the weak bosons involved. All four vertices are of high relevance for the analysis presented in this Thesis.

The last line demonstrates that cubic and quartic self-interactions of the H boson are predicted. The BEH potential can be rewritten to render explicit such self-coupling of the H:

$$\begin{aligned} V(H) &= \frac{1}{2} 2\mu^2 H^2 + \frac{\mu^2}{v} H^3 + \frac{\mu^2}{4v} H^4 - \frac{1}{4}\mu^2 v^2 \\ &= \frac{1}{2} m_{\mathrm{H}}^2 H^2 + \lambda_{\mathrm{HHH}} v H^3 + \lambda_{\mathrm{HHHH}} H^4 - \frac{1}{8} m_{\mathrm{H}}^2 v^2 \end{aligned} \tag{1.41}$$

where we defined the tri-linear and quadri-linear coupling constants as:

$$\lambda_{\mathrm{HHH}} = 4\lambda_{\mathrm{HHHH}} = \frac{m_{\mathrm{H}}^2}{2v^2} \tag{1.42}$$

This strong relation of the self-couplings to m_{H} and the v.e.v. is one of the cornerstones of the SM. Precision measurements of these quantities provide one of the most stringent tests of the SM, giving direct access to the BEH potential and probing the EWSB mechanism detailed here. Therefore, the self-coupling property of the H boson is the primary object of investigation of the analysis presented in this Thesis.

While at first glance there appear to be two free parameters to the BEH Lagrangian, only m_{H} is such, as the v.e.v is fully determined by the measurement of the charged current interaction in the muon decay $\mu \to \mathrm{e}\overline{\nu}_{\mathrm{e}}\nu_{\mu}$, as follows:

$$\frac{G_F}{\sqrt{2}} = \frac{g^2}{8m_{\mathrm{W}}^2} = \frac{1}{2v^2} \quad \longrightarrow \quad v = \frac{1}{\left(\sqrt{2}G_F\right)^{\frac{1}{2}}} \approx 246\,\mathrm{GeV} \tag{1.43}$$

Thus far, we have solved only one of the two mass problems of the SM. The BEH mechanism and ESWB ensure that the weak bosons gain mass, but it does not affect the fermions. Therefore, an additional mechanism needs to be introduced: the Yukawa interaction. This is a direct interaction of the BEH field with the fermions, which couples to both left- and right-handed chirality fields as follows:

$$\mathcal{L}_{\text{Yukawa}} = -y_{f'}\left(\overline{\Psi}_{\text{L}}\phi\psi'_{\text{R}} + \overline{\psi}'_{\text{R}}\phi^{\dagger}\Psi_{L}\right) - y_{f}\left(\overline{\Psi}_{\text{L}}\tilde{\phi}\psi_{\text{R}} + \overline{\psi}_{\text{R}}\tilde{\phi}^{\dagger}\Psi_{L}\right), \tag{1.44}$$

where y_f and $y_{f'}$ are the coupling constants for the components of the doublets with the third component of the isospin $I_3 = 1/2$ and $I_3 = -1/2$, respectively, and we introduced:

$$\tilde{\phi} = i\sigma_2\begin{pmatrix} \phi_0^* \\ -\phi_+^* \end{pmatrix} \quad \longmapsto EWSB \quad \frac{1}{\sqrt{2}}\begin{pmatrix} v + H(x) \\ 0 \end{pmatrix} \tag{1.45}$$

Therefore, after EWSB, the Yukawa Lagrangian can be explicitly written as:

$$\mathcal{L}_{\text{Yukawa}} = -\sum_{f} m_f\left(\overline{\psi}_{\text{L}}\psi_{\text{R}} + \overline{\psi}_{\text{R}}\psi_{\text{L}}\right)\left(1 + \frac{H}{v}\right) \tag{1.46}$$

where the sum runs on all fermions and the fermion mass terms are defined as:

$$m_f = y_f\frac{v}{\sqrt{2}} \tag{1.47}$$

Fermion masses are thus explained in the SM as the interaction of the fermion fields with the BEH field. The strengths of the interactions are directly proportional to the fermion masses, which are free parameters of the theory. An important remark is that the SM does not explain the origin of these couplings nor the hierarchy of the three fermion families.

One important final observation is that the SM thus obtained is a renormalizable theory as demonstrated by 't Hooft and Veltman [18], meaning that the SM is consistent up to the Planck scale $M_{\text{P}} \sim 10^{19}$ GeV, at which point gravity is no longer negligible. As we will further discuss in the following, while the renormalizability of the SM was always regarded as one of its strong points, it is nowadays being regarded as a possible limitation. Nevertheless, the self-consistency of the SM up to such large scales provides a solid theoretical basis for describing the early Universe, as detailed below.

1.1.3 Phenomenology of the Higgs Boson and Experimental Status

On July 4th 2012, the ATLAS and CMS Collaborations jointly announced the discovery of a scalar boson with a mass of about 125 GeV compatible with the one predicted by the BEH mechanism [3, 4]. The discovery was performed by exploiting the data collected during the first operational run of the LHC (2010-2013) at a centre-of-mass energy of $\sqrt{s} = 7 - 8\,\text{TeV}$, effectively measuring the last important free parameter of the SM.

Since then, the focus of the high-energy physics community has been to thoroughly characterise the H boson and to fully probe the EWSB mechanism and its central role in shaping the fabric of the Universe. Central measurements are the determination of its mass and width, its production cross section in the different production mechanisms, its decay branching ratios, its spin-parity, and its most important self-coupling property.

In proton-proton collisions at the LHC, the main production mechanisms are:

- **Gluon Fusion (ggF)**: it's the dominant production at the LHC, with a cross section of $\sigma(\text{ggF}|m_{\text{H}} = 125.09\,\text{GeV}, \sqrt{s} = 13\,\text{TeV}) = 48.61^{+4.27\%}_{-6.49\%}(\text{theory}) \pm 1.85\%$ $(\text{PDF})^{+2.59\%}_{-2.62\%}(\alpha_s)\text{pb}$ [19]. The ggF cross section is enhanced by the large Parton Density Function (PDF) of the gluons, and it is mediated by a heavy quark loop. Due to the typical shape of the Feynman diagram, this process is generally referred to as *triangle diagram*.
- **Vector Boson Fusion (VBF)**: is the second most prominent production at the LHC with a cross section of $\sigma(\text{VBF}|m_{\text{H}} = 125.09\,\text{GeV}, \sqrt{s} = 13\,\text{TeV}) = 3.766^{+0.43\%}_{-0.33\%}$ $(\text{scale}) \pm 2.1\%(\text{PDF} + \alpha_s)\text{pb}$ [19]. The VBF process involves two massive vector bosons being radiated by two quarks from the protons that interact to produce a H boson. This process has a typical signature with two very forward jets that can be used for event tagging.
- **Higgs-strahlung (VH)**: is the third most frequent mechanism and has the following two components, $\sigma(\text{WH}|m_{\text{H}} = 125.09\,\text{GeV}, \sqrt{s} = 13\,\text{TeV}) = 1.358 \pm 0.51\%$ $(\text{scale}) \pm 1.35\%(\text{PDF} + \alpha_s)\text{pb}$ and $\sigma(\text{ZH}|m_{\text{H}} = 125.09\,\text{GeV}, \sqrt{s} = 13\,\text{TeV}) = 0.880^{+3.50\%}_{-2.68\%}(\text{scale}) \pm 1.65\%(\text{PDF} + \alpha_s)\text{pb}$ [19]. In the VH process, a quark interacts with an antiquark to produce a massive vector boson which radiates a Higgs boson, hence the name of the mechanism.
- **Associated production** (ttH, bbH, tH): is the least prominent mechanism of the four and has the following four contributions, $\sigma(\text{ttH}|m_{\text{H}} = 125.09\,\text{GeV}, \sqrt{s} = 13\,\text{TeV}) = 0.5065^{+5.8\%}_{-9.2\%}(\text{scale}) \pm 3.6\%(\text{PDF} + \alpha_s)\text{pb}$, $\sigma(\text{bbH}|m_{\text{H}} = 125.09\,\text{GeV}, \sqrt{s} = 13\,\text{TeV}) = 0.4863^{+20.1\%}_{-23.9\%}\text{pb}$, $\sigma(\text{tH}, t\text{-channel}|m_{\text{H}} = 125.09\,\text{GeV}, \sqrt{s} = 13\,\text{TeV}) = 0.07426^{+6.5\%}_{-14.7\%}\text{pb}$, and $\sigma(\text{tH}, s\text{-channel}|m_{\text{H}} = 125.09\,\text{GeV}, \sqrt{s} = 13\,\text{TeV}) = 0.02875^{+2.4\%}_{-1.8\%}\text{pb}$ [19]. All four processes can be initiated by either two incoming gluons or a quark and antiquark pair; unlike in the ggF case, there are no loops involved in these processes.

The decay modes of the H boson involve all direct decays to massive particles and loop-induced decays to massless particles. At a value of $m_{\text{H}} = 125.09\,\text{GeV}$ the most prominent branching ratio ($\mathcal{B}$) is that in a pair of b quarks, followed by a pair of W bosons. The specific value m_{H} of the observed Higgs boson is particularly interesting as it allows for the simultaneous probe also of $\text{H} \rightarrow \tau^+\tau^-$, $\text{H} \rightarrow \mu^+\mu^-$, and $\text{H} \rightarrow \text{c}\bar{\text{c}}$.

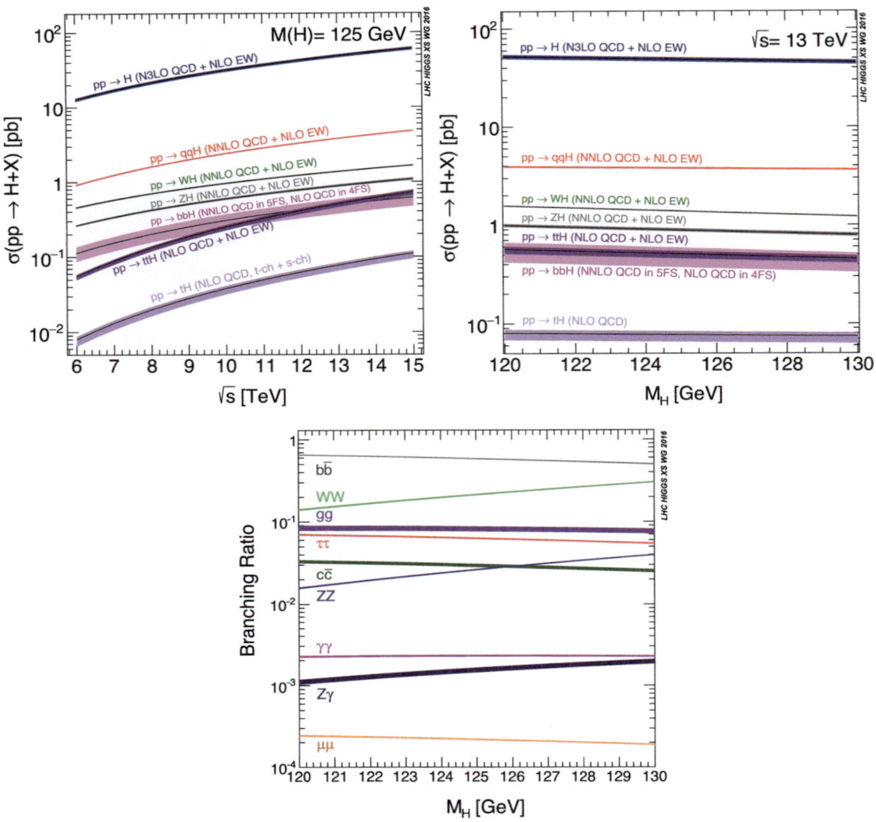

Fig. 1.4 Higgs boson production cross section as a function of the centre-of-mass energy $\sqrt{s}$ (top left), and as a function of the m_{H} (top right), for different production mechanisms. Branching fractions of the decay of a Higgs boson as a function of m_{H} (bottom) [19]

A complete summary of the production cross sections and decay branching ratios of the H boson are given in Fig. 1.4, where they are depicted as a function of the centre-of-mass energy and the H mass hypothesis. Table 1.2 reports the H branching ratios for $m_{\mathrm{H}} = 125.09\,\mathrm{GeV}$.

The discussion reported above represents the theoretical foundation for the experimental probe of the H at the LHC. At the time of writing this Thesis, the most precise measurement of the H mass results from the CMS Collaboration's analysis of the $\mathrm{H} \to \mathrm{ZZ} \to \ell\ell\ell\ell$ decay channel combining the full Run-1 and Run-2 data [20], yielding:

$$m_{\mathrm{H}} = 125.08 \pm 0.10(\mathrm{stat.}) \pm 0.05(\mathrm{syst.})\ \mathrm{GeV} \tag{1.48}$$

The most precise measurement of the H decay width, obtained from the ratio of the on-shell and off-shell H production, has been measured by the ATLAS Collaboration

Table 1.2 Branching fractions of the major H boson decay modes for $m_{\mathrm{H}} = 125.09\,\mathrm{GeV}$. Theoretical uncertainties combine the uncertainties on the H boson partial width, on the value of α_{S}, and on the quark masses [19]

Decay mode	$\mathcal{B}$ [%]
$\mathrm{H} \rightarrow \mathrm{b}\bar{\mathrm{b}}$	$58.09^{+0.72}_{-0.73}$
$\mathrm{H} \rightarrow \mathrm{W}^{\pm}\mathrm{W}^{\mp *}$	21.53 ± 0.33
$\mathrm{H} \rightarrow \mathrm{gg}$	$8, 18 \pm 0.42$
$\mathrm{H} \rightarrow \tau^{+}\tau^{-}$	6.27 ± 0.10
$\mathrm{H} \rightarrow \mathrm{c}\bar{\mathrm{c}}$	$2.88^{+0.16}_{-0.06}$
$\mathrm{H} \rightarrow \mathrm{ZZ}^{*}$	2.641 ± 0.040
$\mathrm{H} \rightarrow \gamma\gamma$	0.2270 ± 0.0047
$\mathrm{H} \rightarrow \mathrm{Z}\gamma$	0.1541 ± 0.0090
$\mathrm{H} \rightarrow \mu^{+}\mu^{-}$	$0.02171^{+0.00036}_{-0.00037}$

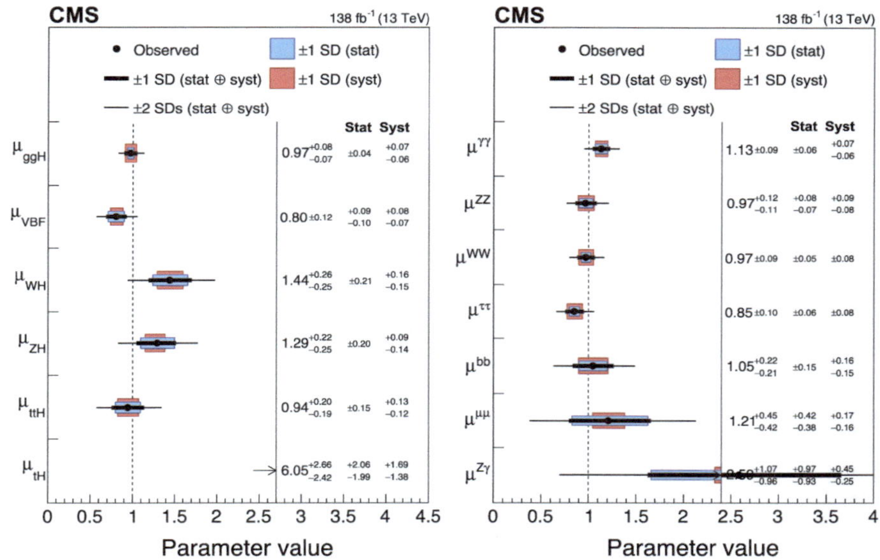

Fig. 1.5 Signal strength parameters extracted for various production modes μ^i, assuming $\mathcal{B}^f = \mathcal{B}^f_{\mathrm{SM}}$ (left), and decay channels μ^f, assuming $\sigma^i = \sigma^i_{\mathrm{SM}}$ (right). The thick (thin) black lines indicate the 1 (2) standard deviation (s.d.) confidence intervals, with the systematic and statistical components of the 1 s.d. interval indicated by the red and blue bands, respectively. The vertical dashed line at unity represents the values of μ^i and μ^f in the SM [33]

at 95% CL [21] to be:

$$\Gamma_{\mathrm{H}} = 4.5^{+3.3}_{-2.5}\,\mathrm{MeV} \tag{1.49}$$

As detailed above, both the production cross section and the decay branching ratios of the H boson are dependent on its mass. The plethora of production mechanisms and decay modes have been extensively probed, and the most recent results from the CMS Collaboration are summarized in Fig. 1.5. The two panels report the agreement

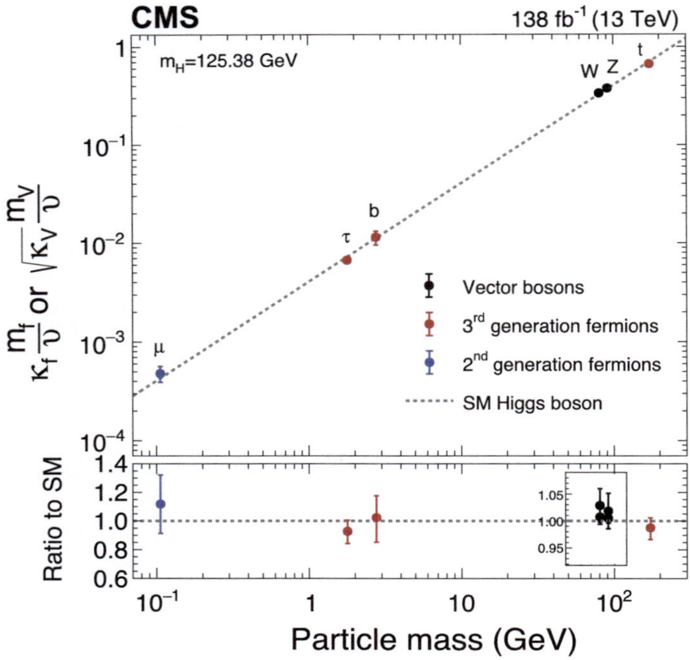

Fig. 1.6 Best fit estimates of the normalized Higgs boson coupling constants as a function of the boson or fermion mass, obtained from the combination of the data collected by the CMS experiment in Run 2, compared to the prediction from the SM [33]

of the experimental measurements with respect to the SM prediction in the form of signal strength modifiers μ. The left panel shows the signal strength modifier for six production mechanisms when the decay branching ratios are fixed at the values predicted by the SM; the right panel presents the signal strength modifier for seven decay modes when the production cross section is fixed at the values predicted by the SM. It is worth highlighting in this picture the outstanding results obtained during Run-2, in which the first measurement of the $\mathrm{H} \to \tau\tau$ decay has been performed [22–24], and the first evidence $\mathrm{H} \to \mu\mu$ decay has been reached [25].

The picture is further enriched by the direct measurement of the Higgs boson couplings to fermions and vector bosons. Their precise measurement constitutes a stringent test of the SM, especially considering the huge number of orders of magnitude that these couplings span. The state-of-the-art measurements performed by the CMS Collaboration are reported in Fig. 1.6.

Two final measurements fully characterise the Higgs boson: its spin-parity and its self-coupling. The first one has been measured to be $J^P = 0^+$ [26, 27] and is thus in agreement with the SM hypothesis. The second represents one of the most challenging measurements at the LHC, and it is one of the topics of this Thesis. A dedicated introduction to it is given in Sect. 1.2.

1.1.4 Weaknesses of the Standard Model

Our confidence in the SM is amply justified on the basis of its ability to accurately describe the bulk of our present-day data and, especially, of its enormous success in predicting new phenomena. Nevertheless, many observations do not find a description within the framework of the SM. A brief, non-exhaustive list of some of these *weaknesses* is reported in the following.

- **Gravity**: the SM does not include gravity, which is described in the theory of general relativity, and whose quantization proved to be not perturbatively renormalizable and thus not compatible with the SM,
- **Neutrinos**: the SM predicts the neutrinos to be massless, but the experimental evidence of neutrino oscillations [9, 28, 29] requires the presence of non-null neutrino masses,
- **Matter-antimatter asymmetry**: the Universe presents a striking imbalance of matter and antimatter, which the SM struggles to explain easily and which requires the introduction of new physics (cf. Sect. 1.1.6),
- **Dark matter**: the measurement of the rotational speed of galaxies [30] suggests the possible existence of a large amount of undetected mass in the Universe. The SM is not able to predict any candidate for this so-called *dark matter*.
- **Dark energy**: precise cosmological observations [31] show that the expansion of the Universe is an accelerating process. This can be understood in the presence of the so-called *dark energy*, which creates a repulsive force that compensates for the gravitational attraction; nevertheless, the SM is not able to predict an interaction of this kind.
- **Higgs boson mass**: the mass of the Higgs boson is a free parameter of the SM, and it is not protected by any symmetry. Thus, m_{H} is sensitive to quantum fluctuations involving masses up to the Planck scale, making it inherently unstable or surprisingly fine-tuned.

These experimental observations suggest that the SM is highly performant at the EW scale but, at the same time, a more fundamental theory beyond the SM is needed to address these shortcomings. Several BSM theories have been proposed throughout the decades, from *Higgs-composite models* to *techni-colour*, from *supersimmetry* to *(super) string theory* [32]. As of today, all of these theories have either proven incorrect or impossible to probe with the technology we dispose of.

1.1.5 Effective Field Theory: the multipole Expansion of the SM

As already remarked, the SM is a renormalizable theory, and it is thus consistent up to the Planck scale, at which point gravity is no longer negligible. Renormalizability has been a long-sought property of all quantum field theories developed during the XXth century; nevertheless, this quality somehow hides one of the fundamental properties of physics itself: the possibility to make predictions only at fixed accuracy.

To understand this point, it is worth considering a very simple example: the electrostatic potential at a large distance. From classical physics, we know that the potential at a distance larger than the charge extension can be expressed in a multipole series. In the limit of very large distances, the lowest monopole is sufficient, and we find the point charge approximation. The smaller the distance, the larger the number of multipoles needed to describe the electrostatic potential accurately. In the limit in which the distance is comparable to the charge dimension, the multipole expansion breaks down as an infinite number of multipoles would be required. This kind of expansion is possible because the electrostatic potential is known only at fixed accuracy at each step. What the expansion does is trade an infinite number of physical variables for just the charge and the distance at the expense of calculating the potential at fixed accuracy via an effective long-distance description. While this example is very simple, the use of effective long-distance descriptions is ubiquitous, and it does not only apply to the static case. Other examples are the classical theory of fluids, where the point particle description is traded for a hydrodynamic description, and the calculation of tidal effects in astrophysical systems [34]. The most prominent example in quantum physics is the Born-Oppenheimer approximation of molecule dynamics [35], where the nucleon dynamics is computed in an effective potential of the electron cloud.

Given this short discussion, there is no apparent reason why the SM should be a theory different from any other and why it should not present the same effective understanding under the correct circumstances. Nonetheless, the requirement for its Lagrangian to be renormalizable elevates the SM to be a fundamental theory and not just an effective one. Effective Field Theory (EFT) was developed to break out of the *renormalizability cage*.

The EFT approach is based on one fundamental concept, that of scale separation. This idea recognizes that in particle interactions, different energy scales can give rise to distinct physical phenomena, thus constructing models specific to a particular energy range, effectively capturing the essential physics of interest while neglecting the higher energy phenomena. More than this, EFT brings forward the idea that interactions of arbitrary complexity that act at higher energy scales can be systematically approximated with a finite number of parameters in the Lagrangian.

To construct the EFT formalism, we introduce the physical scale Λ. From a bottom-up perspective, Λ can be interpreted as the scale up to which only SM fields propagate, while from a top-down perspective, it is the energy scale of the BSM physics itself. In this context, short-distance effects are controlled by an infinite but systematic expansion in powers of $1/\Lambda$, to be added to the SM Lagrangian. This approach opens the way to interaction terms with arbitrary large mass dimensions $\mathcal{D}$, which can be classified in orders of $\mathcal{D}$ and suppressed by powers of Λ. In its general formulation, any EFT can be written as:

$$\begin{aligned}\mathcal{L} &= \mathcal{L}_{\mathrm{SM}} + \sum_{n=5}^{+\infty}\sum_{i} \frac{c_i^{(n)}}{\Lambda^{n-4}}\mathcal{O}_i^{(n)} = \\ &= \mathcal{L}_{\mathrm{SM}} + \sum_{i} \frac{c_i^{(5)}}{\Lambda}\mathcal{O}_i^{(5)} + \sum_{i} \frac{c_i^{(6)}}{\Lambda^2}\mathcal{O}_i^{(6)} + \ldots\end{aligned} \tag{1.50}$$

where $\mathcal{O}_i^{(\mathcal{D})}$ is a complete basis of $\mathrm{SU(2)_L} \otimes \mathrm{U(1)_Y} \oplus \mathrm{SU(3)_C}$ invariant operators of dimension $\mathcal{D}$, and c_i are the so-called Wilson coefficients. The leading order term is the SM with the usual $\mathcal{D} = 4$ dimension. The formulation reported in Eq. 1.50 is generally referred to as Standard Model EFT, and it is only one of the possible EFT approaches; more details are discussed later in this Chapter.

The theory thus obtained is not fully renormalizable but only renormalizable order by order in Λ. Nevertheless, this does not constitute an issue as an EFT only represents the lower energy manifestation of a more extended theory at higher scales. Moreover, in the obtained theory, perturbative computations can be consistently performed at any order, and the theory is predictive.

The only important assumption that must be enforced for this approach to be valid is that all new physics and new states are above the scale $\Lambda \gg \sqrt{s}$. If new physics were accessible already at the centre-of-mass energy of the LHC, the EFT approach would not be applicable as the multipole expansion did not work for the electrostatic charge.

Once the EFT is defined, any UV-complete BSM model can be matched to it, i.e. reduced to its lower scale manifestation to derive an expression of the Wilson coefficients in terms of the fundamental model parameters. From an experimental point of view, an EFT provides a generic parametrization to investigate several BSM signatures with a model-independent approach.

1.1.6 Cosmological Importance of the BEH Mechanism

As mentioned above, the SM is a renormalizable theory, and it is self-consistent up to the Planck scale $M_{\mathrm{P}} \sim 10^{19}$ GeV. While there are substantial reasons why the SM should not be regarded as the *ultimate* theory of Nature up to these energies (one among others being the immense extrapolation of ~ 16 orders of magnitude), the predictive power granted by this assumption is extremely important and should still be regarded as a useful feature. Specifically, being renormalizable to the energies where quantum gravity becomes non-negligible, the SM provides a solid theoretical framework for describing the early Universe.

The BEH field represents a somewhat peculiar part of the SM, as it is the only known scalar field in the theory, it is the only field not arising from a local gauge invariance property of the Lagrangian, and it breaks the degeneracy between the three families of fermions. Given these unique properties, the BEH field might have important cosmological implications relative to the evolution of the Universe: it could have played an important role in the inflation [36] and concurrently favoured a flat, homogeneous, and isotropic Universe [37]; it could be at the basis of dark matter production [38]; modifications of the H self-coupling could be central for baryogenesis, impacting the cosmological phase transitions and leading to a charge asymmetric Universe [39, 40]; its shape and v.e.v. could determine the destiny of the Universe through electroweak vacuum instability [41].

These strong connections between cosmology and particle physics gave birth to the fast-growing field of Higgs cosmology, which aims to shed light on some puzzling questions about the Universe and its history. While some of the questions would need direct cosmological observations to be answered, e.g. with Pulsar Timing Arrays (PTAs) or with the current Laser Interferometer Gravitational-Wave Observatory (LIGO) experiment or the future Laser Interferometer Space Antenna (LISA) detector, others are strongly tied to searches at particle accelerators. In this context, we discuss the importance of the BEH field in understanding the electroweak baryogenesis problem.

Empirically, all particle interactions we observe conserve the baryon number $B = \frac{1}{3}(n_{\mathrm{q}} - n_{\bar{\mathrm{q}}})$, thus the same number of baryons and antibaryons needs to be produced or destroyed by any process. From this observation stems one of the main puzzles in cosmology and particle physics: what is the origin of the matter-antimatter asymmetry observed in the Universe? [42–44].

Electro-Weak Baryogenesis (EWBG) is one of the most attractive ways to account for it by considering that the asymmetry originated during the Electro-Weak Phase Transition (EWPT) of the Universe. The primary idea of EWBG is that of a hot early Universe in which the electroweak $\mathrm{SU(2)_L} \otimes \mathrm{U(1)_Y}$ symmetry is conserved and in which a net baryon charge equal to zero is observed. As the cool-down of the Universe progresses, when a critical temperature threshold is reached, the BEH field triggers EWSB, and in the ensuing EWPT, the electroweak symmetry is reduced to the electromagnetic $\mathrm{U(1)_{EM}}$ symmetry and EWBG takes place. For this process to effectively happen, three conditions need to be fulfilled, the so-called Sakharov conditions [45]:

- baryon number B must be violated
- charge conjugation (C) and charge-parity (CP) symmetries must be violated
- the process must take place out of equilibrium

Remarkably, the SM could *a priori* account for all three conditions, although stringent bounds on them have been set as discussed below.

Experiments have never observed baryon number violation; this observation can be understood by noticing that in the current broken symmetry vacuum state in which the Universe is found, B-violating processes are suppressed by the large value of the v.e.v. of the BEH field. Nevertheless, in the early Universe featuring an unbroken $\mathrm{SU(2)_L} \otimes \mathrm{U(1)_Y}$ symmetry, this suppression would not be present and B-violating processes could take place.

In the SM, C is maximally violated by weak interaction; nevertheless, CP violation in the SM induced by the CKM phase does not appear to be sufficient [46, 47] and recent precision measurements of the neutron electric dipole moment [48] set stringent limits on the strong CP-violating phase. At present, this is the most stringent limitation of the Sakharov conditions; nevertheless, some models have been proposed to overcome the CP issue while being compatible with experimental data either by introducing an additional singlet scalar to the theory [49] or by considering the SM Yukawa couplings to be variable at the same time as the BEH field is acquiring its v.e.v [50].

Finally, the last Sakharov condition can be satisfied if the process of EWPT is a strong first-order transition. In this scenario, the EWPT proceeds through the so-called *nucleation* in which bubbles of the BEH vacuum form in a localized space-time point and grow in dimension until they fill the entire Universe; in this expanding motion, the wall of the bubbles would provide the required non-equilibrium condition. Therefore, the early Universe would resemble a boiling medium; the bubble dynamics would entail their expansion and collision within the external medium still in a different vacuum state. This dynamics can be the source of a primordial stochastic gravitational-wave background originating from the superposition of bubble and sound-wave contributions, which can be experimentally tested via cosmological observations. At the time of writing this Thesis, a collection of very interesting papers on cosmological measurements with a PTA has been published in The Astrophysical Journal Letters. This collection focuses on the measurements of the signal from 67 pulsars performed by the North American Nanohertz Observatory for Gravitational Waves (NANOGrav) Collaboration. Two particularly interesting results are the evidence for a stochastic gravitational-wave background [51] and its interpretation in the context of new physics scenarios [52]. In these results, it is stated that the NANOGrav data is compatible with models in which the gravitational-wave background originates from a strong first-order EWPT. In the SM, the need for such a first-order EWPT would be automatically fulfilled if the mass of the H was $m_{\mathrm{H}} \lesssim 70\,\mathrm{GeV}$ [53, 54], but the discovery of the H with a mass about 125 GeV prevents this possibility, and a more complex scenario is needed.

A plethora of BSM theories have been proposed to account for a strong first-order EWPT, generally including an additional bosonic degree of freedom with sizeable coupling to the SM H. Of particular interest for this Thesis are those that feature modifications of the H self-coupling λ_{HHH} to account for the necessary deviation from the bare SM prediction and where the connection between λ_{HHH} and the EWPT is the following. Denote ϕ_c the expectation value of the BEH field at the EWSB critical temperature T_c; the requirement for a strong first-order EWPT can be expressed as $\phi_c/T_c \gtrsim 1$. As a modification of the H self-coupling directly induces a modification of the shape of the BEH potential, the value of ϕ_c is also directly intertwined with the λ_{HHH} value. Two models that follow this line of reasoning are discussed in the following.

A model of particular interest is the extension of the SM with a scalar singlet (S), as this possibility is largely unprobed at the LHC [39]. Such a theory can be categorised under the Z_2 parity of S, which can be symmetric or can spontaneously break Z_2. The latter is a notably suitable scenario as a large relative phase between the SM box diagram and the singlet triangle diagram can have a large impact on HH production. This minimal extension of the SM features a mixing between the H and

S, and three additional parameters are added to the theory: the mass of the singlet m_S, the the H-S mixing angle $\sin\theta$, and the ratio of v.e.v.s $\tan\beta \equiv \frac{v_S}{v}$. The following two dimensionful couplings then characterise the phenomenology of HH production:

$$
\begin{aligned}
\lambda'_{\mathrm{HHH}} &= -\frac{m_{\mathrm{H}}^2}{2v\tan\beta}\left(\tan\beta\cos^3\theta - \sin^3\theta\right) \\
\lambda_{\mathrm{SHH}} &= -\frac{m_{\mathrm{H}}^2}{2v\tan\beta}\sin 2\theta\left(\tan\beta\cos\theta - \sin\theta\right)\left(1 + \frac{m_{\mathrm{S}}^2}{2m_{\mathrm{H}}^2}\right)
\end{aligned}
\tag{1.51}
$$

where λ'_{HHH} is the modified H self-coupling and λ_{SHH} is the coupling between the heavy scalar and two Higgs bosons. The interplay between these two couplings has fundamental implications on the production of HH at hadron colliders, possibly leading to enhancements of the HH production cross section and modifications of the HH invariant mass differential distribution. An example of such effect is reported in Fig. 1.7 where the direct contribution of the S extension is reported (red, brown, and blue) alongside the SM prediction (grey) and the full result for the singlet SM extension (black). As it can be appreciated, the interference between the S and the H can lead to a consistent modification of the m_{HH} spectrum. From an experimental point of view, this model calls for a differential analysis of the HH invariant mass, which could be targeted by the HH searches exploiting the combined Run-2 and Run-3 dataset, and further on exploiting the HL-LHC dataset whose unparalleled statistical power is central to such a study.

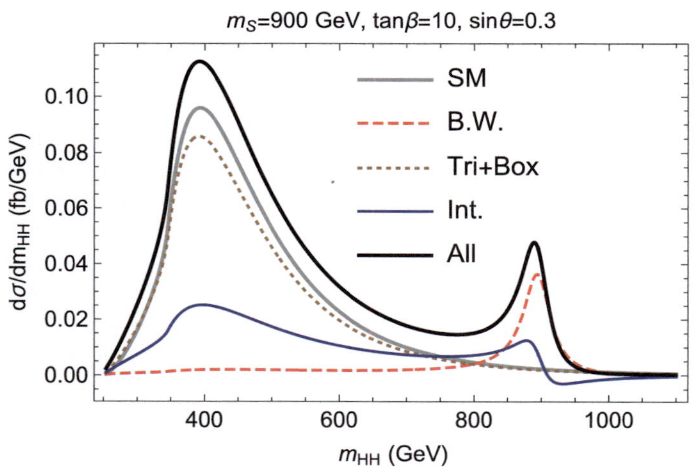

Fig. 1.7 Differential distribution of the di-Higgs invariant mass for a benchmark point of the SM singlet extension. The contributions from the resonant singlet diagram, the non-resonant diagram and the interference between them are shown in red (dashed), brown (dotted) and blue curves, respectively. The SM prediction is shown in grey and the full result for the singlet SM extension is reported in black [39]

A second important extension of the SM to achieve strong first-order EWPT is based on an EFT approach. Assuming that new physics appears only in the scalar sector, several operators can bypass the issue of a Higgs boson mass of about 125 GeV and induce sizeable modifications of the H self-coupling, inducing a first-order EWPT. Some examples of such EFT operators are:

$$\lambda_6\left(\phi^{\dagger}\phi\right)^3/\Lambda^2, \qquad -\lambda_{\ln,2}\frac{\phi^2\Lambda^2}{100}\ln\frac{\phi^2}{2\Lambda^2}, \qquad \lambda_{\exp,4}\phi^4\exp\left\{-\frac{2\Lambda^2}{\phi^2}\right\} \tag{1.52}$$

where λ_j are dimensionless coefficients [40]. The introduction of such extensions to the SM Lagrangian all have an important impact on the value of λ_{HHH}. For example, in the case of the operator of dimension six reported in Eq. 1.52, the H self-coupling would be altered as follows

$$\lambda'_{\mathrm{HHH}} = \frac{m_{\mathrm{H}}^2}{2v}\left(1+\frac{16\lambda_6 v^4}{m_{\mathrm{H}}^2\Lambda^2}\right) = \lambda_{\mathrm{HHH}}\left(1+\frac{16\lambda_6 v^4}{m_{\mathrm{H}}^2\Lambda^2}\right) \tag{1.53}$$

where λ'_{HHH} is the modified H self-coupling, which in the limit $\Lambda \to +\infty$, i.e. no new physics, approaches the SM predicted self-coupling. Figure 1.8 explicitly shows the connection between the strength of the possible self-coupling deviation visible at the LHC and the order of the EWPT as a function of the coupling of the specific operator included in the EFT approach. Red dotted lines represent second-order transitions, while blue lines denote first-order transitions. As it can be appreciated, modifications of the H self-coupling down to $\sim 40-50\%$ can account for a strong first-order EWPT for different operators.

Deviations of such magnitudes might be already accessible with the statistical power available at the end of the LHC Run-3 operations in 2025, at which point an expected $\sim 400\,\mathrm{fb}^{-1}$ will have been collected. Exploiting the complete Run-2 and Run-3 datasets, setting a 95% confidence level limit on $\kappa_\lambda \lesssim 1.5$ should be an achievable goal via the combination of multiple HH analyses within the CMS

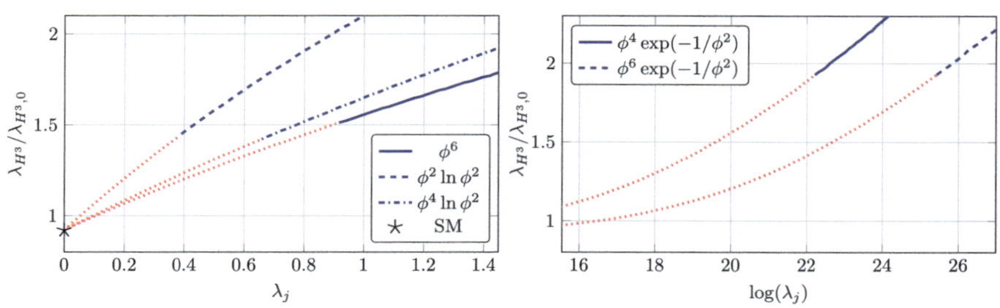

Fig. 1.8 Modification of the H self-coupling (here reported as λ_{H^3}) with respect to the SM prediction (here denoted as $\lambda_{\mathrm{H}^3,0}$) as a function of the λ_j coefficients reported in Eq. 1.52 for different UV potentials. Blue lines represent first-order phase transitions, and red dotted lines depict second-order phase transitions. The new physics cutoff is set to $\Lambda = 2$ TeV [40]

Collaboration alone. Therefore, an extensive effort in such direction should be strived for by all analyses sensible to λ_{HHH} so that stringent limits on such models can be set.

The discussion reported in this Section is not intended as a complete overview of the field of Higgs cosmology, nor of EWBG. Nonetheless, it serves to pinpoint the fundamental importance that the BEH field could play in it and some testable scenarios at the LHC. Performing such interpretations of the LHC data will be an important part of the study of λ_{HHH}, and the recent cosmological observations make the picture even more interesting. While the Run-2 dataset was not exploited to this end, Run-3 represents an immense opportunity to perform such studies. It should also be noted that, most likely, only the statistical power of the $\sim 4000\,\mathrm{fb}^{-1}$ HL-LHC dataset will be sufficient for the rejection of many BSM scenarios relevant to EWBG.

1.2 Higgs Boson Pair Production

The study of Higgs boson pair production is of foremost importance because it is the only direct way to reconstruct the BEH doublet field scalar potential that is responsible for EWSB. Of utmost importance is the measurement of λ_{HHH}, which is the sole parameter controlling the shape of the BEH potential (1.41). Its value is very sensitive to possible variations from BSM physics and could explain some of the open cosmological puzzles. Moreover, the tri-linear H coupling is not a free parameter of the theory but is fully determined once the mass of the H is measured. Therefore, after the precise measurement of its mass, the self-coupling is strictly predicted by the SM to be $\lambda_{\mathrm{HHH}} \sim 0.13$ according to Eq. 1.42, and its measurement is an important consistency test of the SM. It amounts to a closure test of the BEH mechanism and its involvement in the process of EWSB. This examination corresponds to experimentally verifying that the H boson fits into the global picture illustrated in Fig. 1.6 at the ordinate $\sqrt{2\lambda_{\mathrm{HHH}}} = \frac{m_{\mathrm{H}}}{v} \sim 0.51$, thus verifying the profound significance of the H boson in the process of mass generation.

The importance of λ_{HHH} and the study of HH was first highlighted in 1988 when the initial computation of HH production cross section was performed [55]. Nevertheless, the tri-linear self-coupling is not the only interaction that contributes to the HH production; specific production mechanisms involve the Yukawa coupling to the t quark (y_{t}), as well as the coupling of the HH to one and two vector bosons (c_{V} and $c_{2\mathrm{V}}$, respectively). Therefore, a careful disentanglement of the tri-linear contribution from the other production modes is needed, and a simultaneous probe of diverse coupling is possible. In the following, we present the main HH production channels predicted by the SM and the main EFT benchmarks encoding some BSM scenarios.

1.2.1 HH Production in the SM

According to the SM, at the LHC, HH production can happen via five main mechanisms [56–58] that are listed below. For each process, the associated Leading Order (LO) Feynman diagrams are reported alongside the most recent cross section prediction.

- **Gluon Fusion (ggF)**
  The gluon fusion mechanism is mediated by loops of heavy quarks, mainly t with a contribution from the b quark smaller than 1% at LO. Two Feynman diagrams can be drawn at LO, both reported below. The first involves the tri-linear coupling λ_{HHH}, and the HH is produced when an off-shell H splits into two on-shell H. In the second, each H is radiated directly off the heavy quark loop and involves the top Yukawa coupling y_{t}. The two Feynman diagrams make explicit the usual jargon used to address them as *triangle* and *box* diagrams. The calculation of the cross section of this process has considerably improved over the past years, including higher order contributions and shrinking the uncertainty bands (cf. Refs. [59–61]), arriving at the present predicted value [62] of

$$\sigma^{\mathrm{ggF}}_{\text{NNLO-FTapprox}}(\sqrt{s} = 13\,\mathrm{TeV}, m_{\mathrm{H}} = 125\,\mathrm{GeV}) = 31.05^{+2.2\%}_{-5.0\%}(\mathrm{scale})^{+4.0\%}_{-18.0\%}\%(m_{\mathrm{t}}) \pm 2.1\%(\mathrm{PDF}) \pm 2.1\%(\alpha_{\mathrm{s}})\mathrm{fb} \quad (1.54)$$

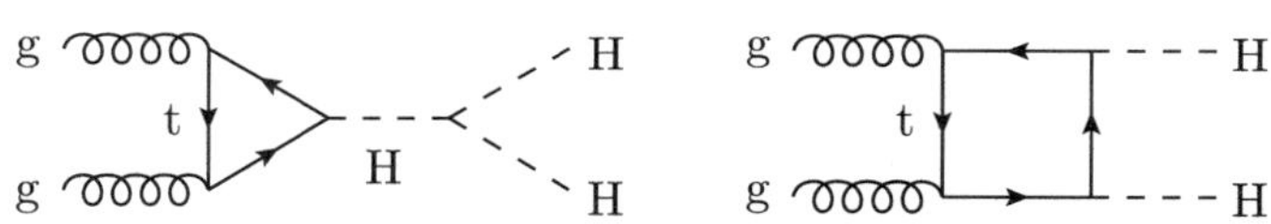

  A fundamental remark needs to be made regarding the triangle and box diagrams: while their amplitude is of similar magnitude, they interfere destructively. This effect, combined with the two H bosons' restricted production phase space, yields the very small cross section reported above. Nonetheless, this destructive interference can be exploited to probe BSM scenarios, which can largely modify the interference behaviour.
- **Vector Boson Fusion (VBF)**
  The vector boson fusion mechanism can be seen as the double elastic scattering of two quarks with two HH radiated off the weak bosons that fuse. At LO, three Feynman diagrams can be drawn and are reported below. The one on the left involves the tri-linear coupling λ_{HHH}, whereas the one in the middle and the one on the right involve the coupling of a H boson with one vector boson (c_{V}) and the coupling of a HH with two vector bosons ($c_{2\mathrm{V}}$), respectively. The two final state jets are generally produced with a very large separation angle and constitute a clean signature for the rejection of background. The calculation of the cross section of this process has also considerably improved during the past years, including higher order contributions and shrinking the uncertainty bands (cf. Ref. [63–67]), arriving at the present predicted value [68] of

$$\sigma^{\mathrm{VBF}}_{\mathrm{N3LO\ QCD}}(\sqrt{s} = 13\,\mathrm{TeV}, m_{\mathrm{H}} = 125\,\mathrm{GeV}) = 1.739^{+0.03\%}_{-0.04\%}(\mathrm{scale}) \pm 2.1\%(\mathrm{PDF} + \alpha_{\mathrm{s}})\mathrm{fb} \tag{1.55}$$

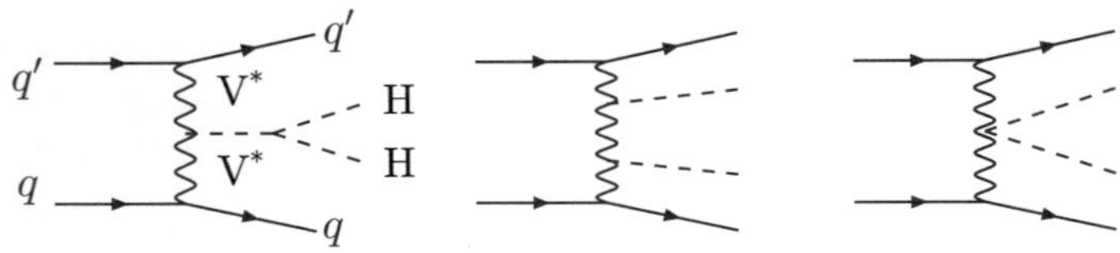

- **Double Higgs-strahlung (VHH)**
  In the double Higgs-strahlung mechanism, a quark interacts with an antiquark to produce a massive vector boson ($\mathrm{W}^{\pm}$ or Z), which radiates the HH. At LO, three Feynman diagrams can be drawn and are reported below. The one on the left involves the tri-linear coupling λ_{HHH}, while the remaining two include the same couplings discussed for the VBF production. The current predicted cross section for this mechanism is decomposed based on the vector boson considered, with values [57] of

$$\begin{aligned}
\sigma^{\mathrm{ZHH}}_{\mathrm{NNLO\ QCD}}(\sqrt{s} = 13\,\mathrm{TeV}, m_{\mathrm{H}} = 125\,\mathrm{GeV}) &= 0.363^{+3.4\%}_{-2.7\%}(\mathrm{scale}) \pm 1.9\%(\mathrm{PDF} + \alpha_{\mathrm{s}})\mathrm{fb} \\
\sigma^{\mathrm{W^{+}HH}}_{\mathrm{NNLO\ QCD}}(\sqrt{s} = 13\,\mathrm{TeV}, m_{\mathrm{H}} = 125\,\mathrm{GeV}) &= 0.329^{+0.32\%}_{-0.41\%}(\mathrm{scale}) \pm 2.2\%(\mathrm{PDF} + \alpha_{\mathrm{s}})\mathrm{fb} \\
\sigma^{\mathrm{W^{-}HH}}_{\mathrm{NNLO\ QCD}}(\sqrt{s} = 13\,\mathrm{TeV}, m_{\mathrm{H}} = 125\,\mathrm{GeV}) &= 0.173^{+1.2\%}_{-1.3\%}(\mathrm{scale}) \pm 2.8\%(\mathrm{PDF} + \alpha_{\mathrm{s}})\mathrm{fb}
\end{aligned} \tag{1.56}$$

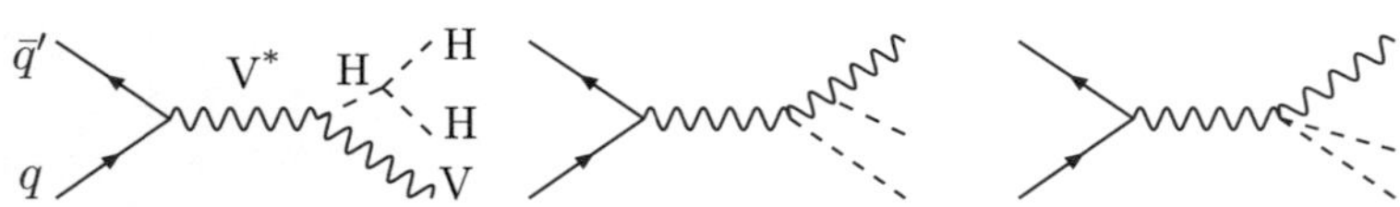

- **tt̄ associated production** (ttHH)
  In the top quark-antiquark pair (tt̄) associated production involves the production of a tt̄ pair that subsequently radiates the HH. At LO, three Feynman diagrams can be drawn and are reported below. This process can either involve the H boson self-coupling or two H bosons can be radiated off the two t quarks independently. Its cross section exceeds that of the VBF mechanism at high centre-of-mass energies and in the high transverse momentum regime of the HH. The current estimate of the cross section at the LHC [58] is

$$\sigma^{\mathrm{ttHH}}_{\mathrm{NLO\ QCD}}(\sqrt{s} = 13\,\mathrm{TeV}, m_{\mathrm{H}} = 125\,\mathrm{GeV}) = 0.775^{+1.5\%}_{-4.3\%}(\mathrm{scale}) \pm 3.2\%(\mathrm{PDF} + \alpha_{\mathrm{s}})\mathrm{fb} \tag{1.57}$$

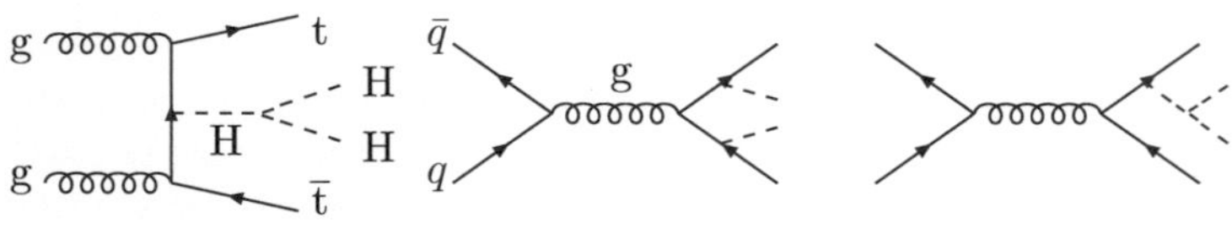

- **Single t quark associated production** tHH
  The single t quark associated production is the least contributing mechanism and

has two contributions from the t- and s-channel production of the t quark, whose LO Feynman diagrams are reported in the top and bottom row of the Figure below. This process is the only one that gives simultaneous access to the HH coupling to both vector bosons and top quarks, allowing to probe the relative phase between c_V, c_{2V}, and y_t. Nevertheless, this process cannot be directly targeted at the LHC due to its small cross section [58] of

$$\sigma^{\mathrm{tqHH}}_{\mathrm{NLO\ QCD}}(\sqrt{s} = 13\,\mathrm{TeV}, m_{\mathrm{H}} = 125\,\mathrm{GeV}) = 0.0289^{+5.5\%}_{-3.6\%}(\mathrm{scale}) \pm 4.7\%(\mathrm{PDF} + \alpha_s)\mathrm{fb} \tag{1.58}$$

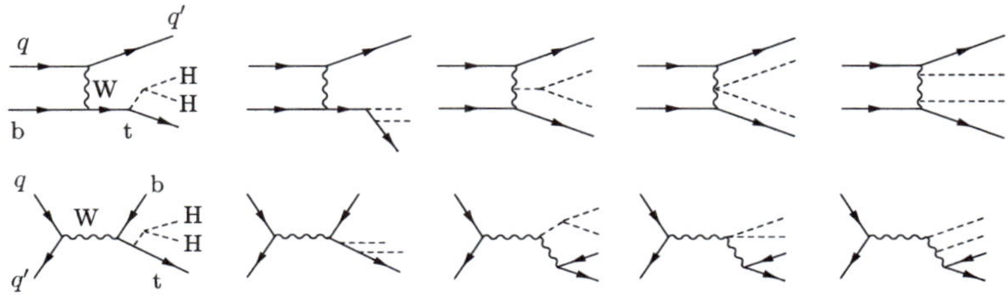

The cross section of the five production mechanisms discussed above is represented as a function of the centre-of-mass energy of the collisions in Fig. 1.9. HH production is a very rare mechanism at the LHC therefore not all processes described are accessible with the dataset collected thus far. Namely, only the ggF and VBF mechanisms are directly targeted in the Run-2 CMS analyses, like that presented in this Thesis. While the former benefits from a larger cross section, the latter is advantaged by the clean signature of the VBF jets. Given these two processes, four couplings are directly accessible: λ_{HHH}, y_t, c_V, and c_{2V}.

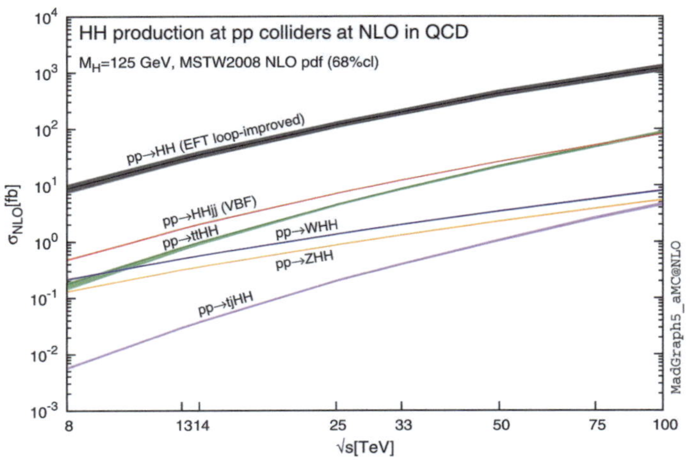

Fig. 1.9 Total cross section for HH production in pp collisions as a function of the centre-of-mass energy for the production modes described in the text. Note that for ggF the figure reports the dependence at Next-to-Leading Order (NLO) FTapprox, but this cross section is now known at Next-to-Next-to-Leading Order (NNLO) with finite top quark mass effects [58]

1.2.2 HH Production Beyond the SM

As already stressed multiple times, the search for HH production is an important channel to probe the presence of physics beyond the SM. Of particular interest is the case in which new physics lies beyond the energy reach of the LHC; in this context, new resonances could not be produced, but their effects could be noticeable through quantum corrections to lower energy processes.

Two separate approaches can be employed to test BSM models. The first and simplest is the introduction of anomalous couplings; the second is the use of EFT. Both methods are described in the following.

Anomalous Couplings

Several BSM models predict diverse modifications of the Higgs boson couplings. Any modification of such couplings effectively leads to a modification of the HH production cross section. All BSM scenarios of this kind can be tested in a model-independent approach referred to as κ-framework, which serves as a parametric approach to looking for hints of new physics.

In the κ-framework, deviations from the standard model are quantified in terms of the so-called κ-modifiers, which are defined as $\kappa_c = c/c^{\mathrm{SM}}$, the ratio between the observed coupling value and its SM prediction. The most important modifiers of interest for HH production are those for the tri-linear coupling $\kappa_\lambda = \lambda_{\mathrm{HHH}}/\lambda_{\mathrm{HHH}}^{\mathrm{SM}}$, for the top Yukawa coupling $\kappa_{\mathrm{t}} = y_{\mathrm{t}}/y_{\mathrm{t}}^{\mathrm{SM}}$, and for the coupling to one or two vector bosons $\kappa_{\mathrm{V}} = c_{\mathrm{V}}/c_{\mathrm{V}}^{\mathrm{SM}}$ and $\kappa_{2\mathrm{V}} = c_{2\mathrm{V}}/c_{2\mathrm{V}}^{\mathrm{SM}}$.

With this approach, the prediction of the production cross section of HH and their kinematics can be obtained as a parametric function of the κ-modifiers. In the case of the gluon fusion and vector boson fusion mechanisms, based on the LO Feynman diagrams reported in Sect. 1.2.1, we can parametrise the deviations of the production cross section from the SM prediction as:

$$\frac{\sigma_{\mathrm{ggF}}}{\sigma_{\mathrm{ggF}}^{SM}} \sim \kappa_\lambda^2\kappa_{\mathrm{t}}^2|\mathcal{A}_\triangle|^2 + \kappa_{\mathrm{t}}^4|\mathcal{A}_\square|^2 + \kappa_\lambda\kappa_{\mathrm{t}}^3\mathcal{I}_{\triangle\square} \tag{1.59}$$

$$\begin{aligned}\frac{\sigma_{\mathrm{VBF}}}{\sigma_{\mathrm{VBF}}^{SM}} \sim\ & \kappa_{2\mathrm{V}}^2|\mathcal{A}_{\mathbf{X}}|^2 + \kappa_{\mathrm{V}}^4|\mathcal{A}_{\mathbf{W}}|^2 + \kappa_{\mathrm{V}}^2\kappa_\lambda^2|\mathcal{A}_{\mathbf{Y}}|^2 + \\ & + \kappa_{2\mathrm{V}}\kappa_{\mathrm{V}}^2\mathcal{I}_{\mathbf{XW}} + \kappa_{2\mathrm{V}}\kappa_{\mathrm{V}}\kappa_\lambda\mathcal{I}_{\mathbf{XY}} + \kappa_{\mathrm{V}}^3\kappa_\lambda\mathcal{I}_{\mathbf{WY}}\end{aligned} \tag{1.60}$$

where $\mathcal{A}$ are the amplitudes associated with the Feynman diagrams and $\mathcal{I}$ denote the interference terms. The subscripts identify the specific diagram to which each term is associated, and they are chosen to resemble the shape of the LO diagram. (A detailed description of this parametrization is discussed in Sect. 5.4.2 when the modelling of the signal is presented as part of the analysis strategy). The dependency of the cross section for all HH production mechanisms is reported as a function of κ_λ in Fig. 1.10. In this specific case, all other couplings are assumed to be those predicted by the SM. As it can be appreciated, modifications of λ_{HHH} have a profound impact on all of them. It is particularly important to notice how the minimum of the cross

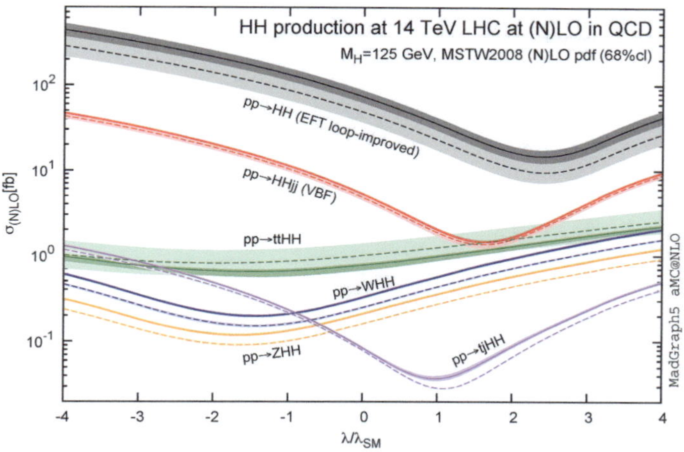

Fig. 1.10 Total cross section for HH production in pp collisions as a function of the κ_λ modifier for the production modes described in the text. Note that for ggF the figure reports the dependence at Next-to-Leading Order (NLO) FTapprox, but this cross section is now known at Next-to-Next-to-Leading Order (NNLO) with finite top quark mass effects [58]

section is found for $\kappa_\lambda = 2.45$. This behaviour is due to the fundamental and large interference between the triangle and box diagrams of the ggF mechanism.

The impact of the κ-modifiers on the production cross sections is but one part of their effect. The kinematics of the Higgs boson pair is also highly affected by them. For this reason, several experimental analyses use the total cross section jointly with the m_{HH} differential distribution to constrain the value of λ_{HHH}.

Figure 1.11 illustrates the effect of different values of κ_λ. The ggF production mechanism dominates these distributions, and the extreme values of κ_λ are easily interpretable. For $\kappa_\lambda = 0$, the box diagram is the only one contributing to produce a broad and hard m_{HH} spectrum; in contrast, when $\kappa_\lambda = 20$, the triangle diagram is the main contributor to a softer m_{HH} spectrum. In the SM configuration, the interference of the two Feynman diagrams produces a broad peak at $m_{\text{HH}} \sim 400$ GeV, while in the maximal interference configuration for $\kappa_\lambda = 2.45$ a huge deep in the distribution is found at $m_{\text{HH}} \sim 345$ GeV.

Figure 1.12 displays the effect of $\kappa_{2\text{V}}$ on the differential distribution of the HH mass produced via the VBF mechanism. As it can be appreciated, a relatively smooth but consistent modification of the kinematics of the HH is expected. This large effect of the $\kappa_{2\text{V}}$ modifier makes the searches for HH production particularly sensible for probing BSM models with a sizeable effect in the VBF mechanism.

While the κ-framework constitutes a simple and effective approach to testing BSM effects in a fully model-independent approach, it suffers from two fundamental drawbacks. The first is purely theoretical and related to the λ_{HHH} coupling not being a free parameter. Having measured the mass of the Higgs boson, the value of the tri-linear self-coupling is fully determined by Eq. 1.42; therefore, a modification of

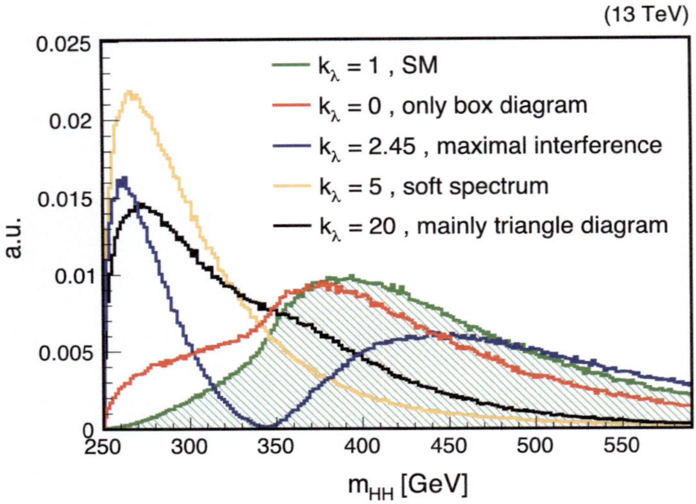

Fig. 1.11 Comparison of the $m_{\rm HH}$ distribution for different values of the κ_λ modifier [69]

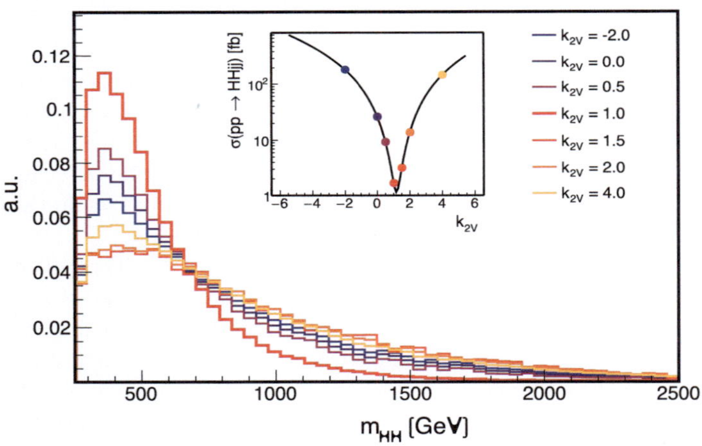

Fig. 1.12 Comparison of the $m_{\rm HH}$ distribution for different values of the $\kappa_{\rm 2V}$ modifier

its value via a multiplicative modifier effectively introduces an inconsistency in the SM itself. The second shortcoming is the complex theoretical calculations needed by such a method. For each specific value of a κ-modifier a series of highly complex, time-consuming, and energy-demanding computations need to be performed. This approach is hardly pursuable as ever-increasing precision is required. The method described in the following Section represents a more consistent and viable approach.

Effective Field Theory

The concept of EFT was introduced in Sect. 1.1.5; in the following, we will discuss the specific use of EFT in the context of HH searches. This formalism is employed to obtain a consistent parametrization of anomalous H properties under the assumption that anomalous H couplings are the dominant effects of BSM physics in the electroweak sector.

Two EFT approaches are currently being discussed: the Standard Model EFT (SMEFT) and the Higgs EFT (HEFT). In the SMEFT framework, the theory is formulated to be $SU(2) \otimes U(1)$ invariant, like the SM, the Higgs field is considered to be a doublet of the theory, and the scalar potential minimization breaks the electroweak symmetry [70, 71]. In the HEFT framework, the theory describes the low-energy dynamics using a non-linear realization of $SU(2) \otimes U(1)$ and the Higgs field is interpreted as a singlet of the theory [72]. This second method to EFT results in a more generic phenomenology at the cost of a larger number of operators to be studied and constrained. It is important to notice that, notwithstanding the different formal approaches of the two EFTs, they still enumerate the same set of operators encoding BSM physics. A detailed overview of both methods can be found in Ref. [72, 73].

In the EFT formalism, there is only one $\mathcal{D} = 5$ operator, and it violates the lepton number [74]. Since lepton number conservation is stringently constrained by experiment, the leading corrections to the Higgs sector are expected from $\mathcal{D} = 6$ operators. In this context, and including only terms relevant for ggF production, the Lagrangian can be written as:

$$
\begin{aligned}
\mathcal{L}_{\mathrm{EFT}}^{\mathrm{HH}} = & \frac{1}{2}\partial^{\mu} H \partial_{\mu} H - \frac{1}{2} m_{\mathrm{H}}^2 H^2 + \kappa_{\lambda} \lambda_{\mathrm{HHH}} v H^3 \\
& - \frac{m_{\mathrm{t}}}{v} \left(v + \kappa_{\mathrm{t}} H + \frac{c_{\mathrm{tt}}}{v} H^2 \right) (\bar{t}_{\mathrm{L}} t_{\mathrm{R}} + t_{\mathrm{R}} \bar{t}_{\mathrm{L}}) \\
& + \frac{\alpha_{\mathrm{s}}}{12\pi v} \left(c_{\mathrm{g}} H - \frac{c_{\mathrm{gg}}}{2v} H^2 \right) G_{\mu\nu}^{a} G_{a}^{\mu\nu}
\end{aligned}
\tag{1.61}
$$

where the dependence on the effective Higgs boson couplings is rendered explicit to provide a simple physics interpretation. The new terms in the Lagrangian predict a point interaction between two t quarks and two H (c_{tt}, or sometimes c_2), between two gluons and two H (c_{gg}), and between two gluons and one H (c_{g}). This amounts to the modification of the already existent triangle and box diagrams and, at the same time, the introduction of new LO Feynman diagrams for the new interaction as reported in Fig. 1.13. While the Lagrangian in Eq. 1.61 stops at $\mathcal{D} = 6$, recent theoretical developments have gone beyond it by also including squared $\mathcal{D} = 6$ operators and double insertions of operators [75].

The formulation in Eq. 1.61, is specific to the HEFT formalism and has some important aspects compared to the one we would obtain from the SMEFT approach. In the SMEFT parametrization, the c_{t} and c_{tt} couplings, and also the c_{g} and c_{gg} couplings, are subject to strict dependency relations; moreover, in the SMEFT approach, a chromo-magnetic interaction (i.e. a point interaction involving a gluon, a t quark,

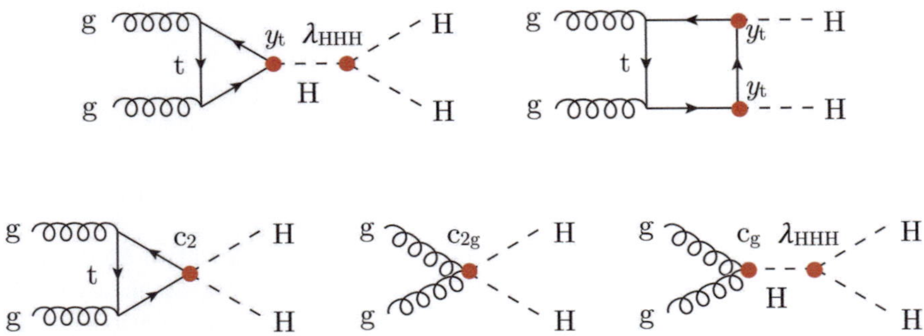

Fig. 1.13 Leading order Feynman diagrams in the HEFT description at dimension $\mathcal{D} = 6$ for the ggF production mechanism [69]

and a H) enters the computation at LO precision. Neither of these features is present in the HEFT Lagrangian, where the chromo-magnetic interaction does not appear at LO but only at Next-to-Leading Order (NLO) precision, and the five couplings are all independent.

The cross section for HH production can then be computed based on the new set of Feynman diagrams introduced by the EFT formulation. Following Refs. [76–78], the cross section normalized to the SM prediction can be parametrized at LO precision in terms of the 15 coefficient $A_j|_{j=1,...,15}$, and can be written as:

$$\begin{aligned} R_{HH} = \frac{\sigma_{HH}}{\sigma_{HH}^{SM}} \overset{LO}{=} &\; A_1\kappa_t^4 + A_2c_2^2 + A_3\kappa_t^2\kappa_\lambda^2 + A_4c_g^2\kappa_\lambda^2 + A_5c_{2g}^2 \\ &+ A_6c_2\kappa_t^2 + A_7\kappa_\lambda\kappa_t^3 + A_8\kappa_t\kappa_\lambda c_2 + A_9^i c_g\kappa_\lambda c_2 \\ &+ A_{10}c_2c_{2g} + A_{11}c_g\kappa_\lambda\kappa_t^2 + A_{12}c_2\kappa_t^2 \\ &+ A_{13}\kappa_\lambda^2 c_g\kappa_t + A_{14}c_{2g}\kappa_t\kappa_\lambda + A_{15}c_gc_{2g}\kappa_\lambda \end{aligned} \tag{1.62}$$

When introducing quantum correction and calculating the same cross section at NLO precision, the coefficient $A_j|_{j=1,...,15}$ are modified, and eight new terms appear, yielding the additional correction:

$$\begin{aligned} \Delta R_{HH} = \frac{\Delta\sigma_{HH}}{\sigma_{HH}^{SM}} \overset{NLO}{=} &\; A_{16}\kappa_t^3c_g + A_{17}\kappa_tc_2c_g + A_{18}\kappa_tc_g^2\kappa_\lambda + A_{19}c_g\kappa_tc_{2g} \\ &+ A_{20}\kappa_t^2c_g^2 + A_{21}c_2c_g^2 + A_{22}c_g^3\kappa_\lambda + A_{23}c_g^2c_{2g} \end{aligned} \tag{1.63}$$

The 15 and 23 coefficients for the LO and NLO calculation, respectively, are reported in Table 1.3. The numerical values are obtained from a simultaneous fit of the cross section determined from LO and NLO simulations [77].

As already noted for the κ-framework, the modification of the cross section is only one part of the use of EFT. Anomalous H couplings significantly modify the kinematic properties of HH events, and diverse combinations of the coupling values can predict

Table 1.3 Results for the coefficients defined in Eqs. 1.62 and 1.63. The uncertainties are obtained from the uncertainties on the total cross sections entering the projections, using error propagation that neglects correlations between these cross sections [77]

Coefficient	LO value	NLO value
A_1	2.08059 ± 0.00163127	2.23389 ± 0.0100989
A_2	10.2011 ± 0.00809032	12.4598 ± 0.0424131
A_3	0.27814 ± 0.00187658	0.342248 ± 0.0153637
A_4	0.314043 ± 0.000312416	0.346822 ± 0.00327358
A_5	12.2731 ± 0.0101351	13.0087 ± 0.0962361
A_6	-8.49307 ± 0.00885261	-9.6455 ± 0.0503776
A_7	-1.35873 ± 0.00148022	-1.57553 ± 0.0136033
A_8	2.80251 ± 0.0130855	3.43849 ± 0.0771694
A_9	2.48018 ± 0.0127927	2.86694 ± 0.0772341
A_{10}	14.6908 ± 0.0311171	16.6912 ± 0.178501
A_{11}	-1.15916 ± 0.00307598	-1.25293 ± 0.0291153
A_{12}	-5.51183 ± 0.0131254	-5.81216 ± 0.134029
A_{13}	0.560503 ± 0.00339209	0.649714 ± 0.0287388
A_{14}	2.47982 ± 0.0190299	2.85933 ± 0.193023
A_{15}	2.89431 ± 0.0157818	3.14475 ± 0.148658
A_{16}	–	$-0.00816241 \pm 0.000224985$
A_{17}	–	$0.0208652 \pm 0.000398929$
A_{18}	–	0.0168157 ± 0.00078306
A_{19}	–	$0.0298576 \pm 0.000829474$
A_{20}	–	$-0.0270253 \pm 0.000701919$
A_{21}	–	0.0726921 ± 0.0012875
A_{22}	–	$0.0145232 \pm 0.000703893$
A_{23}	–	0.123291 ± 0.00650551

very different differential distributions of the HH observables. A generalization of Eqs. 1.62 and 1.63 is used to model the HH kinematics as discussed in detail in Sect. 5.4.3.

The EFT formulation just described presents one drawback: the five-dimensional coupling phase space is extremely wide, making it unfeasible for experimental searches to sample its entirety. One workaround is represented by defining the so-called *shape benchmarks*, which are specific combinations of the five couplings whose predicted kinematics is representative of a large portion of the full phase space. The scheme to define the benchmarks can be generally described as being a scanning procedure of the five-dimensional phase space, followed by the grouping of the points yielding comparable kinematic properties according to some similarity metric. Two sets of benchmarks have been derived.

Table 1.4 Values of the effective Lagrangian couplings for the twelve LO benchmarks defined in Ref. [76] plus the benchmark 8a described in Ref. [77], and the seven NLO benchmarks defined in Ref. [79]

	#	κ_λ	κ_t	c_{tt}	c_g	c_{2g}
LO benchmarks	1	7.5	1.0	−1.0	0.0	0.0
	2	1.0	1.0	0.5	−0.8	0.6
	3	1.0	1.0	−1.5	0.0	−0.8
	4	−3.5	1.5	−3.0	0.0	0.0
	5	1.0	1.0	0.0	0.8	−1.0
	6	2.4	1.0	0.0	0.2	−0.2
	7	5.0	1.0	0.0	0.2	−0.2
	8	15.0	1.0	0.0	−1.0	1.0
	9	1.0	1.0	1.0	−0.6	0.6
	10	10.0	1.5	−1.0	0.0	0.0
	11	2.4	1.0	0.0	1.0	−1.0
	12	15.0	1.0	1.0	0.0	0.0
	8a	1.0	1.0	0.5	0.8/3	0.0
NLO benchmarks	1b	3.94	0.94	−1/3	3/4	−1
	2b	6.84	0.61	1/3	0	1
	3b	2.21	1.05	−1/3	0	−3/2
	4b	2.79	0.61	1/3	1/4	−1/2
	5b	3.95	1.17	−1/3	−3/4	3/2
	6b	5.68	0.83	1/3	1/2	−1
	7b	−0.1	0.94	1	−1/4	1/2

Twelve benchmarks have been defined at LO precision in Ref. [76], where the similarity between two shapes is quantified through a metric defined from a binned likelihood ratio test statistics. The values of the five anomalous couplings are reported in Table 1.4 and the corresponding m_{HH} differential distributions are shown in Fig. 1.14. One additional LO benchmark ($8a$), whose couplings are reported in the same Table, has been defined in Ref. [77].

Seven benchmarks have been defined at NLO precision in Ref. [79], where the similarity between two shapes is quantified through an unsupervised learning technique based on an autoencoder neural network. The values of the associated five couplings are reported in Table 1.4 and the corresponding m_{HH} differential distributions are shown in Fig. 1.15.

As can be appreciated from the Figures, the different benchmarks result in largely diverse kinematics of the HH. Certain benchmarks exhibit prominent interference patterns, while others predominantly occupy the lower energy spectrum; additionally, few benchmarks are characterized by important tails extending far beyond 1 TeV. Their exploration represents a way to perform a model-independent study of BSM models, which can be subsequently reinterpreted in the context of specific

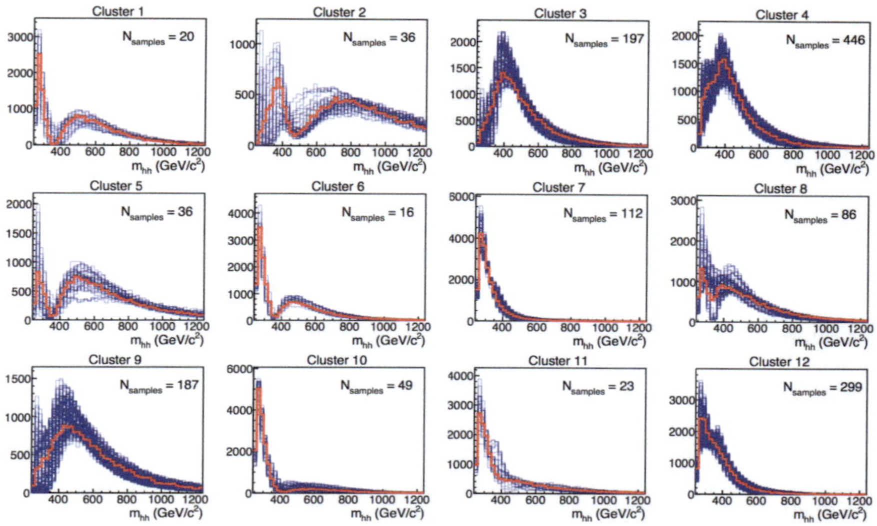

Fig. 1.14 Higgs boson pair invariant mass distributions corresponding to the LO benchmark points devised in Ref. [76]. The red distributions correspond to the benchmark sample in each cluster, while the blue ones describe the other members of each cluster

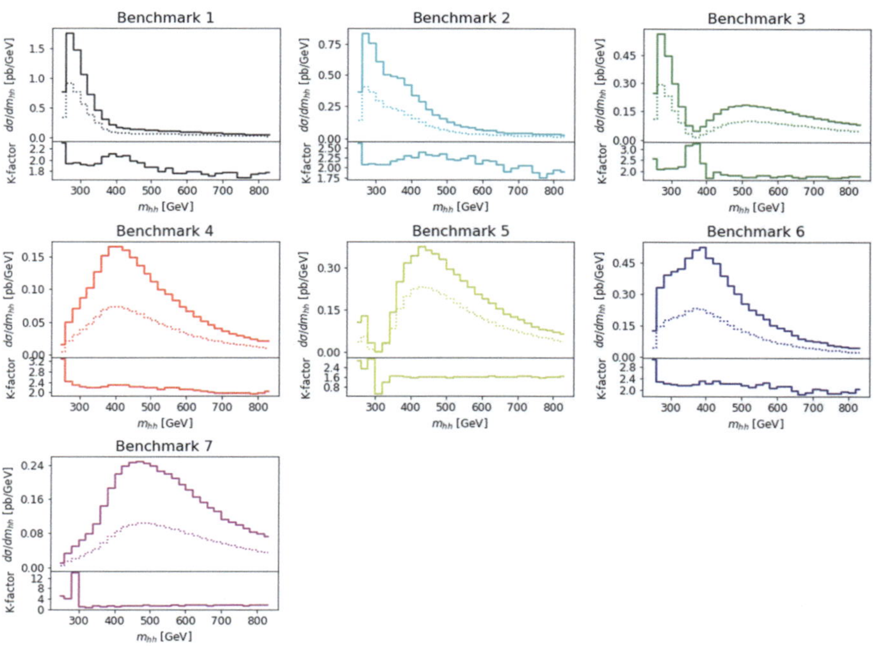

Fig. 1.15 Higgs boson pair invariant mass distributions corresponding to the NLO benchmark points devised in Ref. [79]. The solid curves denote the NLO result, and the dotted curves are the LO result. The lower panels show the K-factor, defined as dσNLO/dσLO (bottom)

UV-complete scenarios. Some examples of such theories are 2HDM models [80], composite Higgs models [81], vector-like leptons [82], and vector-like quarks [83].

In the analysis presented in this Thesis, the results are interpreted within the EFT framework described above. A complete description of the EFT benchmarks modelling is given in Sect. 5.4.3, whereas the results are presented in Sect. 6.2.3.

1.3 Searching for HH at the LHC

The phenomenology of Higgs boson pair production is highly rich, with several combinations of production mechanisms and decay final state accessible at the LHC. Moreover, if BSM physics were to manifest itself, an even richer phenomenology would be at hand, with specific benchmarks predicting highly boosted HHs as well as very low momentum H bosons. Therefore, the HH searches at the LHC explore multiple decay channels to obtain a series of complementary and redundant measurements that guarantee the highest possible sensitivity.

Figure 1.16 presents the branching ratios of HH to a selected group of final states, representing the decay channels with the highest predicted branching fraction. Ideally, the searches for HH production would explore the whole decay phase space to achieve the highest sensitivity; nevertheless, at the current levels of integrated luminosity, only some of the accessible final states guarantee a useful sensitivity level. The choice of which channel is worth investigating is based on a trade-off between the number of expected signal events, background contamination, and selection purity. While the theory fully predicts the former point, the latter two are largely affected by the performance of the detector and the analysis tools. In this context, throughout the latest operational run of the LHC, huge improvements have been achieved by the ATLAS and CMS Collaborations in terms of analysis tools performance, thus opening the way to new studies and searches in decay channels with smaller and smaller branching ratios.

In both the ATLAS and CMS Collaborations, the sensitivity of HH searches is driven by the following three channels:

- HH $\rightarrow$ $\mathrm{b\bar{b}b\bar{b}}$: it is the decay channel with the highest branching ratio $\mathcal{B}(\mathrm{b\bar{b}b\bar{b}}) = 34\%$ but it is affected by a very large QCD-induced multijet background. It was originally considered an unfeasible avenue for HH searches, but the large improvement in jet taggers' performance has brought it forward as one of the most competitive channels during Run-2. Particularly important has been the development of machine learning techniques to identify jets from b quarks and to reject instrumental background from the misidentification of gluon or light flavour quark jets [84]. This channel is particularly instrumental in exploring the H bosons high-boost regime, for which dedicated analyses have been performed. It is used by both Collaborations to set stringent limits on the inclusive HH production cross section and to target the exclusive VBF production phase space [85, 86].

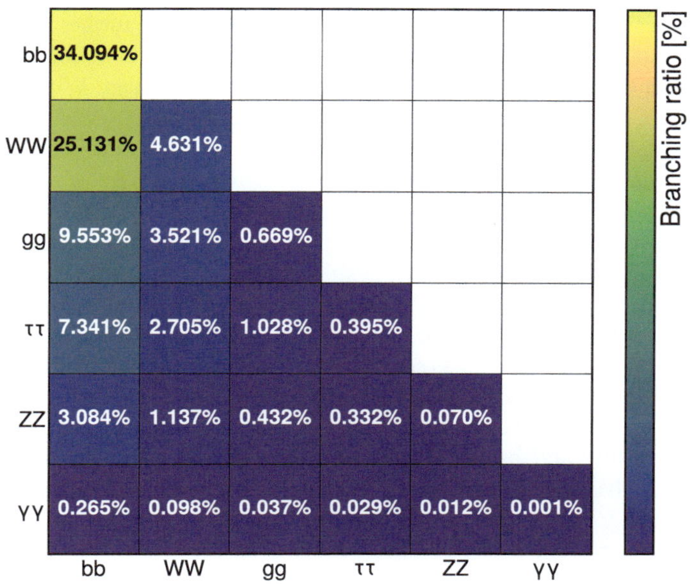

Fig. 1.16 Branching fractions for the decay of a Higgs boson pair to a selected group of final states. The decays of the two Higgs bosons are indicated in the two axes of the figure. The percentage values already account for the Bose-Einstein statistics factor and are computed based on the single H boson branching ratios [19]

- HH $\rightarrow$ $\mathrm{b\bar{b}}\tau^+\tau^-$: the main strength of this final state is the optimal compromise between a sizeable branching ratio $\mathcal{B}(\mathrm{b\bar{b}b\bar{b}}\tau^+\tau^-) = 7.3\%$ and the selection purity of the $\tau\tau$ pair. Its main background contributions are $\mathrm{t\bar{t}}$, Drell-Yan, and QCD multijet, depending on the selected τ final state. It was originally developed as an exploratory avenue; still, during Run-2, it has become one of the most sensitive channels both in CMS and ATLAS, setting tight limits on the inclusive HH cross section. Within the CMS Collaboration, this channel is employed to set the most stringent limit to date on the exclusive VBF production cross section [87, 88].
- HH $\rightarrow$ $\mathrm{b\bar{b}}\gamma\gamma$: it is the purest of the three channels detailed here, but suffers from the tiny branching ratio of $\mathcal{B}(\mathrm{b\bar{b}}\gamma\gamma) = 0.265\%$. It was the first decay channel to be explored in both Collaborations and profits from the precise measurement of the di-photon invariant mass as the main tool for background rejection. The main background is the irreducible di-photon continuum analogous to that of the single H measurements, which can be statistically suppressed only by exploiting the kinematic properties of the selected events. This final state is used to set limits on the inclusive HH cross section and, within the CMS Collaboration, it is also exploited to set limits on the exclusive VBF production cross section [89, 90].

For simplicity of notation, in the remainder of this manuscript, the specification of the charges will be omitted whenever not ambiguous.

Many other final states are explored at the LHC to improve further the sensitivity of the combined HH searches. Each final state presents diverse challenges to be experimentally overcome, creating a great array of techniques. Within the CMS Collaboration, the HH $\rightarrow$ WWWW, HH $\rightarrow$ WW$\tau\tau$, and HH $\rightarrow$ $\tau\tau\tau\tau$ channels are explored simultaneously in the so-called *multilepton* final state [91]; the HH $\rightarrow$ bbZZ [92] search is also performed in the four lepton final state of the Z pair; the HH $\rightarrow$ bbWW channel is explored in the single- and di-lepton W final states [93, 94]; the HH $\rightarrow$ WW$\gamma\gamma$ channel completes the picture [95]. The ATLAS Collaboration has also performed additional searches exploring the HH $\rightarrow$ bbWW channel in the di-lepton final state of the W [96]; the HH $\rightarrow$ WWWW has also been explored [97].

In this wide spectrum of analyses, the primary goal is always that of setting upper limits on the HH production cross section and confidence intervals on κ_λ. Nevertheless, these searches can also be exploited to probe anomalous Higgs boson couplings to fermions and gauge bosons. Particularly interesting, as highlighted above, are the y_t Yukawa coupling, and the vector boson couplings c_V and c_{2V}. A complete overview of the HH searches and the combination of the results have been performed by both the ATLAS and CMS Collaborations in Ref. [33, 98].

The analysis presented in this Thesis focuses on the HH $\rightarrow$ bb$\tau\tau$ search. This final state is one of the most difficult at the LHC, requiring the reconstruction of several different particle candidates and the definition of several τ final states. Moreover, neutrinos in the τ lepton decay prevent the exact reconstruction of the H, thus requiring dedicated fitting approaches. This search is affected by large irreducible and reducible backgrounds that need to be disentangled from the signal, either exploiting the kinematic properties of the signal or employing sophisticated machine learning techniques. This complexity is rewarded by one of the best sensitivities over several HH signal hypotheses, as discussed in several phenomenological studies [57, 73, 99, 100]. Within the CMS Collaboration, it is the second most sensitive channel to the inclusive production cross section, and it is the one setting the current most stringent limit on the exclusive VBF production cross section. The description of the bb$\tau\tau$ final states, its properties, the background contamination, and the analysis strategy are treated in detail in Chap. 5, whereas the results of the search are discussed in Chap. 6.

References

1. Weinberg S (1967) A model of Leptons. Phys Rev Lett 19:1264–1266. https://doi.org/10.1103/PhysRevLett.19.1264
2. Herrero MJ (1998) The standard model. NATO Sci Ser C (1998). https://doi.org/10.48550/arXiv.hep-ph/9812242. arXiv: hep-ph/9812242
3. ATLAS Collaboration (2012) Observation of a new particle in the search for the Standard Model Higgs boson with the ATLAS detector at the LHC. Phys Lett B716:1–29. https://doi.org/10.1016/j.physletb.2012.08.020. arXiv: 1207.7214 [hep-ex]
4. Collaboration CMS (2012) Observation of a new boson at a mass of 125 GeV with the CMS experiment at the LHC. Phys Lett B 716:30–61. https://doi.org/10.1016/j.physletb.2012.08.

021. arXiv:1207.7235 [hep-ex]
5. CMS Collaboration (2013) Observation of a new boson with mass near 125 GeV in pp collisions at $\sqrt{s} = 7$ and 8 TeV. J High Energy Phys 2013.6. issn: 1029-8479. https://doi.org/10.1007/jhep06(2013)081
6. Galbraith D, Burgard C, Standard model of physics. https://texample.net/tikz/examples/model-physics/
7. Iliopoulos J (2014) Introduction to the standard model of the electro-weak interactions. In: 8th CERN–Latin-American School of high-energy physics, pp 1–30. https://doi.org/10.5170/CERN-2014-008.1. arXiv:1305.6779
8. Tanabashi M, others (Particle Data Group) (2018) Review of particle physics. Phys Rev D 98:030001. https://doi.org/10.1103/PhysRevD.98.030001
9. Fukuda Y et al (1998) Evidence for oscillation of atmospheric neutrinos. Phys Rev Lett 81.8:1562–1567. https://doi.org/10.1103/physrevlett.81.1562
10. Maki Z, Nakagawa M, Sakata S (1962) Remarks on the unified model of elementary particles. Prog Theor Phys 28.5:870–880. issn: 0033-068X. eprint:https://academic.oup.com/ptp/article-pdf/28/5/870/5258750/28-5-870.pdf. https://doi.org/10.1143/PTP.28.870
11. Cabibbo (1963) Unitary symmetry and leptonic decays. Phys Rev Lett 10:531–533. https://doi.org/10.1103/PhysRevLett.10.531. https://link.aps.org/doi/10.1103/PhysRevLett.10.531
12. Kobayashi M, Maskawa T (1973) CP-Violation in the renormalizable theory of weak interaction. Prog Theor Phys 49.2:652–657. issn: 0033-068X. https://doi.org/10.1143/PTP.49.652. https://academic.oup.com/ptp/article-pdf/49/2/652/5257692/49-2-652.pdf
13. Glashow SL, Iliopoulos J, Maiani L (1970) Weak interactions with lepton-hadron symmetry. Phys Rev D 2:1285–1292. https://doi.org/10.1103/PhysRevD.2.1285. https://link.aps.org/doi/10.1103/PhysRevD.2.1285
14. Englert F, Brout R (1964) Broken symmetry and the mass of gauge vector mesons. Phys Rev Lett 13:321–323. https://doi.org/10.1103/PhysRevLett.13.321. https://link.aps.org/doi/10.1103/PhysRevLett.13.321
15. Higgs PW (1964) Broken symmetries and the masses of gauge bosons. Phys Rev Lett 13:508–509. https://doi.org/10.1103/PhysRevLett.13.508. https://link.aps.org/doi/10.1103/PhysRevLett.13.508
16. Strocchi F (2008) Symmetry breaking in the ising model. In: Symmetry breaking. Springer, Berlin, Heidelber, pp 131–138. isbn: 978-3-540-73593-9. https://doi.org/10.1007/978-3-540-73593-9_21
17. Goldstone J (1961) Field theories with superconductor solutions. Nuovo Cim 19:154–164. https://doi.org/10.1007/BF02812722
18. 't Hooft G (1971) Renormalizable Lagrangians for massive Yang-Mills fields. Nucl Phys B 35.1:167–188. issn: 0550-3213. https://doi.org/10.1016/0550-3213(71)90139-8. https://www.sciencedirect.com/science/article/pii/0550321371901398
19. de Florian D et al (2017) Handbook of LHC higgs cross sections: 4. deciphering the nature of the higgs sector. CERN yellow reports: monographs. Geneva: CERN. https://doi.org/10.23731/CYRM-2017-002. https://cds.cern.ch/record/2227475
20. Measurement of the Higgs boson mass and width using the four leptons final state (2023) Technical report, Geneva: CERN. https://cds.cern.ch/record/2871702
21. ATLAS Collaboration (2023) Evidence of off-shell Higgs boson production from ZZ leptonic decay channels and constraints on its total width with the ATLAS detector. arXiv: 2304.01532
22. Collaboration CMS (2018) Observation of the Higgs boson decay to a pair of τ leptons with the CMS detector. Phys Lett B 779:283–316. https://doi.org/10.1016/j.physletb.2018.02.004
23. CMS Collaboration (2022) Analysis of the CP structure of the Yukawa coupling between the Higgs boson and τ leptons in proton-proton collisions at $\sqrt{s} = 13$ TeV. J High Energy Phys 2022.6. https://doi.org/10.1007/jhep06(2022)012
24. CMS Collaboration (2023) Measurements of Higgs boson production in the decay channel with a pair of τ leptons in proton–proton collisions at $\sqrt{s} = 13$ TeV. Eur Phys J C 83.7. All the figures and tables, including additional supplementary figures and tables, can be found at http://cms-results.web.cern.ch/cms-results/public-results/publications/HIG-19-

010 (CMS Public Pages), p 562. https://doi.org/10.1140/epjc/s10052-023-11452-8. arXiv: 2204.12957. https://cds.cern.ch/record/2807752
25. CMS Collaboration (2021) Evidence for Higgs boson decay to a pair of muons. J High Energy Phys 2021.1. https://doi.org/10.1007/jhep01(2021)148
26. CMS Collaboration (2015) Precise determination of the mass of the Higgs boson and tests of compatibility of its couplings with the standard model predictions using proton collisions at 7 and 8 TeV. European Phys J C 75.5. https://doi.org/10.1140/epjc/s10052-015-3351-7
27. ATLAS Collaboration (2016) Measurements of the Higgs boson production and decay rates and coupling strengths using pp collision data at $\sqrt{s} = 7$ and 8 TeV in the ATLAS experiment. European Phys J C 76.1. https://doi.org/10.1140/epjc/s10052-015-3769-y
28. Davis R, Harmer DS, Hoffman KC (1968) Search for neutrinos from the sun. Phys Rev Lett 20:1205–1209. https://doi.org/10.1103/PhysRevLett.20.1205. https://link.aps.org/doi/10.1103/PhysRevLett.20.1205
29. Ahmad QR et al (2001) Measurement of the rate of $\nu_e + d \rightarrow p + p + e$ interactions produced by ^{8}B of solar neutrinos at the Sudbury NeutrinoObservatory. Phys Rev Lett 87.7. https://doi.org/10.1103/physrevlett.87.071301
30. Sofue Y, Rubin V (2001) Rotation curves of spiral galaxies. Ann Rev Astron Astrophys 39.1:137–174. https://doi.org/10.1146/annurev.astro.39.1.137
31. Planck Collaboration (2020) Planck 2018 results. VI. Cosmological parameters. Astron Astrophys 641:A6. https://doi.org/10.1051/0004-6361/201833910
32. Giudice GF (2013) Naturalness after LHC8. arXiv:1307.7879
33. CMS Collaboration (2022) A portrait of the Higgs boson by the CMS experiment ten years after the discovery. Nature 607.7917:60–68. https://doi.org/10.1038/s41586-022-04892-x
34. Endlich S, Penco R (2016) Effective field theory approach to tidal dynamics of spinning astrophysical systems. Phys Rev D 93.6. https://doi.org/10.1103/physrevd.93.064021
35. Born M, Oppenheimer R (1927) Zur Quantentheorie der Molekeln. Annalen der Physik 389.20. PDF of an english translation by Blinder SM is available, pp 457–484. issn: 1521-3889. https://doi.org/10.1002/andp.19273892002
36. Bezrukov F, Shaposhnikov M (2008) The standard model higgs boson as the inflaton. Phys Lett B 659.3:703–706. issn: 0370-2693. https://doi.org/10.1016/j.physletb.2007.11.072. https://www.sciencedirect.com/science/article/pii/S0370269307014128
37. Shaposhnikov M (2015) The Higgs boson and cosmology. Phil Trans Roy Soc Lond A 373.2032:20140038. https://doi.org/10.1098/rsta.2014.0038
38. Asaka T, Shaposhnikov M (2005) The νMSM, dark matter and baryon asymmetry of the universe. Phys Lett B 620.1:17–26. issn: 0370-2693. https://doi.org/10.1016/j.physletb.2005.06.020. https://www.sciencedirect.com/science/article/pii/S0370269305008087
39. Carena M, Liu Z, Riembau M (2018) Probing the electroweak phase transition via enhanced di-Higgs boson production. Phys Rev D 97.9. https://doi.org/10.1103/physrevd.97.095032
40. Reichert M et al (2018) Probing baryogenesis through the Higgs boson self-coupling. Phys Rev D 97.7. https://doi.org/10.1103/physrevd.97.075008
41. Rajantie A (2018) Higgs cosmology. Philosoph Trans R Soc Math Phys Eng Sci 376:20170128. https://doi.org/10.1098/rsta.2017.0128
42. Kuzmin VA, Rubakov VA, Shaposhnikov ME (1985) On anomalous electroweak baryon-number non-conservation in the early universe. Phys Lett B 155.1:36–42. issn: 0370-2693. https://doi.org/10.1016/0370-2693(85)91028-7. https://www.sciencedirect.com/science/article/pii/0370269385910287
43. Shaposhnikov ME (1986) Possible appearance of the baryon asymmetry of the universe in an electroweak theory. JETP Lett 44:465–468
44. Shaposhnikov ME (1987) Baryon asymmetry of the universe in standard electroweak theory. Nuclear Phys B 287:757–775. issn: 0550-3213. https://doi.org/10.1016/0550-3213(87)90127-1. https://www.sciencedirect.com/science/article/pii/0550321387901271
45. Sakharov AD (1991) Violation of CP invariance, C asymmetry, and baryon asymmetry of the universe. Soviet Phys Uspekhi 34.5:392. https://doi.org/10.1070/PU1991v034n05ABEH002497. https://dx.doi.org/10.1070/PU1991v034n05ABEH002497

46. Gavela MB et al (1994) Standard Model CP-violation and Baryon asymmetry. Mod Phys Lett A 09.09:795–809. https://doi.org/10.1142/s0217732394000629
47. Huet P, Sather E (1995) Electroweak baryogenesis and standard model CP violation. Phys Rev D 51:379–394. https://doi.org/10.1103/PhysRevD.51.379. https://link.aps.org/doi/10.1103/PhysRevD.51.379
48. Pendlebury JM et al (2015) Revised experimental upper limit on the electric dipole moment of the neutron. Phys Rev D 92:092003. https://doi.org/10.1103/PhysRevD.92.092003. https://link.aps.org/doi/10.1103/PhysRevD.92.092003
49. Cline JM (2018) Is electroweak baryogenesis dead? Philosoph Trans R Soc A: Math Phys Eng Sci 376.2114:20170116. https://doi.org/10.1098/rsta.2017.0116. https://doi.org/10.1098%2Frsta.2017.0116
50. Géraldine Servant (2018) The serendipity of electroweak baryogenesis. Philosoph Trans R Soc A: Math Phys Eng Sci 376.2114:20170124. https://doi.org/10.1098/rsta.2017.0124
51. Agazie G, The NANOGrav Collaboration et al (2023) The NANOGrav 15 yr data set: evidence for a gravitational-wave background. Astrophys J Lett 951.1:L8. https://doi.org/10.3847/2041-8213/acdac6. https://dx.doi.org/10.3847/2041-8213/acdac6
52. Afzal A, The NANOGrav Collaboration et al (2023) The NANOGrav 15 yr data set: search for signals from new physics. Astrophys J Lett 951.1:L11. https://doi.org/10.3847/2041-8213/acdc91. https://dx.doi.org/10.3847/2041-8213/acdc91
53. Bochkarev AI, Shaposhnikov ME (1987) Electroweak production of baryon asymmetry and upper bounds on the higgs and top masses. Mod Phys Lett A 2:417. https://doi.org/10.1142/S0217732387000537
54. Kajantie K et al (1996) The electroweak phase transition: a non-perturbative analysis. Nucl Phys B 466.1:189–258. issn: 0550-3213. https://doi.org/10.1016/0550-3213(96)00052-1. https://www.sciencedirect.com/science/article/pii/0550321396000521
55. Glover EWN, van der Bij JJ (1988) Higgs boson pair production via gluon fusion. Nucl Phys B 309.2:282–294. issn: 0550-3213. https://doi.org/10.1016/0550-3213(88)90083-1. https://www.sciencedirect.com/science/article/pii/0550321388900831
56. Djouadi A et al (1999) Production of neutral Higgs-boson pairs at LHC. European Phys J C 10.1:45–49. issn: 1434-6052. https://doi.org/10.1007/s100529900083. http://dx.doi.org/10.1007/s100529900083
57. Baglio J et al (2013) The measurement of the Higgs self-coupling at the LHC: theoretical status. J High Energy Phys 2013.4. issn: 1029-8479. https://doi.org/10.1007/jhep04(2013)151
58. Frederix R et al (2014) Higgs pair production at the LHC with NLO and parton-shower effects. Phys Lett B 732:142–149. issn: 0370-2693. https://doi.org/10.1016/j.physletb.2014.03.026
59. de Florian D, Mazzitelli J (2015) Higgs pair production at next-to-next-to-leading logarithmic accuracy at the LHC. JHEP 09:053. https://doi.org/10.1007/JHEP09(2015)053. arXiv:1505.07122
60. Borowka S et al (2016) Full top quark mass dependence in Higgs boson pair production at NLO. J High Energy Phys 2016.10. issn: 1029-8479. https://doi.org/10.1007/jhep10(2016)107. http://dx.doi.org/10.1007/JHEP10(2016)107
61. de Florian D et al (2016) Differential Higgs boson pair production at next-to-next-to-leading order in QCD. J High Energy Phys 2016.9. issn: 1029-8479. https://doi.org/10.1007/jhep09(2016)151. http://dx.doi.org/10.1007/JHEP09(2016)151
62. Grazzini M et al (2018) "Higgs boson pair production at NNLO with top quark mass effects. J High Energy Phys 2018.5. issn: 1029-8479. https://doi.org/10.1007/jhep05(2018)059. http://dx.doi.org/10.1007/JHEP05(2018)059
63. Maltoni F et al (2014) Top-quark mass effects in double and triple Higgs production in gluon-gluon fusion at NLO. J High Energy Phys 2014.11. issn: 1029-8479. https://doi.org/10.1007/jhep11(2014)079. http://dx.doi.org/10.1007/JHEP11(2014)079
64. Ling L et al (2014) NNLO QCD corrections to Higgs pair production via vector boson fusion at hadron colliders. Phys Rev D 89.7. issn: 1550-2368. https://doi.org/10.1103/physrevd.89.073001. http://dx.doi.org/10.1103/PhysRevD.89.073001

65. Dolan MJ et al (2014) Further on up the road: hhjj production at the LHC. Phys Rev Lett 112.10. issn: 1079-7114. https://doi.org/10.1103/physrevlett.112.101802
66. Dolan MJ et al (2015) hhjj production at the LHC. European Phys J C 75.8. issn: 1434-6052. https://doi.org/10.1140/epjc/s10052-015-3622-3. http://dx.doi.org/10.1140/epjc/s10052-015-3622-3
67. Nakamura J, Baglio J (2017) Jet azimuthal angle correlations in the production of a Higgs boson pair plus two jets at hadron colliders. European Phys J C 77.1. issn: 1434-6052. https://doi.org/10.1140/epjc/s10052-017-4593-3
68. Dreyer FA, Karlberg A (2018) Vector-boson fusion Higgs pair production at N3LO. Phys Rev D 98.11. issn: 2470-0029. https://doi.org/10.1103/physrevd.98.114016
69. Brost E, Cadamuro L (2022) Searching for Pairs of higgs bosons in the LHC run 2 dataset. Symmetry 14.7. issn: 2073-8994. https://doi.org/10.3390/sym14071467
70. Heinrich G, Lang J (2023) SMEFT truncation effects in Higgs boson pair production at NLO QCD. J Phys Conf Ser 2438.1:012153. https://doi.org/10.1088/1742-6596/2438/1/012153. arXiv:2212.00711 [hep-ph]
71. Heinrich G, Lang J, Scyboz L (2022) SMEFT predictions for gg $\rightarrow$ hh at full NLO QCD and truncation uncertainties. JHEP 08:079. https://doi.org/10.1007/JHEP08(2022)079. arXiv:2204.13045
72. Alasfar L et al (2023) Effective field theory descriptions of Higgs boson pair production. arXiv:2304.01968
73. Di Micco B et al (2020) Higgs boson potential at colliders: status and perspectives. Rev Phys 5:100045. https://doi.org/10.1016/j.revip.2020.100045
74. Jenkins EE, Manohar AV, Stoffer P (2018) Low-energy effective field theory below the electroweak scale: operators and matching. J High Energy Phys 2018.3. https://doi.org/10.1007/jhep03(2018)016
75. Heinrich G, Lang J, Scyboz L (2022) Beyond dimension six in SM effective field theory: a case study in Higgs pair production at NLO QCD. PoS LL2022:009. https://doi.org/10.22323/1.416.0009. arXiv:2207.08790
76. Carvalho A et al (2016) Higgs pair production: choosing benchmarks with cluster analysis. J High Energy Phys 2016.4:1–28. https://doi.org/10.1007/jhep04(2016)126
77. Buchalla G et al (2018) Higgs boson pair production in non-linear Effective Field Theory with full mt-dependence at NLO QCD. JHEP 09:57. https://doi.org/10.1007/JHEP09(2018)057. arXiv:1806.05162
78. Azatov A et al (2015) Effective field theory analysis of double Higgs boson production via gluon fusion. Phys Rev D 92.3. https://doi.org/10.1103/physrevd.92.035001
79. Capozi M, Heinrich G (2020) Exploring anomalous couplings in Higgs boson pair production through shape analysis. JHEP 03:91. https://doi.org/10.1007/JHEP03(2020)091. arXiv:1908.08923
80. Bélusca-Maïto H et al (2017) Higgs EFT for 2HDM and beyond. European Phys J C 77.3. https://doi.org/10.1140/epjc/s10052-017-4745-5
81. del Aguila F, Santiago J, Pérez-Victoria M (2000) Observable contributions of new exotic quarks to quark mixing. J High Energy Phys 2000.09. https://doi.org/10.1088/1126-6708/2000/09/011
82. Giudice GF et al (2007) The strongly-interacting light Higgs. J High Energy Phys 2007.06. https://doi.org/10.1088/1126-6708/2007/06/045
83. del Aguila F, de Blas J, Pérez-Victoria M (2008) Effects of new leptons in electroweak precision data. Phys Rev D 78.1. https://doi.org/10.1103/physrevd.78.013010
84. Bols E et al (2020) Jet flavour classification using DeepJet. J Instrum 15.12:P12012–P12012. https://doi.org/10.1088/1748-0221/15/12/p12012
85. CMS Collaboration (2022) Search for Higgs Boson Pair Production in the Four b Quark Final State in Proton-Proton Collisions at $\sqrt{s} = 13\,\text{TeV}$. Phys Rev Lett 129:081802. https://doi.org/10.1103/PhysRevLett.129.081802. https://link.aps.org/doi/10.1103/PhysRevLett.129.081802

86. ATLAS Collaboration (2023) Search for nonresonant pair production of Higgs bosons in the $b\bar{b}b\bar{b}$ final state in pp collisions at $\sqrt{s} = 13$ TeV with the ATLAS detector. arXiv:2301.03212
87. CMS Collaboration (2023) Search for nonresonant Higgs boson pair production in final state with two bottom quarks and two tau leptons in proton-proton collisions at $\sqrt{s} = 13$ TeV. Phys Lett B 842:137531. issn: 0370-2693. https://doi.org/10.1016/j.physletb.2022.137531. https://www.sciencedirect.com/science/article/pii/S0370269322006657
88. ATLAS Collaboration (2023) Search for resonant and non-resonant Higgs boson pair production in the decay channel using 13 TeV pp collision data from the ATLAS detector. J High Energy Phys 2023.7. https://doi.org/10.1007/jhep07(2023)040
89. CMS Collaboration (2021) Search for nonresonant Higgs boson pair production in final states with two bottom quarks and two photons in proton-proton collisions at $\sqrt{s} = 13$ TeV. J High Energy Phys 2021.3. https://doi.org/10.1007/jhep03(2021)257
90. ATLAS Collaboration (2023) "Search for Higgs boson pair production in the two bottom quarks plus two photons final state in pp collisions at $\sqrt{s} = 13$TeV with the ATLAS detector. Phys Rev D 106:052001. https://doi.org/10.1103/PhysRevD.106.052001. https://link.aps.org/doi/10.1103/PhysRevD.106.052001
91. CMS Collaboration (2023) Search for Higgs boson pairs decaying to WW*WW*, WW*$\tau\tau$, and $\tau\tau\tau\tau$ in proton-proton collisions at $\sqrt{s} = 13$ TeV. J High Energy Phys 2023.7. https://doi.org/10.1007/jhep07(2023)095
92. CMS Collaboration (2023) Search for nonresonant Higgs boson pair production in the four leptons plus twob jets final state in proton-proton collisions at $\sqrt{s} = 13$ TeV. J High Energy Phys 2023.6. https://doi.org/10.1007/jhep06(2023)130
93. CMS Collaboration (2022) Search for heavy resonances decaying to a pair of Lorentz-boosted Higgs bosons in final states with leptons and a bottom quark pair at $\sqrt{s} = 13$ TeV. J High Energy Phys 2023.5. https://doi.org/10.1007/JHEP05(2022)005. https://doi.org/10.1007/JHEP05(2022)005
94. Search for HH production in the bbWW decay mode (2023). Technical report. Geneva: CERN. https://cds.cern.ch/record/2853597
95. Search for nonresonant Higgs boson pair production in the WW $\gamma\gamma$ **channel in pp collisions at** $\sqrt{s} = 13$ TeV (2022). Technical report Geneva: CERN. https://cds.cern.ch/record/2840773
96. ATLAS Collaboration (2020) Search for non-resonant Higgs boson pair production in the bb$\ell\nu\ell\nu$ final state with the ATLAS detector in pp collisions at $sqrts = 13$ TeV. Phys Lett B 801:135145. issn: 0370-2693. https://doi.org/10.1016/j.physletb.2019.135145. https://www.sciencedirect.com/science/article/pii/S0370269319308676
97. ATLAS Collaboration (2019) Search for Higgs boson pair production in the WW*WW* decay channel using ATLAS data recorded at $\sqrt{s} = 13$ TeV. J High Energy Phys 2019.5. https://doi.org/10.1007/jhep05(2019)124
98. ATLAS Collaboration (2023) Constraints on the Higgs boson self-coupling from single- and double-Higgs production with the ATLAS detector using pp collisions at s=13 TeV. Phys Lett B 843:137745. issn: 0370-2693. https://doi.org/10.1016/j.physletb.2023.137745. https://www.sciencedirect.com/science/article/pii/S0370269323000795
99. Barr AJ et al (2014) Di-Higgs final states augMT2ed—Selecting hh events at the high luminosity LHC. Phys Lett B 728:308–313. issn: 0370-2693. https://doi.org/10.1016/j.physletb.2013.12.011. https://www.sciencedirect.com/science/article/pii/S0370269313009866
100. Dolan MJ, Englert C, Spannowsky M (2012) Higgs self-coupling measurements at the LHC. J High Energy Phys 2012.10. https://doi.org/10.1007/jhep10(2012)112

Chapter 2
The Compact Muon Solenoid at the Large Hadron Collider

Located near Geneva, Switzerland, the Conseil Européen pour la Recherche Nucléaire, or European Organization for Nuclear Research (CERN), represents a pioneering hub of scientific inquiry, characterized by its profound contributions to particle physics and fundamental research. Established in 1954, CERN has since become a pre-eminent institution in the exploration of the Universe's fundamental constituents and the underlying principles governing their interactions. Beyond its scientific achievements, CERN has also upheld a profound commitment to peaceful collaboration and technological progress that benefits society as a whole, as stated in the Convention of its establishment:

> The Organization shall provide for collaboration among European States in nuclear research of a pure scientific and fundamental character, and in research essentially related thereto. The Organization shall have no concern with work for military requirements and the results of its experimental and theoretical work shall be published or otherwise made generally available. [1]

The cornerstone of CERN's pursuits is the Large Hadron Collider (LHC), a monumental underground particle accelerator that spans an impressive 26.7 km in circumference. The LHC's primary objective is to study high-energy particle collisions, enabling physicists to investigate the properties and behaviour of matter at the smallest scales. Designed to accelerate protons to nearly the speed of light and collide them at a centre-of-mass energy of 14 TeV (the design value has actually never been reached, and the maximum of 13.6 TeV was achieved in 2022), the LHC is the largest and most powerful particle accelerator ever built. The LHC can generate conditions akin to the early Universe, affording scientists the opportunity to probe the fundamental particles and forces that shape our cosmos, and it represents today the frontier of research in high-energy physics.

At CERN, the pursuit of scientific discovery is driven by the physical infrastructure and the exceptional collaboration between a diverse global community of scientists, engineers, and researchers. This collaboration transcends borders, languages, and disciplines, forging a dynamic ecosystem for knowledge exchange and breakthroughs

J. Motta, *Development of Machine Learning τ Trigger Algorithms and Search for Higgs Boson Pair Production*, Springer Theses,
https://doi.org/10.1007/978-3-031-96288-2_2

in our understanding of the Universe. Currently, CERN counts more than 10000 researchers from over 100 nationalities representing more than 500 universities and institutes.

Within this context, the Compact Muon Solenoid (CMS) experiment stands as a testament to CERN's commitment to precision and innovation. It is a general-purpose detector designed to study the Standard Model (SM), hunt for the Higgs boson (H), and possibly unveil Beyond the SM (BSM) processes; it is situated at one of the LHC's collision points and employs a sophisticated array of detectors to capture and analyse particle interactions. These interactions are then studied to unravel the mysteries of particle physics, including searching for new particles, investigating their properties, and validating established theories.

This Chapter is divided into four main Sections that set the experimental context of this Thesis. The discussion starts with an overview of the LHC accelerator, its design and operations in Sect. 2.1; this is followed by the description of the CMS experiment and its subdetectors in Sect. 2.2. Given the contents of this Thesis and my contributions, particular attention is given to detailing the trigger and data acquisition system in Sect. 2.3. The Chapter is closed by a discussion of the algorithms used for offline objects' reconstruction and identification in Sect. 2.4.

2.1 The Large Hadron Collider

The LHC has been designed as a double-purpose accelerator, guaranteeing the study of proton-proton as well as heavy-ion collisions. In the former configuration, it was designed to collide protons at a centre-of-mass energy of $\sqrt{s} = 14\,\text{TeV}$ with instantaneous luminosity $\mathcal{L} \sim 10^{34}\,\text{cm}^{-2}\,\text{s}^{-1}$, while in the latter setup, it was devised to collide heavy nuclei at a centre-of-mass energy of $\sqrt{s} = 2.76\,\text{TeV}$ per nucleon with instantaneous luminosity $\mathcal{L} \sim 10^{27}\,\text{cm}^{-2}\,\text{s}^{-1}$ [2, 3]. This multipurpose conception guarantees the possibility for physicists to probe a vast scientific program, encompassing the search for the Higgs boson, as well as the test of BSM scenarios and the search for particles yet to be theorized.

The LHC is situated in the Large Electron-Positron (LEP) collider tunnel, of which it inherits the dimension, and it is located between 45 m and 170 m below ground level. The journey from the LHC conception to its operational start has been an endeavour spanning more than two decades, with its first proposal in 1984, its official recognition in 1994, and the start of data-taking in 2008. This journey is currently ongoing with the third operational run of the accelerator, and it is bound to continue with the machine upgrade to its high luminosity specifications and its operation foreseen until the early 2040s. A detailed description of the accelerator complex and operations is given in the following.

2.1.1 Accelerator Complex

The LHC serves as the terminal component of a complex series of particle accelerators, which were established prior to the LHC and subsequently enhanced to satisfy its strict requirements. The initial phase of this accelerator sequence involves the extraction of protons from a hydrogen gas reservoir through the application of a strong electric field. These protons are subsequently directed into a Radio Frequency Quadrupole (RFQ), where they are bunched and accelerated to an energy of 750 keV (3 MeV during Run-3). Following this, the protons are conveyed to the Linear Accelerator (LINAC 2 in Run-2, and LINAC 4 in Run-3), designed to elevate the proton beam's energy to approximately 50 MeV for LINAC 2 and 160 MeV for LINAC 4. Subsequently, the particles advance into the inaugural circular collider, the Proton Synchrotron Booster (PSB), encompassing a 150 m circumference, which boosts the beam's energy to 1.4 GeV and enhances the intensity of the proton bunches. Thereafter, the beam enters the Proton Synchrotron (PS) and, finally, the Super Proton Synchrotron (SPS), characterized by circular configurations spanning 620 m and 6912 m, respectively. These components elevate the beam's energy to 26 GeV and 450 GeV, sequentially. At this stage, the proton bunches are ready to be injected into the LHC accelerator. The complete accelerator complex is illustrated schematically in Fig. 2.1.

The proton beams are transferred to the two LHC beam pipes via fast kicker magnets, which effectively divide the beam into counter-rotating parallel beamlines. Once in the LHC, the beams experience a progressive augmentation of their energy up to 7 TeV (this is the design energy; energies of 6.5 TeV and 6.8 TeV were used during Run-2 and Run-3, respectively), achieved through high-frequency accelerating cavities operated at 400 MHz and positioned within eight linear sections, each spanning 545 m along the ring's circumference. The proton bunches are kept in orbit by means of 1232 superconducting dipole magnets, each measuring 15 m of length and 35 tonnes of weight, systematically distributed across eight arcs, each extending 2.45 km. The magnets are custom-designed and built in a Niobium-Titanium (NbTi) alloy, which exhibits superconductivity when cooled to a temperature of 1.9 K (−271.25 °C) utilizing superfluid He-4. This ensures the production of an 8.3 T magnetic field generated by a current of about 11 kA. The beam optics is then regulated by 392 quadrupole magnets, measuring 5–7 m in length, that focus the particles and reduce the transversal section of the beams; additional octupole magnets are finally installed to control the beam's chromaticity (i.e. its energy spread). Upon stabilizing the beam dynamics and attaining the nominal energy, the proton bunches are further focused by special quadrupoles installed in front of the Interaction Points (IP) to squeeze the beams and increase the proton density at collision. The LHC presents four IPs, each equipped with a sophisticated particle detection apparatus to probe a vast array of physics processes.

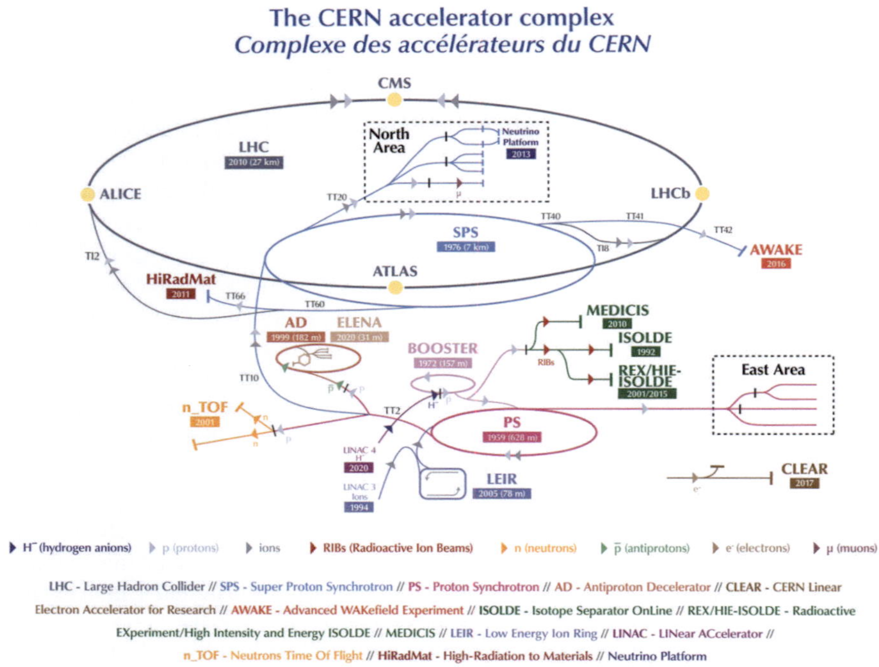

Fig. 2.1 Schematic illustration of the CERN accelerator complex. The LHC is the last ring (dark blue line) in a complex chain of particle accelerators. The smaller machines are used in sequence to accelerate the proton beams that collide in the centre of the four main detectors (yellow dots) [4]

2.1.2 *Design and Specifications*

The LHC has been designed for the collision of protons to overcome LEP's largest limitation: synchrotron radiation, which is the electromagnetic radiation emitted by particles moving on a curved path. The power dissipated by synchrotron radiation scales with the inverse of the fourth power of the particle's mass, i.e. m^{-4}, reducing its effect on protons by a factor $\sim 10^{13}$ compared to electrons. This ensures the ability to reach a nominal proton beam energy of 7 TeV and a centre-of-mass energy of $\sqrt{s} = 14\,\text{TeV}$ at IP (this is the design energy; energies of 13 TeV and 13.6 TeV were used during Run-2 and Run-3, respectively). This comes at the cost of the proton being a composite particle and entailing a more complicated collision dynamic, which is largely dominated by the fraction of gluons in each proton; for this reason, the LHC is sometimes improperly referred to as a *gluon collider*.

A central parameter of the LHC machine is the instantaneous luminosity $\mathcal{L}$ at which collisions take place; it depends on the beam properties as [5]:

$$\mathcal{L} = F \cdot \frac{N_{\text{b}}^2 n_{\text{b}} f_{\text{rev}} \gamma}{4\pi \epsilon_n \beta^*} \tag{2.1}$$

Table 2.1 Nominal design parameters of the LHC machine in proton-proton collisions configuration

Parameter	Description	Value
$\sqrt{s}$	Centre-of-mass energy	14 TeV
N_b	Particles per bunch	$1.15 \cdot 10^{11}$
n_b	Number of bunches per beam	2808
f_{rev}	Revolution frequency	11.2 kHz
ϵ_n	Transverse beam emittance	3.75 μm
β^*	Beta function (focal length)	0.55 m
Δt_b	Bunches spacing	25 ns
θ_c	Collision angle	285 μrad
σ_z	Bunches transverse r.m.s at IP	7.55 cm
σ_{xy}	Bunches longitudinal r.m.s at IP	16.7 μm

where N_b is the number of particles in each of the n_b bunches per beam that revolve in the tunnel with a frequency f_{rev}, and γ is the relativistic factor. The transverse emittance and the focal length of the beam at the IP are quantified by ϵ_n and the beta function β^*, respectively. Finally, F is a factor defined as:

$$F = \left(1 + \frac{\theta_c \sigma_z}{2\sigma_{xy}}\right)^{\frac{-1}{2}} \tag{2.2}$$

which accounts for the reduction of the instantaneous luminosity due to the IP geometry; it depends on the beam crossing angle θ_c, and the longitudinal and transverse r.m.s. bunch sizes σ_z and σ_{xy} at collision. By definition $F \leq 1$. The nominal design values of the LHC parameters are summarized in Table 2.1 alongside their description. In the list, one additional parameter is specified: the spacing of the bunches Δt_b; this parameter fixes the interaction (or bunch crossing) rate to 40 MHz.

The instantaneous luminosity then regulates the number of events per unit of time as

$$\frac{\partial N}{\partial t} = \mathcal{L} \cdot \sigma \tag{2.3}$$

where σ is the cross section of any given process; this renders it evident that a large instantaneous luminosity is essential to produce low cross section processes such as Higgs boson pair (HH) production. When integrating the instantaneous luminosity over time, the so-called *integrated luminosity* $L = \int \mathcal{L} dt$ is obtained, which measures the total amount of collisions produced. While instantaneous luminosity is measured in $\mathrm{cm}^{-2}\,\mathrm{s}^{-1}$, integrated luminosity is generally measured in units of inverse *barn* (b), with $1\,\mathrm{cm}^2 \equiv 10^{24}\,\mathrm{b}$, and typical units being the inverse picobarn (pb^{-1}) and femtobarn (fb^{-1}).

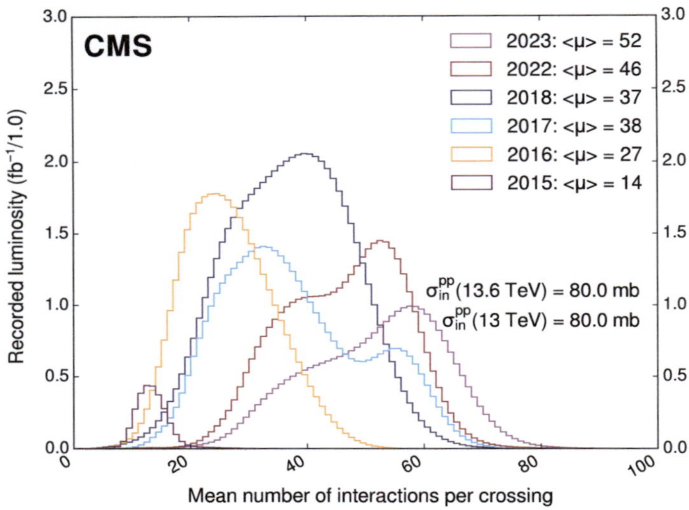

Fig. 2.2 Distribution of the average number of interactions per crossing (pileup) for proton-proton collisions in 2015 (purple), 2016 (orange), 2017 (light blue), 2018 (navy blue), 2022 (brown), and 2023 (light purple); the overall mean values and the minimum bias cross sections are also shown. These plots use only data that passed the *golden* certification (i.e., all CMS sub-detectors were flagged to meet the requirements for any use in physics analysis), and the *LHC standard* values for the minimum bias cross sections, which are taken from the theoretical prediction from Pythia and should be used to compare to other LHC experiments [7]

The achievement of a high instantaneous luminosity is crucial for probing rare processes, but at the same time, it introduces the possibility of several interactions happening simultaneously at each bunch crossing. This effect is referred to as *pileup* (PU), and its average value can be computed from the cross section of inelastic proton-proton scattering $\sigma_{\text{pp}}^{\text{inel.}}$ to be:

$$\langle \text{PU} \rangle = \frac{\mathcal{L} \cdot \sigma_{\text{pp}}^{\text{inel.}}}{n_{\text{b}} f_{\text{rev}}} \tag{2.4}$$

At the centre-of-mass energy of 13 TeV, the inelastic proton-proton scattering has a measured cross section $\sigma_{\text{pp}}^{\text{inel.}} = 68.6 \pm 0.5(\text{syst}) \pm 1.6(\text{lumi})$ mb [6], leading to an average PU $\sim$22 in LHC design conditions. As further detailed in the following, the operational conditions of the LHC machine have steadily increased over the years of operation; the LHC nominal instantaneous luminosity has been largely exceeded, with the current values being as high as 2–2.5 times the design one, leading to the current average PU level of 52, with peak PU exceeding 80. The evolution of the PU differential distributions over the past data-taking years is presented in Fig. 2.2.

This progressive increase in the harshness of the collision conditions requires a continuous improvement of the online data selection system, which is performed by the Level-1 and High-Level triggers detailed in Sect. 2.3; important work has been conducted as part of this Thesis to attain this for the 2022 and 2023 data-taking periods, as detailed in Chap. 3.

The number of simultaneous collisions per bunch crossing is foreseen to largely increase at the High-Luminosity LHC (HL-LHC) where average PU is foreseen to reach levels as high as $\langle PU \rangle = 200$. This PU growth will pose great challenges for the trigger system, which will have to identify interesting signal events among the overwhelming amount of PU. As part of this Thesis, extensive work has been done in view of these conditions, as described in Chap. 4.

2.1.3 Schedule of Operations

On September 10th 2008, after 24 years since its proposal, of which 10 were dedicated to installation and commissioning, the first proton beam circulated in the LHC. Regrettably, a mere week later, an incident impaired progress due to a flawed electrical connection between two magnets, which resulted in mechanical damage and the subsequent release of helium into the tunnel. Swift and thorough rectification procedures ensued, leading to the LHC's resumption of operations in November 2009. After an extensive phase of machine commissioning and preliminary collisions executed at reduced energy levels, a historic milestone materialized on March 30th 2010, marked by the commencement of the first high-energy collisions at 7 TeV centre-of-mass energy. After the inaugural unexpected events, the established operational schedule started.

Figures 2.3 and 2.4 give an overview of the past and future LHC operations and the milestones attained in terms of integrated and instantaneous luminosity up to the present day. The LHC operations are expected to cover a period of almost

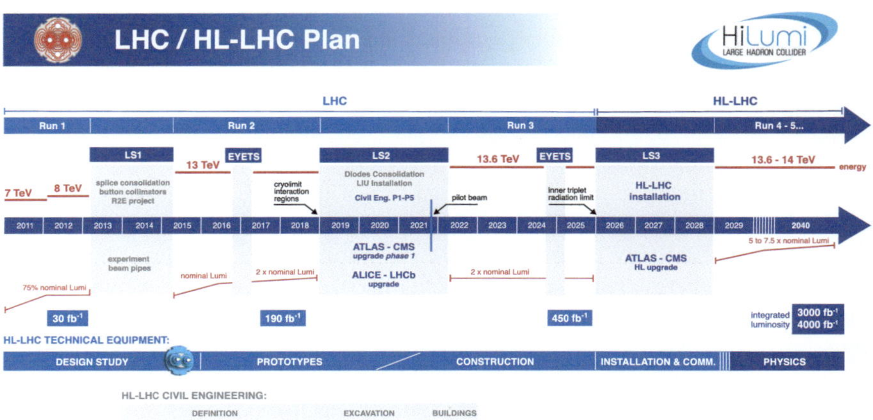

Fig. 2.3 Schedule of the full LHC and HL-LHC operations. These two projects are planned to span a total of four decades of data-taking

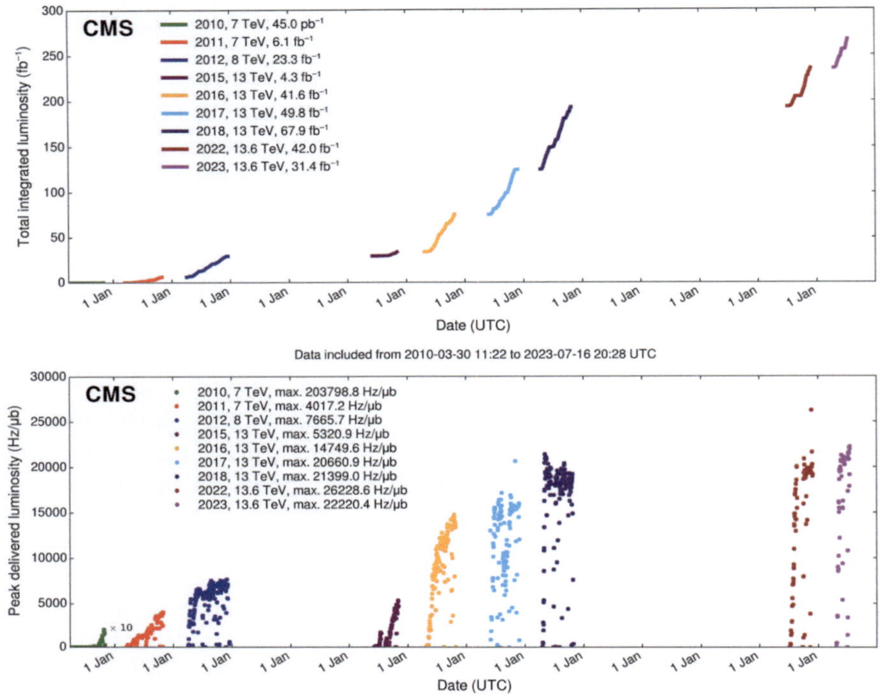

Fig. 2.4 Cumulative luminosity (top) and peak instantaneous luminosity (bottom) versus day delivered to CMS during stable beams for proton-proton collisions at nominal centre-of-mass energy. This is shown for data-taking in 2010 (green), 2011 (red), 2012 (blue), 2015 (purple), 2016 (orange), 2017 (light blue), 2018 (navy blue), 2022 (brown), and 2023 (light purple). These plots use the best available offline calibrations for each year. (Hz/¯b ≡ 10^{30} cm^{-2} s^{-1}) [7]

35 years, divided into two main operational phases: the so-called *Phase-1* (2011–2025) and *Phase-2* (2029–2041). If this schedule is respected, the LHC and its upgraded HL-LHC version will become the longest-lived particle accelerator in the history of physics. Within each Phase of operation, the data-taking years are organized into so-called *Runs*.

Run-1 is the first completed operational run of the LHC, comprising the 2011 and 2012 data-taking years. These two years corresponded respectively to 6.1 fb^{-1} of data collected at 7 TeV centre-of-mass energy and 23.3 fb^{-1} of data registered at 8 TeV collision energy, with a bunch spacing of 50 ns. The dataset accumulated in Run-1 has been the one that ensured the achievement of the historic milestone of the Higgs boson discovery announced on July 4th 2012, and the first measurement of its properties. After the completion of Run-1, the LHC entered its period of First Long Shutdown (LS1), which lasted two years. During this time, a series of maintenance and upgrade efforts were undertaken to bring the LHC toward its intended operational parameters, primarily focusing on reinforcing the superconducting magnets to withstand higher currents for an increase of the energy per beam to 6.5 TeV. During the LS1, the experiments also strategically utilized the opportunity to implement significant upgrades to their detectors in order to accommodate the heightened luminosity requirements. Notably, the CMS trigger electronics underwent substantial revisions as elaborated in Sect. 2.3.

The second completed operational run of the LHC is Run-2; the data collected in this period is used for the analysis presented in this Thesis. Run-2 comprised the data-taking years from 2015 to 2018, all characterized by a collision energy of $\sqrt{s} = 13\,\text{TeV}$ with the nominal 40 MHz bunch crossing frequency. During the four years of Run-2, the LHC delivered $4.2\,\text{fb}^{-1}$, $41.0\,\text{fb}^{-1}$, $49.8\,\text{fb}^{-1}$, and $67.9\,\text{fb}^{-1}$ of integrated luminosity, respectively. While the operations in 2015 were oriented to the commissioning of the LHC in its new configuration, the dataset accumulated in the following three years allowed the achievement of great milestones like the first observation of the direct coupling of the H to the τ lepton [8] and the first evidence of its direct coupling to muons [9]. After the fulfilment of Run-2, the LHC's Second Long Shutdown (LS2) started, lasting three years. Analogously to LS1, important consolidation and renovation works were performed to push the LHC performance. During LS2, the experiments also consolidated their subdetectors and, in some cases, started the upgrade program toward the start of Phase-2. Notably, the CMS hadronic calorimeter readout has been replaced as discussed in Sect. 2.2.2.

Run-3 is the last operational run of the LHC; it started in 2022 and will continue until 2025. At the time of writing this Thesis, the 2023 proton-proton data-taking is on halt due to a helium leak; operations are foreseen to restart in September 2023 for one month of heavy-ion collisions before the customary winter shutdown. In Run-3, a record-breaking centre-of-mass energy $\sqrt{s} = 13.6\,\text{TeV}$ is reached, with instantaneous luminosity as high as $2.6 \cdot 10^{34}\,\text{cm}^{-2}\,\text{s}^{-1}$, and peak PU exceeding 80. In the two years of data-taking of Run-3, the LHC delivered a luminosity of $42.0\,\text{fb}^{-1}$ and $31.4\,\text{fb}^{-1}$. By the end of Run-3, a total of $250\,\text{fb}^{-1}$ is expected to be delivered to the experiments, supplying a large enough dataset to better probe rare processes like HH production and reduce the uncertainties associated with the existing measurements.

The end of Run-3 will celebrate the end of Phase-1 and the LHC accelerator and, at the same time, herald the start of Phase-2 and the HL-LHC machine. In 2025, the Third Long Shutdown (LS3) will begin and last for three years. During this period, the LHC will undergo a profound upgrade towards the HL-LHC specifications [10]. Novel Niobium-Tin (Nb_3Sn) alloy superconducting quadrupole magnets, capable of yielding magnetic fields up to 12 T, will be integrated at the interaction points of the bigger experiments to refine the beam focus. Employing compact superconducting cavities, often referred to as *crab cavities*, precise rotation of the proton bunches prior to collision will be achieved, facilitating a reduction in the crossing angle and increasing the factor F defined in Eq. 2.2. These enhancements are anticipated to amplify the instantaneous luminosity to a value of $5 \cdot 10^{34}\,\text{cm}^{-2}\,\text{s}^{-1}$, which could be further boosted to $7.5 \cdot 10^{34}\,\text{cm}^{-2}\,\text{s}^{-1}$, exceeding more than seven-fold the nominal design value. At the same time, the centre-of-mass energy will be elevated to the original design value of 14 TeV. These improvements also bring exceptional data-taking conditions, which require extensive upgrade programs for all the experiments, some of which already started. Notably, the CMS Collaboration is planning a series of major upgrades of its subdetectors' hardware and software systems [11, 12]. Over the currently foreseen 12 years of operation, the HL-LHC should deliver a total of 3000–4000 fb^{-1}, thus opening the way to unprecedented studies of exceedingly rare phenomena and possibly unveiling new physics never observed.

2.1.4 *Experiments at the LHC*

As reported in Fig. 2.1, the LHC accelerator presents four interaction points where the beamlines converge and the proton beams collide. Being placed in the tunnel facility excavated for the LEP collider, the LHC IPs inherit the numbering of the previous accelerator, i.e. IP1, IP2, IP5, and IP8. In all four IPs, collisions happen at the heart of detectors, which can be seen as the cathedrals of modern high-energy physics. As detailed in the following, four main experiments are placed in the LHC underground caverns, complemented by five additional detectors at IP 1, 5, and 8.

- **A Toroidal LHC ApparatuS (ATLAS)**
  The ATLAS experiment [13] is located at IP1; it is one of the two general-purpose detectors used in the discovery of the Higgs boson, and it was designed to perform a large array of physics studies and searches. The conceptual layout of the ATLAS detector draws an analogy with the layers of an onion, encompassing a series of subdetectors organized in sequential cylindrical strata, each designed to detect different types of particles. The innermost layers are the inner tracker and the transition radiation tracker, which feature mixed silicon- and gaseous-based technology. The following layer is a solenoid magnet providing a 2 T axial field parallel to the beam line. The subsequent layers are the two sampling calorimeters: the inner liquid-argon electromagnetic calorimeter, with its characteristic accordion structure, and the outer hadronic steel and scintillating tiles calorimeter. The final layer is constituted by the muon gaseous chambers, which cover the entire detector surface. The whole detector is encapsulated by several toroidal magnets, which provide a magnetic field up to 4 T around the solenoid and give ATLAS its name. The ATLAS detector is the largest of those at the LHC, tallying a total length of 46 m, a diameter of 25 m, and a weight of $7 \cdot 10^3$ tonnes.
- **Large Hadron Collider forward (LHCf)**
  The LHCf experiment [14] is located at IP1 and shares the experimental cavern of ATLAS; its goal is the perfection of the hadron interaction models used in the study of extremely high-energy cosmic rays. It comprises two identical detectors located ± 140 m from the collision point. This setup ensures the collection of data at zero-degree collision angle via the two imaging calorimeters made of tungsten plates, plastic scintillators, and position-sensitive sensors. This experiment is the smallest on the LHC accelerator, with each detector weighting $\sim$40 kg for a total volume of $30 \times 80 \times 10\,\text{cm}^3$.

- **ForwArd Search ExpeRiment (FASER)**
  The FASER experiment [15] is positioned on the beam collision axis line-of-sight 480 m from IP1 in a service tunnel, and has been installed during the LS2. It is designed to search for new light and very-weakly-interacting particles. The detector is composed of a two-fold scintillator veto system, an interface tracker, a decay volume immersed in a 0.57 T magnetic field generated by a dipole magnet, a timing scintillator station, a tracking spectrometer surrounded by two dipole magnets generating a 0.57 T magnetic field, and an electromagnetic calorimeter system. The FASER detector is complemented by the FASERν [16] emulsion sub-detector, which extends the angular coverage and can detect neutrinos of all flavours produced at the LHC. The FASER detector has a 10 cm radius aperture, the FASERν sub-detector has a $25 \times 30\,\text{cm}^2$ transverse surface, and the total length of the two together is 7 m.
- **Scattering and Neutrino Detector at the LHC (SND@LHC)**
  The SND@LHC experiment [17] is positioned in a service tunnel 480 m from IP1, slightly off-axis from the beam collision axis, and has been installed during the LS2. It is designed to profit from the high flux of energetic neutrinos of all flavours from the LHC. The detector is composed of a hybrid system based on a target made of tungsten plates, interleaved with emulsion and electronic trackers, also acting as an electromagnetic calorimeter, and followed by a hadronic calorimeter and a muon identification system. The SND@LHC detector measures $1 \times 1 \times 2.6\,\text{m}^3$ with a total weight of almost 1 tonne.
- **A Large Ion Collider Experiment (ALICE)**
  ALICE experiment [18] is located at IP2; it is the main experiment dedicated to the study of heavy ion collisions to disclose the nature of the state of the matter expected to have been present in the primordial Universe: quark-gluon plasma. The design of the ALICE detector is very different from that of the other main detector due to the stringent design constraint of coping with the extreme particle multiplicity in heavy-ion collisions. ALICE consists of a central part, much like the ATLAS detector, complemented by a forward muon spectrometer on one side of the experiment to probe decay products of heavy quarkonium states. In the central section, ALICE presents an inner silicon- and gaseous-based tracker, followed by a time projection chamber and time-of-flight identification arrays. The ring imaging Cherenkov and transition radiation detectors further enhance the particle identification capabilities of ALICE. The outer subdetectors are two electromagnetic calorimeters. The forward muon arm consists of an elaborated arrangement of absorbers, dipole magnets, and gaseous tracking chambers. The ALICE detector measures $16 \times 16 \times 26\,\text{m}^3$ with a total weight of approximately 10^4 tonnes.
- **Compact Muon Solenoid (CMS)**
  The CMS experiment [19] is located at IP5; together with its companion detector ATLAS, it is a general-purpose detector pivotal to the discovery of the Higgs boson, and it was designed to probe a large spectrum of physics phenomena. While presenting the similar onion-like structure of ATLAS, the design choices of the CMS detector are largely different from those of its partner experiment. These

choices lead to a substantially smaller detector with a diameter of 14.6 m and a length of 21.6 m, but a weight of $12.5 \cdot 10^3$ tonnes, making it the heaviest detector at the LHC. Section 2.2 gives a detailed description of the CMS detector.

- **TOTal Elastic and diffractive cross section Measurement (TOTEM)**
  The TOTEM experiment [20] is located at IP5 and shares the experimental cavern of CMS; it is designed to exploit a luminosity-independent method for the measurement of the total proton-proton cross-section and the study of elastic and diffractive proton scattering. The detector has a mirrored geometry with respect to the collision point; it comprises two tracking *telescopes*: a cathode strip chamber telescope and a gas electron multiplier telescope, respectively placed at ± 9 m and ± 13.5 m from the IP. These are complemented by Roman Pots, which are silicon-based detectors, placed at ± 147 m and ± 220 m from IP5 for the detection of leading protons a few mm from the beam line.
- **LHC beauty (LHCb)**
  The LHCb experiment [21] is located at IP8; it is designed to perform precision measurements of charge-parity (CP) properties of the SM and to study rare decays of B-hadrons, which could point to the strong violation of the CP symmetry. The LHCb detector is unique in its layout, as it does not present an array of concentric subdetectors like the other three major experiments. Conversely, a single-arm forward spectrometer exploits the property of forward production of B-hadrons. Given the asymmetric geometry of LHCb, to maximally exploit the volume of the underground cavern, the LHC optics is modified at IP8 with a displacement of the collision point by 11.25 m from the centre. Ordered by distance from the collision point, the LHCb experiment presents an array of semi-circular silicon-based detectors composing the Vertex Locator (VELO), followed by a first imaging Cherenkov counter, and the silicon trigger tracker. The fourth component is a warm saddle-shaped magnet, followed by the additional layers of the tracking system and a second imaging Cherenkov counter for particle identification. Moving further away from the IP, we find the Shashlik electromagnetic calorimeter, and the iron and scintillator tiles hadronic calorimeter. The muon detectors complete the design at the opposite end of the IP. The LHCb detector measures 20 m in length and has an angular acceptance ranging from 10 to 300 mrad.
- **Monopole and Exotics Detector At the LHC (MoEDAL)**
  The MoEDAL experiment [22] is located at IP8 and shares the experimental cavern of LHCb; as the name suggests, it is designed to directly search for magnetic monopoles and other exotic particles like highly ionizing stable (or pseudo-stable) massive particles. The MoEDAL detector is composed of an array of plastic nuclear track detectors positioned around the VELO of LHCb for a maximum surface area of 25 m^2.

2.2 The Compact Muon Solenoid Experiment

The CMS experiment stands as one of the pinnacles of scientific ingenuity within the realm of particle physics. Situated in the 100 m deep experimental cavern of interaction point 5 near the French village of Cessy, the CMS detector was designed to explore the physics at the TeV scale in many different signatures and final states to achieve a far-reaching physics program. To attain this, the CMS detector was conceived as a general-purpose experiment that, with its cylindrical structure, hermetically surrounds the bunch crossing point.

The name of the CMS experiment is deeply intertwined with its design choices. The term *compact* stems from the considerably smaller but denser design of the CMS detector, which is six times smaller in volume compared to ATLAS but twice as heavy. The word *muon* originates from the large volume dedicated to the muon tracking chambers, which compose about 80% of the detector's total volume. Finally, the word *solenoid* arises from the central feature of the CMS experiment: a superconducting solenoidal magnet surrounding a large fraction of the detector.

Based on the need to detect different signatures and final states, the CMS experiment comprises several concentric subdetectors that complement each other in characterising the diverse particles resulting from the proton-proton and heavy-ion interactions. Moreover, to cope with the harsh collision environment, the detector design was conceived to be highly granular, fast in response, and highly resistant to radiation. A complete description of the CMS detector structure is given in Sect. 2.2.2 after the introduction of the coordinate system.

2.2.1 Coordinate System

The coordinate system used in the CMS experiment for the description of the detector and the collision products is a right-handed Cartesian system centred on the nominal bunch crossing point. The x and y axes from a plane perpendicular to the proton beamline (generally referred to as *transverse plane*), with the x-axis pointing toward the geometrical centre of the LHC and the y-axis pointing upwards in the direction perpendicular to the LHC plane, which is 1.41% tilted with respect to the gravitationally horizontal plane. The z-axis is the longitudinal coordinate that matches the anticlockwise proton beam direction. Based on this Cartesian system, and given the cylindrical symmetry of the detector, a polar coordinate system is also defined in which the radial coordinate r is measured from the nominal IP, the polar angle θ is defined as the angle formed by r with the z-axis, and the azimuthal angle ϕ is expressed as the angle between r and the x-axis in the transverse plane. A schematic illustration of the CMS Cartesian and polar coordinate systems is given in Fig. 2.5.

While the Cartesian coordinate system is well suited for representing the detector and computing macroscopic quantities, it is not well suited for describing the hard scattering interactions. This problem stems from the composite nature of the colliding

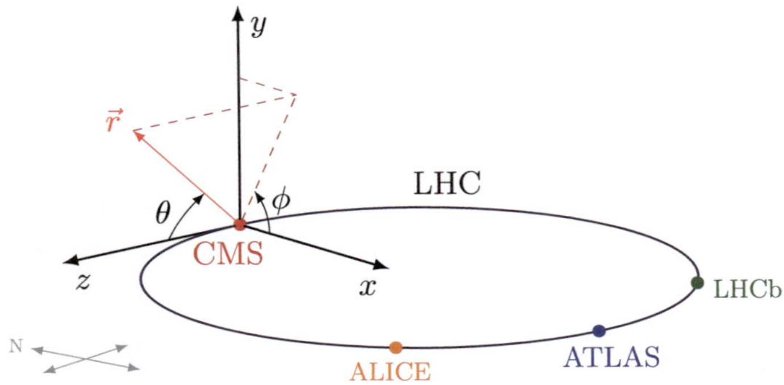

Fig. 2.5 Schematic illustration of the standard coordinate system at the CMS detector, represented relative to the LHC and the other three main LHC experiments [23]

protons. At the LHC, the collisions occur at the level of the fundamental constituents of the proton, whose momentum fraction is unknown. Therefore, the longitudinal boost of the collision's rest frame along the z-axis is also an unknown observable. This leads to the need for the definition of quantities that are either Lorentz boost-invariant or, although not invariant, have transformation properties that are easy to handle and useful for physics analysis.

The colliding partons within the protons carry an unknown fraction of the proton momentum in the longitudinal direction; nevertheless, their momentum in the transverse plane is negligible. Therefore, the boost of the collision rest frame can only be along the z-axis, and the simplest Lorentz-invariant variables involve the projection of the momentum onto the transverse plane for the definition of the transverse momentum (p_{T}) and the transverse mass (m_{T}):

$$p_{\mathrm{T}}^2 = p_x^2 + p_y^2 \tag{2.5}$$

$$m_{\mathrm{T}}^2 = m^2 + p_x^2 + p_y^2 = E^2 - p_z^2 \tag{2.6}$$

A second important quantity is the so-called *rapidity* of a particle, which at colliders is defined as

$$y = \frac{1}{2}\left(\frac{E + p_z}{E - p_z}\right) \tag{2.7}$$

As it contains the longitudinal momentum p_z, the rapidity is not a Lorentz-invariant quantity, but upon a Lorentz boost parallel to the beam axis with velocity $v = \beta c$, the rapidity transforms as

$$y' = y - \tanh^{-1}\beta \tag{2.8}$$

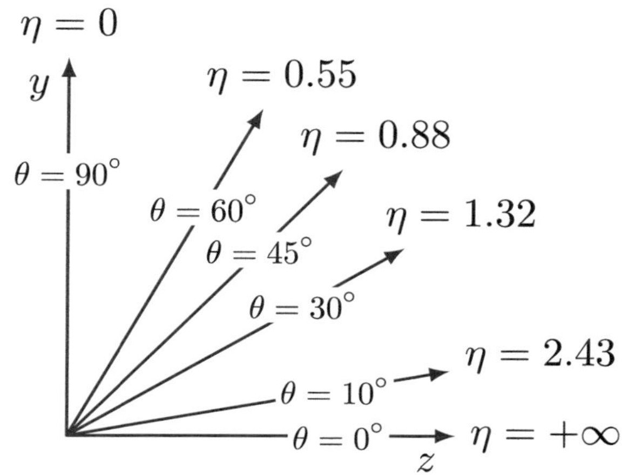

Fig. 2.6 Relation between polar coordinate θ from the standard coordinate system of CMS and the more generally used pseudorapidity η [23]

This particularly simple transformation law for y has an important consequence: the rapidity difference between two particles is a Lorentz-invariant variable. Nevertheless, the use of rapidity still presents a problem in its measurement for highly relativistic particles or for particles quasi-collinear to the beam line. By noticing that at colliders the momentum of a final state particle is generally much larger than its mass, we can approximate the rapidity of ultrarelativistic particles, i.e. $m/E \ll 1$, as

$$y \sim \eta \equiv \frac{1}{2} \log\left(\frac{p + p_z}{p - p_z}\right) = -\log\left[\tan\left(\frac{\theta}{2}\right)\right] \tag{2.9}$$

where η is the so-called *pseudorapidity*, which ranges $\eta \in [-\infty, +\infty]$ and assumes null value for $\theta = \frac{\pi}{2}$. The relation between the pseudorapidity and the polar angle θ is visually represented in Fig. 2.6.

Having defined the pseudorapidity, we can define the following Lorentz-invariant quantity

$$\Delta R^2 \equiv \Delta\eta^2 + \Delta\phi^2 \tag{2.10}$$

which encompasses the angular (or spatial) distance between two particles.

2.2.2 Detector Sub-structure

The CMS detector architecture can be appreciated in Fig. 2.7. It is a cylindrical detector with a length of 21.6 m and a diameter of 14.6 m, constituted of a central section, or *barrel*, and two forward regions, or *endcaps*. The exact boundaries between the two regions depend on the specific subdetectors considered.

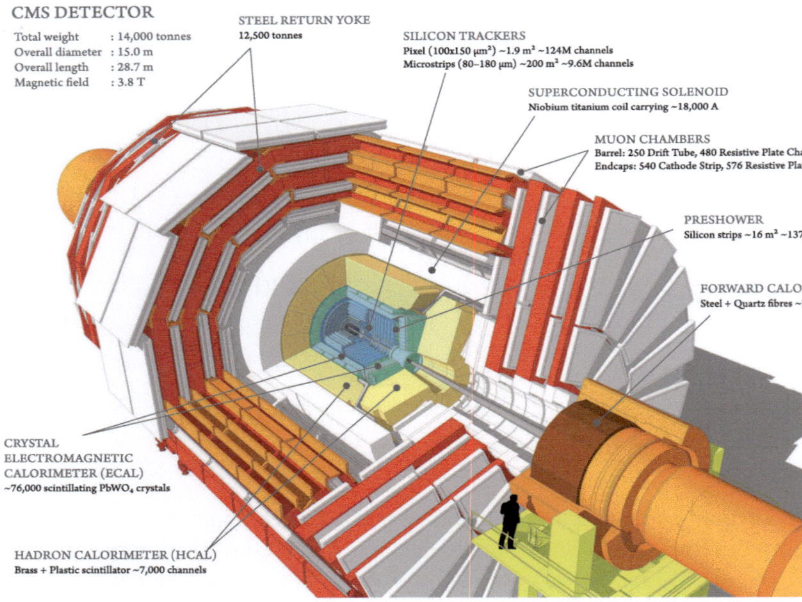

Fig. 2.7 Cutaway 3D model of the CMS detector. This perspective makes visible all the subdetectors at a glance [24]

To serve its multi-purpose scope, the CMS apparatus is equipped with a sophisticated arrangement of concentric detection layers; each specialized in detecting the diverse particles arising from collision events. At the core of this arrangement, the position of the collision vertices is determined by the pixel and strip tracker detectors encircling the interaction point; these detectors also trace the paths and momenta of charged particles generated by the interactions. Surrounding these, the electromagnetic and hadron calorimeters measure the energy deposited in their active material volume by electrons, photons, and hadrons, ensuring their accurate energy measurements. Around the calorimeters, the CMS superconducting solenoid constitutes the most distinctive feature of the experiment. Finally, the outermost layer of the detector is constituted by the muon tracking chambers that encircle the entire CMS volume.

The information from the various subdetectors is often redundant to ensure the highest precision in the measurement of all kinds of final-state particles. The following details the diverse components of CMS; the use of their information in the offline reconstruction of physics objects is discussed in Sect. 2.4.

Superconducting Solenoidal Magnet

The fundamental component of the CMS detector, around which the design of the whole experiment revolves, is the superconducting NbTi solenoidal magnet [25]. The

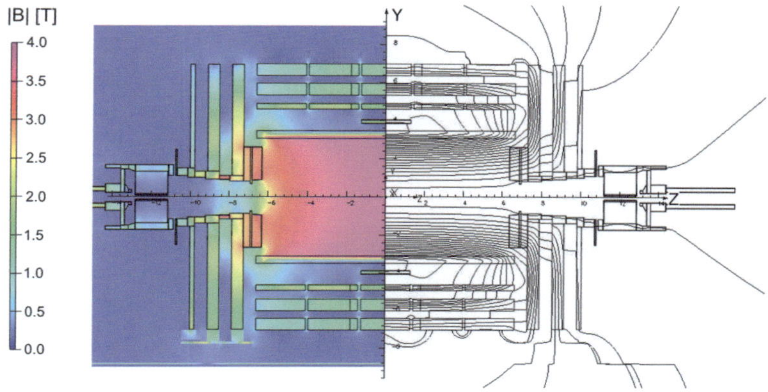

Fig. 2.8 Longitudinal section of the CMS detector superimposed with the heatmap of the predicted absolute value of the magnetic field (left) and field lines (right); each field line corresponds to a flux increment of 6 Wb [26]

CMS magnet is the largest of its kind, with a weight of about 220 tonnes, a diameter of 6 m, and a length of 12.5 m.

Like the dipoles of the LHC accelerator, the magnet is superconducting owing to a liquid He-4 cryostatic system that maintains it at its operational temperature of 4.5 K (−268.3 °C). This ensures the production of a quasi-uniform 3.8 T magnetic field parallel to the z-axis. To minimize the non-active material between the IP and the tracker and calorimeters, the magnet is designed to encircle both, thus posing tight design constraints on them. Surrounding the magnet are the iron flux return yokes, which ensure a quasi-uniform 2 T magnetic field in the volume of the muon detectors. The original design of the magnet foresaw a 4 T magnetic field, which was subsequently lowered to the current value of 3.8 T due to its unique design and the unknown ageing it would experience over operations.

The magnetic field $\vec{B}$ is at the basis of the charged particles' momentum measurement in the tracker. Given a particle of charge q and speed $\vec{v}$ moving inside the CMS solenoid, it will experience a force $\vec{F} = q \cdot \left(\vec{v} \times \vec{B}\right)$. Hence its momentum can be inferred by the bending of its trajectory. To achieve this, a precise mapping of the magnetic field produced by the CMS magnet is performed with a precision of less than 0.1% in the tracker volume [26], and it is shown in Fig. 2.8.

Silicon Tracking System

The innermost part of the CMS detector is the tracking system [27, 28], which is composed of two sections: the inner pixel tracking system and the outer strip tracking system. The former was originally designed to operate under nominal LHC conditions; as the instantaneous luminosity largely exceeded the planned conditions,

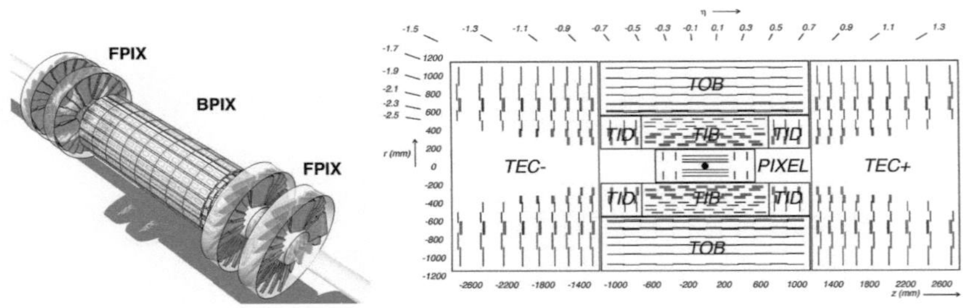

Fig. 2.9 Schematic representation of the CMS pixel tracking subsystems and their separation into Barrel Pixel (BPix) and Forward Pixel (FPix) sections (left). Schematic longitudinal view of the CMS inner tracking system layout (right). The pixel detector is surrounded by the strip tracker detector, which is composed of the Tracker Inner Barrel (TIB) and Tracker Inner Disks (TID), surrounded by the Tracker Outer Barrel (TOB) and the Tracker Endcaps (TEC) [29]. In both panels, the pixel detector is represented in its original three-layer design

the pixel detector has undergone a substantial upgrade during the 2016–2017 Run-2 year-End Technical Stop (YETS) [29]. The specifications given in the following are those of the pixel detector after the upgrade.

The tracking volume has a length of 5.6 m and a diameter of 2.4 m, which is permeated by the uniform 3.8 T magnetic field produced by the CMS magnet. It is instrumented with finely segmented silicon active material and equipped with fast readout to cope with the high particle multiplicity. This ensures the capability of measuring the particles' momentum and charge by combining their position information at the passage in each silicon detector. This information is further used for the reconstruction of the hard scatter interaction points, the so-called *primary vertex*, and its discrimination against PU. Moreover, it guarantees the reconstruction of in-flight decays such as those of B-hadrons or τ leptons (*secondary vertices*).

Figure 2.9 gives a schematic view of the tracking system. The left panel gives a picture of the pixel detector, which is divided into three sections: one barrel section denoted BPix, and two endcap sections denoted FPix. The right panel gives a longitudinal view of the whole tracking system, including the strip tracker. In both cases, the pixel subdetector is shown in its original three-layer design, which was used until the 2016 data-taking; the current upgraded design, in use since 2017, can be appreciated in Fig. 2.10.

In its current configuration, the CMS pixel tracker surrounds the IP for a total pseudorapidity coverage of $|\eta| < 2.50$. The BPix section has a coverage of $|\eta| < 2.16$, and it is constituted by 79 million silicon pixel sensors of dimension $100 \times 150\,\mu\text{m}^2$; the FPix completes the pseudorapidity coverage with an additional 45 million pixel sensors. This corresponds to almost a doubling of readout channels compared to the original design of the subdetector, which had a total of 66 million channels; the BPix and FPix have increased the channel count by 1.6 and 2.5 times, respectively.

The BPix sensors are arranged on four concentric layers at radii 3.0, 6.8, 10.2, and 16 cm, while the FPix is composed of three disks located at 29.1, 39.6, and

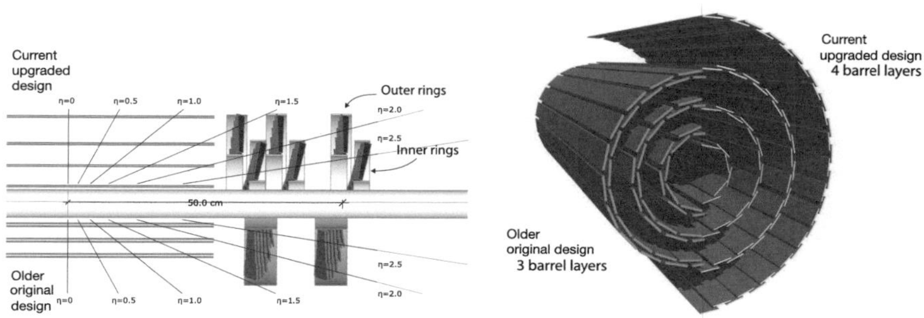

Fig. 2.10 Conceptual representation of the current Phase-1 upgraded pixel tracking system compared to the original design in a longitudinal (left) and transverse-oblique (right) view of the pixel system [29]

51.6 cm from the IP. This amounts to one additional layer in both regions compared to the original design, thus ensuring a four-point track reconstruction over the entire η coverage. Each pixel in this configuration ensures a spatial resolution of 10 µm in the transverse plane and 20 µm in the longitudinal direction.

The high segmentation of the detector bears a corresponding demand for readout electronics and power supply components; the upgrade to a four-layer system further increases this need. However, this augmentation in infrastructure introduces passive materials into the particle trajectory, causing parasitic phenomena such as multiple scattering, pair production, and nuclear interactions. These phenomena can significantly influence the accurate reconstruction of particle trajectories. Therefore, the subdetectors upgrade also meant a full redesign of the CO_2 cooling system, the pixels' supports, and the shift from on-board to off-board electronics. This extended effort resulted in notable achievements: the novel barrel and endcap pixel detectors now weigh 40% and 80%, respectively, compared to their predecessors. The material budget of the pixel tracker is shown in Fig. 2.11, where a substantial reduction of the dead material has been achieved, notwithstanding the presence of an additional layer.

As the particle flux rapidly decreases with the radial distance, the detector occupancy decreases accordingly with an r^{-2} dependence. This allows for lower granularity at increasing distances from the IP. Therefore, the outer tracker is composed of silicon micro-strips of varying dimensions depending on their positioning: from the closest to the IP having a pitch of 60 µm and a length of 7 cm, to those furthest from the IP having a pitch of 270 µm and a length of 12.5 cm. This segmentation totals about 9.3 million readout channels with 15 different geometries. This ensures a single-point resolution between 20 and 50 µm in the transverse plane and between 200 and 500 µm in the longitudinal direction.

The strip tracker is subdivided into several components. The innermost comprises the four layers of the Tracker Inner Barrel (TIB) covering the region $20 < r < 55$ cm, and the three Tracker Inner Disks (TID) located at $58 < |z| < 124$ cm with radii up to

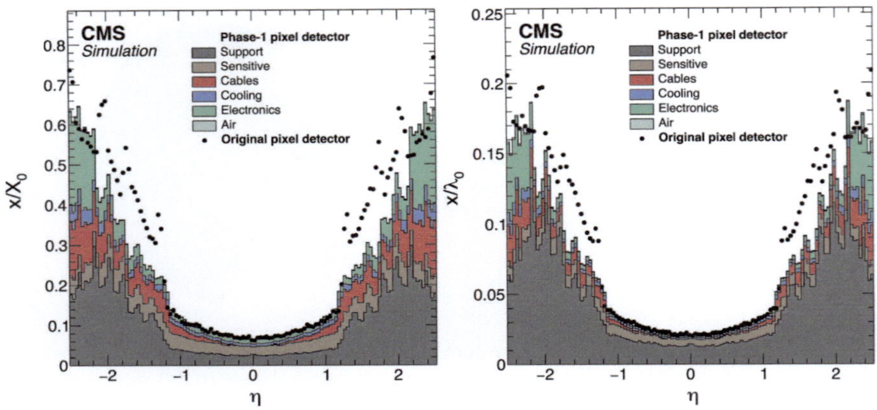

Fig. 2.11 Material budget in the pixel tracker detector in units of radiation lengths (left) and nuclear interaction length (left) as a function of the pseudorapidity, as obtained from simulation. The material budget of the original pixel detector design is reported with black dots, while the current upgraded design is shown with stacked histograms of the different contributions. A considerable reduction of the material budget is obtained while introducing an additional layer [30]

55 cm. These are encapsulated by the Tracker Outer Barrel (TOB) layers covering the region $55 < r < 116$ cm. The architecture is completed by the nine layers of the Tracker Endcaps (TEC) located at $124 < |z| < 282$ cm. The entire strip tracker ensures the same pseudorapidity coverage of the pixel detector $|\eta| < 2.5$ and has a total extension of 5.6 m along the z-axis.

The right panel of Fig. 2.9 also shows an additional feature of the strip tracker: to allow for the measurement of an additional coordinate, several disks and layers are complemented by a second micro-strip sensor mounted back-to-back with the main one and tiled with 100 mrad. This guarantees the measurement of the endcap's radial coordinate and the barrel's longitudinal coordinate.

Given the high fluence region in which they are immersed, to minimize the damage caused by ionizing radiation to the silicon active material, the pixel and the strip detectors are operated at a temperature of about $-10\,^{\circ}$C and $-15\,^{\circ}$C, respectively.

Electromagnetic Calorimeter

In the space between the CMS magnet and the tracker system lies the Electromagnetic Calorimeter (ECAL) [31], mainly devoted to measuring final-state electrons and photons. The CMS ECAL is a hermetic, homogeneous and highly granular calorimeter constituted of Lead-Tungstate ($PbWO_4$) crystals. The measurement of energy relies on the destructive transformation of incoming particles into electromagnetic showers that interact with the crystal material and initiate the emission of scintillation light. Notably, the crystals serve a dual purpose: functioning as both the dense medium for interactions and the active scintillating medium. To achieve this dual role, $PbWO_4$

was chosen for its high 8.28 g/ cm^3 density that ensures a small electromagnetic radiation length of $X_0 = 0.89$ cm and a contained $R = 2.19$ cm Molière radius. The combination of these factors guarantees an excellent containment of the electromagnetic shower within the crystals, which have a length of approximately $25X_0$ and a 22×22 mm^2 section. Moreover, lead-tungstate is radiation-hard and produces 80% of the scintillating light within the 25 ns spacing of the bunch crossings, making it sufficiently fast in response to disentangle particles from separate events. Nevertheless, the radiation hardness of $PbWO_4$ comes at the price of the light yield being limited to a few photons per MeV, necessitating photodetectors with internal amplification.

A schematic representation of the ECAL architecture is given in Fig. 2.12, where the volumetric view of the design is given in the top panel, while the longitudinal section of one single quadrant is given in the bottom panel. The ECAL detector is divided into three sections: one ECAL Barrel (EB) with pseudorapidity coverage $|\eta| < 1.479$ and two ECAL Endcap (EE) sections closing the cylindrical design for a coverage of $1.479 < |\eta| < 3.0$. The EB is segmented into 144 *modules* housing 425 crystals each; the modules are organized into 36 *supermodules*, each spanning half of the length of the EB, covering an azimuthal sector of 20°, and weighting about 1500 kg. In this configuration, a total of 61200 crystals cover the entire EB surface, with each crystal having a frontal surface of $\eta \times \phi = 0.0174 \times 0.0174$ and a depth of 23 cm. In each EE, the crystals are housed in two semi-circular supports called *dees* and arranged into 5×5 group of crystals called *super-crystals*. In both EB and EE, the crystals are placed in a quasi-radial geometry, with their axes being tilted up to 3° with respect to the direction that points to the IP. This is done to achieve minimal shower sharing between crystals and the best hermetic coverage possible. Nevertheless, few gaps remain between the modules, especially at $\eta = 0$ and $\eta = 1.479$ in the transition between EB and EE.

The crystal scintillation signals from the EB and EE are read out by silicon Avalanche Photo-Diodes (APDs) and Vacuum Photo-Triodes (VPTs), respectively. These light detectors are designed to operate in the high magnetic field to which they are exposed and to be resistant to high radiation doses. The ECAL front-end electronics perform the amplification and shaping of the signals, and then sample them at a 40 MHz frequency with a 12-bit Analog-to-Digital Converter (ADC). To achieve stable operational conditions and control the light yield changes as a function of the temperature of the crystals and of the photodetectors, the ECAL modules temperature is maintained at 18 °C with a precision of 0.05 °C in the EB and 0.1 °C in the EE, by a water-based cooling system. This temperature also favours the natural recovery of the crystals' radiation damage. Given the substantial radiation doses to which they are exposed, crystals undergo a slow reduction of their transparency that is partially recovered during the inter-fill operations of the LHC. This effect is monitored by injecting a 440 nm laser light in each crystal and automatically deriving time-dependent correction factors that are applied to the response.

As depicted in Fig. 2.12b , in the region $1.653 < |\eta| < 2.6$ the EE is complemented by an Endcap Preshower (ES) detector with the aim of better disentangling double-photon signals of $\pi^0 \rightarrow \gamma\gamma$ decays form high energy forward photons. The ES is a sampling calorimeter made of two lead absorbers followed by a silicon sensors

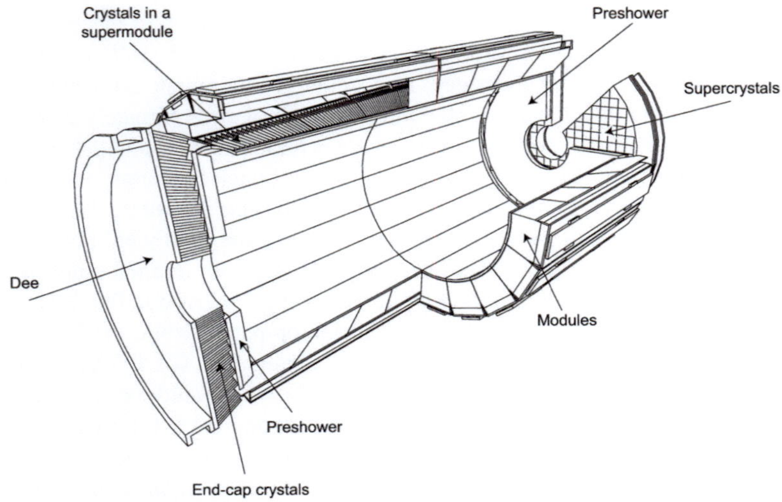

(a) Conceptual representation of the ECAL mechanical structure. The lead-tungstate crystals are housed in the modules and supermodules of the barrel, while in the endcap they are arranged between the preshower and the support dees [31]

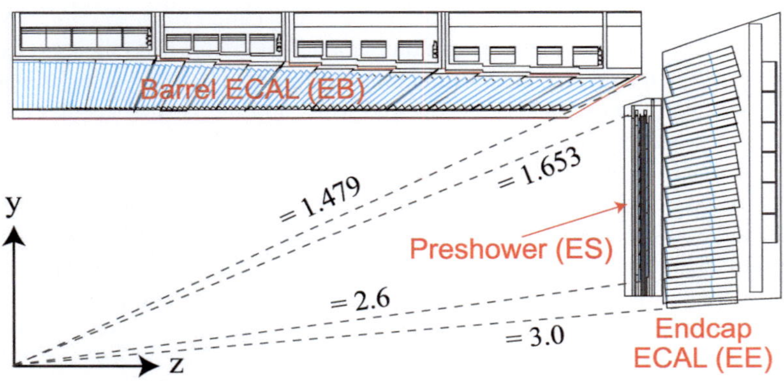

(b) Schematic longitudinal view of the layout of a quarter of the electromagnetic calorimeter. The ECAL Barrel (EB), the Endcap Preshower (ES), and the ECAL Endcap (EE) are shown, and the pseudorapidity envelopes are specified. The lead-tungstate crystals are arranged to obtain a quasi-radial segmentation of the ECAL surface [32]

Fig. 2.12 Figures summarising the structure of the CMS electromagnetic calorimeter plane made of 2 mm-wide silicon strips that measure the deposited energy and the transverse profile of the shower shape, for a total of around $1X_0$. As for the tracker, also the ES is operated at $-10\,^\circ$C temperature to increase the silicon resilience in the high radiation area of the endcap. As approximately 6% to 8% of the shower energy is deposited in the ES, this subdetector is a fundamental piece in the reconstruction of electromagnetic showers in the endcap.

The ECAL calorimeter design ensures the achievement of an outstanding energy resolution that, as for any calorimeter, comprises three terms regulating it. The first contribution is the so-called *stochastic term*, which is due to the statistical fluctuations of the number of photons produced in the scintillation light around the Poissonian mean; being the number of photons proportional to the energy of the impinging particle, the stochastic term scales as $\sqrt{E}$. The second contribution to the resolution is the so-called *noise term*, which accounts for the detector and electronics noise, and thus contributes as the inverse of the particle's energy. The third term is the so-called *constant term*, which encompasses constant losses due to detector inhomogeneities and does not depend on the energy of the particle. The intrinsic energy resolution of ECAL was measured on 3 $\times$ 3 crystals matrices in an electron test-beam environment [33] to be

$$\left(\frac{\sigma_{\text{ECAL}}}{E}\right)^2 = \left(\frac{2.8\%}{\sqrt{E}}\right)^2 + \left(\frac{12\%}{E}\right)^2 + (0.3\%)^2 \tag{2.11}$$

The ECAL subdetector will undergo a substantial upgrade in view of the HL-LHC project starting in 2029. The upgrade aims to considerably increase the granularity of the calorimeter both in the barrel and endcap regions. In the former, this will be achieved by a replacement of the readout electronics; in the latter, the calorimeter will be fully replaced by the High-Granularity Calorimeter (HGCAL). The upgrades are fully detailed in Chap. 4 alongside the development of a novel trigger algorithm that exploits the upgraded Phase-2 calorimeter.

Hadronic Calorimeter

In the space between the CMS magnet and ECAL lies the Hadronic Calorimeter (HCAL) [34], which is composed of five distinct compartments: the HCAL Barrel (HB), the two HCAL Endcaps (HE), the HCAL Outer barrel detector (HO), and the Hadron Forward calorimeter (HF). As was the case for the tracker, the HCAL detector was originally designed to operate under nominal LHC conditions; as the average PU largely exceeded the planned conditions, the HCAL detector has undergone a substantial upgrade started during LS1 and concluded during the LS2 [35, 36], the so-called HCAL *Phase-1 upgrade*.

The HCAL detector is mainly devoted to measuring final-state hadrons and hadronic showers that, while depositing ~30% of their energy in ECAL, are not contained by its active material. Hadronic showers develop by interacting mostly via the strong force; thus, it is especially important to manage the containment of

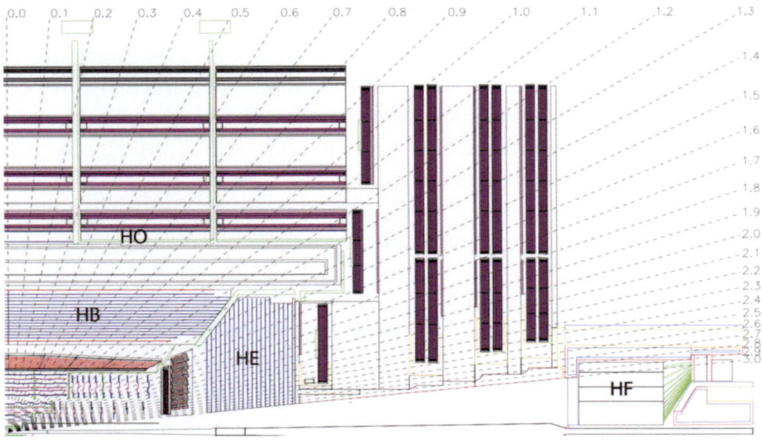

Fig. 2.13 Schematic longitudinal view of the layout of a quarter of the hadronic calorimeter. The barrel HCAL (HB), outer HCAL (HO), endcap HCAL (HB), and forward HCAL (HF) are shown; a pseudorapidity position reference is given with the dashed lines [19]

the neutral component of the shower, which develops in the form of neutrons and $\pi^0 \to \gamma\gamma$ decays. Moreover, as the nuclear interaction length (λ_0) is much larger than X_0, showers present larger fluctuations in terms of spatial development and energy loss. For these reasons, HCAL is designed as a sampling calorimeter with alternating layers of brass absorber and plastic scintillator tiles, and has a much deeper section than ECAL. Nevertheless, the presence of the CMS superconducting magnet places important constraints on the HCAL design.

A schematic representation of the HCAL detector is given in Fig. 2.13 in the form of a longitudinal view of one quadrant. The HB and HE detectors have a similar design; the former has a pseudorapidity coverage of $|\eta| < 1.3$ while the latter complements it with $1.3 < |\eta| < 3.0$. They are both sampling calorimeters consisting of interspersed brass absorbers and plastic scintillator tiles, which amount to a total depth (including the previous subdetectors) of $\sim 7\lambda_0$ at $\eta = 0$, steadily increasing up to $\sim 12\lambda_0$ at $\eta \sim 1.2$, and then stabilized at $\sim 10\lambda_0$ in the HE. Similarly to the EB, the HB is mechanically segmented into 36 wedges, each spanning half of the length of the HB, covering an azimuthal sector of 20°, and weighting about 25.7 tonnes; inside each wedge, the plastic scintillators are organized into 16 η regions. Conversely, the HE follows a different mechanical design, with absorber plates bolted together to form a single 18-sided polyhedric monolithic structure with gaps for scintillator insertion; each HE monolith weighs about 300 tonnes and is fastened to CMS to achieve a quasi-hermetic coverage at the HB-HE edge.

To facilitate the combined processing of the HCAL and ECAL information for event selection in the trigger system, the readout geometry of the HB and HE is designed to closely and easily match the one of ECAL. Therefore, the HB is segmented into so-called *projective towers* of extension $\eta \times \phi = 0.087 \times 0.087$, which match 5×5 ECAL crystals arrays and ensures a high granularity in both the η and ϕ directions. The HE is segmented, with a coarser granularity, in towers of dimension $\eta \times \phi \sim 0.17 \times 0.17$ (the exact extension of towers slightly varies depending on the pseudorapidity position). For each tower, the scintillating light is collected by Wavelength Shifter (WLS) fibres embedded in the tiles and read out by photodetectors.

As the total material budget from the IP to the outer edge of the HB is not enough to guarantee full containment of hadronic showers from highly energetic particles, the HB is complemented by the HO in the pseudorapidity region $|\eta| < 1.4$. The HO is composed of scintillating tiles installed between the magnet's iron flux return yokes, which are exploited as absorber material. The scintillating tiles are organized in trays of the same ϕ extension of the HB wedges and length 2.53 m for a weight of 25–30 kg each. The HO ensures the addition of roughly $1 - 2\lambda_0$ interaction lengths over its pseudorapidity coverage. The readout of the HO is performed with WLS fibres, analogous to the one of the HB/HE, coupled with photodetectors.

The final component of HCAL is the HF, which completes the pseudorapidity coverage in $3.0 < |\eta| < 5.0$. The HF is placed at $|z| = \pm 11.1$ m from the IP, and it is a Cherenkov sampling calorimeter composed of steel absorbers alternating with quartz fibres active material. The selection of this active material is bound to the hostile environment in which the HF is placed, where particle fluxes can be as high as $6 \cdot 10^6\,\text{cm}^{-2}\,\text{s}^{-1}$, thus requiring a highly radiation-hard material that necessitates low maintenance. Fibres of two different lengths are installed to estimate the electromagnetic and hadronic components of the shower, and the Cherenkov light is read out by Photo-Multiplier Tubes (PMTs).

In the original HCAL design, Hybrid Photo-Diodes (HPDs) were employed as photodetectors for all compartments up to $|\eta| < 3.0$, while PMTs were used in the HF. HPDs were chosen based on their magnetic field tolerance and high gain; nevertheless, unexpected problems of electric discharges from the high-voltage supplies were encountered, thus highly increasing the noise contribution to the HCAL resolution. Recent developments have led to an ideal replacement for the HPDs: the Silicon Photo-Multiplier (SiPM). Thus, over an extensive Phase-1 upgrade program that took place in three stages (LS1, 2016-2017 YETS, and LS2), all photodetectors of each HCAL section have been replaced. The current readout is performed with SiPMs for all compartments up to $|\eta| < 3.0$, while new state-of-the-art PMTs are used in the HF. The implementation of SiPMs is complemented by the replacement of the readout electronics and the introduction of newer, more performant data linking. This upgrade renders possible a significant increase in depth segmentation in the HB and HE calorimeters, with the possibility to achieve three depth segments in the barrel and four to five in the endcap, as shown in Fig. 2.14. This segmentation ensures better tracking of hadronic shower development, which can be exploited to boost the sensitivity to BSM signatures with displaced and delayed jets.

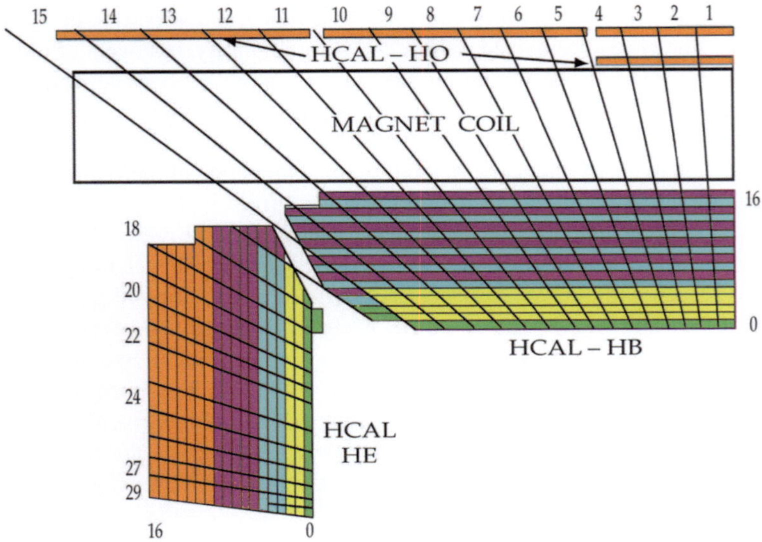

Fig. 2.14 Depth segmentation of the HCAL detector achievable after the Phase-1 upgrade. With the improved gain of the SiPM, a segmentation with three depth segments in the barrel and four to five in the endcap is attained [35]

Muon Chambers

The outermost part of the CMS detector is the muon-tracking system [37, 38], which is devoted to detecting muons for reconstructing their tracks. With typical energies ranging from a few to hundreds of GeV, muons at the LHC are close to Minimum Ionizing Particles (MIPs); thus, they lose minimal energy when traversing the CMS detector and easily escape it. For this reason, the muon system is placed outside the CMS magnet volume, using the flux return yokes as mechanical support. The presence of the 2 T return filed ensures a measurement of the muons' charge and momentum, which is complementary to that of the tracker. Due to the importance of muons in the CMS physics program, the muon system has been designed with high redundancy to ensure the highest performance and trigger selectivity. Moreover, considering the large surface and volume it covers, a compromise between performance and cost was found in the gaseous chamber technology.

Figure 2.15 gives a schematic representation of the muon system in the form of a longitudinal view of one quadrant. A total of 1400 muon chambers are organized into three groups of detectors: the central pseudorapidity region is covered by Drift Tubes (DTs) up to $|\eta| < 1.2$, the coverage is completed in the endcaps by the Cathode Strip Chambers (CSCs) with $0.9 < |\eta| < 2.4$, and measurement redundancy is achieved with the Resistive Plate Chambers (RPCs) up to $|\eta| < 2.1$. These elements combined add up to a total detection surface of about 25000 m^2.

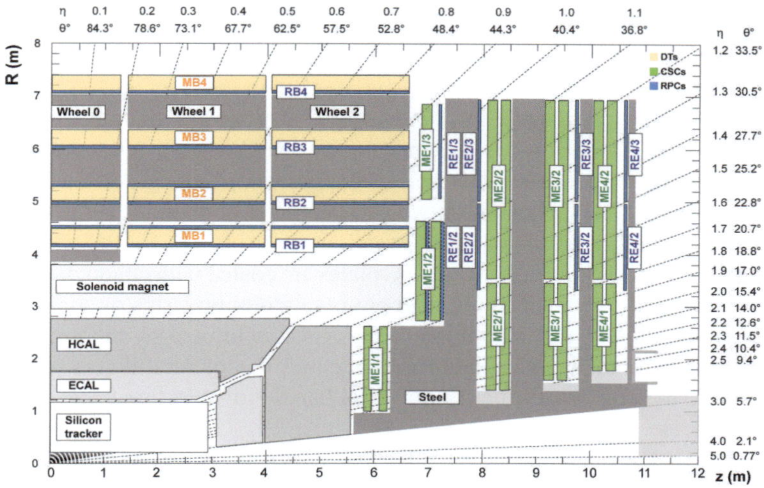

Fig. 2.15 Schematic longitudinal view of the layout of a quarter of the muon detection system. The Drift Tubes (DT, yellow), the Cathode Strip Chambers (CSC, green), and Resistive Plate Chambers (RPCs, blue) are shown; a pseudorapidity position reference is given with the dashed lines [38]

In the barrel region, 250 DTs of $2 \times 2.5\,\mathrm{m}^2$ surface are deployed with a lateral arrangement in five sections, the so-called *wheels*, and a depth segmentation of four stations, ensuring a four-point detection. Each chamber is segmented into 12 planes, and each plane is segmented into tubes of $1.3 \times 4.2\,\mathrm{cm}^2$ section and 2.4 m length; this relatively low granularity is allowed by the low expected occupancy in this region. Each DT consists of an anode wire spanning the entire length of the tube and two cathode strips, all immersed in a gaseous mixture of Ar (85%) and CO_2 (15%). To achieve the highest time and position resolution, DT layers are stacked with an offset of half of a tube width, thus ensuring a time resolution lower than 6 ns, a spatial resolution of 100 μm in the transverse plane and 150 μm in the longitudinal direction, and station efficiency larger than 98%.

In the endcaps, 540 CSC chambers (270 per endcap) are implemented. Each chamber has a trapezoidal shape with a length ranging between 1.7 and 3.3 m, and a width ranging between 60 cm and 1.5 m, depending on the pseudorapidity position. Each unit is divided into 72 or 36 chambers, each with six alternating planes of anode wires and cathode strips, the former providing the η measurement and the latter providing the $r - \phi$ coordinates. Strip width varies from 3 to 16 mm for different chambers, or from about 2 to 5 mrad in ϕ coordinate. This higher granularity compared to the barrel region is needed to cope with the higher background rate and the stronger, non-uniform magnetic field in which they are immersed. The chosen gaseous mixture is composed of Ar (45%), CO_2 (50%) and CF_4 (10%). The CSCs ensure a time resolution similar to that of the DTs, a spatial resolution of 45–150 μm, and station efficiency larger than 92%.

To achieve the highest trigger performance, 612 RPC chambers are deployed to complement the DT and CSC measurements. In the barrel, six RPC stations are mounted as follows: four RPC stations are mounted on the front and on the back of the first two DT stations, while the last two RPC stations are mounted on the front of the remaining DT stations. In the endcap, four RPC stations are mounted in the CSCs inter-space. The full architecture can be appreciated in Fig. 2.15, where RPCs are depicted in blue. RPCs consist of two parallel plates of phenolic resin with a separation of a few millimetres in which a gaseous mixture of $C_2H_2F_4$ (96.2%), i-C_4H_10 (3.5%) and SF_6 (0.3%) is placed. The RPC plates are coated with conductive graphite paint to form electrodes, which are read out by aluminium strips on the outside of the plate. While the spatial resolution of the RPCs is limited by the segmentation size of about 1 cm, their time resolution challenges that of scintillators with values as low as 2–3 ns. This makes them a powerful handle for selecting events incoming at a 40 MHz rate.

Like the pixel detector and HCAL, also the muon detection system has undergone an upgrade with the installation of the Gas Electron Multiplier (GEM) detectors in the region $1.6 < |\eta| < 2.2$ [39]. This GEM implementation is an anticipation of the Phase-2 upgrade program, whose benefit can also be experienced in Run-3. A total of 144 (72 in each endcap) chambers have been installed during LS2; each chamber has a trapezoidal form with an opening angle of 10.15°, filled with an Ar (70%), CO_2 (30%) gaseous mixture. Each chamber is segmented in eight η sections with 384 readout strips of 230 µrad angular width and radially placed with a 463 µrad pitch. These specifications ensure a spatial resolution of 30–100 µm and a timing resolution of about 5 ns.

2.3 The CMS Trigger System

The full information from all CMS subdetectors amounts to $\sim$1 Mb per event; therefore, if read out at the nominal LHC bunch crossing rate of 40 MHz, they would produce a total throughput of $\sim$40 Tb/s. At the present day, technology falls short of efficiently reading and storing such formidable data quantities. However, a substantial portion of these collisions yields low-energy proton-proton interactions, which hold no relevance to the CMS physics program, which targets hard scattering processes. Figure 2.16 shows the summary of the cross section measurements of SM processes at CMS; as it can be appreciated, the process with the highest cross section is single W boson production with $\sigma(W|\sqrt{s} = 13\,\text{TeV}) = 1.8 \cdot 10^5$ pb. This value stands six orders of magnitude below the inclusive proton-proton interaction cross section that towers at $\sigma(pp) \sim 10^{11}$ pb. The knowledge of this huge discrepancy can be exploited to perform an online event selection with the goal of reducing the data acquisition rate by $\sim 10^5$. This procedure is the so-called *triggering process*, and the CMS Trigger system performs it. After the trigger selection, the data is sent to storage by the Data Acquisition System (DAQ). The Trigger and DAQ are generally jointly referred to as the TriDAS project [40].

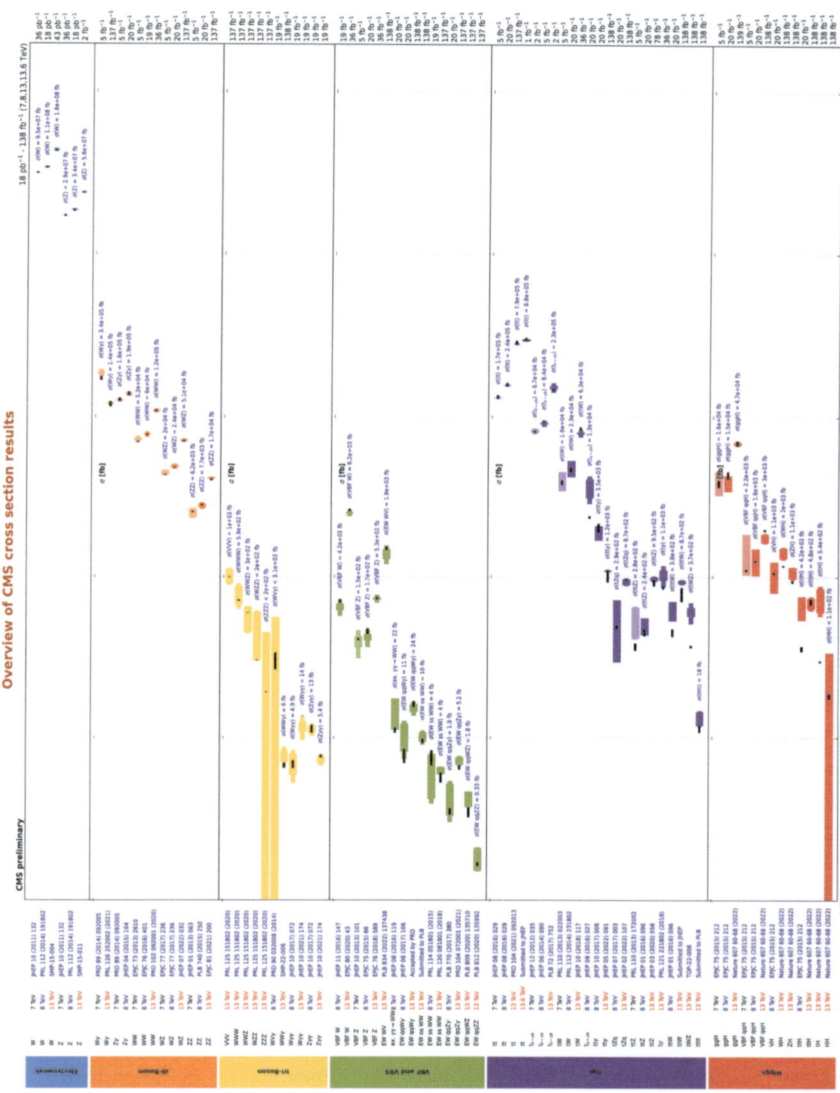

Fig. 2.16 Summary of the cross section measurements of Standard Model processes at CMS. The process, centre-of-mass energy of the measurement, and the associated publication are reported on the left of the panel; the integrated luminosity used for each result is reported on the right [41]. Values are to be compared to the total proton-proton interaction cross section of about 10^{11} pb

The trigger system acts as the bridge linking online data-taking and offline data analysis, the latter being impossible without the former. Therefore, to fulfil the ambitious CMS physics program, the trigger system must adhere to both the technical constraints set by the online hardware system and the stringent efficiency benchmarks and background suppression expected on the side of the analyses. Moreover, adaptability to varied data conditions and resilience against the instantaneous luminosity and PU challenges posed by the LHC are paramount prerequisites for the system. These are the all-important and exacting guidelines that underpin the trigger system design.

To achieve the best flexibility of the trigger system, the CMS experiment adopts a two-tiered approach in which the event selection is based on the kinematic properties of the particles produced in an LHC bunch crossing. The two steps have to fulfil very different requirements and are implemented in different kinds of hardware and with different levels of sophistication. The first selection is performed by the Level-1 (L1) trigger, which is composed of dedicated hardware that processes the information from calorimeters and muon systems only with reduced granularity; the L1 has at its disposal a maximum processing time (the so-called *latency*) of 3.8 μs and selects the most interesting events for a rate up to 100 kHz. The second selection is performed by the High-Level Trigger (HLT), which exploits the full detector granularity on commercial CPU and GPU processors; the HLT has a latency of $\sim$200 ms and selects the most interesting events for a rate up to 1 kHz. The events thus selected are acquired by the DAQ system and sent for permanent storage in the tapes of the Tier-0 and Tier-1 sites; the two core levels of the so-called *WLCG grid* (Worldwide LHC Computing Grid) [42]. As it can be appreciated, the triggering process needs to perform a real-time reduction of the data by a factor $4 \cdot 10^4$ while retaining the most interesting events for physics analysis.

The TriDAS system is detailed in the following, with particular attention given to the Level-1 trigger, especially its calorimeter-based part, as it is a central topic of this Thesis.

2.3.1 The Level-1 Trigger

The L1 trigger is designed based on two main guidelines: latency and flexibility. The event accept decision must be performed within 3.8 μs. This limitation enforces the use of hardware-implemented processing of the data from the CMS subdetector; this, in turn, sets tight restrictions on the amount of information that can be processed and how it can be processed within the hardware resources limits. Moreover, being the first tier in the data acquisition chain, the L1 trigger must be flexible and scalable to adapt to the challenging and varying LHC running conditions while abiding by the needs of evolving physics searches.

The input to the L1 trigger are the so-called *Trigger Primitives* (TPs), which are produced either by the on-detector electronics, the so-called *front-end* placed in the experimental cavern, or the off-detector electronics, the so-called *back-end* placed in the service cavern. Due to the tracker back-end hardware and latency constraints, the production of tracker TPs for the L1 trigger is not yet possible; therefore, only the information from the calorimeters and the muon detectors is exploited. TPs offer a coarse view of the CMS detector; thus, the precise reconstruction algorithm developed for the offline reconstruction of physics objects cannot be performed. Conversely, the L1 employs the TPs for the production of coarse-granularity and low-resolution physics objects, the so-called *L1 candidates*.

As already detailed for the pixel and HCAL detectors, the L1 trigger system was designed to cope with the nominal LHC running conditions; foreseeing these conditions to be largely exceeded in Run-2, a major upgrade of the L1 trigger [43, 44] was installed and commissioned between 2015 and 2016. The main aspect of the L1 trigger Phase-1 upgrade has been the replacement of the custom ASICs (Application-Specific Integrated Circuits) used in part of the system with more powerful, more flexible, and easier-to-maintain industry standards. The original electronics were replaced by Advanced Mezzanine Cards (AMC) technology, electronics boards that mount powerful Field-Programmable Gate Arrays (FPGAs) and are designed following the μTCA (micro Telecommunications Computing Architecture) industry standard, which ensures additional flexibility and higher bandwidth. The FPGAs are electronic circuits whose runtime functionality can be pre-configured using a Hardware Description Language (HDL). This upgrade brought all the subcomponents of the L1 trigger to the same standard, thus ensuring higher flexibility and scalability of the trigger system. Moreover, state-of-the-art optical serial links with a bandwidth of up to 10 Gb/s replaced the original copper linking between boards to maximise data throughput. In parallel to the hardware replacement, a consistent upgrade of the L1 trigger algorithms was strived for, ensuring the new powerful FPGAs were exploited at their maximum. A schematic representation of the L1 trigger system, in its upgraded Run-2 and Run-3 architecture, is given in Fig. 2.17.

The upgraded trigger system retains the basic division into two subsystems, which run parallelly before flowing into the third subsystem that performs the event accept decision. The TPs from the ECAL and HCAL subdetectors are collected by the *calorimeter trigger*, while those from the DT, RPC, and CSC subdetectors are collected by the *muon trigger*. The TPs from the GEM subdetectors are currently being validated; thus, they are also available in the muon trigger, but they are not yet used for building the muon candidates. In the hardware of these two subsystems are implemented the L1 reconstruction algorithms, which optimally exploit the TPs to produce the L1 candidates. In the calorimeter trigger, the objects are built from local energy deposits to form electrons and photons, which are indistinguishable at this stage and are jointly referred to as e/γ, hadronically decaying τ leptons (τ_h), jets, and energy sums. In the muon trigger, the objects are μ candidates constructed based on tracks built from the hits in the muon chambers. All the objects from the two subsystems are then fed to the Global Trigger (μGT) that combines the information to perform the event accept or reject decision based on the L1 candidates' kinematics and high-level variables, such as invariant masses and angular distances. The decision is based on pre-defined energy, position, and isolation criteria; the energy cutoffs applied at this stage are commonly denoted as *L1 thresholds*. The numerical values of these

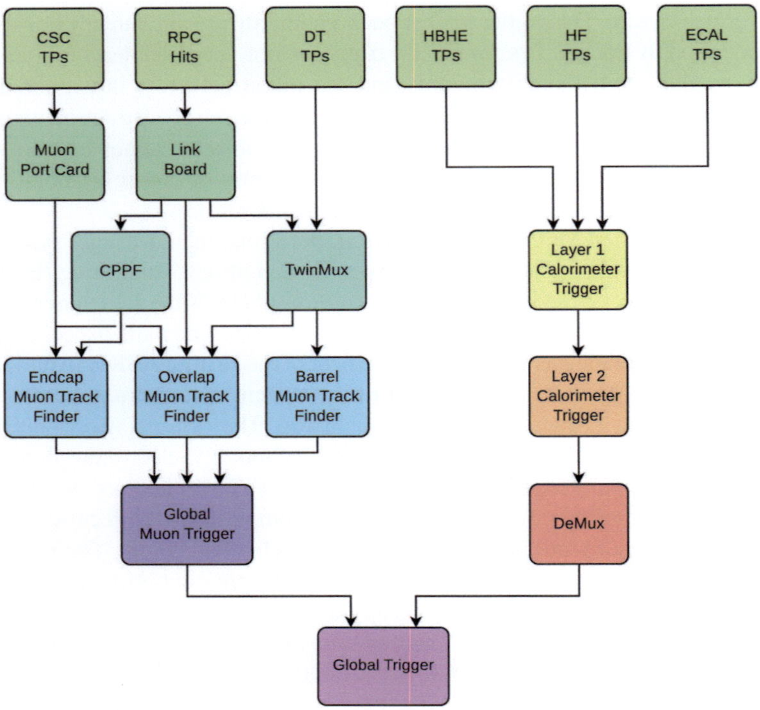

Fig. 2.17 Schematic representation of the Level-1 trigger system architecture in Run-2 and Run-3. The information from the CSC, RPC, and DT detectors are used to reconstruct the μ candidates; the energy deposits in the ECAL, HCAL, and HF detectors are combined to build e/γ, τ, and jet candidates, as well as global HT and p_T^{miss} quantities. The global trigger collects the output of the two independent trigger lines and takes the event accept or reject decision [45]

thresholds are chosen to find a compromise between the L1 rate and the phase space available for physics analyses: the lower the threshold, the wider the latter and the higher the former.

Level-1 Calorimeter Trigger

One of the key technological changes brought by the L1 trigger upgrade is the implementation of the original Time-Multiplexed Trigger (TMT) architecture [43, 46, 47], which allows for a global view of the detector per each event. The TMT architecture and the conceptual choices that lead to it are discussed in the following.

The experience accumulated during Run-1 showed that the main constraining factor in achieving higher trigger performance was the limited view of the detector; improved granularity of the input and global view of the detector are the leading factors to achieve the efficient reconstruction and identification of e/γ, τ_h, jets and

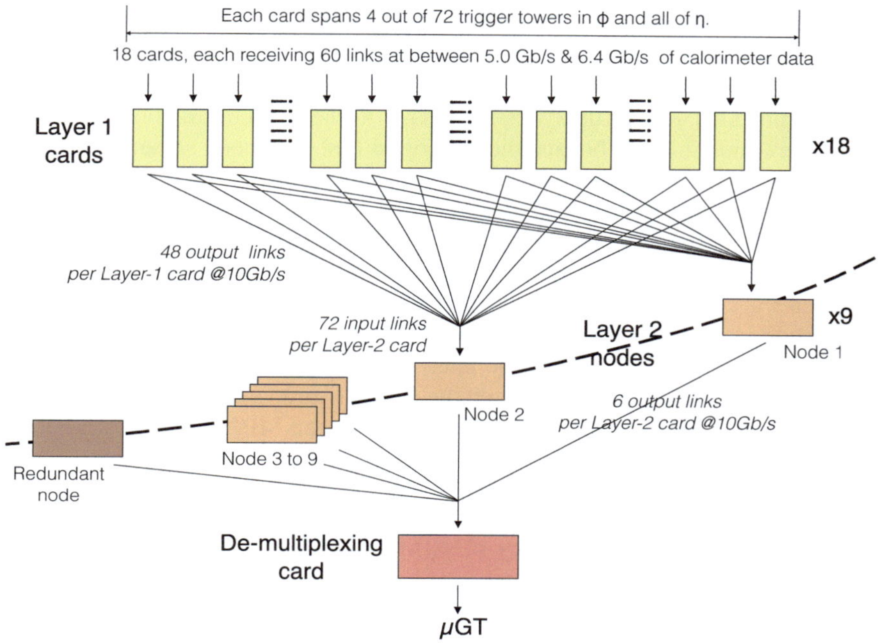

Fig. 2.18 Layout of the time-multiplexed architecture of the Run-2 and Run-3 calorimeter trigger. The information from the calorimeters is collected by the 18 CTP7 card constituting the Layer-1 of the CT, which calibrates and sorts the trigger towers in decreasing E_T. The Layer-1 output is sent to the 10 MP7 cards of Layer-2, where the e/γ, τ_h, jet, and sums algorithms are implemented. The objects built at Layer-2 are sent to the μGT to be used for the event accept or reject decision [43]

sums. Therefore, accessing the whole calorimeter information at the same time, with improved granularity, was paramount in view of Run-2 and Run-3. Nevertheless, the transmission of the totality of the TPs of a specific bunch crossing to a single electronic board is not possible given the 25 ns bunch crossing spacing. The TMT architecture answers this issue by analysing two consecutive events in separate boards rather than sequentially in one board.

A schematic representation of the TMT calorimeter trigger architecture is given in Fig. 2.18. The calorimeter trigger presents a two-layer architecture, with the second tier fully dependent on the first one. The highly granular TPs from the calorimeters, which have a typical size $\eta \times \phi \sim 0.087 \times 0.087$ over most of the subdetectors acceptance, are first processed into the 18 Calorimeter Trigger Processor (CTP7) cards [48] of the Layer-1 system. The CTP7 cards are AMCs that implement a Xilinx Virtex-7 FPGA optimized for fast data sharing with the subsequent layer. At Layer-1, the TPs are calibrated and pre-processed to build the so-called *Trigger Towers* (TTs), which encode the sum of the ECAL and HCAL TPs that lie one behind the other in the physical world. Each CTP7 processes a $\Delta\phi = 20°$ sector and, after calibration, sorts the TTs in order of decreasing energy deposit. The Layer-1 then dispatches the full $\Delta\phi = 360^\circ$ information to one of the nine Layer-2 processing nodes where the reconstruction and identification algorithms are implemented. Layer-2 is constituted by 10 Master Processor (MP7) cards [49], the 10th card being a redundant safety node, each instrumented with a Xilinx Virtex-7 FPGA. As the L1 reconstruction algorithms are implemented in these boards, they are optimized as generic stream-processing engines to provide the best flexibility in algorithm embedding. The use of nine boards, each processing the information from consecutive events, effectively introduces an *artificial latency* of $9 \cdot 25$ ns with respect to a non-multiplexed system, thus providing the possibility of implementing more sophisticated algorithms. Finally, the output of Layer-2 is collected by the so-called demultiplexer node, which reorganizes the L1 candidates, converts their energy and position coordinates to the μGT specific format, and transmits them to it. The interface with the CMS data acquisition and system synchronization is ensured with the AMC13 card [50].

The information from the CTP7 card is sent to the MP7 card over 72 optical links, four for each Layer-1 board; as each CTP7 processes only a portion of the calorimeter information, the MP7 card needs to perform a reorganization of the inputs. This process introduces a sizeable latency, which is partially recovered by running the MP7 at a clock frequency of 240 MHz; in this way, the data takes less than seven bunch crossing to be sent from Layer-1 and Layer-2. Moreover, algorithms are designed to start the processing as soon as a minimal amount of data is received, thus further reducing the latency.

The introduction of the TMT architecture ensured an enhancement of the spatial granularity by a factor of four compared to Run-1, and the use of powerful FPGAs opened the road to the firmware embedding of complex algorithms to precisely cluster relevant energy deposits into identifiable L1 candidates, ultimately improving the trigger resilience to PU. Moreover, the improved granularity gives the possibility to compute higher-level variables like the spatial correlation between objects or their invariant mass, allowing for the implementation at L1 of analysis-specific triggers. A brief overview of the L1 calorimeter candidates' reconstruction algorithms is given in the following.

In the L1 trigger, e/γ candidates can be initiated either by electrons or photons, which are indistinguishable at this stage due to the inability to access the tracking information at L1. The candidate is initiated by a local energy deposit in a fixed-dimension region, the so-called *seed*; the TTs around the seed are dynamically clustered to the seed based on basic position and energy deposit rules. As electromagnetic showers are known to have a small lateral dimension, trimming rules are implemented after clustering to exclude soft PU contributions better; the resulting shape is then required to pass a *shape veto* requirement that targets cluster profiles from genuine electrons of photons. As discussed above, the ECAL subdetector has a depth of $23X_0$, making longitudinal containment of e/γ showers highly probable; thus, an energy-dependent threshold on the ratio between ECAL and HCAL energy deposit is enforced to reduce the misidentification rate of hadronic showers. The resulting candidates are calibrated based on their energy, η position, and shape. Finally, as electrons and photons tend to induce narrower showers compared to QCD-induced background, an isolation criterion based on the activity surrounding the e/γ candidate is enforced to increase further the true positive rate. The excellent efficiency achieved for inclusive L1 e/γ candidates in 2022 is shown in Fig. 2.19a for typical L1 thresholds.

The τ_h candidates are reconstructed with an algorithm similar to that used for the e/γ candidates. Specific modifications are implemented to account for the particle multiplicity in the hadronic decays of τ leptons, while calibration and isolation requirements are tailored to the specific needs of these candidates. As the optimization of the L1 τ_h algorithm in view of Run-3 has been a part of this Thesis work, a fully detailed description of the algorithm is given in Sect. 3.2.

The L1 calorimeter trigger builds jet candidates employing a fixed dimension clustering approach. A local energy maximum seed initiates the clustering, and the 9×9 array of TTs around it is grouped as a single candidate; this dimension corresponds to an angular opening $\Delta R \sim 0.4$, which is the same extension used for the offline reconstruction of jets, thus facilitating interpretation. Given the relatively large extension of the 9×9 array, the PU contribution needs to be dealt with optimally; two techniques are available at L1: the *chuncky doughnut* and the *phi ring* subtractions, both correcting the candidate energy on a jet-by-jet basis. While the former estimates the PU contribution from the energy deposit in the flat approximately toroidal area surrounding the candidate, the latter estimates it on rings of TTs with the same η coordinate. Both approaches have been implemented in the Layer-2 firmware; the chunky doughnut approach has been used as the default technique during Run-2 and the start of Run-3, while extensive studies have been performed on the phi ring method as a possible substitute for the remaining period of Run-3 data-taking. Before being sent to the μGT, jet candidates are calibrated based on their energy and η position. The great efficiency achieved for L1 jet candidates in 2022 is shown in Fig. 2.19b for two typical L1 thresholds. An important new feature introduced for Run-3 by the HCAL Phase-1 upgrade is a finer segmentation of the detector's depth, which can be exploited in conjunction with the ECAL timing information to design L1 trigger algorithms that target delayed and displaced jets. Such objects are relevant for the search for Long-Lived Particles (LLP) and are thus referred to as *LLP triggers*. Such triggers exploit the better shower tracking ensured by the SiPMs to target LLPs decaying into jets within the HCAL volume. The efficiency of L1 delayed jets was evaluated during the 2023 HCAL phase scan, where an artificial delay in the HCAL TPs signals is introduced for time alignment purposes. During this procedure, artificially delayed jets can be simulated and the performance of the L1 trigger tested. The remarkable efficiency achieved for L1 delayed jet candidates in 2023 is shown in Fig. 2.19c for several values of phase delay.

The energy sums are the last type of object constructed in the L1 calorimeter trigger. Two sums are of particular interest: the negative of the magnitude of the vector sum of the transverse energy over all the TTs ($E_\mathrm{T}^\mathrm{miss}$) and the total scalar transverse energy of all jets (H_T). These objects significantly profit from the TMT architecture as the full calorimeter granularity can be used for their computation. A dedicated PU subtraction and calibration is applied to $E_\mathrm{T}^\mathrm{miss}$ candidates to remove large contributions from soft, diffuse PU-induced energy deposits; these techniques exploit the activity in the central part of the barrel to estimate the PU on an event-by-

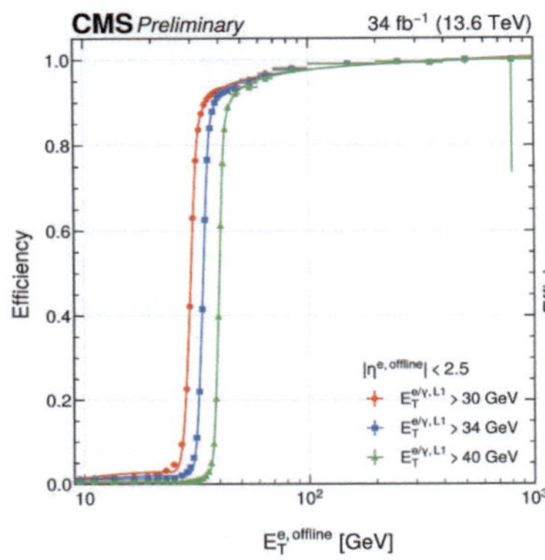

(a) e/γ candidates

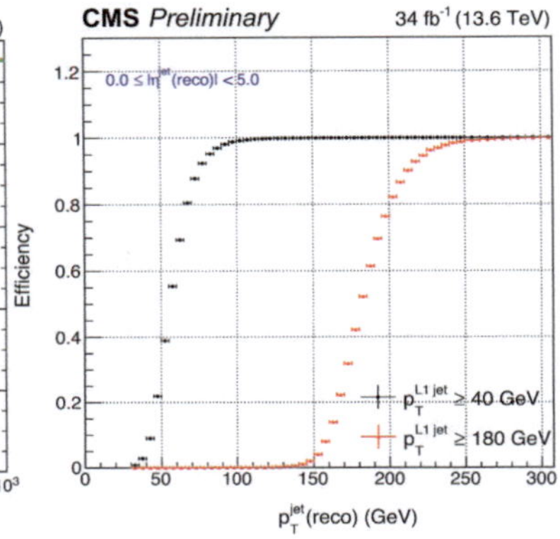

(b) Jet candidates

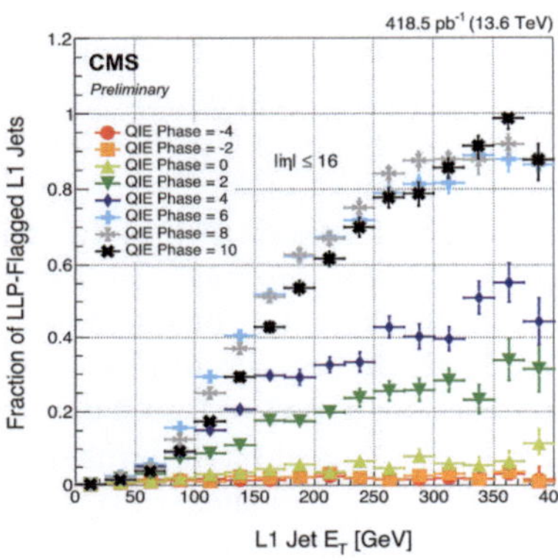

(c) LLP jet candidates

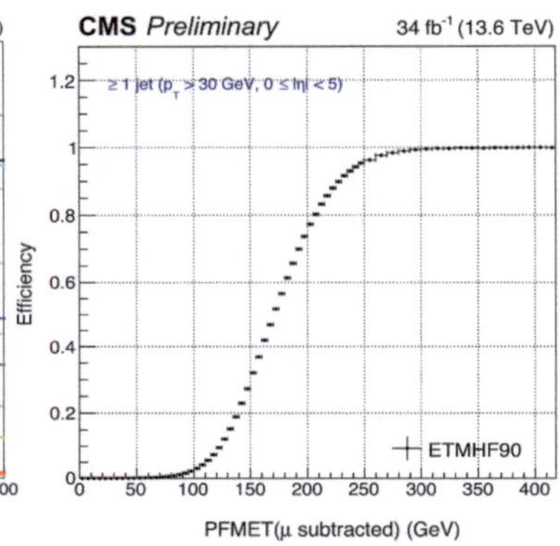

(d) $E_{\mathrm{T}}^{\mathrm{miss}}$ candidates

Fig. 2.19 Level-1 performance of selected objects relevant for the calorimeter trigger. **a** The L1 e/γ trigger efficiency as a function of the offline reconstructed E_{T} for three different high E_{T} thresholds on the L1 trigger e/γ candidate; the functional form of the fits consists of a cumulative Crystal Ball function with a polynomial tail in the low E_{T} region. **b** L1 jet efficiency as a function of the offline jet p_{T}, for two L1 p_{T} thresholds. Low $p_{\mathrm{T}}^{\mathrm{L1}}$ jets are observed to show a response below unity, resulting in a 50% efficiency for the $p_{\mathrm{T}}^{\mathrm{L1}} > 40\,\mathrm{GeV}$ condition reached above the corresponding offline p_{T} value. This feature is primarily due to the limited accuracy of the jet energy corrections derived from simulations [51]. **c** The HCAL LLP-flagged L1 delayed jet fraction as a function of jet E_{T} during the 2023 HCAL phase scan demonstrates that the delayed jet fraction reaches unity as the phase delay increases. The delayed jet fraction begins decreasing at the largest delays (10 ns and above), as at these large delay settings, the total hadronic jet energy is reduced due to a significant amount of the jet energy being pushed into the subsequent bunch crossing. No direct selection is made with respect to jet E_{T}: the implicit requirement for a jet to have at least two cells with $E_{\mathrm{T}} > 4\,\mathrm{GeV}$ sculpts this distribution [52]. **d** L1 ETMHF90 efficiency, evaluated as the fraction of events for which the magnitude of the vector sum of all TT is $E_{\mathrm{T}} > 90\,\mathrm{GeV}$, as a function of the offline transverse momentum of the event, removing the contribution from muons. In order to mitigate the impact of PU collisions, low E_{T} TTs are excluded from the ETMHF TT sum. Moreover, jet-level energy corrections are not propagated. These two effects result in a sizeable shift of the ETMHF response with respect to the offline measurement of the missing transverse energy [51] event basis and define the energy threshold that TTs need to pass to be considered. The remarkable efficiency achieved for L1 $E_{\mathrm{T}}^{\mathrm{miss}}$ candidates in 2022 is shown in Fig. 2.19d for a typical L1 threshold.

Level-1 Muon Trigger

As shown for the calorimeter trigger, the ability to have a global view per event can highly enhance the L1 trigger performance. In the context of the muon trigger, the L1 upgrade ensures a quasi-global view of the detectors, whose redundancy is fully exploited. During Run-1, the muon trigger performed the candidates' reconstruction separately for each muon chamber type, i.e. DTs, CSCs, and RPCs; the upgraded system introduced a shift in this approach by using all subdetectors information at the same time in three separate regions of pseudorapidity where the tracks of the muons are reconstructed. These three regions, schematically reported in Fig. 2.17, are the Barrel Muon Track Finder (BMTF), the Overlap Muon Track Finder (OMTF), and the Endcap Muon Track Finder (EMTF).

The BMTF uses the information from the DT and RPC muon detectors and has the pseudorapidity coverage $|\eta| < 0.83$. The TPs from the two detector back-ends are collected by the TwinMUX system, which combines them into *super-primitives*; compared to TPs, super-primitives have an improved precision in the reconstruction of the muon hit position owing to the combination of the spatial resolution of the DT and the precise timing of the RPC. In Run-2, the BMTF used this information to perform a track extrapolation from the inner to the outer muon stations, defining distinct hit acceptance windows for each layer based on the extrapolated track from previous steps. In Run-3, the BMTF reconstruction has been upgraded to a Kalman Filter algorithm [53], which was originally developed for the HL-LHC and subsequently optimized to fit the Run-3 technical constraints. After tracks are built, the quality of the extrapolation and the number of hits are used to define quality criteria, which are used as a handle to increase true positive fractions and decrease the L1 rate. To remain within the imposed latency of about 750 ns, the muon candidates' p_{T} assignment is performed leveraging pattern recognition algorithms based on the track hits. Given the iterative approach to track reconstruction, the BMTF and TwinMUX systems need large computing power; therefore, MP7 cards identical to those used in the calorimeter trigger Layer-2 are employed. The use of the same hardware further increases the flexibility and scalability of the upgraded Phase-1 trigger.

The OMTF uses the information from all muon detectors and has a pseudorapidity coverage of $0.83 < |\eta| < 1.24$, while the EMTF uses the information from CSCs and RPCs, and has a coverage of $1.24 < |\eta| < 2.4$. These subsystems use a different approach to track finding than the BMTF; rather than a pure track extrapolation that would require a long latency in these high occupancy regions, muon tracks are reconstructed using fast pattern recognition, with each pattern encoding information on the hit probability density and the average track propagation between stations for a given energy. The pattern recognition starts from a hit seed, preferably in one of the inner stations, and at the same time performs the track reconstruction and the

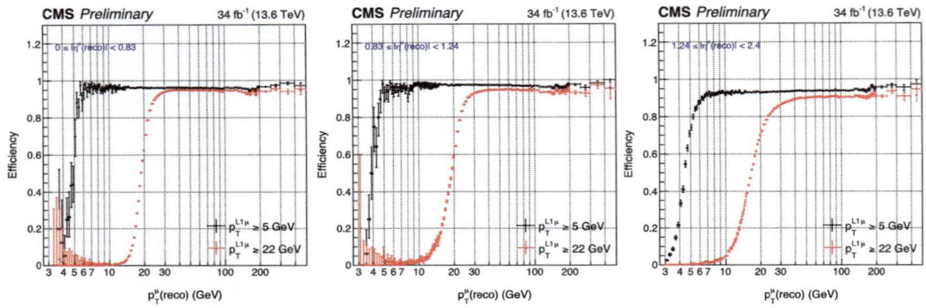

Fig. 2.20 Level-1 muon efficiency in the BMTF (left), OMTF (centre), and EMTF (right) regions as a function of the offline muon p_T, for two L1 p_T thresholds. Only the highest-quality muons are considered. As very high momentum muons can initiate a shower in the muon detectors, a small decrease in efficiency can be noticed [51]

p_T assignment. Similar to the BMTF, the number and topology of hits pertaining to a track are used to establish quality criteria. Given the specific approach to muon track finding employed in the OMTF and EMTF, the hardware is optimized to have a large storage memory for the implementation of the track patterns; the specific boards used are the Modular Track Finder (MTF7) cards [54].

The muon candidates from the BMTF, OMTF, and EMTF are collected by the Global Muon Trigger (μGMT), which performs the candidates' post-processing. Duplicate candidates are removed at the system's boundaries, the muons are converted into the μGT specific format, ranked by p_T and quality, and finally sent to the μGT. Due to the need for large computing power, the μGMT is also implemented in an MP7 card. The excellent efficiency achieved for inclusive L1 muon candidates in 2022 is shown in Fig. 2.20 for the three subsystems separately.

Level-1 Global Trigger

All L1 candidates built in the calorimeter and muon triggers are collected by the global trigger, which is devoted to deciding whether to accept or reject an event. In the μGT, the kinematical variables of the L1 candidates are exploited to either apply basic selections or compute higher-level correlations between objects, like angular distances and invariant masses; this is achievable owing to the μGT instrumentation with six MP7 boards with 72 10 Gb/s optical links each (to be compared with a single μGT board available in Run-1). The final selection of the events is based on the so-called *L1 trigger menu*, which is a collection of algorithms (or *seeds*) that apply a diverse array of kinematics and quality criteria. The most simple seeds perform basic cutoff selection on candidates' energy, momentum, pseudorapidity, and quality; these seeds are generally employed as single- or double-candidate selection (with the two objects being of the same kind). More sophisticated selections are available in the form of cross-triggers, which simultaneously target candidates of different types; these are generally analysis-oriented seeds that exploit the introduction of an additional handle in the form of a second object to loosen the requirements on the first object.

In both Run-2 and Run-3, the L1 trigger μGT implemented a menu of 400–500 seeds, a more than three-fold increase compared to Run-1, where the integration of only 128 seeds was possible. To any seed is assigned an adjustable *prescale* factor P that reduces the trigger rate of $1/P$ by retaining only one event accept decision every P occurrences. The set of prescales is pre-computed as a function of the instantaneous luminosity conditions and embedded in the μGT firmware. The set of prescale values is changed during data-taking to maintain a constant ~100 kHz trigger rate when the instantaneous luminosity is reduced, thus maximizing signal acceptance. Both the L1 menu and the associated prescale set undergo constant optimization and redefinition to cope with the changing running conditions of the LHC; this effort is particularly intensive during the period of the Run-3 restart.

It should be noted here that the rate value of 100 kHz is not a hard threshold, and larger rates can be accommodated. Namely, during Run-3 the L1 trigger rate is stably increased at ~110 kHz to favour the introduction of the new *parking* techniques detailed in the next Section.

2.3.2 The High-Level Trigger

The second triggering level of CMS is the HLT, which takes as input the events marked with a L1 accept signal at ~100 kHz and reduces the information rate to ~1 kHz, which is a rate compatible with the maximum speed of writing to permanent storage. The HLT, like other subsystems, has undergone an upgrade during Phase-1; in this case, the upgrade is part of the Phase-2 upgrade program [55] and has been deployed in LS2; thus, its benefits are already advantageous during Run-3, as discussed below.

The HLT is implemented in a software computing farm located in the surface building of IP5. The available input at HLT is the full granularity information of all the CMS subdetectors, including the tracker not exploitable at L1; this load of information allows a more sophisticated reconstruction close to the offline standard. This is performed with a streamlined version of the CMS Software (CMSSW), optimized to be executed within the HLT latency of ~200 ms per event. Moreover, to lower the computational burden, the HLT reconstruction is performed only locally around the L1 objects that triggered the event accept, thus making evident the naming *seed* for the L1 algorithms. The HLT implements a menu of algorithms, the so-called *paths*, which are defined based on the set of reconstruction and filtering instructions of increasing complexity through which the candidates pass. To meet timing and hardware resources constraints, the HLT performs an initial selection based only on the calorimeter and muon detectors information; for those events that pass this pre-filtering, the tracker information is fully exploited and matched to the calorimeter and muon system information.

The HLT menu consists of around 600 independent algorithms, by construction closely related to the L1 seeds. These algorithms generally employ highly sophisticated algorithms based on Machine Learning (ML) techniques for the reconstruction of jets, their identification as originating from gluon, light-flavour or heavy-flavour quark, and the specific tagging of b-quark jets; additional ML methods are also employed for the identification of hadronically decaying τ leptons. Like the L1 menu, the HLT menu is regularly updated to cope with the ever-changing collision conditions of the LHC. Moreover, new triggering techniques have been implemented during Run-2, like the so-called *parking*, which consists of writing events directly to tape during the latest stages of an LHC fill when the rate is substantially decreased due to the lower instantaneous luminosity. Therefore, the parked events consist of the complete raw data information that can be exploited to increase signal acceptance and perform quasi-unbiased physics analyses.

All events accepted for storage by the HLT trigger are finally sent to permanent storage in the tapes of the CERN Tier-0, where events are classified into *primary datasets* according to the HLT paths that accepted the event.

During Run-2, the HLT farm was composed of Intel Skylake Gold 6130 CPUs allowing for high-performance running of CMSSW. As the use of ML techniques is becoming wider within the high-energy physics community, the number of HLT algorithms based on neural networks is increasing rapidly. To cope with this need, as part of the Phase-2 upgrade, the HLT computing farm has been upgraded during LS2 by deploying 200 dual-processor servers, each equipped with two AMD EPYC Milan 7763 CPUs and two NVIDIA Tesla T4 GPUs [56]. This improvement called for extensive work of software portability, i.e. conversion of CPU-developed CMSSW algorithms to GPU-friendly formats. The improvement in performance granted by this upgrade is reported in Figs. 2.21 and 2.22; the former reporting the latency needed for HLT processing under CPU-only and CPU+GPU conditions, and the latter showing the throughput improvement achieved with the GPUs compared to a CPU-only based system. It is also worth noticing that the introduction of GPU acceleration ensures a $\sim$30% reduction in power consumption at the same throughput level.

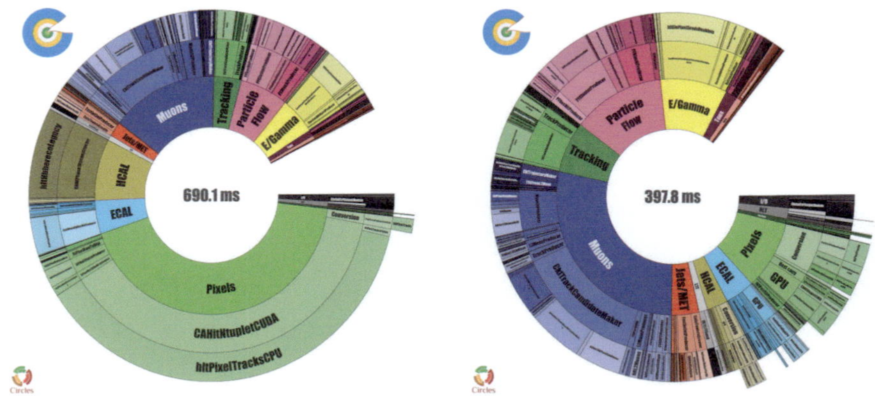

(a) CPU-only (b) CPU+GPU

Fig. 2.21 Distribution of CPU time in different instances of CMSSW modules (outermost ring), their corresponding C++ class (one level inner), grouped by physics object or detector (innermost ring). In the right panel, the extra outer wedges indicate the time spent converting GPU-friendly data to CPU data formats. The empty slice indicates the time spent outside of the individual algorithms. The timing is measured at $\langle PU \rangle = 56$ in Run-3 data processed on one AMD EPYC 7763 Milan CPU with eight concurrent jobs, each running with 32 CPU threads and 24 concurrent events (left) and with the same setup plus acceleration on one NVIDIA Tesla T4 GPU (right). The GPU is used to accelerate the pixel local reconstruction, the track and vertex reconstruction, the HCAL local reconstruction, and the ECAL unpacking and local reconstruction [57]

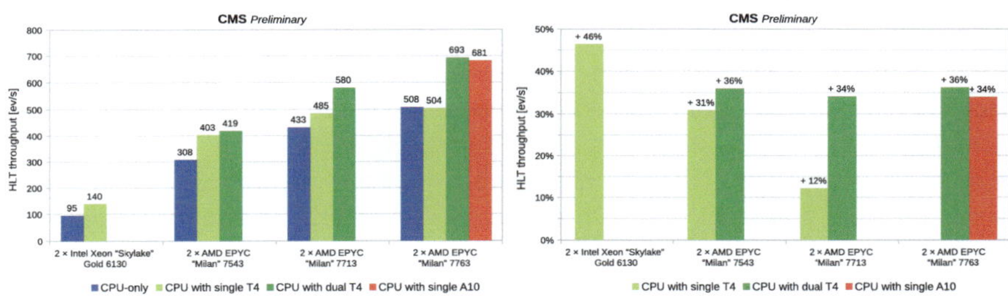

Fig. 2.22 Absolute (left) and relative (right) performance of the HLT on different machines, running only on the CPU (blue bars) and offloading to different GPUs: a single NVIDIA T4 (light green), a pair of NVIDIA T4 (dark green), and a single NVIDIA A10 (red) [56]

2.4 Physics Objects Reconstruction and Identification

The hard scattering events that are selected by the trigger system are recorded for permanent storage in the CERN grid; at this stage, the raw detector information can be exploited for the reconstruction of the so-called *physics objects*, i.e. particles and other quantities that are used to perform physics analyses.

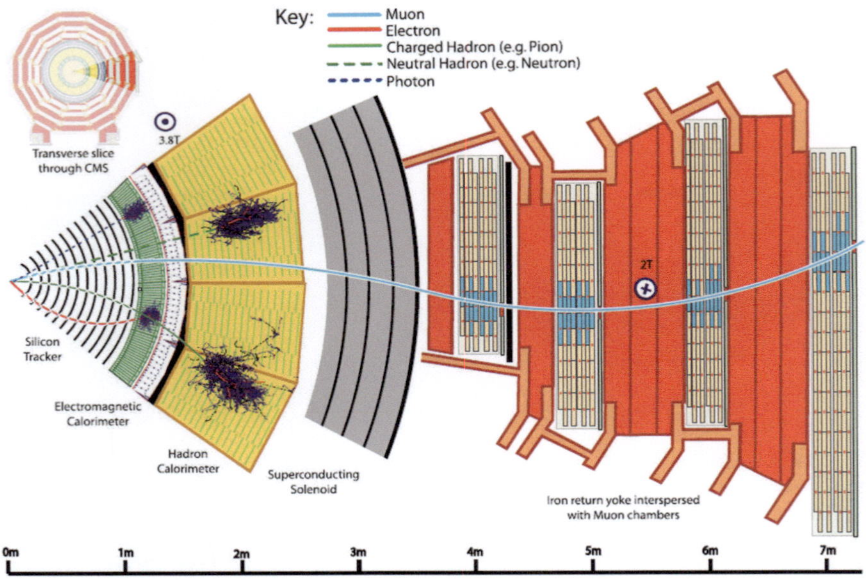

Fig. 2.23 Typical final-state particle interactions and signatures in a transverse slice of the CMS detector, from the interaction point to the muon detectors [58]

The offline reconstruction is performed based on each particle's distinct signature inside the detector. Figure 2.23 gives a schematic view of the interactions and signatures in a transverse slice of the CMS detector. From the IP, particles travel outward, encountering the various subdetectors and depositing energy in them depending on their nature. Muons are MIP at the energies at which they are produced at the LHC; they traverse the entire detector with minimal loss of energy, leaving detectable hits in the tracker and in the muon chambers; these hits are combined to reconstruct the offline muon candidates. Hadrons traverse the ECAL calorimeter with mild energy loss and are absorbed by the HCAL calorimeter, which is exploited to estimate the offline hadron candidate energy; in the case of charged hadrons, the hits in the tracker add a further handle to their identification. The ECAL active material absorbs electrons and photons, and their energy is evaluated from it; in the case of electrons, as they are charged, their track can be reconstructed and used for the computation of auxiliary variables. Finally, neutrinos escape the CMS detector and are not directly detectable; nevertheless, their momentum fraction can be estimated in the form of a transverse momentum imbalance.

This approach to reconstructing particles is the Particle Flow (PF) approach [58]. The PF algorithm aims to obtain a global event view by taking advantage of the high granularity and precision of the CMS detector components to achieve an accurate reconstruction of each particle. While the PF technique was originally developed for the reconstruction of low occupancy events in the ALEPH experiment at LEP, the CMS Collaboration has been the first to successfully employ this method at hadron colliders and use it as its golden spear to physics object reconstruction. A detailed description of the PF algorithm and how it is employed in CMS is given in the following.

2.4.1 Particle-Flow for Global Event Reconstruction

The fundamental principle underlying the PF approach is the utilization of information from multiple subdetectors to optimally reconstruct and identify individual particles within an event. The process commences by exploiting the high granularity of the CMS detector to build the two basic elements of the PF algorithm: clusters and tracks. These components are then *linked* with one another in a likelihood-based algorithm that takes into account a wide array of information, encompassing energy measurements, particle trajectories, and response patterns within the various detector components.

The clusters are built by grouping neighbouring energy deposits in the ECAL and in the HCAL detectors. The clustering is performed separately in each subdetector: EB and EE, HB and HE, and the two ES layers. No clustering is performed in the HF, as the quartz fibres of different lengths automatically build electromagnetic and hadronic clusters. This separation into subdetectors is chosen to achieve high efficiency also at low p_T and to be able to disentangle overlapping showers. The clustering starts with seeding cells, i.e. energy deposits above a threshold, then *topological clusters* are built by grouping the seed's neighbouring cells if their energy deposit is larger than twice the expected cell noise. Finally, a Gaussian-mixture expectation-maximization algorithm is used to reconstruct the clusters within a topological cluster. The Gaussian-mixture model hypothesizes that the energy deposits in the M individual cells of the topological cluster arise from N Gaussian energy deposits where N is the number of seeds [58]. This approach results in an iterative technique that closely reproduces the typical lateral shower profile and generally converges in less than five iterations.

The tracks are built employing a Combinatorial Track Finder (CTF) based on a Kalman Filter (KF) method [59–61]. The CTF reconstructs tracks in three stages: seed generation with few hits compatible with a charged-particle trajectory; pattern recognition to gather hits from all tracker layers; and final fitting to determine the track/charged-particle properties. The complete reconstruction follows an iterative approach in which the CTF is repeatedly applied. In the first iteration, only tracks with high compatibility to the primary vertex are constructed, and their hits are removed; in the following iterations, the quality criteria are progressively loosened as the combinatorial complexity decreases. This approach increases the tracking efficiency while keeping the misreconstructed track rate at acceptable levels. This technique allows for the reconstruction of tracks with $p_T \gtrsim 0.1$ GeV and produced as far as 60 cm for the primary vertex while generally converging in less than 12 iterations.

As schematically shown in Fig. 2.23, any given particle is generally expected to produce several PF elements in the various CMS subdetectors. Therefore, the reconstruction of a particle necessitates the definition of a linking algorithm that connects the PF elements from different subdetectors. Ideally, the linking algorithm would test all possible combinations of PF elements; still, to avoid the computing time to grow quadratically with the number of elements, only the nearest neighbours are considered, as obtained with a k-dimensional tree in the (η, ϕ) plane. The specific linking conditions are diverse and depend on the elements' nature; a quality variable is assigned to the links depending on a specific metric. In a simplified description: muon tracks are independently reconstructed based on tracker tracks and muon system tracks; electrons employ a specialized tracking methodology designed to account for bremsstrahlung radiation and are matched to PF clusters; charged hadrons follow from a fusion of track and PF cluster information; upon assigning all tracks to their respective candidates, the remaining PF clusters are interpreted as photons or neutral hadrons depending on the cluster being from ECAL of HCAL. The information in each PF block is finally exploited to estimate the candidates' momentum, energy, and quality.

After all PF blocks have been processed and all particle candidates have been identified, a global event description is ultimately achieved. At this stage, a post-processing of the reconstructed event is performed to correct for the small residual probability of particle misidentification and misreconstruction. Particular care is given to the momentum imbalance calculation, which can be highly affected by genuine cosmic muons. After this last step, the final output of the PF algorithm is an event global reconstruction in the form of a PF candidates list: muons, electrons, photons, charged hadrons, and neutral hadrons. The PF candidates can then be exploited to build higher-level objects like jets, hadronically decaying τ leptons, and missing transverse momentum.

2.4.2 Muons Reconstruction

The CMS detector was designed to achieve the best muon reconstruction performance owing to the clean signature in the muon chambers. For this reason, muons are the first particles to be identified, with an approach partially independent of the PF algorithm.

Muon tracks are reconstructed independently in the silicon tracker and in the muon chambers; while the former are built using the PF approach and are referred to as *tracker tracks*, the latter employ a standalone procedure and are referred to as *standalone tracks*. These two types of tracks are then combined to form muon candidates in two approaches: the *inside-out* and the *outside-in* method. This leads to a final collection of three different muon types:

- **Standalone muons**
  Standalone muons are built relying solely on the hits in the DT, RPC, and CSC stations. The hits are first combined to form track segments, used as seeds for

pattern recognition, and an initial estimation of the muon's direction and momentum is performed. A KF method that accounts for the muon energy loss in the detector materials is then used to build full tracks from the track segments. Purely standalone muons show high contamination from cosmic rays background, thus they are seldom used for physics analyses.

- **Tracker muons**
  Tracker muons are built inside-out: PF tracks with transverse momentum $p_T >$ 0.5 GeV and total momentum $p > 2.5$ GeV are extrapolated from the tracker to the muon system; if at least one standalone muon segment matches the extrapolated track, the PF track qualifies as a tracker muon. The track-to-segment matching is performed in a local (x, y) coordinate system defined on the CMS transverse plane. Matching is achieved either if $\Delta x < 3$ cm or if $\Delta x/\sigma_{\Delta x} < 4$ (where $\sigma_{\Delta x}$ is the uncertainty on Δx).
- **Global muons**
  Global muons are built outside-in: standalone tracks are matched to the tracker tracks based on the compatibility of their parameters when extrapolated to a common surface. The hits from the standalone and tracker tracks are fitted jointly to form a global muon.

As both the tracker and muon systems guarantee high reconstruction efficiency, about 99% of muons are reconstructed either as tracker or global muons; candidates that share the same inner tracks are merged into a single object. For muons of $p_T <$ 200 GeV, the charge and momentum assignments are computed solely from the PF tracker measurement; for muons of $p_T > 200$ GeV, the momentum is evaluated from the fit of the global muons candidate trajectory. This guarantees a muon transverse momentum resolution ranging between 1% and 6% for muons with $p_T < 100$ GeV and better than 10% for central muons of $p_T \sim 1$ TeV.

Finally, several criteria are derived to rank muons based on their quality. These criteria involve factors such as the track fit χ^2, the count of hits per track, and the alignment between the tracker and standalone tracks. Diverse thresholds for these criteria lead to the definition of Working Points (WPs) for muon identification, categorized as loose, medium, and tight. These WP exhibit an ascending purity and diminishing efficiency. Moreover, to discern between prompt muons and those stemming from weak decays within jets, a PF isolation variable is defined. The relative PF isolation is computed as the ratio between the cumulative p_T of all PF candidates contained within a cone of size $\Delta R < 0.4$ centred around the muon and the p_T of the muon itself.

Muons are an essential part of the HH $\rightarrow$ bb$\tau\tau$ analysis, as they offer a very clean signature for events where a τ lepton decays to a muon. The specific isolation and identification criteria applied to muons for the search of Higgs boson pairs are presented in Chap. 5.

2.4.3 Electrons Reconstruction

Unlike muon, electrons are subject to significant interactions in both the active and dead material they traverse. This energy loss mainly comes in the form of bremsstrahlung photons, which carry on average 33% of the total electron energy at $\eta \sim 0$ and up to up to 86% at $\eta \sim 1.4$. The picture is complicated even further when considering the large probability of high p_{T} bremsstrahlung photons to convert into $\mathrm{e}^{+}\mathrm{e}^{-}$ pairs. These factors make reconstructing electrons quite challenging and require a dedicated approach that merges an ECAL-driven method with a devoted track-finding technique.

The track-finding technique for electrons is independent of the PF approach and is based on the Gaussian-Sum Filter (GSF) algorithm [62]. The GSF has been developed to optimally account for possible electron energy losses and consequent sudden changes in their path, for which a KF approach is not well suited. This is achieved by employing looser track criteria than KF and a Gaussian weighted sum rather than a single Gaussian to model the electron energy loss. This approach enhances the overall tracking efficiency but requires considerable computing power and time.

The traditional ECAL-driven approach defines a seed PF cluster in the ECAL detector and merges it with neighbouring clusters to recover the radiated energy. The merging window is defined to be narrow in η and extended in ϕ to account for the azimuthal bending of the electron in the magnetic field. The resulting merged object is referred to as *supercluster*. This approach is well suited for high energetic electrons but falls short of performance on low p_{T} candidates.

These two approaches can then complement each other for global electron reconstruction. To reduce the combinatorial burden of the GSF algorithm, the supercluster energy and position can be used to infer the expected position of the tracker hits under the assumption that the cluster is produced either by an electron or a positron. Hence, the GSF reconstruction can be performed based on ECAL-driven seeds, achieving excellent performance on high p_{T}, isolated electrons. Conversely, to recover soft electrons, the standalone GSF tracks with $p_{\mathrm{T}} > 2\,\mathrm{GeV}$ can be used as seeds for ECAL clusters. Tracks are propagated to the ECAL surface and matched to the nearest PF clusters to form electron seeds if the ratio between the track transverse momentum and the cluster energy is compatible with unity. GSF tracks and PF superclusters are finally associated with an electron candidate if they satisfy loose requirements on their qualities and matching.

Owing to the high granularity of the tracker and the ECAL detector's outstanding energy resolution, electrons with $p_{\mathrm{T}} \sim 45\,\mathrm{GeV}$ reconstructed with this mixed approach have an energy resolution ranging between 2 and 5%, depending on electron η and energy loss in the detector material. In the same range of energies, the energy scale uncertainty is smaller than 0.1% and 0.3% in the barrel and endcaps, respectively [63].

Finally, an identification criterion based on a Boosted Decision Tree (BDT) is derived to distinguish the genuine electrons from the misidentified jets, namely pions. The input variables to the BDT are related to the shower shape from the ECAL clusters, the GSF track parameters, the track-cluster ($\Delta\eta$, $\Delta\phi$) matching distance, the fraction of momentum lost due to bremsstrahlung evaluated with a Bethe-Heitler function, and the electron isolation. The output of the BDT represents the probability of the candidate being a genuine electron; diverse thresholds are defined on the BDT output in ascending purity and diminishing efficiency to define several WPs.

Analogously to muons, electrons are a very important part of the HH $\rightarrow$ bb$\tau\tau$ analysis, as they guarantee a very good selectivity of events where a τ lepton decays to an electron. The specific isolation and identification criteria applied to electrons for the HH search are presented in Chap. 5.

2.4.4 *Hadronic τ Reconstruction*

The τ lepton lifetime is $\tau_\tau = (290.3 \pm 0.5) \cdot 10^{-15}$ s [64]; thus, a τ decays within a few millimetres from its production point for the typical Lorentz boosts at the LHC. This results in the possibility of detecting only its visible decay products in the detector. Moreover, the τ is the heaviest lepton in the SM, with a mass $m_\tau = 1776.86 \pm 0.12$ MeV [64]; hence, it is the only one that can decay both leptonically and hadronically with the branching fractions reported in Table 2.2. The fully leptonic decays account for $\sim 1/3$ of the total decay width and are reconstructed using the muon and electron reconstruction algorithm detailed above. Conversely, for the $\sim 2/3$ of the cases in which the τ decays hadronically, a dedicated algorithm is needed. The hadronic decay can happen directly into charged and neutral hadrons or mediated by a ρ770 or $a_1$1260 resonance; notwithstanding the decay resonance, the final state particles are in ~98% of the cases pions and in the remaining ~2% kaons [64]. Due to charge conservation, the decay can happen into any odd number of single charge hadrons; nevertheless, the most likely decays involve one or three charged hadrons,

Table 2.2 τ lepton branching fractions [64]

Decay mode	Meson resonance	$\mathcal{B}$ [%]
$e\nu_e\nu_\tau$		17.8
$\mu\nu_\mu\nu_\tau$		17.4
All leptonic decays		**35.2**
$h^\pm\nu_\tau$		11.5
$h^\pm\pi^0\nu_\tau$	ρ770	26.0
$h^\pm\pi^0\pi^0\nu_\tau$	$a_1$1260	9.5
$h^\pm h^\mp h^\pm\nu_\tau$	$a_1$1260	9.8
$h^\pm h^\mp h^\pm\pi^0\nu_\tau$		4.8
Other hadronic decays		3.2
All hadronic decays		**64.8**

generally referred to as *prongs*. As the neutrinos in the decay remain undetected, it is impossible to evaluate exactly the momentum of the τ lepton before its decay.

Given the particle multiplicity in the τ_h decay, the reconstruction algorithm should be able to identify the PF candidates associated with the charged hadrons and the photons issued from the $\pi^0 \rightarrow \gamma\gamma$ decay, regroup them, determine the decay mode, and estimate the τ_h kinematic properties. Therefore, the Hadrons-Plus-Strips (HPS) algorithm [65–67] has been designed to perform this procedure. The reconstruction is performed in four steps. As τ_h decays result in narrow jets, the first step is the seeding based on PF jets; all PF particles within $\Delta R < 0.5$ from the seeding jet are considered for the following steps. The second stage is the reconstruction of the π^0 candidates using an iterative procedure to build the so-called *strips*:

- *1st iteration*: initiate strip with a PF electron or photon candidate within a strip of predetermined $\eta \times \phi$ dimension
- *Nth iteration*:
  - cluster photon and electron PF candidates of $p_T > 0.5$ GeV
  - evaluate strip position as the p_T-weighted average of the clustered candidates
  - re-computed the $\eta \times \phi$ window size as functions of the strip p_T itself to ensure optimal collection of energy and minimize the impact of background.

This approach optimally reconstructs the $\pi^0 \rightarrow \gamma\gamma$ candidates by including neutral pion decays appearing either directly as photon PF candidates or as electron candidates because of the large $\gamma \rightarrow e^+e^-$ conversion probability. The third step consists of creating all possible combinations of PF charged hadrons and strips to build one of seven decay mode hypotheses. Compared to the algorithm documented in Reference [67], three additional hypotheses are currently available to target the higher multiplicity decay; the seven hypotheses are:

- $h^\pm$, targeting $\tau^\pm \rightarrow h^\pm \nu_\tau$ decays
- $h^\pm\pi^0$, targeting $\tau^\pm \rightarrow h^\pm\pi^0\nu_\tau$ decays
- $h^\pm\pi^0\pi^0$, targeting $\tau^\pm \rightarrow h^\pm\pi^0\pi^0\nu_\tau$ decays
- $h^\pm h^\mp h^\pm$, targeting $\tau^\pm \rightarrow h^\pm h^\mp h^\pm \nu_\tau$ decays
- $h^\pm h^\mp h^\pm \pi^0$, targeting $\tau^\pm \rightarrow h^\pm h^\mp h^\pm \pi^0 \nu_\tau$ decays (new hypothesis)
- $h^\pm h^{\mp/\pm}$, targeting $\tau^\pm \rightarrow h^\pm h^\mp h^\pm \pi^0 \nu_\tau$ decays (new hypothesis)
- $h^\pm h^{\mp/\pm}\pi^0$, targeting $\tau^\pm \rightarrow h^\pm h^\mp h^\pm \pi^0 \nu_\tau$ decays (new hypothesis).

Additional quality and invariant mass requirements are imposed on the PF candidates in each of the seven decay mode hypotheses to ensure compatibility with a genuine τ_h. For the decay modes hypotheses compatible with a resonance-mediated decay, loose compatibility with the resonance's mass is enforced; the τ_h charge must be unity unless the hypothesis is one with only two charged hadrons; all reconstructed decay products must lie within $\Delta R < 3.0/p_T(\tau_h)$ from the τ_h momentum. Finally, the τ_h candidate with the highest p_T is selected among all the candidates satisfying the aforementioned requirements.

The HPS algorithm ensures a misreconstruction rate between 11% for 1-prong decays and 25% for 1-prong+π^0 decays, with the overall reconstruction efficiency mostly limited by the ~90% efficiency in charged hadrons tracks reconstruction [58]. The charge assignment is 99% correct in the decay mode hypotheses without missing charged hadrons for an inclusive $Z \rightarrow \tau\tau$ sample, 98% for τ_h with $p_T \sim 200$ GeV, and 92% for $p_T \sim 1$ TeV. The decay modes with missing charged hadrons guarantee the recovery of 19% of the 3-prong decays and 13% of the 3-prong+π^0 decays; however, owing to the missing charge hadron, the charge assignment is correct only ~70% of the times. Nevertheless, since physics analyses apply requirements on the reconstructed τ_h charge to suppress background events, these decay modes are only useful for analyses not limited by background events. In the case of the $HH \rightarrow bb\tau\tau$ search presented here, these decay modes are not exploited.

Ultimately, the τ_h reconstruction is complemented with a ML-based identification algorithm called DEEPTAU [68]. This algorithm combines high-level features of the reconstructed τ_h and low-level information from the PF candidates within the τ_h cone to perform an identification against light leptons and QCD-induced jets. A detailed description of the DEEPTAU algorithm is given in Chap. 5 in the context of the $HH \rightarrow bb\tau\tau$ analysis.

Hadronically decaying τ leptons are the most important objects in the $HH \rightarrow bb\tau\tau$ analysis because they capture the largest fraction of HH decays in the selected final state. The specific isolation and identification criteria applied to muons for the search of Higgs boson pairs are presented in Chap. 5.

2.4.5 *Jets Reconstruction*

Quarks and gluons produced in the hard scatterings manifest themselves through their hadronization in the CMS detector; this results in several particles produced in a cone-shaped jet. Jets are reconstructed from charged and neutral PF candidates employing the anti-k_T algorithm [69, 70]. The anti-k_T algorithm is particularly well suited to disentangle hits corresponding to isolated particles from those belonging to the jet cone, thus optimally dealing with the jets' large occupancy in the detector. The following two distance parameters are defined in this approach:

$$d_{ij} = \min(k_{T,i}^{-2}, k_{T,j}^{-2})\frac{\Delta R_{ij}^2}{R} \qquad d_{iB} = k_{T,i}^{-2} \tag{2.12}$$

where d_{ij} is the distance between the algorithm entities i and j, i.e. PF candidate and anti-k_T pseudojets, and d_{iB} is the distance between entity i and the beam. These distances are defined to be inversely proportional to the entities' transverse momentum to ensure that the jet is built around the hardest particle of the event and the soft radiation or collinear parton splitting at the borders is suppressed. The algorithm proceeds iteratively by identifying the smallest of the two distances between all available entities. If the smallest distance is of the d_{ij} type, entities i and j are combined to form a new pseudojet; if the smallest distance of the d_{iB} type, the clustering is ended and pseudojet i is defined as the output hadronic jet and removed from the list of entities. This procedure is repeated until no entities are left and a list of jets has been produced. In this approach, the jet angular extension is encoded in the $\Delta R_{ij}^2/R$ term, in which the R parameter governs the radius of the clustering.

This PF-based approach allows the reconstruction of up to 90% of the jet components, substantially improving the performance of a calorimeter-based reconstruction. The jet four-momentum is computed as the vector sum of the clustered PF candidates four momenta, and a set of simulation-based corrections, known as Jet Energy Corrections (JECs), are applied to account for contribution from pileup, non-linearity in the detector response and residual data-simulation differences. Typically achieved jet resolutions are of about 15 − 20% for $p_T \sim 30$ GeV, 10% at $p_T \sim 100$ GeV, and 5% at $p_T \sim 1$ TeV [71].

According to the topology of the events studied in a particular analysis, additional identification criteria can be introduced to target specific jets. Particularly interesting is the case of identifying jets originating from b quarks, which are essential in the HH → bb$\tau\tau$ search presented in this Thesis. The latest and best performing b-jet tagger is a ML-based identification algorithm called DEEPJET [72]. This algorithm combines 650 high-level features of the reconstructed jet and low-level information from the PF candidates within the jet cone to identify the typical signatures of b-jets. A detailed description of the DEEPJET algorithm is given in Chap. 5 in the context of the HH → bb$\tau\tau$ analysis.

The reconstruction and identification of b-jets represent a crucial point of the HH → bb$\tau\tau$; the specific identification criteria applied to b-jets for the HH search are presented in Chap. 5.

2.4.6 Missing Transverse Momentum Reconstruction

After all detectable particles have been reconstructed, the presence of undetectable particles like neutrinos or BSM weakly-interacting particles can be inferred from the appearance of an apparent non-conservation of momentum. As discussed in Sect. 2.2.1, the projection of the momentum on the transverse plane is a conserved quantity; therefore, the missing transverse momentum ($\vec{p}_T^{\text{miss}}$) can be defined as minus the vectorial sum of the transverse momenta of all the reconstructed PF objects

$$\vec{p}_T^{\text{miss}} = -\sum_{i=0}^{N} \vec{p}_T^{\,i} \tag{2.13}$$

The high granularity and hermetic design of the CMS detector is of central importance for the precise measurement of $\vec{p}_T^{\text{miss}}$. Nevertheless, in some rare cases, an artificially large missing transverse momentum is reconstructed in the event, most often caused by a misidentified or misreconstructed high-p_T muon. The main cause of this is to be found in the presence of genuine cosmic muons that traverse the CMS detector in coincidence with the LHC bunch crossing. To tackle this, a dedicated post-processing is employed in the PF approach.

The inefficiencies arising from the tracking or clustering algorithms and the non-linearities in the calorimeter response for hadronic particles might introduce biases in the determination of $\vec{p}_{\mathrm{T}}^{\mathrm{miss}}$. Therefore, the JECs are propagated also to the $\vec{p}_{\mathrm{T}}^{\mathrm{miss}}$ computation as follows:

$$\vec{p}_{\mathrm{T}}^{\mathrm{miss,corr}} = \vec{p}_{\mathrm{T}}^{\mathrm{miss}} - \sum_{\mathrm{jets}} \left(\vec{p}_{\mathrm{T}}^{\mathrm{corr}} - \vec{p}_{\mathrm{T}}\right) \tag{2.14}$$

where the superscript *corr* stands for the JEC corrected variables. In this way, the relative energy resolution of the missing transverse energy is about 20% for all energy ranges [58].

References

1. European Organization for Nuclear Research (1954) Convention for the Establishment of a European Organization for Nuclear Research. https://council.web.cern.ch/en/content/convention-establishment-european-organization-nuclear-research. Accessed 12th Aug 2023
2. Baconnier Y et al (1993) LHC: the Large Hadron Collider accelerator project. Geneva: CERN. https://cds.cern.ch/record/257706
3. Pettersson TS, Lefèvre P (1995) The large Hadron Collider: conceptual design. Technical report. https://cds.cern.ch/record/291782
4. Lopienska E (2022) The CERN accelerator complex, layout in 2022. Complexe des accélérateurs du CERN en janvier 2022. In: CERN-GRAPHICS-2022-001. General Photo. https://cds.cern.ch/record/2800984
5. Evans L, Bryant P (2008) LHC machine. J Instrum 3.08:S08001–S08001. https://doi.org/10.1088/1748-0221/3/08/s08001
6. CMS Collaboration (2018) Measurement of the inelastic proton-proton cross section at $\sqrt{s} = 13\,\mathrm{TeV}$. J High Energy Phys 2018.7. https://doi.org/10.1007/jhep07(2018)161
7. CMS Collaboration (2023) Public luminosity results 2023. https://twiki.cern.ch/twiki/bin/view/CMSPublic/LumiPublicResults. Accessed 1st Aug 2023
8. Collaboration CMS (2018) Observation of the Higgs boson decay to a pair of τ leptons with the CMS detector. Phys Lett B 779:283–316. https://doi.org/10.1016/j.physletb.2018.02.004
9. CMS Collaboration (2021) Evidence for Higgs boson decay to a pair of muons. J High Energy Phys 2021.1. https://doi.org/10.1007/jhep01(2021)148
10. Aberle O et al (2020) High-Luminosity Large Hadron Collider (HL-LHC): technical design report. CERN yellow reports: monographs. Geneva: CERN. https://doi.org/10.23731/CYRM-2020-0010. https://cds.cern.ch/record/2749422
11. Contardo D et al (2015) Technical proposal for the Phase-II upgrade of the CMS detector. Technical report Upgrade Project Leader Deputies: Lucia Silvestris (INFN-Bari), Jeremy Mans (University of Minnesota) Additional contacts: Lucia.Silvestris@cern.ch, Jeremy.Mans@cern.ch. Geneva. https://doi.org/10.17181/CERN.VU8I.D59J. https://cds.cern.ch/record/2020886
12. Butler J et al (2015) CMS Phase II upgrade scope document. Technical report Geneva: CERN. https://cds.cern.ch/record/2055167
13. The ATLAS Collaboration (2008) The ATLAS experiment at the CERN Large Hadron Collider. J Instrum 3.08:S08003–S08003. https://doi.org/10.1088/1748-0221/3/08/s08003

14. The LHCf Collaboration (2008) The LHCf detector at the CERN Large Hadron Collider. J Instrum 3.08:S08006–S08006. https://doi.org/10.1088/1748-0221/3/08/s08006
15. FASER Collaboration (2022) The FASER detector. arXiv:2207.11427
16. FASER Collaboration (2020) Detecting and studying high-energy collider neutrinos with FASER at the LHC. European Phys J C 80.1. https://doi.org/10.1140/epjc/s10052-020-7631-5
17. SND@LHC Collaboration (2023) SND@LHC: the scattering and neutrino detector at the LHC. arXiv:2210.02784
18. The ALICE Collaboration (2008) The ALICE experiment at the CERN LHC. J Instrum 3.08:S08002–S08002. https://doi.org/10.1088/1748-0221/3/08/s08002
19. The CMS Collaboration (2008) The CMS experiment at the CERN LHC. J Instrum 3.08:S08004–S08004. https://doi.org/10.1088/1748-0221/3/08/s08004
20. The TOTEM Collaboration (2008) The TOTEM experiment at the CERN Large Hadron Collider. J Instrum 3.08:S08007–S08007. https://doi.org/10.1088/1748-0221/3/08/s08007
21. The LHCb Collaboration (2008) The LHCb detector at the LHC. J Instrum 3.08:S08005–S08005. https://doi.org/10.1088/1748-0221/3/08/s08005
22. Pinfold J et al (2009) Technical design report of the MoEDAL experiment. Technical report. https://cds.cern.ch/record/1181486
23. Neutelings I, CMS Wiki Pages, how to draw diagrams in LaTeX with TikZ. https://wiki.physik.uzh.ch/cms/latex:tikz
24. Sakuma T (2019) Cutaway diagrams of CMS detector. https://cds.cern.ch/record/2665537
25. CMS Collaboration (1997) The CMS magnet project: technical design report. Technical design report. CMS. Geneva: CERN. https://doi.org/10.17181/CERN.6ZU0.V4T9. https://cds.cern.ch/record/331056
26. CMS Collaboration (2010) Precise mapping of the magnetic field in the CMS barrel yoke using cosmic rays. J Instrum 5.03:T03021–T03021. https://doi.org/10.1088/1748-0221/5/03/t03021
27. Karimäki V et al (1997) The CMS tracker system project: technical design report. Technical design report. CMS. Geneva: CERN. https://cds.cern.ch/record/368412
28. Karimäki V et al (2000) The CMS tracker: addendum to the technical design report. Technical design report. CMS. Geneva: CERN. https://cds.cern.ch/record/490194
29. Dominguez A et al (2012) CMS technical design report for the pixel detector upgrade. Technical report. https://cds.cern.ch/record/1481838
30. The Tracker Group of the CMS Collaboration (2020) The CMS Phase-1 pixel detector upgrade. arXiv:2012.14304
31. CMS Collaboration (1997) The CMS electromagnetic calorimeter project: technical design report. Technical design report. CMS. Geneva: CERN. https://cds.cern.ch/record/349375
32. Bayatian GL et al (2006) CMS physics: technical design report volume 1: detector performance and software
33. Ingram Q (2007)Energy resolution of the barrel of the CMS electromagnetic calorimeter. J Instrum 2.04:P04004–P04004. https://doi.org/10.1088/1748-0221/2/04/p04004
34. CMS Collaboration (1997) The CMS Hadron Calorimeter project: technical design report. Technical design report. CMS. Geneva: CERN. https://cds.cern.ch/record/357153
35. Mans J et al (2012) CMS technical design report for the Phase 1 Upgrade of the Hadron Calorimeter. Technical report 2012. https://cds.cern.ch/record/1481837
36. Lutz B (2012) Upgrade of the CMS Hadron Outer Calorimeter with SiPM sensors. J Phys: Conf Ser 404:012018. https://doi.org/10.1088/1742-6596/404/1/012018
37. Layter JG (1997) The CMS muon project: technical design report. Technical design report. CMS. Geneva: CERN. https://cds.cern.ch/record/343814
38. Sirunyan AM et al (2018) Performance of the CMS muon detector and muon reconstruction with proton-proton collisions at $\sqrt{s} = 13$ TeV. JINST 13:P06015. https://doi.org/10.1088/1748-0221/13/06/P06015. arXiv:1804.04528
39. Colaleo A et al (2015) CMS technical design report for the Muon Endcap GEM upgrade. Technical report 2015. https://cds.cern.ch/record/2021453

40. CMS Collaboration, CMS TriDAS project: technical design report, volume 1: the trigger systems. Technical design report. CMS. http://cds.cern.ch/record/706847
41. CMS Collaboration (2023) Summaries of CMS cross section measurements. https://twiki.cern.ch/twiki/bin/view/CMSPublic/PhysicsResultsCombined. Accessed 14th Aug 2023
42. Bird I et al (2014) Update of the computing models of the WLCG and the LHC experiments. Technical report 2014. https://cds.cern.ch/record/1695401
43. Tapper A, Acosta D (2013) CMS technical design report for the Level-1 trigger upgrade. Technical report. Additional contacts: Jeffrey Spalding, Fermilab, Jeffrey.Spalding@cern.ch Didier Contardo, Universite Claude Bernard-Lyon I, didier.claude.contardo@cern.ch. https://cds.cern.ch/record/1556311
44. Zabi A et al (2017) The CMS Level-1 Calorimeter Trigger for the LHC Run II. J Instrum 12.01. https://doi.org/10.1088/1748-0221/12/01/c01065
45. CMS Collaboration (2020) Performance of the CMS Level-1 trigger in proton-proton collisions at $\sqrt{s} = 13$ TeV. J Instrum 15.10:P10017–P10017. https://doi.org/10.1088/1748-0221/15/10/p10017
46. Frazier R et al (2012) A demonstration of a Time Multiplexed Trigger for the CMS experiment. J Instrum 7.01:C01060. https://doi.org/10.1088/1748-0221/7/01/C01060
47. Hall G et al (2014) A time-multiplexed track-trigger architecture for CMS. J Instrum 9.10:C10034. https://doi.org/10.1088/1748-0221/9/10/C10034
48. Svetek A et al (2016) The Calorimeter Trigger Processor Card: the next generation of high speed algorithmic data processing at CMS. J Instrum 11.02. https://doi.org/10.1088/1748-0221/11/02/c02011
49. Imperial College of London (2023) MP7. http://www.hep.ph.ic.ac.uk/mp7/. Accessed 15th Aug 2023
50. Hazen E et al (2013) The AMC13XG: a new generation clock/timing/DAQ module for CMS MicroTCA. J Instrum 8.12. https://doi.org/10.1088/1748-0221/8/12/c12036
51. CMS Collaboration (2023) Performances of Muons, Jets and MET Level 1 trigger algorithms in Run 3. In: CMS detector performance summaries.https://cds.cern.ch/record/2853675
52. CMS Collaboration (2023) Performance of long lived particle triggers in Run 3. In: CMS detector performance summaries. https://cds.cern.ch/record/2865844
53. Bachtis M (2019) Upgrade of the CMS barrel muon track finder for HL-LHC featuring a Kalman Filter algorithm and an ATCA Host Processor with Ultrascale+ FPGAs. Technical report. Geneva: CERN. https://doi.org/10.22323/1.343.0139. https://cds.cern.ch/record/2648953
54. Acosta D et al (2013) The CMS modular track finder boards, MTF6 and MTF7. J Instrum 8.12. https://doi.org/10.1088/1748-0221/8/12/c12034
55. CMS Collaboration (2021) The Phase-2 upgrade of the CMS data acquisition and high level trigger. Technical report. This is the final version of the document, approved by the LHCC. Geneva: CERN. https://cds.cern.ch/record/2759072
56. Bocci A (2023) CMS High Level Trigger performance comparison on CPUs and GPUs. J Phys: Conf Ser 2438.1:012016. https://doi.org/10.1088/1742-6596/2438/1/012016
57. CMS Colaboration (2022) Commissioning CMS online reconstruction with GPUs. In: CMS detector performance summaries. https://cds.cern.ch/record/2851656
58. CMS Collaboration (2017) Particle-flow reconstruction and global event description with the CMS detector. J Instrum 12.10:P10003–P10003. https://doi.org/10.1088/1748-0221/12/10/p10003
59. Frühwirth R (1987) Application of Kalman filtering to track and vertex fitting. In: Nuclear instruments and methods in physics research section a: accelerators, spectrometers, detectors and associated equipment, vol 262.2-3, pp 444–450. https://doi.org/10.1016/0168-9002(87)90887-4 <error l="308" c="Invalid command: paragraph not started." />
60. Billoir P (1989) Progressive track recognition with a Kalman-like fitting procedure. Comput Phys Commun 57.1-3:390–394. https://doi.org/10.1016/0010-4655(89)90249-x
61. Adam W, Mangano B, Speer (2006) Track reconstruction in the CMS tracker. Technical report. Geneva: CERN. https://cds.cern.ch/record/934067

62. Adam et al (2005) Reconstruction of electrons with the Gaussian-sum filter in the CMS tracker at the LHC. J Phys G: Nucl Part Phys 31.9:N9–N20. https://doi.org/10.1088/0954-3899/31/9/n01
63. CMS Collaboration (2021) Electron and photon reconstruction and identification with the CMS experiment at the CERN LHC. J Instrum 16.05:P05014. https://doi.org/10.1088/1748-0221/16/05/p05014
64. Tanabashi M, others (Particle Data Group) (2018) Review of particle physics. Phys Rev D 98:030001. https://doi.org/10.1103/PhysRevD.98.030001
65. CMS Collaboration (2012) Performance of τ-lepton reconstruction and identification in CMS. J Instrum 7.01:P01001–P01001. https://doi.org/10.1088/1748-0221/7/01/p01001
66. CMS Collaboration (2016) Reconstruction and identification of τ lepton decays to hadrons and ν_τ at CMS. J Instrum 11.01:P01019–P01019. https://doi.org/10.1088/1748-0221/11/01/p01019
67. CMS Collaboration (2018) Performance of reconstruction and identification of τ leptons decaying to hadrons and ν_τ in pp collisions at $\sqrt{s} = 13$ TeV. J Instrum 13.10:P10005–P10005. https://doi.org/10.1088/1748-0221/13/10/p10005
68. CMS Colaboration (2022) Identification of hadronic tau lepton decays using a deep neural network. J Instrum 17.07:P07023. https://doi.org/10.1088/1748-0221/17/07/p07023
69. Cacciari M, Salam GP, Soyez G (2008) The anti-k_T jet clustering algorithm. J High Energy Phys 2008.04. https://doi.org/10.1088/1126-6708/2008/04/063
70. Cacciari M, Salam GP, Soyez G (2012) FastJet user manual. European Phys J C 72.3. https://doi.org/10.1140/epjc/s10052-012-1896-2
71. CMS Collaboration (2020) Jet energy scale and resolution performance with 13 TeV data collected by CMS in 2016-2018. CMS detector performance summaries. https://cds.cern.ch/record/2715872
72. Bols E et al (2020) Jet flavour classification using DeepJet. J Instrum 15.12:P12012–P12012. https://doi.org/10.1088/1748-0221/15/12/p12012

Chapter 3
The Level-1 τ_h Trigger: From the Past, to the Present

The Phase-1 upgrade of the Level-1 (L1) trigger [1] took place during the first long shutdown of the LHC (2012–2015). In this instance, the calorimeter trigger was upgraded to its current Run-2 and Run-3 specification, allowing for sophisticated triggering algorithms to be implemented in powerful Field Programmable Gate Arrays (FPGAs). As discussed in Sect. 2.3.1, the L1 calorimeter trigger is constituted of two processing tiers, the Layer-1 and Layer-2. In the former, the calibration of the calorimeter triggers primitive is performed. In the latter are implemented the reconstruction algorithms for e/γ, τ, jet, and sum objects, which optimally exploit the input they receive from Layer-1 in order to reduce the event rate to an adequate level while retaining the potentially interesting physics events with the highest efficiency possible.

As part of my Thesis work, I have been the leading contributor to developments targeting both Layer-1 and Layer-2. I have been the main developer for the optimization of the L1 τ algorithm for the restart of the data-taking in Run-3. I have introduced a simple yet more informative approach to optimizing the algorithm's parameters. The same techniques have also been successfully implemented in the workflow for optimizing the L1 e/γ algorithm, yielding excellent performance. Moreover, I have been coordinating the development and measurement in data of the performance of both the L1 τ and e/γ algorithms. This work is fundamental to the ambitious CMS Run-3 physics program's success, which involves, among others, the development of multiple trigger paths dedicated to the collection of Higgs boson pair (HH) signal events. Moreover, I have been the leading contributor to developing the first algorithm for calibrating L1 Trigger primitives (TPs) that exploits a fully data-driven Machine Learning (ML) technique. This novel method exploits offline reconstructed electrons and jets to calibrate single calorimeter TPs optimally, leading to promising performance. This innovative technique, based on a neural network architecture, is still being advanced and improved, and it is presented for the first time in this Thesis.

This Chapter is structured into four main Sections. The first one details the L1 τ algorithm implemented in Run-2 and still used in Run-3; the second and third

J. Motta, *Development of Machine Learning τ Trigger Algorithms and Search for Higgs Boson Pair Production*, Springer Theses,
https://doi.org/10.1007/978-3-031-96288-2_3

describe my contribution to the preparation and commissioning of the τ and e/γ L1 trigger algorithms for the Run-3 data-taking; while the fourth details the development and performance evaluation of a novel technique for calibrating L1 TPs with a data-driven machine learning technique.

3.1 Physics Goals Meet Experimental Challenges

The τ is the heaviest lepton in the Standard Model (SM) and showcases a large Yukawa coupling to the Higgs boson (H); nevertheless, the fundamental reason behind the huge difference of the τ Yukawa coupling with respect to lighter leptons is still unknown. These properties make τ leptons a very interesting probe to test the SM and to perform searches of several Beyond the SM (BSM) models. Run-2 yielded the first observation and subsequent measurement of the $H \rightarrow \tau\tau$ channel [2, 3] constituting the first-ever observation of the direct coupling between the H and fermions, as well as the first measurement of the charge-parity (CP) structure of the Yukawa coupling between the τ lepton and the H boson [4]. In the context of the searches for HH production, the $HH \rightarrow bb\tau\tau$ decay channel is one of the most sensitive to the gluon fusion production mechanism, and it is the most sensitive to the vector boson fusion mechanism, as detailed in this Thesis. The physics goals of Run-3 further expand the ones of Run-2, targeting heavy scalar resonances decaying to τ leptons and charged Higgs bosons decays $H^+ \rightarrow \tau^+\nu_\tau$ predicted by the Minimally Super Symmetric SM (MSSM). Finally, in Run-3, HH searches will benefit from the implementation of the novel *Higgs parking* technique that largely exploits the τ L1 trigger in the definition of the High-Level Trigger (HLT) paths.

The sensitivity of the analyses mentioned above is crucially impacted by the efficient reconstruction and identification of τ leptons. This starts with the first selection tier in CMS: the Level-1 trigger. Given their sizeable mass, τ leptons can decay to lighter leptons or hadrons, with the branching ratios reported in Table 2.2. The leptonic decays of the τ are easily targeted at L1 using the clean muon and electron signatures in the CMS detector. Conversely, in roughly $2/3$ of the cases when the τ decays hadronically (τ_h), the reconstruction at L1 is extremely challenging due to the particle multiplicity of the decay that makes it very similar to jet and gluon hadronization.

The first experimental challenge to implementing an L1 τ_h trigger is striking a balance between achieving high signal efficiency to optimize the sensitivity of the physics analyses and implementing substantial background rejection to manage the trigger rate effectively. The decay products of τ_h involve charged and neutral hadrons, leading to energy depositions in both the Electromagnetic Calorimeter (ECAL) and Hadronic Calorimeter (HCAL), which generally lie within a relatively narrow angular distance. In the Phase-1 L1 trigger, not having access to the tracker detector highly constrains the exploitable information, which represents a second challenge. Thus requiring the identification of τ_h to be based solely on different footprints left in the calorimeter by the narrow τ_h jets and the broader QCD-induced jets. A third demanding task is the implementation of the algorithm within the boundaries set by the hardware limitations. The L1 trigger is an electronic system fully synchronous with the LHC clock that has to make the accept/reject decision within a fixed latency, which is limited in Phase-1 to just 3.8 μs, of which roughly 1 μs is reserved to the calorimeter algorithms execution. Additionally, the used FPGAs have a finite number of logic gates and available memory, the so-called *hardware resources*, for implementing the algorithm. The combination of these factors imposes limitations on the number and complexity of operations that can be performed, necessitating a highly optimized τ algorithm to address these constraints effectively. The final challenge lies in effectively reducing the background while keeping high signal acceptance. Given that the majority of proton bunch crossings result in energy deposits in the calorimeters, it is imperative to devise efficient strategies to maintain the trigger rate within manageable limits for the CMS data acquisition system. While applying a minimal energy threshold on the τ_h helps suppress some background, it is essential to develop complementary rejection criteria to keep these thresholds sufficiently low, thus ensuring high signal efficiency. This becomes particularly demanding in the context of the LHC Run-3, where instantaneous luminosity conditions can reach 2.6×10^{34} cm^{-2}s^{-1} with peak pileup (PU) as high as 80 and average PU exceeding 50. This challenging collision environment increases energy and jet counts, necessitating a robust τ algorithm capable of withstanding the impact of PU and maintaining effective background rejection.

For all of these reasons, the L1 τ_h algorithm needs a delicate optimization to cope with the harsh experimental conditions. Such optimization is particularly crucial at the start of new LHC data-taking operations like Run-3, which brings additional challenges due to both the accelerator and the CMS detector's three years of inactivity. Ensuring that the accelerator and detector systems are brought back to optimal performance requires extended commissioning. Moreover, in the fast-changing environment of the accelerator ramp-up to nominal running conditions, all L1 algorithms need to be re-optimized and commissioned at the same fast pace. This is the context in which my contribution, detailed in the following, has taken place.

3.2 The τ_h Algorithm from the Past

The τ_h reconstruction procedure incorporates elements from the e/γ and jet algorithms while being customized to account for the distinctive topology of τ_h decays. During this process, relevant energy deposits are clustered dynamically to minimize the impact of PU effects, and secondary clusters arising from multiple objects in the final state can be merged based on their proximity. However, L1 τ_h candidates face notable background contamination from QCD-induced jets, which can escalate the L1 trigger rate when coupled with the challenging PU environment. To effectively address this issue, the background is efficiently removed by evaluating the isolation of the candidate. QCD-induced jets typically yield broader energy deposits, making their isolation stand out from genuine τ_h decays. This isolation criterion proves instrumental in mitigating the impact of the QCD background and refining the τ_h candidate selection process. In this Section, the algorithm summarised above is detailed. The discussion then moves to the optimization improvements introduced to better cope with the harsher environment of Run-3; finally, the performance measured in the Run-3 2022 data-taking is presented.

3.2.1 Calorimeter Inputs

All algorithms implemented in the calorimeter trigger Layer-2 for the reconstruction of e/γ, τ, jets, and sums are built from the same inputs, the Trigger Towers (TTs), which in turn are built from the TPs generated in the subdetectors back-ends.

The information from the ECAL, HCAL, and Hadron Forward (HF) detectors is sent to the L1 trigger system in the form of TPs, which are digital quantities corresponding to the 40 MHz samplings of the calorimeter pulses. In each subdetector, TPs carry the information on localized energy deposits in confined detector regions. In the barrel of the CMS detector, each TP corresponds to the extension $\eta \times \phi = 0.087 \times 0.087$, which covers a square of 5×5 ECAL crystals and one HCAL readout unit. Given the extension of the CMS detector, each half barrel is sectioned into 17 and 72 TPs in the η and ϕ directions, respectively; therefore, a discrete two-indices Cartesian notation is employed to identify each TP using the pair $(i\eta, i\phi)$. In this configuration, for the barrel, $i\phi \in [1, 72]$ while $i\eta \in [-17, 17] \setminus \{0\}$, therefore position 0 does not exist in either of the two directions. It should be noted at this point that the $(i\eta, i\phi)$ position of the TPs does not need to be encoded into digital quantities with a specific number of bits as it is fully determined by the linking of the detector readout to the calorimeter trigger Layer-1. A schematic representation of the geometry of the TPs is reported in Fig. 3.1.

A more complex definition of the TPs is needed for the endcaps and HF due to the different geometry of the detector components. In the ECAL endcap, where the crystals are arranged in the specific pattern reported in Fig. 3.2 on the left, the trigger towers do not follow exact (η, ϕ) boundaries. In this region, the η-extension of the TPs increases as η grows while the number of crystals per TP varies between 25 at $|\eta| \approx 1.5$ and 10 at $|\eta| \approx 2.8$. This complex geometry is selected based mainly on the necessity to match the HCAL physical tower geometry while keeping a homogeneous segmentation of the detector. In the HCAL endcap, where the physical towers have an extension of $\Delta\phi = 0.174$, each HCAL readout is evenly split into two TPs of extension $\Delta\phi = 0.087$. In this configuration, the discrete Cartesian coordinates of the TPs in the endcap are in the following ranges: $|i\eta| \in [18, 28]$ and $i\phi \in [1, 72]$. For the HF detector, the same tower-to-TPs ϕ-splitting geometry of the HCAL endcap detector is used, while the η direction is segmented into 12 splittings, therefore $|i\eta| \in [29, 41]$ and $i\phi \in [1, 72]$, as reported on the left of Fig. 3.2.

For each TP detailed above, the energy deposit is computed as the projection onto the transverse plane of the momentum vector originating in the detector centre and pointing to the calorimeter cells, which is denoted as E_T. This value is encoded for

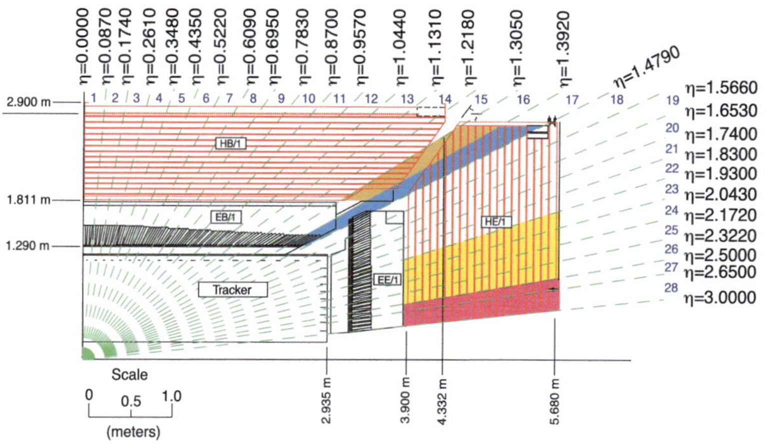

Fig. 3.1 Layout of the boundaries between trigger towers in the r-z projection. Each trigger tower regroups inputs from both the ECAL and HCAL subdetectors [5]

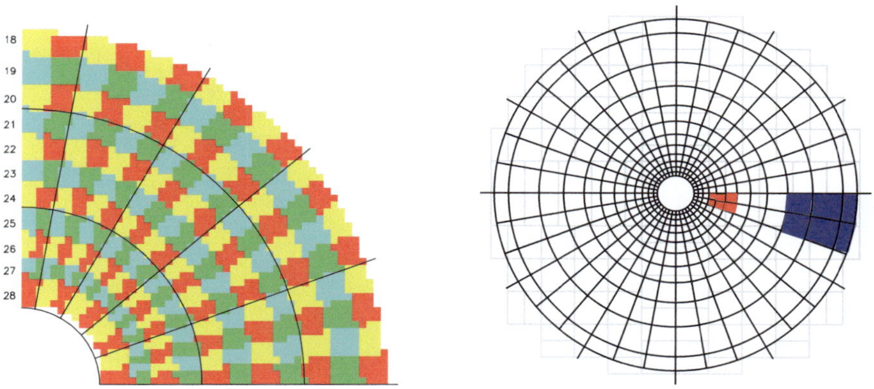

Fig. 3.2 Calorimeter Trigger Tower (TT) layout in the x-y projection of the ECAL endcap (left); each square denotes an ECAL crystal, and regions with the same colour represent one TT. Calorimeter TT layout in the x-y projection of the HF detector (right); each square denotes an HF readout unit, and regions with the same colour represent one TT [5]

each TP in an eight bits digital quantity using a linear scale with a 0.5 GeV unit (the minimal unit is often referred to as the *Least Significant Bit* or LSB); therefore, each TP can contain up to 127.5 GeV.

The TPs are transmitted to the calorimeter trigger Layer-1, where a calibration factor is applied to each of them based on position and energy deposit; a detailed description of this step and a novel approach to it are detailed in Sect. 3.5. After calibration, all TPs are arranged into TTs by merging the information of ECAL and HCAL TPs that geometrically lie one behind the other in the physical world. The HF

Table 3.1 Geometrical characteristics of the Level-1 calorimeter Trigger Towers (TTs). HCAL physical towers in the endcaps ($1.74 < |\eta| < 3.0$) have a $\Delta\phi = 0.174$ extension and are split into two trigger primitives each of $\Delta\phi = 0.087$. ECAL crystals in the endcaps are composed into TTs with the layouts depicted on the left of Fig. 3.2, while HF physical towers are composed into TTs with the layout depicted on the right of the same Figure

$\|i\eta\|$	$\|\eta_{min}\|$	$\Delta\eta$	$\Delta\phi$	ECAL crystals	HCAL physical towers
1–17	(n − 1) × 0.087	0.087	0.087	5 × 5	1
18–20	(n − 1) × 0.087	0.087	0.087	Endcap layout	1
21	1.740	0.090	0.087	Endcap layout	1 with ϕ-splitting
22	1.830	0.100	0.087	Endcap layout	
23	1.930	0.113	0.087	Endcap layout	1 with ϕ-splitting
24	2.043	0.129	0.087	Endcap layout	
25	2.172	0.150	0.087	Endcap layout	1 with ϕ-splitting
26	2.322	0.178	0.087	Endcap layout	
27	2.500	0.150	0.087	Endcap layout	1 with ϕ-splitting
28	2.650	0.350	0.087	Endcap layout	
29	3.000	0.500	0.348	–	HF
30	3.500	0.500	0.348	–	HF
31	4.000	0.500	0.348	–	HF
32	4.500	0.500	0.348	–	HF

detector information does not take part in the τ_h algorithm, but a seamless TTs coverage in the L1 trigger is needed for the jet and sums algorithms. Therefore, a coarser granularity is used, as reported in Fig. 3.2 on the right. The forward calorimeter is thus split into 72 TTs, with four and 18 TTs in the η and ϕ direction, respectively. In this geometry, the ϕ divisions are exactly four times the size of those of the barrel and endcap, and the η splittings are approximately the size of the last endcap TT. A summary of the geometry of the TTs is reported in Table 3.1.

At this stage, each TT is represented in digital form by 16 bits that encode the total E_T (ECAL and HCAL E_T sum) on nine bits, the ratio of the ECAL and HCAL energies on five bits and HCAL and ECAL quality flags on the two remaining two bits. At this stage, similarly to TPs, TTs can contain up to 127.5 GeV, and any TT with $E_T \geq 0.5$ GeV is referred to as an *active tower*. Finally, the Layer-1 sorts the TTs in descending E_T order and transmits them to Layer-2, where any of the Master Processor (MP7) cards receives the complete set of TTs from a specific bunch crossing (cf. Sect. 2.3.1 for the technical specification of the MP7 cards). In Layer-2, the L1 reconstruction algorithms are dedicated to e/γ, τ_h, jet, and sums reconstructions that optimally exploit the TTs information for efficient high-level object definition.

3.2.2 Algorithm Steps

In this Section, the Level-1 τ_h trigger algorithm is described corresponding to the architecture implemented in 2016 [6, 7] and which is unchanged at the start of Run-3. The algorithm is implemented in MP7 boards of the calorimeter trigger Layer-2, embedded in FPGAs firmware, and it optimally exploits the inputs from Layer-1 in four steps:

1. **clustering**: consisting of the selection of TTs into localized energy deposits referred to as *clusters*,
2. **merging**: involving the potential merging of multiple clusters into single L1 τ_h candidates,
3. **calibration**: performing energy calibration to improve the L1 τ_h scale and resolution,
4. **isolation**: comprising the application of an isolation criterion to reject QCD-induced jet background.

A full description of the original conception and development of the L1 τ_h trigger can be found in Ref. [7]. As mentioned, the algorithm is implemented in hardware; therefore, for the study and the optimization of its performance, a so-called *emulator* is employed, i.e. a C++ code that replicates with 100% accuracy the VHDL (Very High-speed integrated circuits Hardware Description Language) implementation of the firmware.

Clustering

The initial stage of the algorithm is the identification of local calorimetric energy deposits compatible with the particle products of the τ_h decay. Any odd number of hadrons in association with any number of neutral pions can be produced in the τ_h decay. Nonetheless, for the sake of this discussion, we can limit ourselves to consider $\tau \rightarrow h^{\pm}h^{\mp}h^{\pm}\pi^0\nu_\tau$ as the most complex τ_h decay, while neglecting the contribution from other higher multiplicity decay modes. Moreover, it should be noted that while any light hadron or meson can be a product of a τ_h decay, the most probable products are by far charged pions. Therefore, a τ_h decay produces charged pions, which can interact with the detector material and leave energy deposits in both the ECAL and HCAL subdetectors, as well as neutral pions that decay into photon pairs, leading to energy deposits either directly in ECAL or through conversion into electrons as they traverse the detector material. Furthermore, given the presence of the CMS magnetic field as well as bremsstrahlung radiation, the τ_h energy deposits tend to have elongated shapes in the ϕ direction.

Discriminating the τ_h signature amidst the overwhelming presence of PU, which generates low-energy diffused deposits in the calorimeter, poses significant challenges. As the τ_h is a collimated object, a dynamic clustering procedure is employed, selecting small groups of TTs to mitigate the impact of PU. As illustrated in Fig. 3.3, the L1 τ_h candidate begins with a seed, a local energy maximum within a region spanning three TTs along the η direction and nine TTs along the ϕ direction. Valid seed towers possess $E_T \geq 2$ GeV and are confined to either the barrel or endcaps, excluding the forward calorimeter, i.e. $|i\eta| < 29 = |\eta| < 3.0$. All the eight TTs within one unit in $i\eta$ and $i\phi$ of the seed are first incorporated into the so-called *proto-cluster* if they satisfy $E_T \geq 1$ GeV; additionally, the two TTs with the same $i\eta$ position as the seed but situated at two units of distance in $i\phi$ can be included in the proto-cluster, provided the TT in between is also part of the proto-cluster. A proto-cluster can comprise a total of up to 11 TT, extending over a maximum of three TT in the η direction. Nonetheless, the energy deposits from τ_h decays are generally more compact than this and tend to be asymmetric along η; therefore, a *lateral trimming* procedure is implemented to remove the proto-cluster side, i.e. TTs (2, 3, 4) or $(2', 3', 4')$ in the notation from Fig. 3.3, with the lowest E_T deposit sum, sacrificing only a fraction of the total candidate energy.

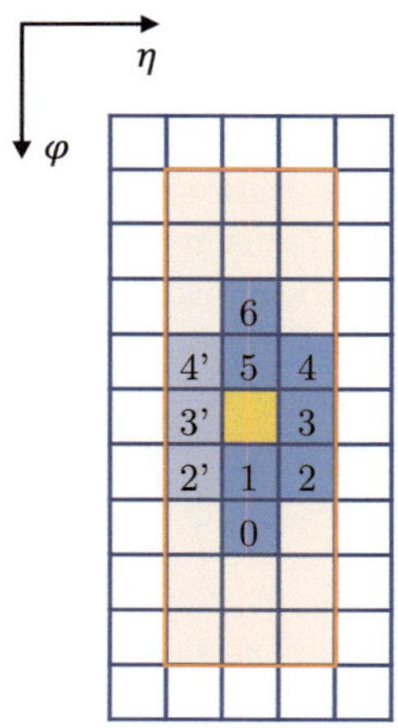

Fig. 3.3 Schematic representation of the clustering step of the Level-1 τ_h algorithm. Each square identifies a Trigger Tower (TT): in yellow the *seed* TT (local E_T maximum) initiating the clustering, in shades of blue the 10 TTs that can be clustered. The two sides labelled with primed and non-primed indices are considered for the creation of the proto-cluster but the lateral section with the lowest energy deposit is dropped during the trimming stage [7]

Finally, the proto-cluster position corresponds to the centre of the seed TT with possible adjustment based on the energy deposit pattern. Along both the ϕ axis, an offset of 1/4 of the TT size is applied in the direction of the highest energy deposit, i.e. $E_T(0, 1, 2/2') \lessgtr E_T(4/4', 5, 6)$; in the η direction, a shift of the same magnitude is applied toward the side remaining after the lateral trimming.

Merging

The clustering procedure described above is shared between the L1 τ_h and e/γ algorithm; therefore, it is well suited to collect the energy deposit of a single particle, being it a charged hadron, an electron, or a photon. As previously discussed, the τ_h decay targeted by this algorithm can produce up to three charged pions and one neutral pion, leading to a higher multiplicity and broader energy deposits for which the simple proto-clusters obtained in the first step are not well suited. The simplest approach to solving this issue would be the enlargement of the area considered for clustering, but this would lead to a large loss in the performance of L1 e/γ reconstruction due to its sensitivity to PU contribution; thus, a different approach needs to be implemented. Owing to the presence of the CMS magnetic field, the most important problem to overcome is the ϕ extension of the proto-clusters, which being $\Delta i\phi = \pm 2$ is not wide enough to collect the totality of the energy; the *merging* procedure is designed exactly for this purpose.

At the same time as the proto-clusters creation, *secondary clusters* are built with the same approach described above but with two differences: first, a smaller window

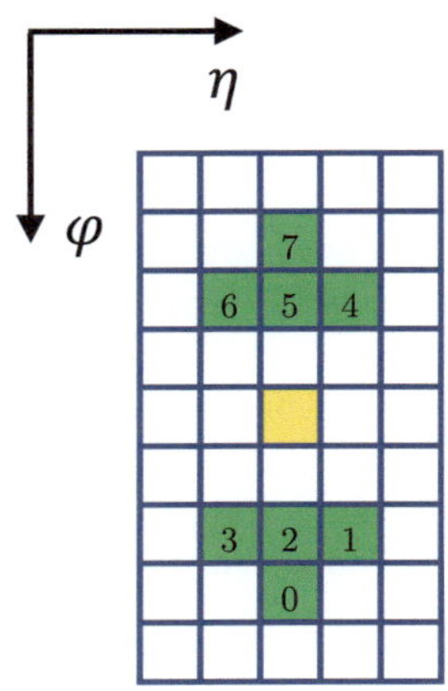

Fig. 3.4 Schematic representation of the merging step of the Level-1 τ_h algorithm. Each square identifies a trigger tower (TT): in yellow the seed TT (local E_T maximum) of the primary proto-cluster, in green the seven TTs that can seed secondary clusters that get merged to the primary one. The numbering scheme reported in the figure matches the one adopted in the firmware implementation [7]

spanning only three TTs in both the η and ϕ directions is employed; second, no lateral trimming is applied. The association rule between primary proto-clusters and secondary clusters is represented in Fig. 3.4: a secondary cluster will be merged with the primary proto-cluster, whose seed is shown in yellow, if its seed is found in one of the eight positions highlighted in green. Given the simultaneous construction of primary and secondary clusters, an overlap removal procedure is implemented to avoid the double-counting of TTs; moreover, the secondary clusters that are seeded in positions number 2 or 5, using the numbering scheme of Fig. 3.4, are not considered if their seed TT is already included in the primary proto-cluster. Finally, if multiple secondary clusters are found, the one with the largest E_T sum is selected; if multiple secondary clusters are found with the same energy deposit, the one with the smallest $\Delta i\phi$ distance from the primary seed is chosen. In this way, each L1 τ_h candidate is reconstructed by a single cluster, be it merged or not merged.

The total fraction of merged clusters is about 15%; however, this value is strongly dependent on the τ_h decay mode and its p_T. In the low p_T regime, i.e. $p_T \lesssim 40$ GeV, more than 30% of $\tau \rightarrow h^{\pm}h^{\mp}h^{\pm}\nu_{\tau}$ decays result in merged clusters while only 15 and 10% of respectively $\tau^{\pm} \rightarrow h^{\pm}\pi^0$ and $\tau^{\pm} \rightarrow h^{\pm}$ undergo merging. Conversely, in the high p_T regime, i.e. $p_T \gtrsim 80$ GeV, only 10% of 3-prong decays have merged clusters, while about 5% of 1-prong(+π^0) decays present merged clusters [7].

Calibration

As discussed earlier, the TTs used in the first two steps of the algorithm undergo calibration in the calorimeter trigger Layer-1; therefore, a first evaluation of the energy of the L1 τ_h candidate can be obtained with reasonable precision by summing the energy deposit in each TTs; this quantity is referred to as raw energy (E_T^{raw}). Nevertheless, the raw energy does not fully encompass the actual energy of the τ_h candidate owing to multiple reasons. Firstly, the energy deposit is shared between ECAL and HCAL, while the Layer-1 calibration does not account for any overlap of the two quantities, thus an artificial total non-linearity can be induced by the different ECAL and HCAL energy responses. This problem has become particularly relevant in Run-3 due to the missing calibration of HCAL TTs at Layer-1 (this will be fully detailed in Sect. 3.5). Secondly, the clustering procedure induces unavoidable losses due to the limited number of TTs that can be considered in each L1 τ_h candidate, this issue becoming particularly acute in the large $|\eta|$ regions where the TTs themselves have a complex geometrical definition. Thirdly, the merging procedure induces systematic differences between merged and non-merged clusters that need to be addressed and corrected.

Following the reasoning of the previous paragraph, the energy calibration of the clusters associated with the L1 τ_h candidates is derived as a function of four variables: the E_T^{raw} energy deposit, the $i\eta$ position of the cluster, the presence or absence of ECAL energy deposit identified by the flag i^{EM}, and the cluster being issued by the merging procedure or not according to the flag i^{merged}. The calibrated L1 τ_h energy can thus be expressed as:

$$E_T = c(E_T^{raw}, i\eta, i^{EM}, i^{merged}) \cdot E_T^{raw} \tag{3.1}$$

where c is the calibration factor given the value of the four quantities on which it depends. The calibration constants are derived as the inverse of the correction factor computed with a Boosted Decision Tree (BDT) having the ratio $E_T^{raw}/p_T^{offline}$ as a target. Given the need for hardware implementation of such calibration, the calibration constants are derived in exclusive intervals of the input variables and are stored in a Look-Up Table (LUT), which is an array replacing runtime computations with a simpler indexing operation. At the same time, the hardware resources availability constrains the number of entries of the LUT; therefore, the input quantities are compressed, i.e. converted with a non-linear transformation, into a lower number of bits: E_T^{raw} is reduced from 13 to five bits, $i\eta$ is compressed from five to two bits, while i^{EM} and i^{merged} are kept as single-bit variables. Finally, the calibration factor itself is encoded on 10 bits and represents a decimal value in a linear scale between 0.0 and 2.0.

Isolation

The first three steps of the algorithm are designed to reconstruct L1 τ_h candidates with the highest possible efficiency, taking into account only the hardware capabilities of the Phase-1 Level-1 trigger. Therefore, the algorithm described up to now reconstructs also background contributions with high efficiency, leading to very high L1 rates. The isolation step of the algorithm is designed precisely to reject background candidates and meet the rate requirements of the L1 trigger.

Jets originating from quarks and gluons constitute the largest contamination to the L1 τ_h signal. Given the higher multiplicity of particles produced by a quark or gluon hadronization, compared to the τ_h decay, hadronic jets result in wider energy deposits, which are more spread out in both the η and ϕ directions. This allows for the definition of the isolation energy (E_T^{iso}), which encodes the amount of calorimeter activity around the candidate that is the most suitable handle to identify genuine L1 τ_h candidates against QCD-induced jets.

The isolation energy is computed as the difference between the total energy deposit in a local region of dimension $i\eta \times i\phi = 6 \times 9$ around the τ_h candidate seeding TT and the uncalibrated energy of the candidate itself as:

$$E_T^{iso} = E_T^{6\times 9} - E_T^{raw} \tag{3.2}$$

where the uncalibrated E_T^{raw} is used for two reasons: to ensure that homogeneous quantities are subtracted to avoid biases in E_T^{iso} due to the application of the calibration factors and to meet latency constraints in the algorithm implementation by allowing the two quantities to be computed in parallel rather than in series. It should be noted that, given the asymmetry of the L1 τ_h clusters, the isolation region is also asymmetric, extending one TT more in the direction of the proto-cluster side that

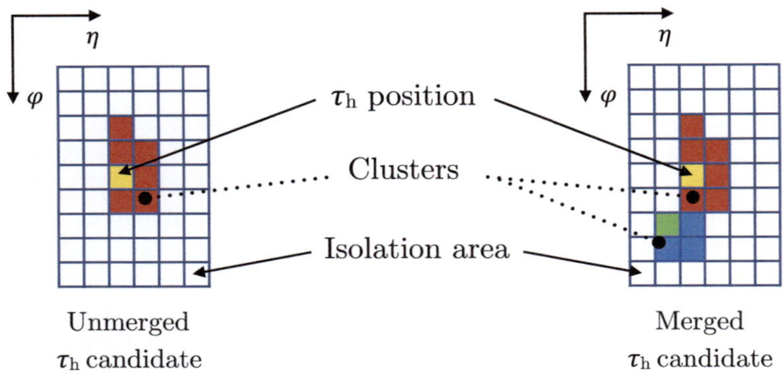

Fig. 3.5 Schematic representation of the isolation step of the Level-1 τ_h algorithm. Each square identifies a trigger tower: the isolation energy is computed as the difference between the energy deposit in an $i\eta \times i\phi = 6 \times 9$ region and the Level-1 τ_h candidate [7]

survives the trimming process. The choice of the extension of the isolation area is a trade-off between the optimal rejection of QCD-induced background and the sharing of this step with the L1 e/γ algorithm. A schematic representation of the isolation region is reported in Fig. 3.5.

The separation power of E_T^{iso} can be appreciated in Fig. 3.6, where the differential distribution of the isolation energy for the τ_h signal and jet background. The signal is obtained from Monte Carlo (MC) simulated Vector Boson Fusion (VBF) H $\to \tau\tau$ events, while the background is obtained from data collected in high PU runs during the 2016 data-taking. As expected, due to the larger extension of the hadronic activity, the isolation energy assumes larger values for the jet background, and the application of an isolation criterion ensures their rejection.

The isolation criterion can be defined via the application of a simple threshold on E_T^{iso}, thus defining an L1 τ_h to be isolated if $E_T^{iso} < \Theta$, where the numerical value of Θ needs to be defined as a trade-off between signal efficiency and background rejection. Moreover, the definition of the cut-off point must ensure uniform performance over the entire geometry of the detector and E_T range considered, as well as over the range of expected simultaneous interactions per bunch crossing. The threshold on the L1 τ_h isolation energy can thus be expressed as:

$$\Theta \equiv \Theta(E_T^{raw}, i\eta, n_{TT}) \tag{3.3}$$

where n_{TT} is a PU estimator defined as the number of active TTs in the most central region of the barrel, i.e. $|i\eta| \leq 4$, and whose value is on average proportional to the number of primary vertices reconstructed in the event. The dependency on the energy of the L1 τ_h candidates arises from the non-uniform collection of the energy deposited in the calorimeter by the clustering procedure at different energies, while the dependency on the pseudorapidity position stems from the expected difference of QCD-induced jets multiplicity in the endcaps, as well as the different geometrical

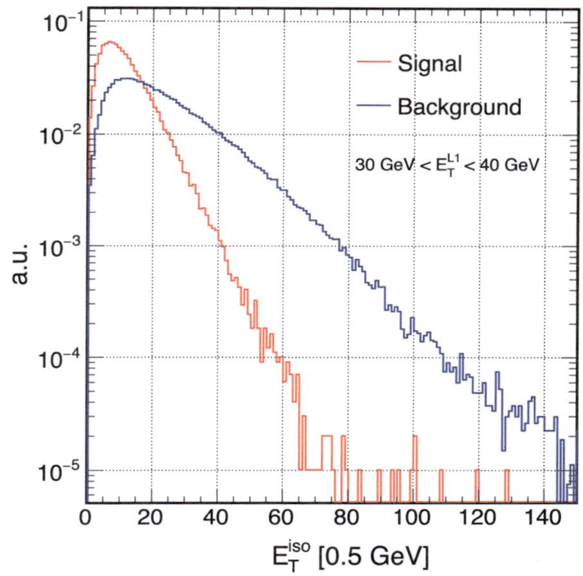

Fig. 3.6 Distributions of the isolation energy (E_T^{iso}) of Monte Carlo simulated Level-1 τ_h candidates from the vector boson fusion $H \to \tau\tau$ process (red) and Level-1 jet candidate from data recorded in 2016 (blue). The simulated pileup conditions of the signal are the same as those measured for the background [7]

definition of the TTs. The use of E_T^{raw}, for the definition of Θ, instead of the calibrated energy, follows the same considerations outlined above for the computation of E_T^{iso}.

The reason for using n_{TT} as a pileup estimator in the computation of E_T^{iso} can be appreciated in Fig. 3.7. The left panel shows how n_{TT} is linearly proportional to the number of offline reconstructed vertices, which is itself an estimate of the PU. Moreover, the right panel showcases the analogous linear dependency of E_T^{iso} on n_{TT}. The right panel has considerably larger error bars because its entries are only well-identified τ_h candidates; in contrast, the left panel uses the entire dataset available as no selection is applied to these events. The robust dependency of n_{TT} on the number of vertices is a consequence of its definition, which makes this variable range between 0 and 576, thus making it almost insensitive to the presence of a possible τ_h candidate in the central region given its very localized interaction. Furthermore, as n_{TT} is defined as the simple counting of active TTs in a certain area, its firmware implementation is straightforward, and its computation latency is contained.

The numerical values of Θ are derived in MC simulated events in separate intervals of the input variables to ensure flat efficiency across the $i\eta$ and n_{TT} ranges while keeping a dependence on the raw energy E_T^{raw}. This decision is taken to fully exploit the steeply falling E_T spectrum of the background by applying tighter isolation thresholds in the low energy region, which are progressively relaxed until 100% L1 τ_h rigger efficiency is reached. This procedure is referred to as *isolation relaxation* and necessitates the definition of a parametrization of the isolation as a function of E_T^{raw}. Such parametrization can be defined as follows: constant efficiency ϵ_{min} is targeted below a certain energy threshold $E_T^{raw,\,min}$, linearly increased until the maximal efficiency is reached at another energy threshold $E_T^{raw,\,max}$, after which the

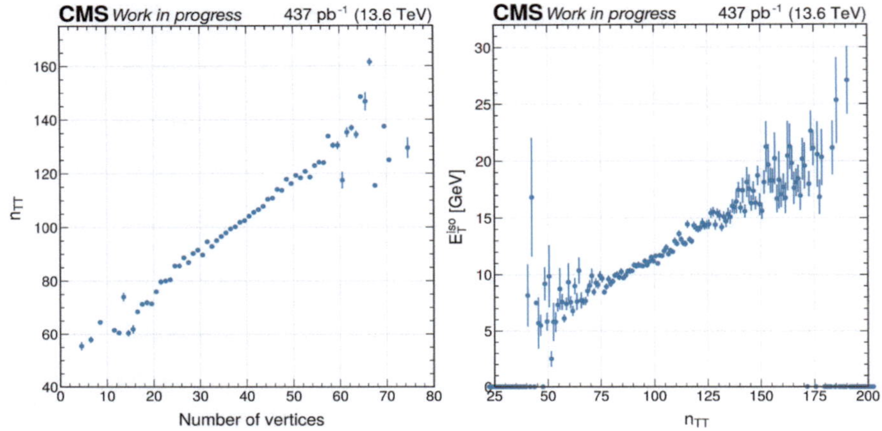

Fig. 3.7 Distribution of the n_{TT} pileup (PU) estimator as a function of the number of offline reconstructed vertices (left), and distribution of the isolation energy ($E_{\mathrm{T}}^{\mathrm{iso}}$) as a function of n_{TT} (right), both measured in a sample from the Run-3 2022 data-taking. A robust linear dependency can be appreciated in both panels, showcasing the effectiveness of n_{TT} as an event PU estimator

efficiency is maintained maximum. The values of the parameters ϵ_{min}, $E_{\mathrm{T}}^{\mathrm{raw,\,min}}$ and $E_{\mathrm{T}}^{\mathrm{raw,\,max}}$ constitute the so-called *isolation option*, and they are optimized according to the expected PU and instantaneous luminosity profiles.

In Run-2, increasingly tight isolation schemes were deployed as the experimental conditions turned harsher, and the approach to selecting the isolation option was based on a trial-and-error approach: various combinations of isolation parameters fully conceived by the analyzer were tested and modifications, again directly chosen by the analyzer, were implemented until satisfactory performance was reached. This approach suffers from two main drawbacks: it is highly inefficient in terms of time, as it requires the analyzer to take all the isolation optimization steps actively, and at the same time only points in the parameter space that are selected by the analyzer are investigated, resulting in possible loss of performance. Moreover, being the numerical values of Θ derived in separate intervals of the input variables, the determination of Θ can suffer from the large statistical fluctuations of the number of simulated candidates in such bins. To tackle these three points, for the beginning of Run-3, a new systematic approach to the optimization, which at the same time is resilient to the statistical power of the samples used for the optimization, has been designed and implemented within this Thesis work as detailed in Sect. 3.3.

3.2.3 Main Level-1 τ Seeds

The geometry of the L1 τ_h algorithm described above implies that up to 144 L1 τ_h candidates can be constructed per bunch crossing. Nevertheless, only a few L1 τ_h

candidates at most per event are expected to be relevant to take a trigger decision meaningful for physics analyses. Moreover, this very large number of candidates cannot be transmitted to the global trigger due to the huge bandwidth they would require. Therefore, only the six highest energy candidates in the positive and negative η regions of the detector are selected and sent from the calorimeter trigger Layer-2 to the μGT. Similarly, the μGT receives the L1 μ, e/γ, jet, and sums candidates to be exploited for the event accept or reject decision based on the L1 Menu seeds. In Run-3, the L1 seeds using τ_h candidates can be grouped in three categories:

- **Single-τ_h L1 triggers**: they require the presence of one L1 τ_h candidate satisfying specific E_T and η requirements. They are denoted as `L1_SingleTauXerY` or `L1_SingleIsoTauXerY`, depending if the isolation requirement is enforced or not, where `X` denotes the L1 E_T threshold applied and `Y` represents the η restriction applied (which for τ_h is always $|\eta| < 2.1$). These inclusive triggers offer a broad physics acceptance, rendering them suitable for exploring various physics phenomena, including the identification of highly boosted H bosons. However, due to the selection of a single object in these triggers, a higher trigger rate is observed, necessitating the application of very stringent thresholds ($\gtrsim$100 GeV) on the energy of the τ_h candidate.
- **Double-τ_h L1 triggers**: they require the presence of two L1 τ_h candidates satisfying the same E_T and η requirements. They are denoted as `L1_DoubleIsoTauXerY`, following the same naming scheme of the single-τ_h triggers, and the isolation requirement is always applied to lower the energy threshold while keeping an acceptable rate. This type of trigger targets more specific signal topologies for searches like H $\rightarrow \tau\tau$ and HH $\rightarrow$ bb$\tau\tau$. The simultaneous presence of two L1 τ_h candidates ensures the lowering of trigger thresholds to $\approx 32 - 34$ GeV.
- **Cross τ_h + X triggers**: they require the presence of one or two L1 τ_h candidates and the simultaneous presence of other L1 objects (X). These seeds are denoted by a name including both objects, e.g. `L1_DoubleIsoTauXerY_JetZ_RmOvlp_dR0p5` or `L1_MuZerW_TauXerY`, and follow the same naming convention described above. In the former example, `RmOvlp` indicates the removal of the jet and τ_h objects overlap, while `dR` denotes the angular separation between the objects. Similarly to the double-τ_h seeds, they target specific physics signal topologies with high efficiency and a low rate. The two examples reported above are highly efficient for HH $\rightarrow$ bb$\tau\tau$ and H $\rightarrow \mu\tau$ searches, respectively. The simultaneous presence of multiple objects ensures the lowering of τ_h trigger thresholds to $\approx$ 24–26 GeV.

3.3 The Present: A Systematic and Resilient Optimization

The L1 τ_h algorithm described above performed excellently through the 2016, 2017, and 2018 data-taking, reaching high efficiency and stability levels. During the three years of Run-2, the algorithm has been optimized in several instances to cope with the changes in the LHC running conditions. In the same way, the start of Run-3 brought major changes in the LHC conditions, making further development needed to maintain the performance of Run-2 in the considerably harsher collision environment of Run-3. In particular, the instantaneous luminosity is increased reaching peak values of $\sim 2.6 \cdot 10^{34}\,\text{cm}^{-2}\,\text{s}^{-1}$ in 2022 and $\sim 2.2 \cdot 10^{34}\,\text{cm}^{-2}\,\text{s}^{-1}$ in 2023 (cf. Fig. 2.4). At the same time, both the average and peak PU are largely enhanced, with an average of respectively 46 and 52 in 2022 and 2023, to be compared with 38 (37) for 2017 (2018) data-taking and a peak PU exceeding 80 in Run-3 while seldom values surpassing 70 were observed in Run-2 (cf. Fig. 2.2). Concurrently with the increase in instantaneous luminosity, the centre-of-mass energy has been increased from 13 TeV to 13.6 TeV, therefore increasing the production cross section of QCD-induced jets, which are the main background to the L1 τ_h reconstruction.

Owing to the changes in running conditions detailed above, the L1 τ_h algorithm has been fully optimized for the 2022 and 2023 data-taking. The changes in the algorithm are aimed at maintaining the adequate energy thresholds and L1 rates achieved in Run-2. They are mostly concentrated on the derivation of new calibration constants and the optimization of isolation options adapted to the harsher running conditions. Nevertheless, as noted above, the Run-2 approach to the optimization was highly time-consuming, not systematic, and suffered from the MC simulation's statistical power. Therefore, a new approach to optimization was implemented to be resilient to the statistical limitations of MC samples and, simultaneously, more systematized in the search for the optimal isolation parameters.

The improvements to the algorithm are detailed in this Section, while the resulting performance is reported in Sect. 3.4.

3.3.1 *Derivation of Calibration Factors*

The calibration of calorimeter objects in the L1 trigger is a two-tier process: first, the trigger primitives are calibrated in Layer-1; subsequently, the objects-specific calibration is applied in Layer-2. This stacked approach implies that any change in the calibration performed at Layer-1 requires an update of the one at Layer-2.

At the beginning of Run-3, due to the large noise levels expected in the HCAL detector, the derivation of the associated Layer-1 calibration factors proved extremely challenging; therefore, the decision was taken to remove the Layer-1 calibration of HCAL TPs. This corresponds to a major change for the L1 τ_h algorithm as the Layer-2 calibration needs to compensate for the missing Layer-1 adjustment of the energy deposits. Moreover, during the 2022 data-taking, a large instability of the HCAL noise levels was noticed in the barrel region. This instability highly affected the n_{TT} PU estimator used by the L1 τ_h and e/γ algorithm, making the definition of a stable isolation working point extremely challenging. To cope with this, the decision was taken to introduce a dedicated *zero-suppression* scheme for the HCAL barrel TPs, i.e. $|i\eta| \leq 15$. In this approach, all TPs with $E_T = 0.5\,\text{GeV}$ are considered to be activated by noise, and their energy deposit is set to zero. This decision once again highly affects the Layer-2 calibration of L1 τ_h objects.

To counterbalance the changes in the Layer-2 inputs from Layer-1 during the 2022 and 2023 data-taking periods, a new set of calibration factors $c(E_T^{\text{raw}}, i\eta, i^{\text{EM}}, i^{\text{merged}})$ for the L1 τ_h candidates was derived. Their values are reported in Fig. 3.8 where the c constants derived for the 2022 and 2023 Run-3 data-taking are shown in the form of heat maps. The dependence of the calibration factors on the i^{EM} flag is made evident by splitting the calibration factor into two orthogonal sets.

The difference in the magnitude of the calibration for 2022 and 2023 can be appreciated from the numerical values reported in the Figure. Notwithstanding the missing calibration of HCAL TPs in the calorimeter trigger Layer-1, the calibration factors of L1 τ_h candidates are, on average, larger for those candidates with ECAL deposit ($i^{\text{EM}} = 1$); this is a symptom of the high level of noise experienced by HCAL in 2022 data-taking conditions, for which the Layer-2 calibration needs to compensate. The calibration factors in 2023 are mostly smaller with respect to 2022, owing

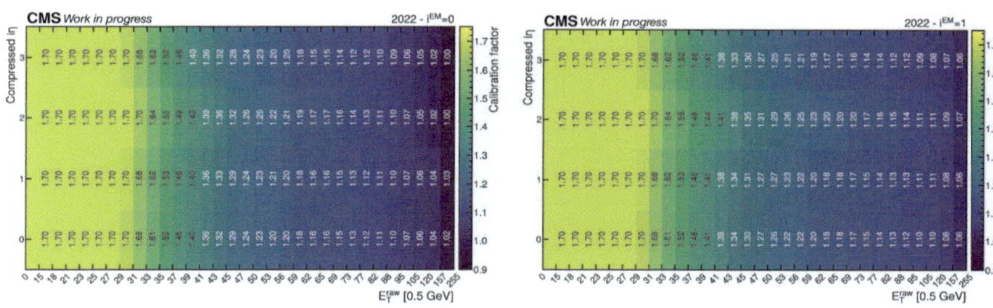

(a) 2022

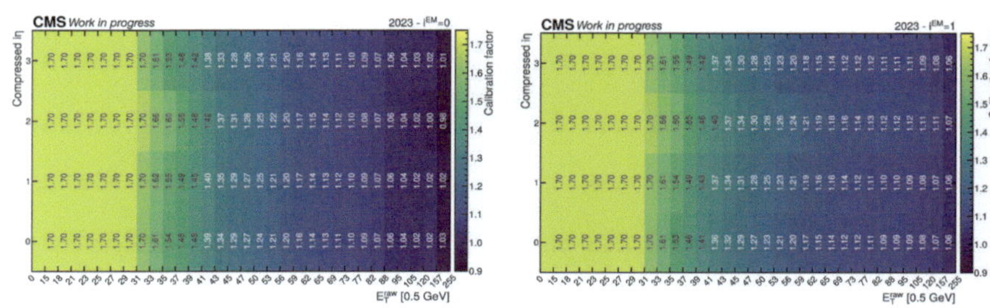

(b) 2023

Fig. 3.8 Calibration factors $c(E_T^{\text{raw}}, i\eta, i^{\text{EM}}, i^{\text{merged}})$ of Level-1 τ_h candidates used during the Run-3 2022 (top row) and 2023 (bottom row) data-taking. The orthogonal cases of $i^{\text{EM}} = 0$ (left column) and $i^{\text{EM}} = 1$ (right column) are displayed in the form of heat maps as a function of the raw energy and the compressed pseudorapidity position of the τ_h candidate, while the inclusive i^{merged} spectrum is considered. The numerical value of the calibration factors is reported per each bin of the heatmap. The calibration factors for $i^{\text{EM}} = 1$ are mostly larger than $i^{\text{EM}} = 0$ to compensate for the high level of noise experienced by HCAL in Run-3 data-taking conditions. The calibration factors in 2023 are mostly smaller with respect to 2022, owing to the HCAL improved calibration scheme to the HCAL back-end improved TPs calibration scheme. This can be appreciated by comparing the heat maps in Fig. 3.8a to those in Fig. 3.8b, where the calibration factors of the latter are mostly smaller compared to the former.

The effect of the new calibration is reported in Fig. 3.9, where the ratios of the L1 τ_h raw and calibrated energies to the offline reconstructed p_T are reported. The 2022 and 2023 calibration conditions are shown separately to compare the different data-taking conditions. The improvement of the raw response in 2023 owes to the introduction of a better calibration of the TPs in the HCAL back-end, which further mitigates the missing Layer-1 calibration, and to the different energy spectrum of the τ_h candidates in the samples used. The Layer-2 calibration improves the response distribution, which assumes a mean value compatible with unity; the improvement is particularly important in 2022, where the scale is corrected by 11%. In achieving a better scale, the RMS of the distribution is slightly degraded because the BDT is trained to obtain a flat scale over the p_T spectrum of the τ_h and a minor smearing of the response is introduced to attain that.

These results highlight that the Layer-2 calibration of the L1 τ_h algorithm can mitigate and compensate for the different Layer-1 calibrations. At the same time, it should be noted that the Layer-1 calibration is indispensable for the proper calibration

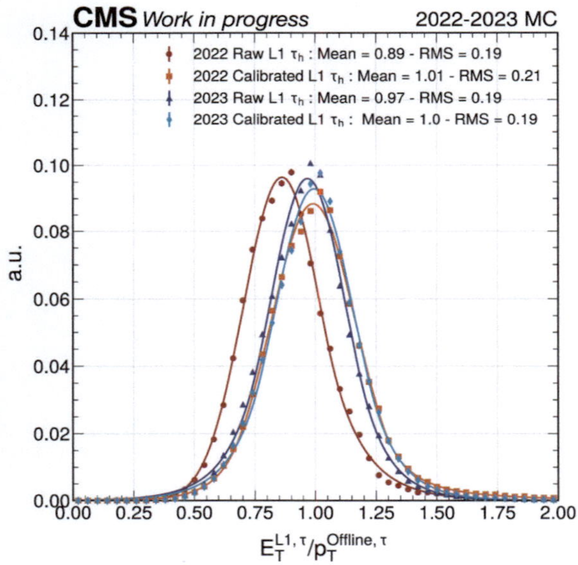

Fig. 3.9 Response of the Level-1 (L1) τ_h defined as the ratio between the energy measured at L1 and the offline reconstructed p_T, separately for 2022 and 2023 Monte Carlo (MC) simulation. These 2022 result is obtained using MC simulated vector boson fusion (VBF) H $\rightarrow \tau\tau$ events with a realistic Gaussian pileup (PU) distribution, while the 2023 result is obtained and tested using MC simulated VBF and gluon fusion H $\rightarrow \tau\tau$ events, and Drell-Yan $Z/\gamma^* \rightarrow \tau\tau$ events, all generated assuming a distribution of the PU according to a Poisson law centred at 70 simultaneous interactions per bunch crossing

of the sum objects that are not subject to Layer-2 calibrations. Section 3.5 presents a novel method developed as part of this Thesis, which tries to answer this need.

3.3.2 τ_h Isolation Resilient to Statistical Limitations

The most extensively used seed in CMS analyses targeting hadronically decaying τ leptons, such as the HH $\rightarrow$ bb$\tau\tau$ and H $\rightarrow \tau\tau$ analyses, is `L1_DoubleIsoTauXerY`. In Run-3, the two unprescaled seeds of this form with the lowest energy threshold are `L1_DoubleIsoTau34er2p1` and `L1_DoubleIsoTau36er2p1`, which require two isolated L1 τ_h candidates with $E_T > 34$ (36) GeV and $|\eta| < 2.1$. The L1 rate of these seeds is expected to have a linear dependence on PU; if the dependency is not reached, it is generally a symptom of too stringent isolation criteria. During the 2022 data-taking, both seeds presented a deviation from the expected linear behaviour, with $\sim$5% less rate than expected at $PU \approx 57$, as reported in Fig. 3.10. These results are obtained by analyzing run number 360927 (from 2022-10-23 at 00:53:11, to 2022-10-23 at 13:23:11), with a recorded integrated luminosity of 657 pb^{-1} and a peak instantaneous luminosity of $2.0 \cdot 10^{34}$ cm^{-2} s^{-1}. This sub-optimal behaviour is understood under the considerations made in the following paragraph.

The numerical values of the isolation threshold Θ defined in Eq. 3.3 can be derived from MC simulated events in separate intervals of the input variables. The range

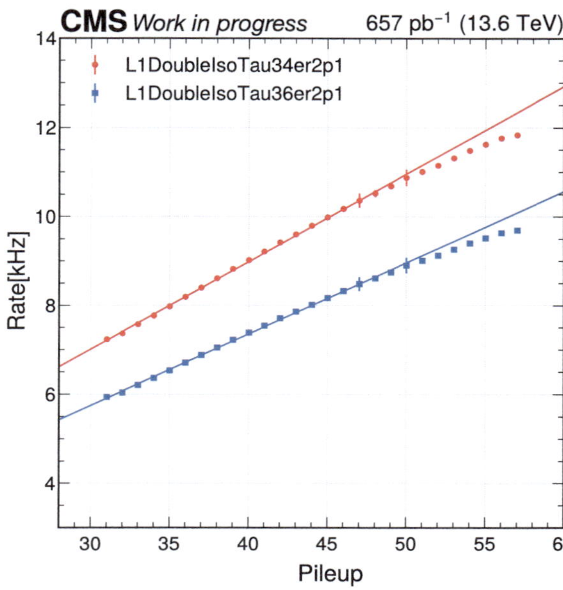

Fig. 3.10 Level-1 trigger rate as a function of pileup, evaluated in a single LHC fill, for two Level-1 τ_h trigger seeds: double isolated τ_h with $E_T > 34$ GeV and $|\eta| < 2.1$ (red), and double isolated τ_h with $E_T > 36$ GeV and $|\eta| < 2.1$ (blue). Rate and pileup are averaged over a time interval of a *luminosity section*, corresponding to 2^{18} LHC orbits or 23.31 s of data-taking. The functional form of the fits consists of a line. Both curves are measured in Run-3 2022 data

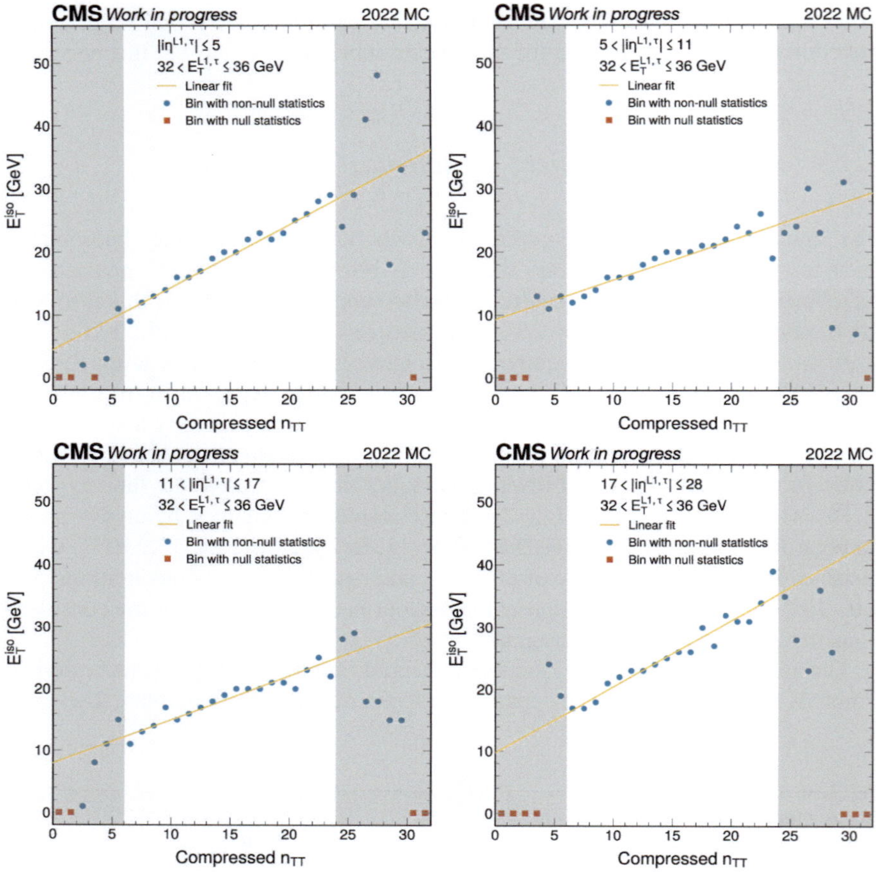

Fig. 3.11 Isolation energy (E_T^{iso}) of the Level-1 τ_h candidate as a function of the compressed value of the n_{TT} PU estimator in four specific bins of $i\eta$ and E_T. The blue points represent the E_T^{iso} values obtained from the Monte Carlo simulated vector boson fusion $H \rightarrow \tau\tau$ sample, while the red squares represent the bins where no entry is found. In yellow is reported the best linear fit of the distribution within the fiducial region $n_{TT} \in [6, 24]$. The grey areas are the ones excluded from the fit. The large statistical fluctuations due to the limited number of events—sometimes being exactly zero—at the extremes of the n_{TT} range are easily overcome by the linear fit

and the boundaries of these bins are partially driven by the statistical power of the simulated sample used to derive the isolation cut-off. Nevertheless, at the edges of the ranges of the input variables, the available number of MC events might still remain highly constrained, making Θ very sensible to statistical fluctuations. In the extreme case where the MC simulated sample does not contain any event pertaining to a specific interval of the E_T^{raw}, $i\eta$, and n_{TT} variables, the numerical value of Θ becomes zero, possibly leading to large inefficiencies. Such behaviour can be appreciated in Fig. 3.11. The blue points represent the E_T^{iso} values obtained from the MC simulated

VBF H $\rightarrow \tau\tau$ sample, while the red squares represent the bins where no entry is found. Two things can be noticed: firstly, there are multiple bins, both at the high and low edges of the n_{TT} range, where the statistical power of the MC sample is null; secondly, in bins where entries are found, large fluctuations of $E_{\mathrm{T}}^{\mathrm{iso}}$ are observed due to the limited statistical power in them. This situation is what was experienced in 2022, where the available number of MC simulated events used for the L1 τ_{h} optimization was very limited. To cope with this undesirable behaviour, a new approach to the evaluation of the isolation threshold Θ was conceived as part of this Thesis and introduced between 2022 and 2023.

The main reason for the missing statistical power in 2022 was due to the too granular definition of the input variables intervals. These orthogonal regions were defined in units of the compressed $i\eta$, $E_{\mathrm{T}}^{\mathrm{raw}}$, and n_{TT} variables, corresponding to four, 32, and 32 bins, respectively. This meant a total of 4096 exclusive intervals in which $E_{\mathrm{T}}^{\mathrm{iso}}$ is computed and the relative threshold Θ selected. The first improvement to the algorithm has been the introduction of a *super-compression* scheme to reduce the number of regions in the inputs phase space: both $E_{\mathrm{T}}^{\mathrm{raw}}$ and n_{TT} are further compressed from 32 to 16 bins with a non-linear transformation, while $i\eta$ is left unchanged. This allows for the definition of coarser intervals less subject to statistical fluctuation. The second modification to the algorithm is aimed at further reducing the number of orthogonal regions needed for the selection of Θ. Therefore, instead of splitting the phase space based on all three input variables, only the energy and position of the L1 τ_{h} candidate are used to build exclusive bins. In each of these subsets of phase space, the MC events are profiled against the n_{TT} variable and a linear fit of the $E_{\mathrm{T}}^{\mathrm{iso}}$ distributions is performed as a function of the PU estimator. This approach is depicted in Fig. 3.11, where a yellow line represents the linear fit, and the grey areas are those excluded from the fit. This approach has been studied using compressed and super-compressed n_{TT} schemes, the latter being better suited to cope with the high $E_{\mathrm{T}}^{\mathrm{raw}}$ region.

The comparison of the numerical values of the Θ thresholds obtained before and after the introduction of the new scheme can be grasped in Fig. 3.12. In both panels, the $E_{\mathrm{T}}^{\mathrm{iso}}$ threshold is reported as a function of the compressed $E_{\mathrm{T}}^{\mathrm{raw}}$ and n_{TT} values in the most central eta region, i.e. $|i\eta| \leq 5$. The panel on the left reports the values of Θ used for the 2022 data-taking, while the panel on the right shows those used in 2023. It can be appreciated how the Run-2 approach to the calculation of the threshold leads to highly unstable values, which are strongly dependent on the statistical power of the MC samples. In contrast, the newly introduced scheme guarantees a smooth behaviour of Θ throughout the whole n_{TT} range, ensuring the appropriate threshold selection.

Owing to the improved stability in the derivation of the isolation threshold, this method has become a stable and fundamental component of the τ_{h} algorithm optimization. At the same time, given the similarity of the two algorithms, this approach has been adopted in its entirety also for the optimization of the L1 e/γ algorithm, which benefits from the same improvements. The performance attained for both the L1 τ_{h} and e/γ candidates is reported in Sect. 3.4.

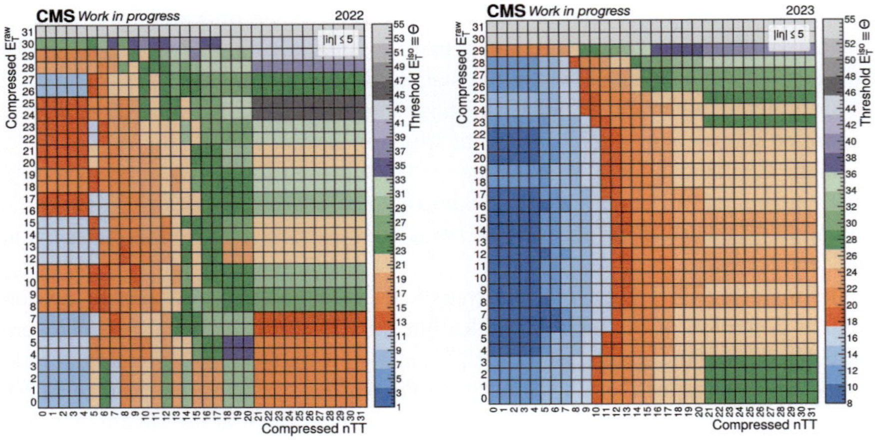

Fig. 3.12 Isolation energy threshold Θ used for the 2022 (left) and 2023 (right) data-taking. In both panels, the threshold Θ is reported as a function of the compressed E_T^{raw} and n_{TT} values in the most central pseudorapidity region, i.e. compressed $i\eta = 0$ corresponding to $i\eta \leq 5$. In both panels, the binning matches that of the firmware look-up table

3.3.3 Systematic τ_h Isolation Optimization

As discussed in Sect. 3.2.2, the process of selecting the isolation option in Run-2 was guided by a non-systematic approach; moreover, it should be noted that the missing Layer-1 calibration highly impacts the optimization of the L1 τ_h isolation. Therefore, methodical exploration of the isolation options phase space is crucial. Such a technique has been designed as part of this Thesis and is detailed below.

The isolation relaxation procedure was conceived to exploit the steeply falling E_T spectrum of the background. This method applies tighter isolation thresholds in the low τ_h raw energy region, which are progressively relaxed until 100% efficiency is reached at high E_T^{raw} values, according to a certain parametrization of the relaxation. The choice of this parametrization is fully arbitrary and has generally been chosen to be linear between a fixed minimum efficiency ϵ_{min} at a fixed minimum energy $E_T^{raw,\,min}$ and the maximum 100% efficiency at a maximum energy $E_T^{raw,\,max}$. In this simple scheme, the slope and intercept of the straight line are fully determined by choice of ϵ_{min}, $E_T^{raw,\,min}$, and $E_T^{raw,\,max}$ as:

$$\text{Linear slope}: m = \frac{1-\epsilon_{min}}{E_T^{raw,\,max} - E_T^{raw,\,min}} \tag{3.4}$$

$$\text{Linear intercept}: q = 1 - \frac{1-\epsilon_{min}}{E_T^{raw,\,max} - E_T^{raw,\,min}} \cdot E_T^{raw,\,max} \tag{3.5}$$

and the piece-wise definition of the linear isolation relaxation reads:

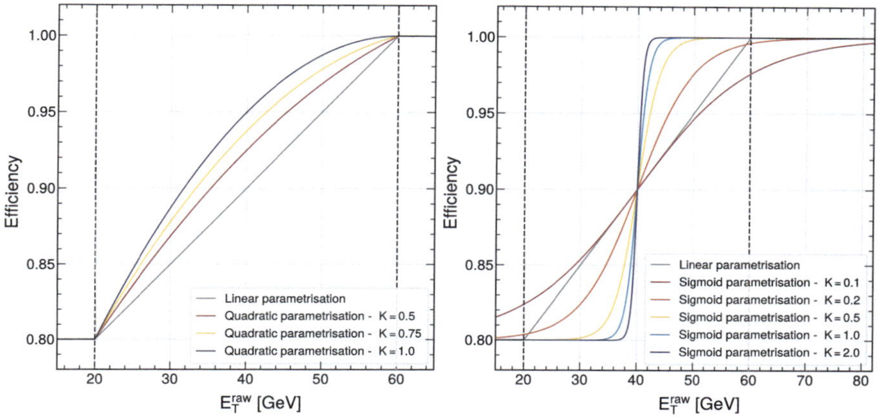

Fig. 3.13 Example of explored quadratic (left) and sigmoid (right) parametrizations of the efficiency as a function of the raw energy (E_T^{raw}) for the identification of Level-1 τ_h candidates. The parameter K reported in the legend is the sharpness factor; in all cases, the other isolation parameters are chosen to be: $\epsilon_{min} = 0.8$, $E_T^{raw,\,min} = 20\,\text{GeV}$, and $E_T^{raw,\,max} = 60\,\text{GeV}$. The vertical dashed lines highlight the boundaries of the piece-wise definition of the efficiency parametrization

$$\epsilon(E_T^{raw}) \equiv \begin{cases} \epsilon_{min} & \text{if } E_T^{raw} < E_T^{raw,\,min} \\ m \cdot E_T^{raw} + q & \text{if } E_T^{raw,\,min} \leq E_T^{raw} \leq E_T^{raw,\,max} \\ 1 & \text{if } E_T^{raw} > E_T^{raw,\,max} \end{cases} \tag{3.6}$$

Given the three relaxation parameters defined above, two other parametrizations can be fully defined without any increase of complication. The first is the parabolic relaxation defined in Eq. 3.7, and the second is the sigmoid relaxation established in Eq. 3.8. In both cases, a sharpness factor $K \in [-\infty, +\infty]$ is introduced to regulate the steepness of the parabola and of the sigmoid.

$$\epsilon(E_T^{raw}) \equiv \begin{cases} \epsilon_{min} & \text{if } E_T^{raw} < E_T^{raw,\,min} \\ m \cdot E_T^{raw} + q & \\ \quad + K \cdot (E_T^{raw} - E_T^{raw,\,min})(E_T^{raw} - E_T^{raw,\,max}) & \text{if } E_T^{raw,\,min} \leq E_T^{raw} \leq E_T^{raw,\,max} \\ 1 & \text{if } E_T^{raw} > E_T^{raw,\,max} \end{cases} \tag{3.7}$$

$$\epsilon(E_T^{raw}) = \frac{1 - \epsilon_{min}}{1 + \exp -(E_T^{raw} - \frac{E_T^{raw,\,max} + E_T^{raw,\,min}}{2}) \cdot K} + \epsilon_{min} \tag{3.8}$$

At this stage, it is easy to appreciate that the selection of the relaxation scheme can be transformed into a systematic sampling of the relaxation parameters, i.e. ϵ_{min}, $E_T^{raw,\,min}$, $E_T^{raw,\,max}$, and K. An example of some of the tested combinations is reported in Fig. 3.13, where $\epsilon_{min} = 0.8$, $E_T^{raw,\,min} = 20\,\text{GeV}$, and $E_T^{raw,\,max} = 60\,\text{GeV}$, while the sharpness factor K is modified to highlight its impact on the parametrization.

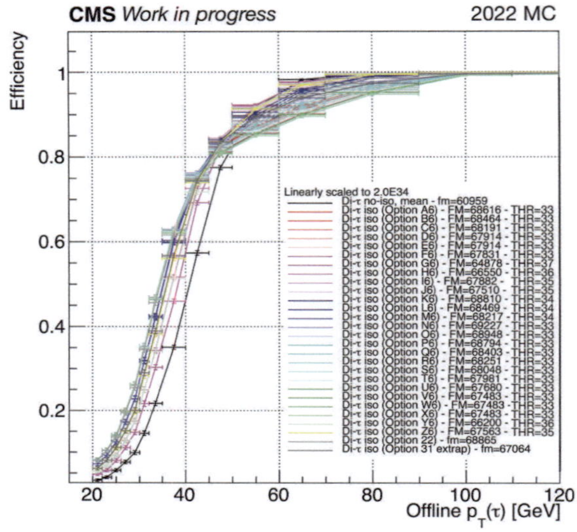

Fig. 3.14 Level-1 τ_h efficiency as a function of the offline τ candidate p_T obtained using some of the relaxation options explored in the grid-search approach described in this Section

The exploration of the relaxation parameters phase space can then be performed with a simple grid search approach. The probed points are defined as follows:

- $\epsilon_{min} \in \{0.1, 0.2.0.3, 0.4, 0.6, 0.6, 0.7, 0.8, 0.9\}$
- $E_T^{raw,\,min} \in \{10, 13, 16, 19, 22, 25, 28, 31, 34, 37, 40, 43, 46\}$ [GeV]
- $\Delta \in \{15, 18, 21, 24, 27, 30, 33, 36, 39, 41, 44, 47, 50, 53, 56, 59, 61, 64\}$ [GeV]
- $K \in \{0.1, 0.2, 0.5, 0.75, 1.0, 2.0\}$

where the third isolation parameter $E_T^{raw,\,max}$ is substituted by the difference $\Delta \equiv (E_T^{raw,\,min} - E_T^{raw,\,max})$ without loss of generality. Considering all the combinations of parameters, the total points explored in the relaxation parameters phase space is 21060, while only a limited number could be explored with the former method. Probing this enormous amount of combinations is performed in roughly 24 h of computing time when the exploration is run sequentially and under six hours when the grid search is parallelized. An example of the L1 τ_h identification efficiency as a function of the offline τ candidate p_T, the so-called *efficiency turnon*, obtained with few of the explored points in the relaxation parameters phase space, is reported in Fig. 3.14, showcasing how the visual identification of the best relaxation configuration is impossible.

In order to fully exploit this new grid-search approach, a last ingredient is required: a systematic evaluation of the performance of one isolation option over the others. This estimation can be done based on the L1 τ_h selection efficiency as a function of the offline p_T, obtained for each explored isolation option. To be considered as a good isolation option, the associated efficiency turnon, computed at a fixed L1 rate, must satisfy the following requirements:

Table 3.2 Parameters corresponding to the different isolation options used for the relaxation of the isolation in the 2022 and 2023 Run-3 data-taking. ϵ_{min} is the minimal efficiency, $E_T^{raw,\,min}$ is the start of the relaxation, $E_T^{raw,\,max}$ is the point where 100% efficiency is reached, and K is the factor regulating the sharpness of the relaxation. The full definitions of the parametrizations are given in Eqs. 3.4–3.8

Year	ϵ_{min} (%)	$E_T^{raw,\,min}$ (GeV)	$E_T^{raw,\,max}$ (GeV)	K	Parametrization
2022	90	22	72	0.0	Linear
2023	90	16	76	0.0	Linear

- $\epsilon > 0.98$ for $p_T > 95$ GeV to ensure high plateau efficiency,
- $\epsilon > 0.95$ for $p_T > 75$ GeV to guarantee fast efficiency onset,
- $\epsilon < 0.05$ for $p_T < 20$ GeV to reduce contamination of miscalibrated objects.

If these three conditions are met, a specific sharpness figure of merit is computed to order the isolation options in terms of performance. This figure of merit is defined as:

$$\Gamma \equiv \left(\int_{thr}^{120} \epsilon(p_T) dp_T - \int_{20}^{thr} \epsilon(p_T) dp_T \right) \cdot \mathcal{E} \sim \left(\int_{thr}^{120} \epsilon(p_T) dp_T \right)^2 - \left(\int_{20}^{thr} \epsilon(p_T) dp_T \right)^2 \tag{3.9}$$

where $\epsilon(p_T)$ is the efficiency turnon function, *thr* is the L1 threshold at fixed rate, and $\mathcal{E}$ is the total efficiency over the whole p_T spectrum. This choice is taken because the simple use of $\mathcal{E}$ would lead to the selection of an isolation scheme with high efficiency in the low p_T region, resulting in non-manageable L1 rates. At the same time, using the difference in integrals alone would give too much importance to minimising the integrated efficiency before the threshold. Finally, the product of these two terms results in a good compromise between the total acceptance and the reduction of the impact on the rate from the low p_T region. A visual representation of the sharpness figure of merit is reported in Fig. 3.15, where the yellow and green areas correspond to the $\int_{20}^{thr} \epsilon(p_T) dp_T$ and $\int_{thr}^{120} \epsilon(p_T) dp_T$ integrals, respectively; the red vertical line highlights the L1 threshold at a fixed rate.

The maximization of the Γ figure of merit allows the selection of the best performing isolation relaxation that satisfies the L1 trigger rate restrictions. Table 3.2 reports the isolation parameters selected via the approach described above. The results obtained in 2022 with this new isolation scheme are reported in Sect. 3.4. This approach has been adopted in its entirety (with minor modifications) also for the optimization of the L1 e/γ algorithm, which benefits from the same improvements.

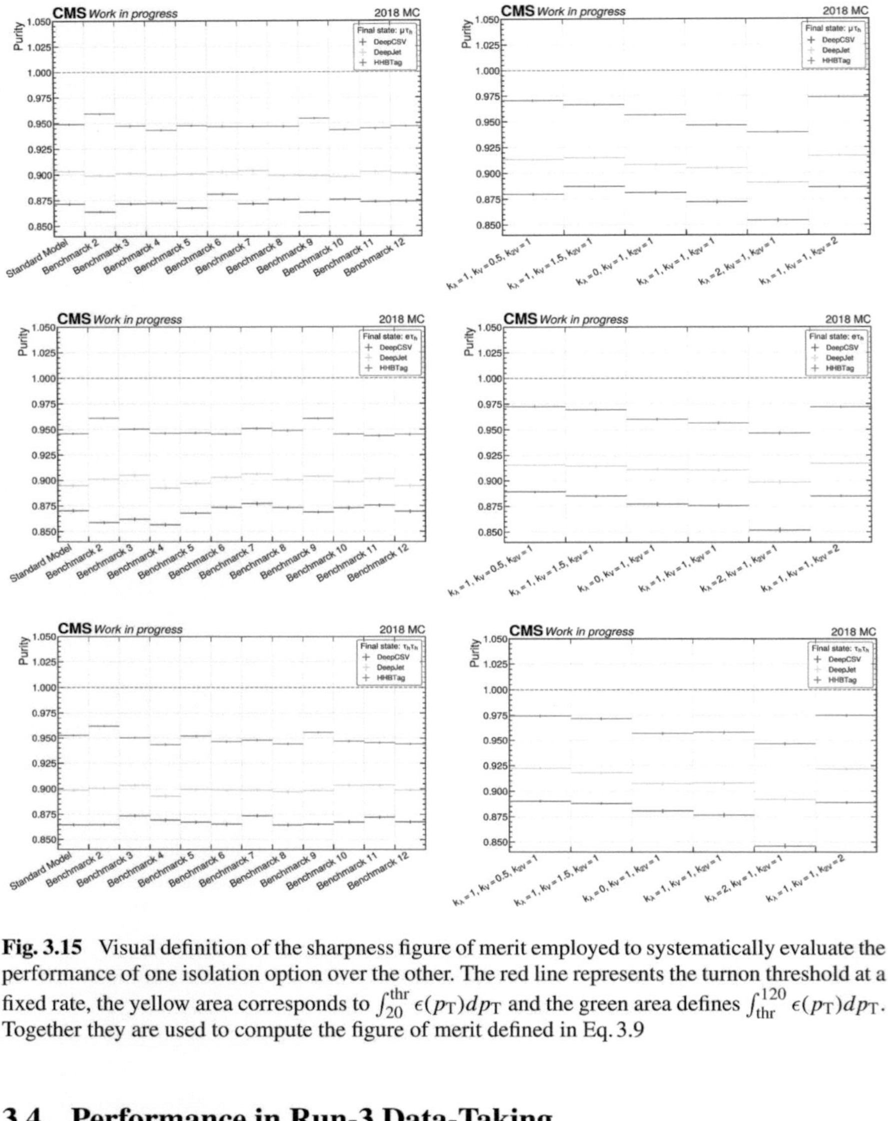

Fig. 3.15 Visual definition of the sharpness figure of merit employed to systematically evaluate the performance of one isolation option over the other. The red line represents the turnon threshold at a fixed rate, the yellow area corresponds to $\int_{20}^{\text{thr}} \epsilon(p_T)dp_T$ and the green area defines $\int_{\text{thr}}^{120} \epsilon(p_T)dp_T$. Together they are used to compute the figure of merit defined in Eq. 3.9

3.4 Performance in Run-3 Data-Taking

The Run-3 data-taking started in 2022, bringing to an end the second long shutdown of the LHC. In this new era of the LHC accelerator, the centre-of-mass energy is increased from 13 TeV to 13.6 TeV, enhancing all physics processes cross section. At the same time, the instantaneous luminosity is substantially increased, reaching peak values of $2.6 \cdot 10^{34}\,\text{cm}^{-2}\,\text{s}^{-1}$. These two factors resulted in an increase in the simultaneous interaction per bunch crossing to a mean value larger than 50 and peak values exceeding 80, reached in 2023. These facts make the data-taking conditions the harshest ever encountered, while proton-proton collisions are performed at an unprecedented centre-of-mass energy. In addition to the evolving LHC running conditions, the restart of the CMS detector has been a critical period in which all subdetectors needed to be switched-on after three years of inactivity. Moreover, the Level-1 trigger has now to operate on an ageing detector whose response degrades over time, in particular in the forward region, where the radiation damage from the first two Runs of LHC is significant.

In these demanding conditions, the high quality of the L1 τ_h algorithm is ensured by the optimization scheme described in the previous Section. To verify the achievement of this goal, the performance of the L1 τ_h trigger was measured with the data collected in 2022, allowing us to monitor the robustness of the algorithm. The results presented in this Section are obtained by analyzing the full dataset collected in 2022 at $\sqrt{s} = 13.6\,\text{TeV}$ passing the *golden* certification, i.e. all CMS sub-detectors were fully operational for all kinds of use in physics analysis, corresponding to $34\,\text{fb}^{-1}$. In the following, the so-called *tag-and-probe* method is presented, followed by the presentation of the position and energy resolution, the L1 τ_h selection efficiency, and the L1 trigger rates.

3.4.1 Level-1 τ_h Tag-and-Probe

The evaluation of the performance of the L1 τ_h trigger is performed on an unbiased sample of τ_h candidates from a sample of $Z \to \tau\tau \to \mu\nu_\mu\nu_\tau\tau_h\nu_\tau$ events selected with the tag-and-probe technique. The decay of the Z boson in this channel is especially appropriate for L1 trigger performance measurements as the clean muon signature can be exploited to select the events, while the well-known kinematics of the $Z \to \tau\tau$ decay can be taken advantage of to reject background contamination. The tag-and-probe approach starts with the selection of a muon (the *tag*) that satisfies identification, isolation, and trigger requirements and then proceeds with the study of an associated τ_h (the *probe*) whose kinematics is compatible with the decay of the Z boson. The ability to select the τ_h candidate without any trigger requirement makes the τ_h sample fully unbiased and suited to performance evaluation.

The event selection is adapted to Run-3 from the basic criteria of the Run-2 $H \to \tau\tau$ analysis [2]. The tag muon is required to be selected via the single-μ HLT path with a threshold of $p_T^{HLT} > 27\,\text{GeV}$, and to pass the offline selection of $p_T > 24\,\text{GeV}$ and $|\eta| < 2.1$. Furthermore, the medium identification criterion is enforced together with a tight isolation requirement. It should be noted that these selections are chosen solely to ensure the quality of the reconstructed muon candidate with a reduced presence of background events, and they are not used in any way to compute the τ_h trigger efficiency. The probe τ_h needs to fulfil the offline threshold $p_T > 20\,\text{GeV}$, it must have opposite charge with respect to the muon, and must satisfy $|\eta| < 2.1$. Moreover, the DEEPTAU identification algorithm is used to reject misidentified objects: the medium Working Point (WP) is used against jets and muons, while the loose WP is used against electrons (a detailed description of the DEEPTAU algorithm and its WPs is given in Chap. 5). In case multiple candidates are found, the most isolated one is chosen. Finally, the requirement $\Delta R(\mu, \tau_h) > 0.5$ is enforced to ensure that the tag and the probe are not reconstructed from the same Particle Flow (PF) objects.

Having selected the candidates for the tag and the probe, their kinematics can be exploited to reject background contamination. First, the invariant mass of the pair ($m_{\mu\tau_h}$) is required to be compatible with that of the Z boson decay, thus imposing the selection $40 < m_{\mu\tau_h} < 80$ GeV. The contribution from $Z \rightarrow \mu\mu$ events is further reduced by applying a lepton veto in cases where an additional lepton is found with $p_T > 20$ GeV and $|\eta| < 2.4$. The top quark-antiquark pair ($t\bar{t}$) background is suppressed by rejecting jets issued by b quarks with $p_T > 20$ GeV and $|\eta| < 2.4$. Finally, the contribution from W boson production in association with jets is reduced with the following requirement on the transverse mass of the muon:

$$m_T^{\mu} = \sqrt{\left(p_T^{\mu} + p_T^{\text{miss}}\right)^2 - \left(\vec{p}_T^{\,\mu} + \vec{p}_T^{\,\text{miss}}\right)} < 30\,\text{GeV}$$

where $\vec{p}_T^{\,\mu}$ and $\vec{p}_T^{\,\text{miss}}$ are the transverse momentum vector of the muon and the imbalance in the sum of the energy of the reconstructed PF objects in the event, respectively.

After applying these selections, a residual contamination of roughly 20% of background events is present in the sample. They mainly originate from QCD production of jets misidentified as hadronically decaying τ leptons which mostly populate the low p_T region [7].

Having selected an unbiased τ_h sample through the tag-and-probe technique, the performance is evaluated by defining a matching criterion between L1 and offline candidates. For all results, an offline τ_h is considered to be successfully reconstructed by the L1 trigger algorithm if an L1 τ_h candidate is found within an angular distance $\Delta R < 0.5$. In those cases where more than one L1 τ_h candidate is found to satisfy this requirement, the most energetic one is chosen. All results presented in the following use this approach, considering either the inclusive dataset of τ_h candidates or only those that are isolated.

3.4.2 Level-1 τ_h Position and Energy Resolution

The first performance figure to be addressed is the position resolution in the η and ϕ directions. When computing the position response, the effect of the magnetic field on the tracks of charged particles must be considered. As the magnetic field bends the trajectory of the charged hadron issued from the τ_h decay, the offline τ_h position is computed as the energy-weighted average of the position of its PF components at the entrance of the ECAL subdetector. This procedure avoids the mismatch between the direction of the offline τ_h emission at its production vertex and the position where the object impacts the ECAL surface. The position resolution, computed as the difference between the L1 and offline positions, measured in 2022 data is presented in Fig. 3.16 split in the barrel and endcap contributions. As expected,

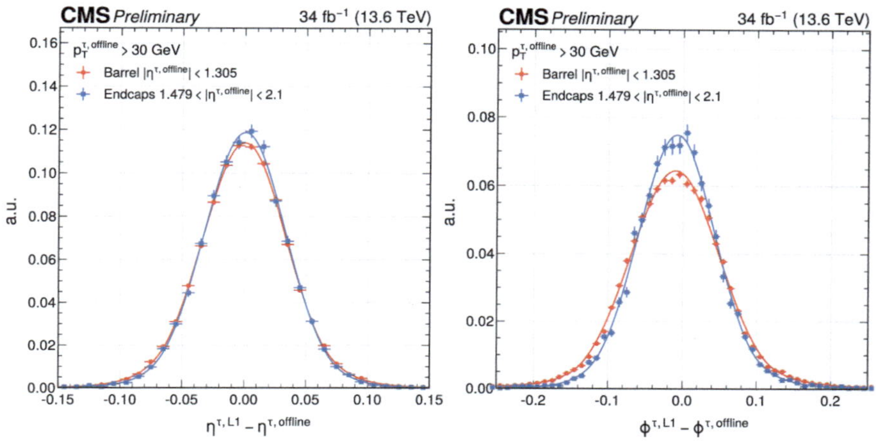

Fig. 3.16 Pseudorapidity (left) and azimuthal angle (right) positions of Level-1 τ_h candidates with respect to the offline reconstructed position, separately for the barrel and endcap regions, measured in Run-3 2022 data. Offline τ_h candidates are required to have $p_T > 30\,\text{GeV}$ to mimic a typical analysis' offline threshold. The functional form of the fits consists of a two-sided tail asymmetric Crystal Ball function

in Run-3, the position response is compatible with Run-2, showcasing a full width at half maximum of about 0.08 and 0.1 rad in η and ϕ directions, respectively. The worse response in the azimuthal angle is understood as the effect of the magnetic field bending of charged hadrons.

Having assessed the good position response, the following performance figures to be addressed are the energy response and resolution. The energy response of the L1 τ_h, defined as the ratio between the online energy and the offline reconstructed p_T, is reported on the left of Fig. 3.17, alongside the energy resolution as a function of the offline p_T. The energy resolution is defined as the root-mean-square of the response distribution divided by its mean in exclusive intervals of p_T, and is reported on the right of Fig. 3.17. The response distribution presents a long tail at low values of $E_T^{\tau,\text{L1}}/p_T^{\tau,\text{offline}}$ due to the missing Layer-1 calibration of the HCAL TPs. This tail is populated mainly by non-isolated candidates with $30 \lesssim p_T \lesssim 40\,\text{GeV}$. Nevertheless, the resolution of τ_h candidates is improved compared to the Run-2 performance; moreover, owing to the enhanced resolution of the ECAL and HCAL subdetectors at high energies, the resolution of the τ_h is reduced in the high p_T regime. The discontinuities in resolution are understood as being the product of the intrinsic discontinuous nature of the calibration LUT, which can present sharp changes in the calibration factor in adjacent bins.

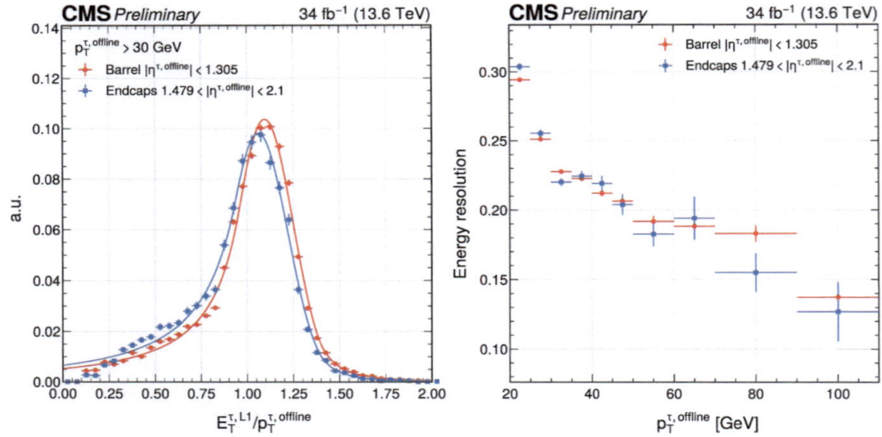

Fig. 3.17 Level-1 τ_h trigger energy response with respect to the offline reconstructed p_T, separately for the barrel and endcap regions, measured in Run-3 2022 data; offline τ_h candidates are required to have $p_T > 30$ GeV to mimic a typical analysis' offline threshold; the functional form of the fits consists of a two-sided tail asymmetric Crystal Ball function (left). The Level-1 τ_h trigger energy resolution, as a function of the $p_T^{\tau,\,\text{offline}}$, estimated by the root-mean-square of the energy response distribution divided by its mean, in bins of $p_T^{\tau,\,\text{offline}}$, measured in Run-3 2022 data (right). The discontinuity in the endcap energy resolution at $p_T \approx 35$ GeV is due to the discontinuous nature of the calibration LUTs

3.4.3 Level-1 τ_h Selection Efficiency

The most important performance figure is the reconstruction efficiency attained by the L1 τ_h algorithm. The efficiency is computed as the ratio of the number of successfully reconstructed L1 τ_h candidates over the total number of offline candidates, and it is generally computed in exclusive intervals of the offline p_T. The typical efficiency shape as a function of p_T, the turnon curve, is reported in Fig. 3.18 for inclusive and isolated τ_h candidates. The efficiency is shown in both panels for three typical energy thresholds used in L1 τ_h seeds. The excellent response resolution shown above ensures a sharp onset of the efficiency, which reaches a flat plateau of 100% efficiency also in the presence of the isolation requirement, as a result of the isolation relaxation at high E_T. At the L1 threshold of 34 GeV, the lowest unprescaled available in 2022, the 90% efficiency point is reached at 52(60) GeV for the inclusive (isolated) candidates.

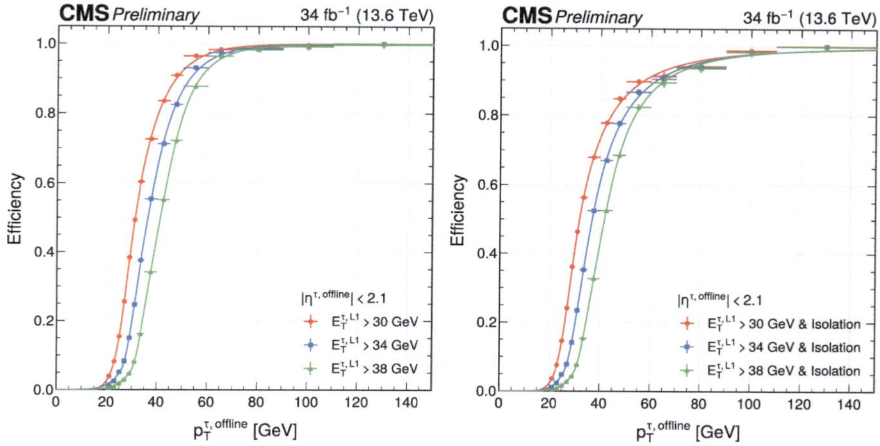

Fig. 3.18 Level-1 τ_h trigger efficiency as a function of the offline reconstructed p_T for three typical thresholds on the Level-1 trigger τ_h candidate (left) and Level-1 isolated τ_h trigger efficiency as a function of the offline reconstructed p_T for the same thresholds (right), measured in Run-3 2022 data; the functional form of the fits consists of a cumulative Crystal Ball function convolved with an arc-tangent in the high p_T region

To test the robustness of the L1 τ_h trigger, the other two important performance figures are the efficiency as a function of the pseudorapidity coverage and as a function of the number of reconstructed vertices. These results are presented in Fig. 3.19. A reasonably flat behaviour is reached in both cases for offline τ_h with $p_T > 40$ GeV. The efficiency exceeds 80% across the whole η range when applying an L1 threshold of $E_T > 30$ GeV, for both inclusive and isolated candidates. The perceivable inefficiency in the positive endcap is due to the missing part of the ECAL TPs caused by the water leak that affected ECAL in the summer of 2022. Analogously, a flat efficiency is attained across the whole range of offline reconstructed vertices, with only a minor inefficiency above 50 vertices corresponding to events with PU $\gtrsim$ 70.

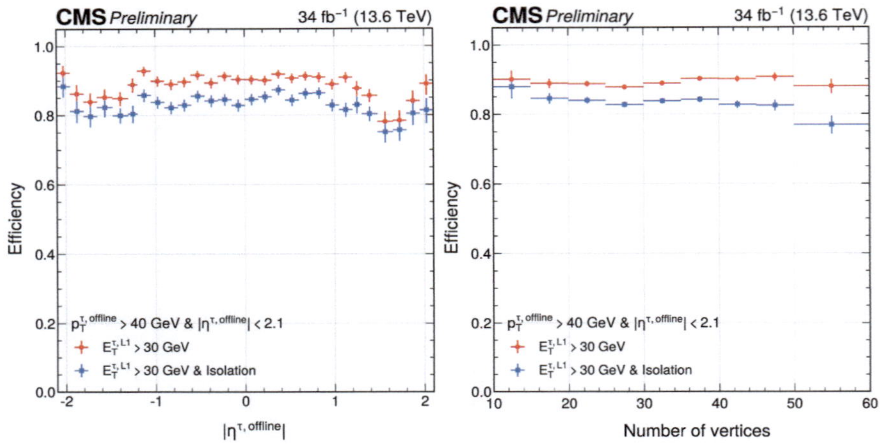

Fig. 3.19 Level-1 τ_h trigger efficiency as a function of the offline reconstructed η for a typical threshold on the Level-1 trigger τ_h candidate (left) and Level-1 τ_h trigger efficiency as a function of the number of offline reconstructed vertices for the same threshold (right), measured in Run-3 2022 data. The efficiency of both isolated and non-isolated candidates is reported in the two panels. The offline τ_h candidates are required to have $p_T > 40\,\text{GeV}$ to mimic a typical analysis' offline threshold

3.4.4 Level-1 τ_h Trigger Rates

The main L1 seeds used during the Run-3 2022 and 2023 data-king are `L1_DoubleIsoTau34er2p1` and `L1_DoubleIsoTau36er2p1`, the latter being fully shadowed by the former. The left panel of Fig. 3.20 reports the rate as a function of PU measured in 2023 data-taking. It is obtained by analyzing run number 367838 (from 2023-05-23 at 03:40:31 to 2023-05-23 at 08:48:29), with a recorded integrated luminosity of 319 pb^{-1} and a peak instantaneous luminosity of $2.0 \cdot 10^{34}\,\text{cm}^{-2}\,\text{s}^{-1}$. As it can be appreciated by comparing this plot to the one in Fig. 3.10, the introduction of the new isolation evaluation method fully solves the issue encountered in 2022, ensuring the proper linear dependence of the rate as a function of PU.

The right panel of Fig. 3.20 presents the double-τ_h rate as a function of the online L1 E_T threshold applied on both τ_h candidates. This is obtained by analyzing run number 367883 (from 2023-05-24 at 00:40:38 to 2023-05-24 at 07:28:02), with a recorded integrated luminosity of 464 pb^{-1} and a peak instantaneous luminosity of $2.1 \cdot 10^{34}\,\text{cm}^{-2}\,\text{s}^{-1}$. Both the inclusive and isolated τ_h rates are reported to highlight the effectiveness of the isolation criterion. This result is obtained from the so-called *zero bias* dataset, which is collected using a set of triggers synchronized to the LHC bunch crossing, constituting an unbiased sample of calorimeter activity.

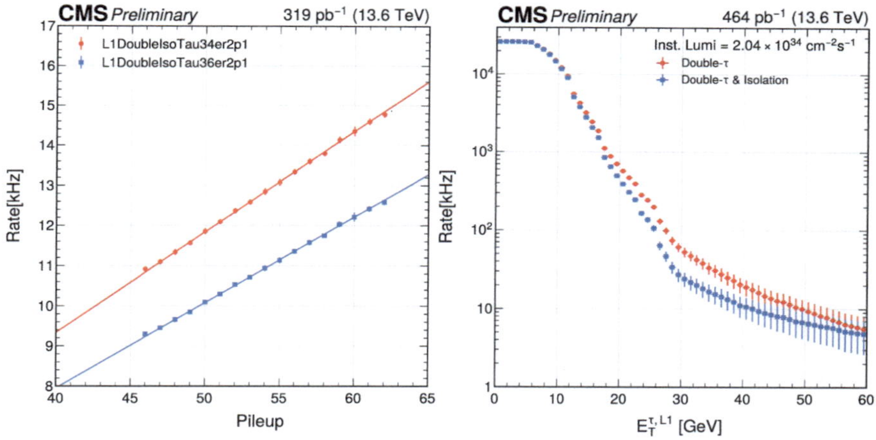

Fig. 3.20 Level-1 trigger rate as a function of pileup, evaluated in a single LHC fill, for two Level-1 τ_h trigger seeds: double isolated τ_h with $E_T > 34$ GeV and $|\eta| < 2.1$ (red), and double isolated τ_h with $E_T > 36$ GeV and $|\eta| < 2.1$ (blue). Rate and pileup are averaged over a time interval of a *luminosity section*, corresponding to 2^{18} LHC orbits or 23.31 sec of data-taking. The functional form of the fits consists of a second-degree polynomial (left). Level-1 double-τ_h trigger rate as a function of the E_T threshold applied on both candidates, for both isolated and non-isolated candidates (right). Both results are measured in Run-3 2023 data

3.4.5 Level-1 e/γ Performance

The L1 τ_h and e/γ reconstruction and identification algorithms are very similar, and their optimization procedure is designed to follow the same steps. Therefore, the improvements designed within this Thesis and detailed above have also been adopted for the derivation of the Run-3 e/γ algorithm conditions. Minor modifications were required to adapt the methods to the e/γ. Namely, the super-compression scheme used for the isolation derivation has been redefined to suit the different bit encoding of the η coordinate of e/γ, and the fiducial fit range has been consequently adapted. Moreover, the figure of merit for the ranking of the relaxation scheme has been adjusted to adapt to the sharper turnon curves of L1 e/γ candidates.

Figure 3.21 presents the performance attained by the L1 e/γ trigger algorithm during the 2022 Run-3 data-taking. The top row reports the efficiency of the e/γ reconstruction as a function of the offline electron candidate for several L1 thresholds and isolation combinations. The bottom row presents the efficiency as a function of the number of vertices and the rate as a function of the offline threshold. These results are obtained using a tag-and-probe technique analogous to the one detailed in Sect. 3.4.1 but targeting $Z \rightarrow ee$ decays. The exceptional performance of the Run-3 L1 e/γ trigger owes to the use of the optimization techniques developed within this Thesis.

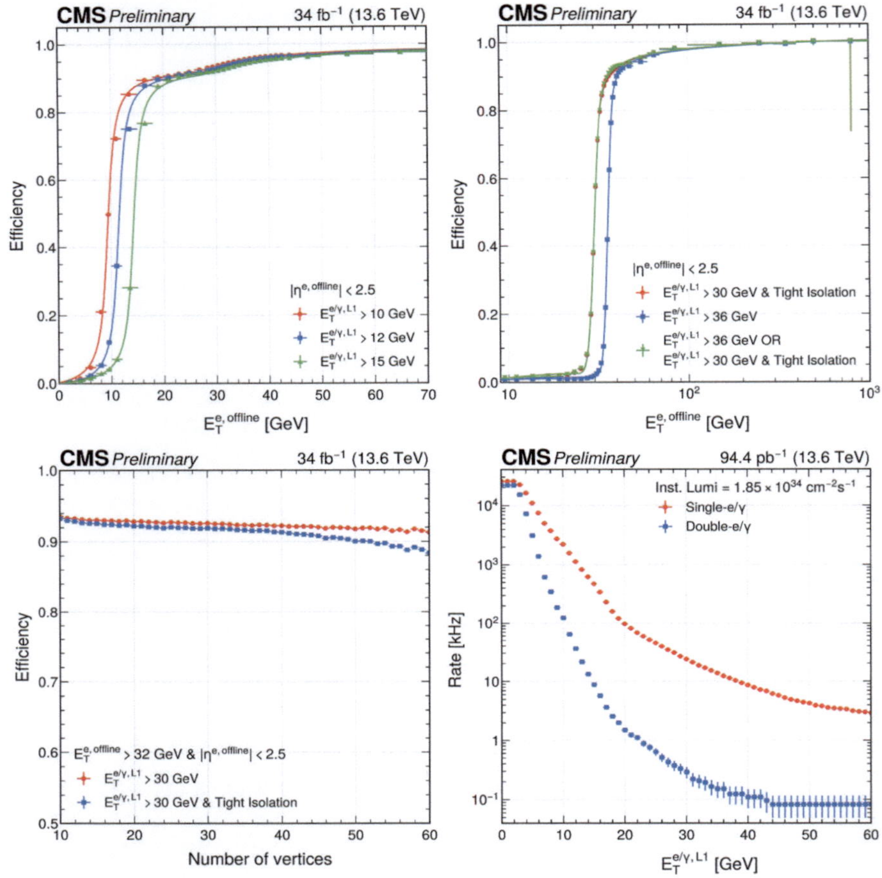

Fig. 3.21 Level-1 e/γ trigger efficiency as a function of the offline reconstructed E_T for three different low E_T thresholds on the Level-1 trigger e/γ candidate, the functional form of the fits consists of a convolution of two arc-tangent functions; the increase in plateau efficiency at $E_\mathrm{T} \approx$ 30 GeV is due to the HCAL over ECAL energy deposit fraction selection which, being E_T dependent, changes considerably in that energy region. (top left). Level-1 e/γ trigger efficiency as a function of the offline reconstructed E_T for two typical unprescaled algorithms, as well as their logical OR, the functional form of the fits consists of a cumulative Crystal Ball function with a polynomial tail in the low E_T region (top right). Level-1 e/γ trigger efficiency as a function of the number of offline reconstructed vertices for a typical threshold on the Level-1 trigger e/γ for both isolated and non-isolated candidates (bottom left). Level-1 e/γ trigger rate as a function of the E_T threshold for non-isolated single and double e/γ candidates; in the double-e/γ case, the same threshold is applied to both candidates (bottom right). All results are measured in Run-3 2022 data

3.5 The Present: A New Approach to Calorimeter Trigger Primitives Calibration

As highlighted above, the start of Run-3 brought a set of major changes in the LHC running conditions. To cope with these new harsher conditions, all physics objects reconstructed in the Level-1 trigger have been further optimized to maintain the excellent Run-2 performance; the optimization of the L1 τ_h objects being carefully detailed in the previous Sections. At the same time, for what concerns calorimeter objects, a lesser effort has been dedicated to the calibration of the inputs to the Layer-2 algorithm, i.e. the trigger primitives and trigger towers.

Given the stacked architecture of Layer-1 and Layer-2, with the latter being fully dependent on the former's output, large performance improvements can be expected for all L1 calorimeter objects if a better performing Layer-1 is achieved. One important example of how a lower-level calibration can largely benefit high-level objects is the PF algorithm; in this approach, the proper calibration of the PF elements ensures that the final PF candidates are well calibrated, and their high-level calibration factors are generally close to unity. For this reason, as part of this Thesis work, I have been the leading contributor to developing a novel technique for calibrating calorimeter TPs. This method is based on a ML approach and exploits offline reconstructed electrons and jets to calibrate single TPs optimally. This ML technique is detailed in the present Section; first, a brief overview of the past approach to TP calibration is given, highlighting its critical points, then the new method is described in detail, followed by the presentation of its performance.

3.5.1 *The Layer-1 Calibration in the Past*

The current approach to deriving the Layer-1 calibration factors was developed at the beginning of Run-2 and further improved in 2018. This method applies separate calibration factors to the ECAL and HCAL TPs arriving in the calorimeter trigger Layer-1 from the detectors' electronics. The calibration factors are derived in exclusive regions of energy deposit and pseudorapidity position of the TP from MC simulated samples, as explained in the following.

The ECAL TPs calibration is derived in MC simulated events of single photon production in the ideal environment of no PU, and with the γ having a flat energy spectrum $p_T \in [0, 200]$ GeV and a flat η distribution. Each generated photon is matched to the closest ECAL TP based on a ΔR angular distance; subsequently, a cluster with extension $i\eta \times i\phi = 3 \times 3$ is built. The requirement is applied to each cluster that 90% of the cluster's energy deposit be in the central TP. Then, clusters are binned based on the energy of the central TP with binning $E_T^i \in [0, 3, 6, 9, 12, 15, 20, 25, 30, 35, 40, 45, 55, 70, 256]$ and in units of hardware absolute pseudorapidity position, i.e. 28 bins for $|i\eta| \leq 28$. In each bin, the differential distribution of the ratio $p_T^\gamma/E_T^{3\times3}$ of the generated photon p_T to the 3×3

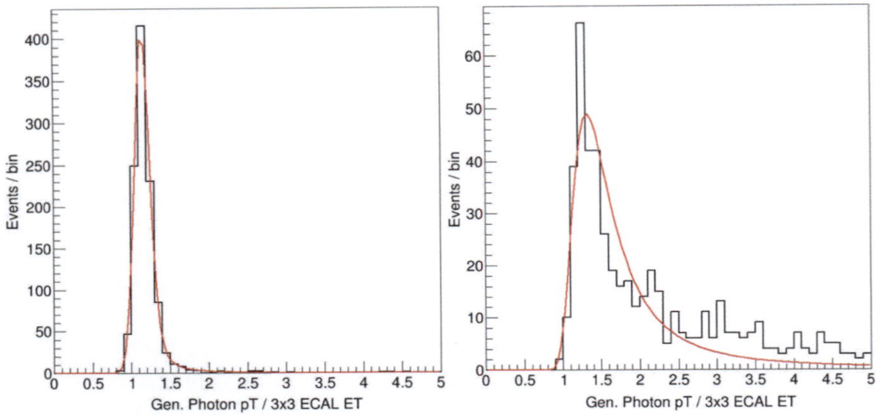

Fig. 3.22 Ratio between the generated photon p_T and the E_T deposited in an $(i\eta \times i\phi) = 3 \times 3$ trigger towers square centred around the photon, evaluated in the ECAL for $|i\eta| = 1$ (left) and $|i\eta| = 18$ (right), obtained from single-γ simulated events. The functional form of the fit (red) is a Landau pdf convolved with a Gaussian

cluster energy deposit (considering only ECAL deposits) is built and fitted with a Landau distribution convolved with a Gaussian distribution. An example of such fit is reported in Fig. 3.22 for $|i\eta| = 1$ and $|i\eta| = 18$. The calibration scale factors were originally designed to be extracted as the mean of the fit. Still, given the possible large tails in the distributions, the method was upgraded in 2018 to use the mode instead.

The calibration factors of HCAL TPs are obtained following the same procedure detailed above, using a MC sample of double charged pion production in the ideal environment of no PU, and with each $\pi^{\pm}$ having a flat energy spectrum $p_T \in [0, 200]$ GeV and a flat η distribution. The HCAL derivation differs from ECAL in the dimension of the clusters being raised to $i\eta \times i\phi = 5 \times 5$, and the central tower is required to hold only 20% of the cluster's energy deposit. Moreover, the calibrated ECAL energy deposit is summed to the uncalibrated HCAL energy deposit. For the HF calibration, the same approach of HCAL is used, but without the ECAL contribution. For both HCAL and HF, the pseudorapidity binning is of unitary width, i.e. 41 bins for $|i\eta| \leq 41 \setminus \{29\}$.

The calibration constants thus derived are stored in three firmware-compatible LUTs, one for each subdetector. In each of these LUTs, the 520 calibration constants stemming from 13 E_T bins and 40 $i\eta$ rings are encoded into 10-bit digital variables. Figure 3.23 reports the scale factors derived with this approach for both ECAL and HCAL TPs, as a function of $i\eta$, for the different bins in E_T. The increase of the calibration factors with η reflects the profile of the detector material in front of the calorimeters.

The method presented here has been used with good achievements throughout Run-2; however, it suffers from some inherent critical problems. The first issue stems from the finite statistical power of the MC samples used in the derivation of the calibrations; having to perform fits in bins of $i\eta$ an E_{T} necessitates large statistics in all exclusive regions, thus requiring the definition of a small number of coarse energy bins. The second issue resides in the use of a linear regression to obtain a single TP calibration factor from clusters of multiple TPs; this approach completely removes any correlation information between TPs. The last critical point is the highly non-trivial association between the $\pi^{\pm}$ momentum and the TP energy deposit; HCAL shower can spread beyond the $i\eta \times i\phi = 5 \times 5$ extension of the cluster, and often a large contribution of the shower in the ECAL detector is found.

These points have become particularly evident at the start of the Run-3 data-taking. The optimization of ECAL, HCAL, and HF calibrations was performed, but in the case of HCAL and HF, the new correction factors did not pass the necessary validation and closure tests. For this reason, for the Run-3 data-taking, the choice was taken to set HCAL and HF Layer-1 calibration factors to unity, thus effectively removing the needed calibration.

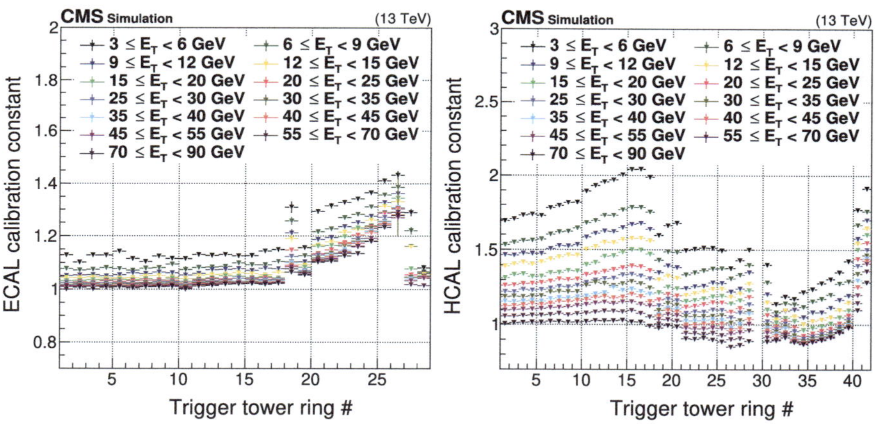

Fig. 3.23 Layer-1 energy scale factors for ECAL (left) and HCAL (right), shown for each constant-$|\eta|$ ring of Trigger Primitives (TPs). As specified in the legend, the colour of each point corresponds to a range of uncalibrated TP transverse energy values received by the Layer-1 calorimeter trigger. Because of the HCAL geometry, the signals of TPs ring 29 are divided between rings 28 and 30, and no scale factors are applied [8]

3.5.2 *New Layer-1 Calibration Using Machine Learning: the* CALIBRATON

In this Section, a proposal for a new ML technique for the derivation of Layer-1 calibration constants is presented: the CALIBRATON[1]; this method aims to resolve the three aforementioned critical points of the current approach.

The first improvement aims at removing the current assumptions on how energy is distributed within the L1 object. This is achieved by defining L1 clusters of dimension $i\eta \times i\phi = 9 \times 9$, referred to as $CL^{9\times9}$, which in the barrel roughly correspond to an angular extension of $\Delta R \sim 0.4$. This choice is taken under two intertwined considerations. Firstly, in the L1 trigger, jets are reconstructed as clusters of the same 9×9 extension to ensure maximal containment of the hadronization shower; thus, exploiting the same shape ensures minimal loss of energy deposit and a minimal bias in the evaluated calibration factors. Secondly, in the offline reconstruction, the AK4 jet algorithm constitutes the CMS standard and jets are built with an angular distance parameter of $\Delta R = 0.4$; hence, the selected L1 cluster dimension makes the online-to-offline mapping of the objects' E_T and p_T very precise.

The second improvement is the design of a dedicated Neural Network (NN) to supersede the current linear regression. The introduction of this NN can solve two issues at the same time: it removes the need for the definition of an energy binning, as all inputs are considered inclusively during the NN training, and it also introduces the ability to exploit correlations between the input TPs, which the NN can optimally learn.

One additional improvement is that this method gives the possibility of using either MC or data to derivate the calibration factors. While in the current method, the use of data is quite arduous due to the large contribution from PU that is not easily factored out in a simple linear fit approach, introducing a more complex NN architecture that can better discern between signal and PU contributions, allows us to exploit data. In this context, the target of the training is represented by offline reconstructed objects, namely electrons for ECAL TPs and hadronic jets for HCAL/HF TPs.

3.5.3 *Technical Interlude: NNs and Datasets*

In this Section, a brief overview of NNs technology is given alongside the summary of the samples used in the design of the CALIBRATON algorithm. Regarding the presentation of NNs, this discussion is not intended as an in-depth course about machine learning techniques but rather as an intuitive exposition of NNs and their power.

---

[1] The name of this method stems from a typo in the name of the folder containing the first tests that were conducted, in which the second *i* of *calibration* was missing; nonetheless, it interestingly reminds of the crunching of *calibration neuron*.

Neural Networks

The main idea behind NNs, which for the correctness of notation should be called artificial NNs, is reproducing in a software implementation the neuronal organization of the biological neural networks constituting animal brains. However, this approach to understanding NNs shadows the basic underlying concepts that were used in the same manner, but without calling them NNs, already 200 years ago. The method of least squares introduced by Legendre and Gauss is, in fact, a one-layer linear NN in which the inputs are the data points, the squared error is the metric to be minimized, and the parameters of the fit function are the quantities to be adjusted. The modern concept of a learning artificial NN (fully intended as such) is attributed to Shun-Ichi Amari's recursive NN model, based on the Lenz-Ising model, and published in 1972 [9].

The fundamental building blocks of NNs are the *artificial neurons* (in the following simply *neurons*), which are organized into layers; in each layer, a number N of neurons can be arranged so that the inputs can be passed through them. Each neuron can be mathematically represented as a unit performing the following operation:

$$y = f\left(\sum_{i=0}^{W} w_i \cdot x_i + b\right) \tag{3.10}$$

where we identify x_i as the set of inputs to the neuron, w_i the set of dimension W of weights of the neuron, b its bias term, and $f(z)$ is the so-called *activation function*. Activation functions can be either linear or non-linear, the latter option generally being preferred as the introduction of non-linearity enables the NN to capture complex relationships in data. The process of feeding the inputs to a neuron and applying Eq. 3.10 is known as *forward pass* (or *forward propagation*).

The NN's architecture is defined by the arrangement of its layers and the number of neurons within each layer. A typical NN consists of an input layer, a certain number of hidden layers, and an output layer. The forward pass computations propagate through these layers, with each neuron in a layer receiving outputs from the previous layer's neurons. The output layer can be chosen to have either one or several output values, depending on the problem at hand.

After defining a NN architecture, its training involves adjusting the weights and biases to minimize a chosen metric, quantifying the discrepancy between the network's predictions and the actual quantities that are trying to be learnt. This metric is generally referred to as *loss function*. Mathematically, the objective of the training is to adjust the weights and biases to minimize the loss function, which following Legendre and Gauss, can be defined as:

$$L(w, b) = \sum_{i=0}^{N} (y_i - \hat{y}_i)^2 \tag{3.11}$$

where N is the number of input samples generally referred to as *batches*, y_i are the truth values being learnt, and $\hat{y}_i$ are the predictions made by the NN. The minimization of the loss function is often accomplished through a dedicated optimizer in what is called the *backward pass* (or *backward propagation*). In this process, the loss function is derived with respect to weights and biases for fixed input-output pairs, and the trainable parameters are updated in the opposite direction to the gradient, thus minimizing the loss.

The process of training a NN is then the iterative repetition of the following schedule: forward pass, loss computation, gradient evaluation, and parameters update in the backward pass; each unfolding of the iterative training schedule is generally referred to as an *epoch*. The more the epochs performed, the better the fitting of the NN to the input dataset. Figure 3.24 gives a schematic overview of a typical NN architecture and its training process.

It is worth noticing at this point that the NN architecture is directly related to the degree of complexity of the NN's *fit function*. To a first approximation, the number of layers of NN corresponds to the degree of the polynomial function, and the number of neurons in each layer regulates the number of monomials of the same degree. The far-reaching power of NNs can be appreciated from this simple introduction. The arbitrary complexity of the NN architecture drives the complexity of the fit function, thus allowing the learning of the most intricate features of a dataset.

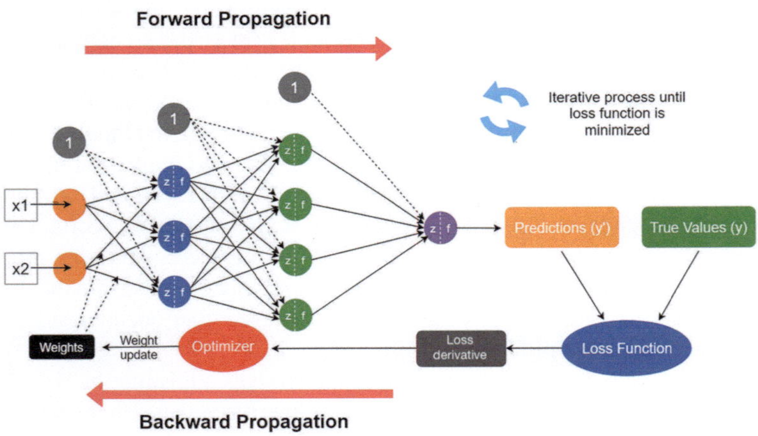

Fig. 3.24 Schematic representation of a typical Neural Network (NN) architecture, alongside its basic training steps. The input data is fed to the NN, which processes it in the forward propagation to give a prediction; the loss function is computed, encoding the discrepancy between prediction and truth values; in the backward propagation, the loss is derived and the trainable parameters are updated according to an optimizer that minimizes the loss function

Dataset Overview

The design and optimization of the CALIBRATON algorithm presented in this Section are performed solely based on data collected during the CMS 2022 Run-3 data-taking. Two separate samples are used for the derivation of the ECAL and HCAL/HF calibration constants; in both cases, the dataset corresponds to samples recorded using specific HLT trigger paths. For the derivation of the ECAL calibration factor, the so-called `EGamma` dataset is used, which is registered using paths requiring the presence of electron or photon candidates. Conversely, for the derivation of the HCAL calibration, the so-called `JetMET` dataset is used, which is recorded by requiring the presence of jet or missing transverse energy candidates. For the `EGamma` dataset, the whole set of data recorded in data-taking periods (Eras) E, F, and G is used; this corresponds to an integrated luminosity of $27.3\,\text{fb}^{-1}$. For the `JetMET` dataset, only the data from run number 362617 (from 2022-11-24 at 11:07:15 to 2022-11-24 at 12:41:49) is used, corresponding to an integrated luminosity of $106\,\text{pb}^{-1}$. In all cases, the `ZeroBias` dataset is used for the evaluation of rates. The validation of the performance is conducted using an orthogonal set of data corresponding to run number 362616 (from 2022-11-24 at 08:45:47 to 2022-11-24 at 11:04:40) amounting to an integrated luminosity of $156\,\text{pb}^{-1}$. Under all circumstances, only data that passed the golden certification is used. A summary of the datasets used for the CALIBRATON training is reported in Table 3.3.

Table 3.3 List of datasets used in the design of the CALIBRATON algorithm, the run intervals considered and corresponding integrated luminosity. The samples are separated between those used for the training of the neural network and those used for the performance evaluation

Use	Dataset	Run range	Int. lumi
Training	`EGamma`		
	Era E	359002–360331	$5.9\,\text{fb}^{-1}$
	Era F	360390–362167	$18.2\,\text{fb}^{-1}$
	Era G	$362433–362760 \setminus \{362616\}$	$3.2\,\text{fb}^{-1}$
	`JetMET`		
	Era G	362617	$106\,\text{pb}^{-1}$
	`ZeroBias`		
	Era G	362433–362760	$3.2\,\text{fb}^{-1}$
	`EGamma`		
Performance	`JetMET`		
	Era G	362616	$156\,\text{pb}^{-1}$

3.5.4 Algorithm Architecture and Training

The calorimeter trigger Layer-1 performs the calibration of the TPs from ECAL, HCAL, and HF. The geometrical scheme of calorimeter TPs and their encoding into digital quantities was detailed in Sect. 3.2.1; a summary of the necessary information useful for the following discussion is reported here. The entire solid angle coverage of the calorimeters is organised into TPs; the pseudorapidity range $0 \leq |\eta| \leq 5.1$ is segmented into 41 partitions, while the azimuthal angle range $-\pi \leq \phi < \pi$ is organised into 72 sections. Each TP is identified by its discrete Cartesian coordinates $|i\eta| \in [0, 41] \setminus \{29\}$ and $i\phi \in [0, 72]$; the missing TP 29 is caused by hardware constraints of the Layer-1, which splits its energy deposit evenly into the two adjacent TPs. In each TP, the energy deposit is encoded into eight bits with LSB 0.5 GeV, therefore in the following, the notation $i E_\mathrm{T}$, with $i E_\mathrm{T} \in [0, 255]$, will identify the hardware-encoded energy deposit in each TP. In the remainder of this Chapter, the nomenclatures *trigger primitive* will be substituted by the more general *trigger tower* for simplicity of treatment.

In the CALIBRATON, the L1 objects are defined as a TT grid of an area extending over $i\eta \times i\phi = 9 \times 9$; the energy of each object is defined as the sum of the $i E_\mathrm{T}$ energy deposit in the array of 81 TTs. The target objects used to derive the calibration factors are the offline reconstructed electrons and AK4 jets. The CALIBRATON strategy is then fully determined by the following equation:

$$E_\mathrm{T}^\mathrm{L1} = \sum_{j=0}^{8}\sum_{k=0}^{8} \chi\left(i\eta^{(j,k)}, i E_\mathrm{T}^{(j,k)}\right) \cdot i E_\mathrm{T}^{(j,k)} \overset{NN}{\longmapsto} p_\mathrm{T}^\mathrm{reco} \tag{3.12}$$

where E_T^L1 is the energy deposit of the L1 object and the sums run on its array of 81 TTs, χ are the calibration factors to be evaluated, $i\eta^{(j,k)}$ and $i E_\mathrm{T}^{(j,k)}$ are respectively the pseudorapidity position and the energy deposit of a single TT, and $p_\mathrm{T}^\mathrm{reco}$ is the offline reconstructed target. The χ factors are by construction a function of $i\eta$ and $i E_\mathrm{T}$, but not of $i\phi$; therefore, all TT lying on an $i\phi$-ring with a specified $i\eta$ value will undergo the same calibration. The CALIBRATON is trained to map the L1 energy deposit onto the offline reconstructed p_T, effectively learning each tower's χ calibration factor.

The architecture of the CALIBRATON is reported in Fig. 3.25. The basic building block is the Trigger Tower Predictor (TTP), a shallow NN taking as input the energy deposit and $i\eta$ position of a single TT. The TTP is cloned 81 times, one for each TT in the $\mathrm{CL}^{9\times9}$, with each clone sharing the same trainable parameters. The outputs of the 81 TTP clones are summed in a non-trainable summation layer to form the Predictive Network Model (PNM) which predicts the calibrated L1 energy $\hat{E}_\mathrm{T}^\mathrm{L1}$. The PNM is finally cloned twice to obtain the full CALIBRATON model; each clone of the PNM receives separate inputs and predicts separate outputs, which are used for the regression of the calibration factors and the computation of a rate proxy, as detailed in the following. The notation $\hat{E}_\mathrm{T}$ is used to identify the prediction made by the NN.

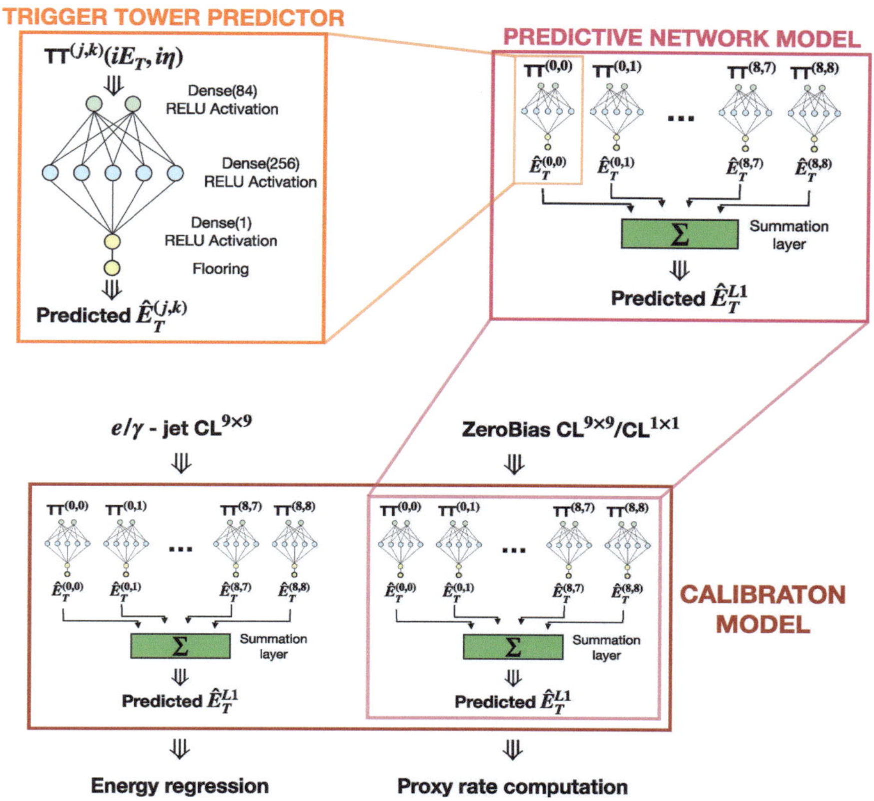

Fig. 3.25 Visual representation of the CALIBRATON algorithm architecture. The basic building block is a shallow neural network with the hyperparameters reported in the figure and called Trigger Tower Predictor (TTP); the TTP is cloned 81 times, one per each TT in the raw L1 object, to build the Predictive Network Model (PNM) which predicts the calibrated $\hat{E}_T^{L1}$; the PNM is cloned twice to obtain the full CALIBRATON model. The $CL^{9\times9}$ is the main input to the model and identifies the clusters from electrons, hard jets, and zero bias jets candidates. The model has two outputs, one used for the regression of the calibration constants and one for the rate evaluation. In the case of the rate evaluation, special $CL^{1\times1}$ clusters are defined for e/γ candidates as described in the text

The CALIBRATON algorithm is implemented in Keras [10] with a TensorFlow backend [11]. The architecture of the NN model entails the following:

- Trigger Tower Predictor
  - ◇ a fully connected (or dense) layer with 84 neurons and Rectified Linear Unit (ReLU) [12] activation function,
  - ◇ a dense layer with 256 neurons and ReLU activation function,
  - ◇ a dense layer with one neuron and ReLU activation function,

- ◇ a custom layer flooring the output of the previous layer in units of $i E_T$ (i.e. 0.5 GeV precision),
- ◇ in all layers, the neurons' bias parameters are inhibited.

- Predictive Network Model
  - ◇ 81 clones of the TTP, all sharing the same trainable parameters,
  - ◇ one non-trainable summation layer, which adds the outputs of the 81 TTP clones.

This unique architecture results from a comprehensive synthesis of considerations drawn from physics, technical constraints inherent to the NN implementation, and the calorimeter trigger Layer-1 hardware restrictions. As detailed in the following, the same considerations also pose the necessity for a specific pre-processing of the $CL^{9\times9}$ inputs.

The most unique architectural choice of the CALIBRATON model is the extensive use of NN clones. This choice stems from the intrinsic problem that is trying to be solved. The goal of the CALIBRATON is learning to predict the best χ calibration factors for each TT. To do so, the NN is actually trained to perform the mapping in Eq. 3.12, where the sum of the predicted energy deposit in an array of 81 TT is mapped to the offline reconstructed p_T. In this process, the NN should learn to calibrate each TT independently rather than the entire $CL^{9\times9}$ object. This is achieved by having 81 clones of the TTP, all sharing the same trainable parameters, and feeding single TT information to each of the clones. This design choice also facilitates the derivation of calibration factors that are compatible with the current calorimeter Layer-1 firmware, as further detailed in the following.

The second consideration regards the discrete nature of the TTP input variables. Each TT carries two tiers of information: the energy deposit in $i E_T$ units and the Cartesian coordinate $i\eta$ (the second coordinate $i\phi$ is dropped as the calibration factors are computed per $i\phi$-ring), which are both discrete quantities by construction. Nevertheless, given the adjacency of the values they can assume, the NN will interpret them more likely as continuous variables rather than discrete ones. This is solved with two independent approaches in the two cases. To ensure that the $i\eta$ coordinate is interpreted as a discrete variable, a *categorical one-hot encoding* is performed. This technique foresees the encoding of a variable into a string of bits, in which only a single bit is allowed to be active, i.e. be set to 1. In the specific case of the CALIBRATON algorithm, the $i\eta$ coordinate is encoded into a string of 40 bits, and only the bit of j-th position is set to 1 if $i\eta = j$. This efficiently covers the range $|i\eta| \in [0, 41] \setminus \{29\}$ and transforms the Cartesian coordinate of TTs into a categorical variable, thus allowing for less smooth inferences at changing values of $i\eta$. While the discreteness of $i\eta$ is enforced at input time, the one of $i E_T$ is implemented at output time: the last layer of the TTP is a custom-built layer which performs the flooring of the predicted TT energy in units of $i E_T$, i.e. $i E_T^{\text{floored}} = \lfloor \hat{E}_T \rfloor$. The process of flooring a quantity is, by definition, a non-derivable operation; therefore, the flooring layer is implemented to transfer the loss derivative without acting on it.

The third consideration regards the type of information that the NN learns. As introduced above, the input to the TTP is a single TT's iE_T and $i\eta$, with the pseudorapidity being on-hot encoded. The CALIBRATON aims to find the optimal calibration factor for each TT's energy deposit; therefore, no learning should be allowed from non-active TTs. Three separate approaches are used and collectively referred to as *zero suppression* to enforce this. The first is the complete removal of information from TTs with no energy deposit by setting the entire input array to null values; this prevents the NN from learning about the position of non-active TTs. The second approach is the removal of the neurons' bias terms; as per Eq. 3.10 even if all inputs are set to 0, a neuron's output can still be different from 0 owing to the bias parameter: removing it effectively suppresses all propagation of information from null-input TTs. The third method is the selection of a proper activation function; from Eq. 3.10, we can appreciate that neurons' output can be different from 0 if the activation function at 0 has a non-null value. The ReLU is an activation function defined as the positive part of its argument, i.e. $f(x) = \max(0, x)$, thus properly ensuring the needed behaviour of each neuron. The use of the ReLU function also presents the advantage of ensuring the strict positiveness of the predicted energy $\hat{E}_T$.

The final consideration is based on how the application of calibration factors affects different physics processes. The achievement of a good calibration of physics objects in the L1 trigger translates into a more precise application of energy thresholds and, at the same time, better management of the trigger rate. These two intertwined factors are the cornerstones to reaching high L1 trigger selection efficiency of interesting events. Nevertheless, a good calibration of physics objects also means a better calibration of the objects originating from PU collisions. If not correctly taken care of, these candidates can cause a steep rise in the L1 trigger rate. The first training iterations of the CALIBRATON showed that, especially in the case of the HCAL calibration factors, attaining excellent calibration of signal objects directly translated into an escalating and unmanageable L1 trigger rate. The presented architecture with the PNM double clone was introduced to cope with this behaviour. While one clone receives inputs from signal $\mathrm{CL}^{9\times9}$s obtained from electron and jet candidates, the other receives $\mathrm{CL}^{9\times9}$s ($\mathrm{CL}^{1\times1}$s) from zero bias events. The outputs are then used in two independent pipelines of the algorithm, the former being employed for the energy regression of the offline objects p_T and the latter for evaluating a rate proxy (R). The value of R is computed in the following way:

$$\mathrm{e}/\gamma \text{ rate proxy} : R = \frac{N\left(\mathrm{CL}^{1\times1}\middle|\hat{E}_T^{\mathrm{L1}} > 25\,\mathrm{GeV} \wedge \frac{\hat{E}_T^{\mathrm{L1,\,ECAL}}}{E_T^{\mathrm{L1,\,HCAL}}} < 2^{-3(4)}\right)}{N} \tag{3.13}$$

$$\text{Jet rate proxy} : R = \frac{N\left(\mathrm{CL}^{9\times9}\middle|\hat{E}_T^{\mathrm{L1}} > 50\,\mathrm{GeV}\right)}{N} \tag{3.14}$$

thus being equal to the fraction of input zero bias L1 clusters passing the selections specified in the equations. The $\mathrm{CL}^{1\times1}$ objects are single TT clusters that mimic e/γ

candidates. The thresholds on the H/E energy fraction are chosen as the ones used in the firmware currently online: 2^{-3} and 2^{-4} for the barrel and endcap, respectively. The obtained values of the rate proxy are used to define a penalty term in the loss function, effectively reducing the PU-induced rate.

Having fully defined the architecture of the CALIBRATON, whose design choices have been comprehensively detailed above, the training of the NN is performed with the following specifications:

- batch size $N = 1024$; testing sample = 20%
- custom loss function:

$$L = \frac{A \cdot 100}{N} \cdot \sum_{i=0}^{N} |\frac{\hat{E}_T^{L1} - p_T^{reco}}{p_T^{reco}}|_i + B \cdot \sum_j w_j^2 + C \cdot \cosh\left(D \cdot \frac{R_{proxy} - R_{target}}{R_{target}}\right) \quad (3.15)$$

  where the first term corresponds to a mean absolute percentage error loss function, the second term is a regularization term computed as the squared sum of all the network's weights, and the third term is a penalty term based on the evaluation of the L1 rate proxy. The values of the parameters regulating the relative importance of the loss terms are $A = 100$, $B = 1$, $C = 1$, and $D = 1.5$. The value of R_{target} is estimated as the rate proxy computed by applying the current Layer-1 calibration constants.
- Adam minimizer [13] with learning rate 10^{-3}
- distributed training on four NVIDIA Tesla V100 Tensor Core GPUs [14] for training acceleration
- three separate training iterations for ECAL, HCAL, and HF calibration factors.

The last ingredient for performing the training of the CALIBRATON is the definition of the training and testing datasets. For the calibration pertaining to ECAL, electrons originating from the Z peak are selected, satisfying the requirement $|\eta| < 3.0$, and having an electromagnetic energy deposit fraction larger than 95%. The $CL^{9\times9}$ are built around the TT with the smallest angular distance from the offline electron and are required to satisfy $0.3 < p_T/E_T^{L1} < 3.0$; this choice prevents the NN from learning features of the small fraction of outliers. This selection leads to a total of $\sim 20 \cdot 10^6$ candidates, of which $\sim 4 \cdot 10^6$ are reserved as a testing sample to keep overfitting under control at training time. For the calibration pertaining to HCAL and HF, jets identified with the PileUp Per Particle (PUPPI) algorithm [15] are used, fulfilling the criterion $|\eta| < 5.1$, and having a hadronic energy deposit fraction larger than 80%. Moreover, jets are required to have transverse momentum $30 < p_T < 1000$ GeV and the associated $CL^{9\times9}$ must satisfy $0.3 < p_T/E_T^{L1} < 3.0$. This selection gives a total of $\sim 12 \cdot 10^6$ candidates, of which $\sim 2.5 \cdot 10^6$ are reserved for testing. The two independent training datasets for the calibration of HCAL and HF are obtained by applying the threshold $i\eta \lessgtr 29$ on the $i\eta$ position of the clusters' central TT.

3.5.5 Firmware-Compatible Calibration Factors

After the successful training of the CALIBRATON model, the χ calibration constants need to be computed. This process is devised to be fully compatible with the current calorimeter trigger Layer-1 architecture so that the use of this new technique would require minimal online implementation effort. As mentioned, the calibration constants for all three subdetectors are stored in LUTs organised on the $i\eta$ position and iE_{T} energy deposit; these LUTs are generally constrained to contain less than $2^{12} = 4096$ entries in order to reduce the hardware resources usage.

The calibration factors from the CALIBRATON model are computed using what we can call a *standard candle* approach. In this method, the trained TTP is extracted from the full model and the scale factors are evaluated as

$$\chi(i\eta, iE_{\mathrm{T}}) = \frac{i\hat{E}_T}{iE_{\mathrm{T}}} \equiv \frac{\mathrm{TTP}(i\eta, iE_{\mathrm{T}})}{iE_{\mathrm{T}}} \tag{3.16}$$

where $i\eta$ and iE_{T} are the position and energy deposit of the standard candle TT, and $i\hat{E}_T$ is the TT energy deposit predicted by the TTP. The calibration factors are derived in bins of width $i\eta \times iE_{\mathrm{T}} = 1 \times 2$ up to a maximum value of $iE_{\mathrm{T}} = 200$, after which one single energy bin is considered. Given the geometrical organization of the TT in Layer-1, this choice leads to 2828 χ constants for ECAL, 2828 for HCAL, and 1212 for HF; in all cases, the number of calibration factors is much lower than the upper limit of 4096. Figure 3.26 reports the χ factors derived with this approach for ECAL and HCAL/HF, as a function of $i\eta$, for the different bins in ET; for simplicity of display, only the calibration constants for energy deposits up to 50 GeV are shown. As already noted, the increase of the calibration factors with $i\eta$ reflects the profile of the detector material in front of the calorimeters.

In the case of ECAL, it can be appreciated how few scale factors largely deviated from the bulk of the distribution. The deviating calibration constants are all associated with low energy deposit TTs and perform a consistent downward calibration of them. This behaviour is understood under one consideration: being trained with data, the CALIBRATON is exposed to the presence of PU and ECAL noise and learns to correct for it. To obtain the best calibration when targeting the p_{T} of the offline reconstructed electron, the NN learns to effectively remove the soft and diffuse contributions from PU and ECAL noise. Moreover, a considerable drop in the calibration constants for TT 28 is observed. This behaviour is well understood since Run-2: it is caused by the large noise due to radiation damage, and already during the 2018 data-taking, a suppression scheme of TT 28 was introduced to remove energy deposits below 9 GeV. The CALIBRATON is confirming this behaviour and showcasing that in Run-3 the situation is possibly worsened, requiring the suppression of large energy deposits too.

In the case of HCAL, the calibration constants have an overall larger value compared to ECAL, understood to be caused by the initiation of the hadronic showers in the electromagnetic calorimeter. Also in this case few scale factors assume values

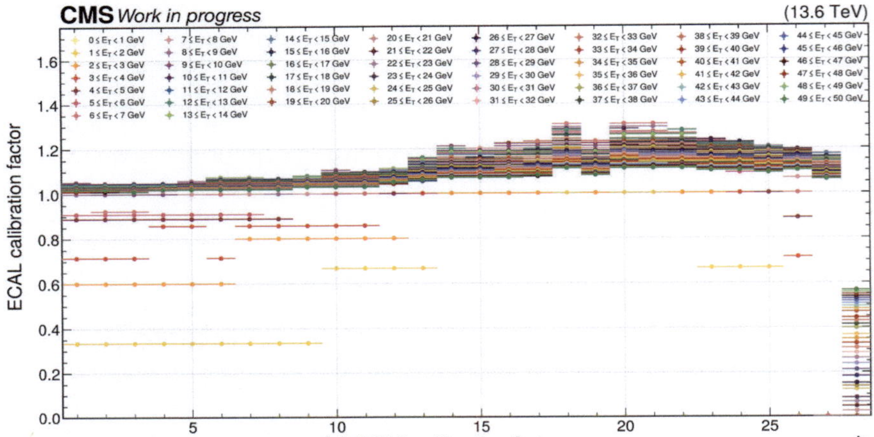

(a) ECAL calibration factors

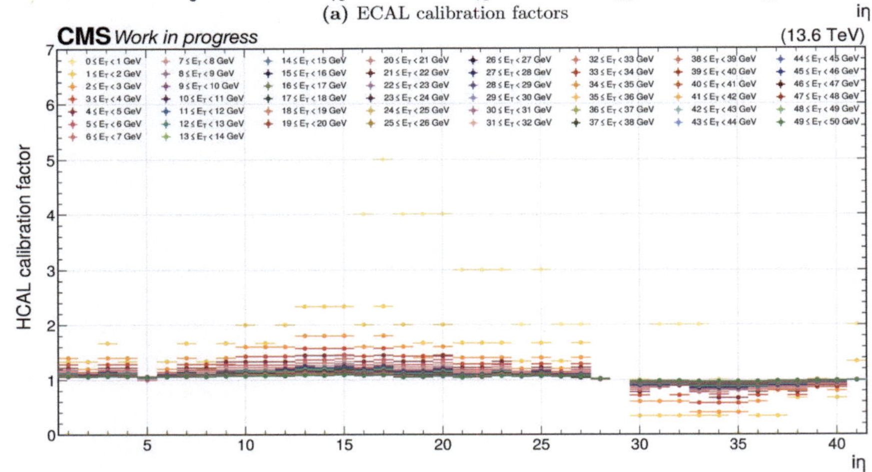

(b) HCAL and HF calibration factors

Fig. 3.26 Layer-1 energy scale factors obtained with the CALIBRATON for ECAL (top) and HCAL (bottom), shown for each constant-$|i\eta|$ ring of trigger towers. As specified in the legend, the colour of each point corresponds to a range of uncalibrated trigger primitive transverse energy values received by the Layer-1 calorimeter trigger. Because of the HCAL geometry, trigger tower ring 29 signals are divided between rings 28 and 30, and no scale factors are applied. For simplicity of display, only the calibration constants for energy deposits up to 50 GeV are shown, but the new calibration factors have a 1 GeV granularity up to 100 GeV very different from the bulk of the distribution and create a particularly interesting spike at $|i\eta| = 17$. This behaviour is understood by noticing that the position $|i\eta| = 17$ corresponds to the edge of the transition region between the barrel and endcap sections of the calorimeters. The CALIBRATON algorithm is effectively learning the presence of this edge and correcting for it. In the HF compartment, a similar situation of noise suppression to that of ECAL TT 28 can be appreciated; in this case, the suppression is of lesser magnitude, with calibration constants averaging around $\sim$0.9.

The use of this novel approach to Layer-1 calibration would represent a sevenfold increase in the granularity of the χ factors with respect to the present method. This would result in a highly increased capability of fine-tuning the energy deposit of TTs to the high benefit of the object reconstruction algorithms implemented in the calorimeter trigger Layer-2.

3.5.6 The CALIBRATON *Performance*

The performance of this new ML technique for the calorimeter trigger Layer-1 calibration of TTs is evaluated by means of the L1 trigger emulator, which replicates in C++ software the exact behaviour of the firmware implemented in the hardware in the CMS service cavern. Data from run number 362616 (from 2022-11-24 at 08:45:47 to 2022-11-24 at 11:04:40) amounting to an integrated luminosity of 156 pb^{-1} are used to this end. The performance of the CALIBRATON is evaluated by emulating the data collected in run number 362616 under three different Layer-1 configurations: one in which no calibration is applied, i.e. all $\chi = 1$; one in which the calibration factors are set to the values obtained with the old method; one in which the calibration constants are derived with the technique presented in this Thesis.

The first performance figure evaluated is the energy response of L1 objects, defined as the differential distribution of the ratio between the L1 e/γ or jet candidate and the offline reconstructed electron or PUPPI jet. The energy response performance is then evaluated by extracting two figures of merit from the response: the energy scale and energy resolution; the former is defined as the mean of the response distribution, while the latter is defined as the RMS of the distribution divided by its mean. The results are reported in Fig. 3.27 for both the L1 e/γ and PUPPI jet candidates. In both cases, the CALIBRATON ensures higher performance in both scale and resolution of the L1 candidate across the whole range of offline object transverse momentum. The L1 e/γ scale and resolution are improved by a consistent 5–10% over the entire p_T spectrum, with peaks of improvement of $\sim$15% in the low p_T regime. This advancement can greatly impact the Run-3 B-parking strategy. The L1 jet scale and resolution are improved by a consistent 10–15% over the entire p_T spectrum, with peaks of improvement of $\sim$25% in the low p_T regime.

The most important performance figure to be evaluated is the L1 trigger efficiency. The efficiency is computed as the ratio of the number of successfully reconstructed L1 candidates over the total number of offline candidates, and it is computed as a

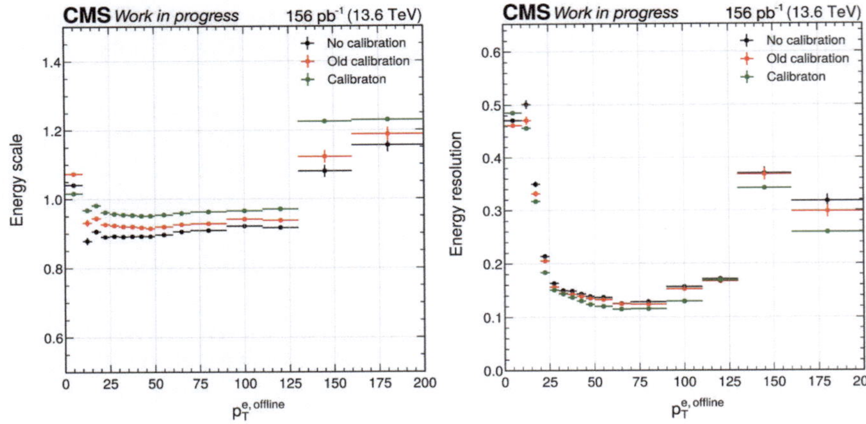

(a) Level-1 e/γ energy scale and resolution

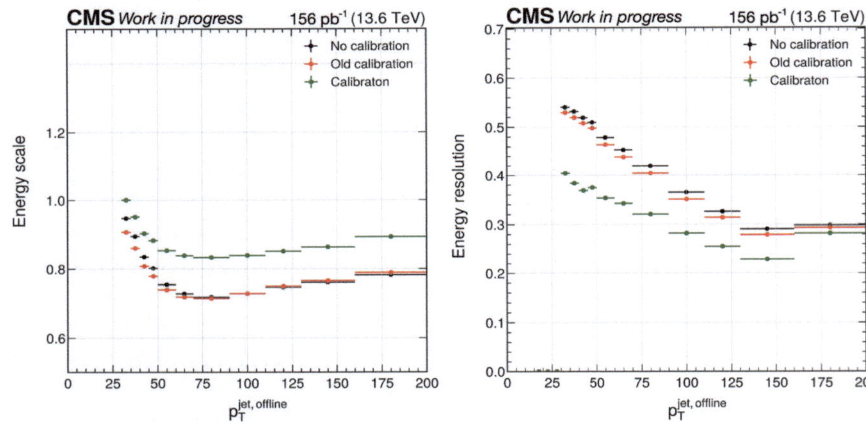

(b) Level-1 jet energy scale and resolution

Fig. 3.27 Level-1 e/γ (**a**) and jet (**b**) candidates energy scale (left) and energy resolution (right) as a function of the offline object p_T. The energy scale is defined as the mean of the response distribution (i.e. $E_T^{L1} / p_T^{offline}$), while the latter is defined as the RMS of the distribution divided by its mean. Three configurations of the Layer-1 calibration factors are shown: no calibration (black), old method (red), CALIBRATON method (green)

function of the offline objects' transverse momentum. The efficiency turnon curves are reported in Fig. 3.28 for both the L1 e/γ and PUPPI jet candidates. The efficiency curves are computed at a fixed L1 rate for the three distinct configurations of the Layer-1 calibration to ensure a fair comparison of the different methods. In the case of e/γ, the new calibration ensures sharper turnons compared to the old calibration. Especially important is the possibility of lowering the threshold in the low energy regime by ~8%, as reported on the left of Fig. 3.28a. This is a direct consequence of the better energy scale attained by the CALIBRATON algorithm over the present

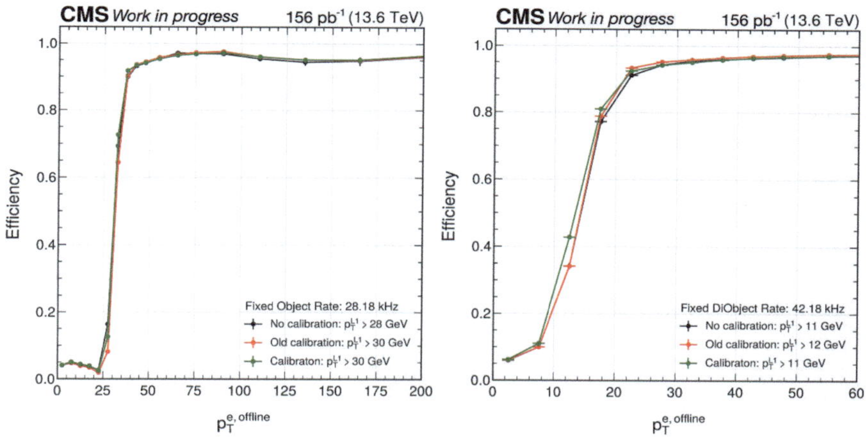

(a) Level-1 e/γ efficiency turnon

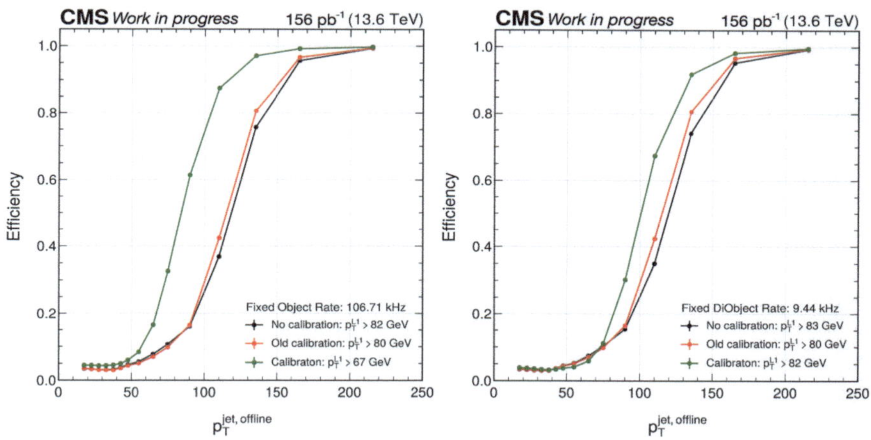

(b) Level-1 jet efficiency turnon

Fig. 3.28 Level-1 e/γ (**a**) and jet (**b**) candidates efficiency as a function of the offline object p_T. The efficiency turnons are evaluated at a fixed rate value to guarantee a fair comparison among different Layer-1 configurations. Three configurations of the Layer-1 calibration factors are shown: no calibration (black), old method (red), CALIBRATON method (green)

method. This type of improvement can be of extensive impact in the Run-3 B-physics program, which highly relies on the low threshold L1 e/γ candidates. An impressive improvement in performance is reached in the case of the L1 jet candidates, owing to the outstanding improvement in energy scale and resolution. In this case, two different efficiency turnon curves are reported: one for fixed single-jet L1 rate in Fig. 3.28b (left) and one for fixed double-jet L1 rate in Fig. 3.28b (right). In both cases, the CALIBRATON method largely outperforms the current approach.

The performance presented in this Section showcases very promising results in terms of energy resolution and efficiency curves for both electromagnetic and hadronic objects. This improvement in the input energy definition at Layer-1 holds immense potential in enhancing the L1 trigger performance during the Run-3 data-taking, thus contributing to the best success of the CMS physics program. Currently, the effect of the CALIBRATON-derived Layer-1 calibration on the object reconstruction algorithms implemented in Layer-2 is being assessed; the final Layer-2 objects' performance will be the ultimate proof or disproof of the need to transition to this new calibration method.

3.5.7 *Possible Future Applications of the* CALIBRATON

The previous Section highlighted how the CALIBRATON substantially impacts the performance of the L1 trigger, ensuring better scale and resolution over the entire p_T range considered for both electrons and jets. The improvement directly translates to a more correct application of the L1 E_T threshold, better rate management, and better efficiency turnons. Notwithstanding this very good performance, the CALIBRATON additionally has a crucial property: it is easily scalable for applications to contexts with much higher granularity. This characteristic makes it particularly well-suited for several future applications.

The first important example is the newly developed High Granularity Calorimeter (HGCAL), which will replace the current endcap calorimeter of the CMS detector at the High-Luminosity LHC (HL-LHC). A detailed description of the HGCAL design is given in Sect. 4.1.2; still, for the scope of this discussion, it is sufficient to highlight that this new calorimeter will feature roughly six million readout channels. The use of traditional techniques in this context can prove extremely challenging. Calibrating six million channels requires a much more detailed approach like the CALIBRATON, which can easily be trained to learn the correlation between the separate channels and correct for possibly faulty readouts in a seamless manner.

A second excellent application will be the Phase-2 barrel calorimeter. An in-depth discussion of the barrel calorimeter upgrade is given in Sect. 4.1.1; nevertheless, for the scope of this argument, it is sufficient to emphasize that the granularity of the trigger primitives generated by the barrel calorimeter will increase by a factor 25. Therefore, using traditional calibration methods can prove difficult due to the large number of parameters to be extracted, and the CALIBRATON can overcome this limitation.

Finally, the most exciting application could be the real-time implementation of the CALIBRATON in the Phase-2 L1 trigger. As presented in Sect. 4.1.3, the L1 trigger at the HL-LHC will undergo a substantial upgrade with the use of state-of-the-art FPGAs, ensuring a highly enhanced computing power with respect to the current system. In this context, ML algorithms will be implemented for real-time inference in FPGA firmware, and the CALIBRATON is a perfect candidate for this type of undertaking. The algorithm could precisely calibrate single detector objects based on the specific running conditions of the LHC, ensuring the best calibration of trigger objects under all PU conditions.

Future applications like those reported above will need to be commissioned. In this context, of particular interest will be the test beam environment, in which the incoming particles have a fixed and well-known energy. This will favour the CALIBRATON training, which could be performed using the known test beam energy as a regression target. Moreover, the particularly clean experimental environment could benefit the development of the algorithm itself, as parameters like the dimension of the inputs or the architecture of the network could be tuned based on the available clean signal.

References

1. Tapper A, Acosta D (2013) CMS Technical Design Report for the Level-1 Trigger Upgrade. Technical report. Additional contacts: Jeffrey Spalding, Fermilab, Jeffrey.Spalding@cern.ch, Didier Contardo, Universite Claude Bernard-Lyon I, didier.claude.contardo@cern.ch. https://cds.cern.ch/record/1556311
2. CMS Collaboration (2018) Observation of the Higgs boson decay to a pair of τ leptons with the CMS detector. Phys Lett B 779:283–316. https://doi.org/10.1016/j.physletb.2018.02.004
3. CMS Collaboration (2023) Measurements of Higgs boson production in the decay channel with a pair of τ leptons in proton–proton collisions at $\sqrt{s} = 13$ TeV. Eur Phys J C 83(7). All the figures and tables, including additional supplementary figures and tables, can be found at http://cms-results.web.cern.ch/cms-results/public-results/publications/HIG-19-010 (CMS Public Pages), p 562. https://doi.org/10.1140/epjc/s10052-023-11452-8. arXiv: 2204.12957. https://cds.cern.ch/record/2807752
4. CMS Collaboration (2022) Analysis of the CP structure of the Yukawa coupling between the Higgs boson and τ leptons in proton-proton collisions at $\sqrt{s} = 13$ TeV. J High Energy Phys 2022(6). https://doi.org/10.1007/jhep06(2022)012
5. CMS Collaboration CMS TriDAS project: technical design report, vol 1: The trigger systems. Technical design report. CMS. http://cds.cern.ch/record/706847
6. Mastrolorenzo L (2015) Search for the Higgs boson decaying into tau lepton pairs with the matrix element method and tau trigger optimization in the CMS experiment at the LHC. Thèse de doctorat dirigée par Beaudette, Florian Physique Palaiseau, Ecole polytechnique 2015. PhD thesis, vol 1, p 296. http://www.theses.fr/2015EPXX0102
7. Cadamuro L (2018) Search for Higgs boson pair production in the $b\bar{b}\tau^+\tau^-$ decay channel with the CMS detector at the LHC. Springer International Publishing. ISSN: 978-3-030-04055-0. https://doi.org/10.1007/978-3-030-04055-0
8. CMS Collaboration (2020) Performance of the CMS Level-1 trigger in proton-proton collisions at $\sqrt{s} = 13$ TeV. J Instrum 15(10):P10017–P10017. https://doi.org/10.1088/1748-0221/15/10/p10017
9. Amari S-I (1972) Learning patterns and pattern sequences by self-organizing nets of threshold elements. IEEE Trans Comput C-21(11):1197–1206. https://doi.org/10.1109/T-C.1972.223477
10. "Keras website" (2023) (last visited: 26 Aug 2023). https://keras.io
11. "TensorFlow website" (2023) (last visited: 26 Aug 2023). https://tensorflow.org
12. Fukushima K (1969) Visual feature extraction by a multilayered network of analog threshold elements. IEEE Trans Syst Sci Cybern 5(4):322–333. https://doi.org/10.1109/tssc.1969.300225
13. Kingma DP, Ba J (2017) Adam: a method for stochastic optimization. arXiv: 1412.6980 [cs.LG]

14. NVIDIA (2023) NVIDIA Tesla V100 Tensor Core GPU. (last visited: 9 Aug 2023). https://www.nvidia.com/en-us/data-center/v100/#:~:text=NVIDIA%C2%AE%20V100%20Tensor%20Core,CPUs%20in%20a%20single%20GPU
15. Bertolini D et al (2014) Pileup per particle identification. J High Energy Phys 2014(10). https://doi.org/10.1007/jhep10(2014)059

Chapter 4
The Level-1 τ_h Trigger: From the Present, to the Future

The High-Luminosity LHC (HL-LHC) is scheduled to start in 2029, and it will constitute the Phase-2 of the LHC operations. It is designed to operate at a centre-of-mass energy of 14 TeV while delivering an instantaneous luminosity in the range of $5-7.5 \cdot 10^{34}\,\text{cm}^{-2}\,\text{s}^{-1}$. These conditions correspond, in the ultimate HL-LHC configuration, to a number of simultaneous collisions per bunch crossing of $\mathcal{O}(200)$, a fluence of up to $10^{16}\text{n}_{\text{eq}}\,\text{cm}^{-2}$, and a dose reaching 2 MGy. In these unprecedented running conditions, a remarkable integrated luminosity of $4000\,\text{fb}^{-1}$ is expected to be collected over the anticipated ten years of data-taking.

This unparalleled dataset will open a unique window on the weak-scale nature of the Universe, providing high-precision measurements of the Standard Model (SM) as well as searches for new physics Beyond the SM (BSM). The study of the Higgs boson (H) self-coupling (λ_{HHH}) represents the most important target of the HL-LHC. Its possible measurement can be strived for via the direct search of Higgs boson pair (HH) production in the HH $\rightarrow$ bb$\tau\tau$, HH $\rightarrow$ bbbb, and HH $\rightarrow$ bb$\gamma\gamma$ channels, which hold the highest sensitivity, and via indirect searches profiting from the electroweak corrections to single H production [1]. Moreover, the study of rare processes like the $\text{t}\bar{\text{t}}\text{t}\bar{\text{t}}$ production in leptonic final states will be of utmost importance to constrain the magnitude and Charge-Parity (CP) properties of the top Yukawa coupling, as well as probe 2HDM BSM models [1]. An important test of CP violation is represented by the *golden channel* $B_s \rightarrow J/\psi\,\phi(1020)$, with an expected uncertainty on CP-violating phase $\phi_s \sim 5-6$ mrad in Phase-2 [1]. Further studies of BSM physics include the search for lepton flavour non-universality in the $\tau \rightarrow \mu\mu\mu$ decay [2] and the search for long-lived particles [1]. This will be complemented by a deep study of the H boson properties [3].

Such precision measurements and searches require information-rich datasets with statistical power that matches the high luminosity provided by the Phase-2 upgrade of the LHC. The achievement of this goal means not only maintaining the current discovery potential of the CMS detector but also extending its physics reach. To attain this, the CMS Collaboration plans a series of major upgrades of its subdetectors'

J. Motta, *Development of Machine Learning τ Trigger Algorithms and Search for Higgs Boson Pair Production*, Springer Theses,
https://doi.org/10.1007/978-3-031-96288-2_4

hardware and software systems [4, 5]. This upgrade has already started during the Second Long Shutdown (LS2, 2018–2022) and will continue in the Third Long Shutdown (LS3, 2025–2029) when the commissioning of the new detector will be performed. The CMS upgrade for the HL-LHC will enable efficient data collection in the 200 pileup (PU) harsh environment. In these new conditions, the already challenging implementation of an efficient τ lepton trigger will become an even more crucial and complicated task; especially interesting will be the hadronically decaying τ leptons (τ_h). To this end, the highly upgraded capabilities of the Phase-2 Level-1 (L1) triggering system (fully detailed in Sect. 4.1.3) can be exploited to design new sophisticated Machine Learning (ML) based triggering algorithms that are not yet implementable in the current Phase-1 system. As part of this Thesis work, I have been the sole developer of a completely new and innovative L1 trigger algorithm for the reconstruction, calibration, and identification of τ_h candidates, which is currently considered one of the baselines for the Phase-2 L1 trigger. This algorithm is based on convolutional neural networks, whose implementation and test in Field Programmable Gate Array (FPGA) firmware have also been carried out as part of this Thesis.

The algorithm described in this Chapter has been published as an article in Proceedings of Science [6], as well as a CMS Detector Performance Summary note [7]. For both publications, I am the sole editor.

This Chapter is structured into four main Sections. The first gives a brief overview of the CMS Phase-2 upgrade, detailing the parts of specific interest for the work documented in the following. The second Section describes the design and firmware implementation of a novel L1 τ_h trigger algorithm. The third reports the ultimate performance attained by the algorithm. The Chapter is closed with a fourth Section giving a perspective view of the future of L1 triggering with ML techniques.

4.1 The Phase-2 CMS Detector

The objective of the CMS Phase-2 upgrade program for the HL-LHC encompasses two key aspects. Firstly, the anticipated challenging conditions at the HL-LHC demand comprehensive renovations of several ageing detector components, necessitating their complete replacement or substantial upgrade. The harsh environment expected during Phase-2 requires enhanced resilience and robustness to ensure optimal performance. Secondly, the HL-LHC will enable groundbreaking precision measurements, facilitating direct searches for rare processes and providing an excellent opportunity to probe potential BSM phenomena. Therefore, it is imperative to maintain the current level of physics performance and possibly extend it, not only addressing technical limitations but also guaranteeing that the detector's capabilities align with the ambitious research goals of the HL-LHC era.

A schematic overview of the CMS Phase-2 upgrade is reported in Fig. 4.1, where the green boxes connected to the coloured parts on the CMS detector model represent

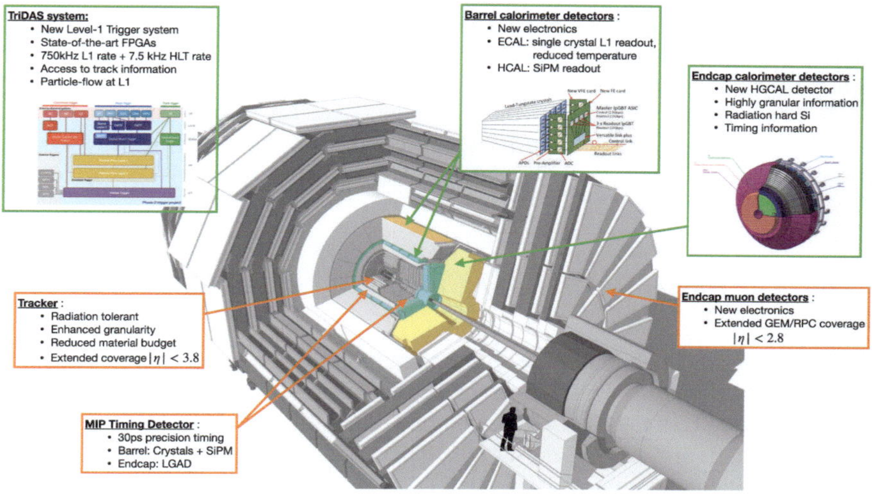

Fig. 4.1 Cutaway 3D model of the CMS detector summarising the Phase-2 upgrade foreseen for the HL-LHC [2, 8–13]. The green boxes, connected to the coloured parts on the CMS detector model, represent the upgraded systems that are significant for the discussion in this Chapter; the orange boxes denote the upgrades that are out of the scope of this Thesis

the upgraded systems that are significant for the discussion in this Chapter, while the orange boxes denote the upgrades that are out of the scope of this Thesis.

The pixel and strip tracking detectors will be entirely replaced [12] to increase the granularity of the detector. This is achieved by substituting the current pixel detectors with ones of smaller size and superseding the present strips with modern ones featuring macro pixels. This upgrade will guarantee, at the same time, a lower material budget and extended pseudorapidity coverage to $|\eta| < 3.8$. The new tracker system will highly benefit the reconstruction and identification capabilities of the CMS detector and will supply the L1 trigger with the track information to be exploited in the first online Particle Flow (PF) implementation. Multiple Minimum Ionising Particle Timing Detectors (MTDs) [13] will be introduced in the gap between the tracker and the calorimeters in both the barrel and endcap regions, thus increasing the timing information on charged candidates. This upgrade will have extensive use in disentangling the approximately 200 PU interactions foreseen per bunch crossing. The muon detectors will be complemented by installing state-of-the-art Resistive Plate Chamber (RPC) detectors and the newly developed Gas Electron Multiplier (GEM) chambers. These additions to the muon detection system will increase the geometrical coverage to $|\eta| \leq 2.8$, ensuring the possibility of extended muon track matching in the forward region. With the aim of further increasing the overall granularity of the detector, and providing additional timing measurements, an upgrade of the Electromagnetic Calorimeter (ECAL) and Hadronic Calorimeter (HCAL) barrel electronic readout is foreseen [9], as further examined in Sect. 4.1.1. The ageing endcap calorimeters, already highly impacted by the radiation dose of the first three runs of the LHC, will be replaced with the High Granularity Calorimeter (HGCAL) [10], a five-dimensional sampling calorimeter whose specifications are further discussed in Sect. 4.1.2. Finally, to efficiently collect data, the trigger and data acquisition systems will be completely replaced [8], redesigning the Level-1 hardware trigger to include tracking and high-granularity calorimeter information exploited via the extensive use of state-of-the-art FPGAs and ML techniques, as detailed in Sect. 4.1.3.

4.1.1 The Barrel Calorimeter Upgrade

The ambitious CMS Phase-2 physics program requires all subdetectors to maintain and possibly extend the Phase-1 physics performance in the higher instantaneous luminosity and PU of the HL-LHC. In the case of the Barrel Calorimeters (CB), multiple technical improvements, most of them being related to the upgraded L1 trigger latency and rate requirements (cf. Sect. 4.1.3), are needed to accomplish the performance enhancement. Moreover, the current electronics of the CB were designed to withstand integrated luminosities up to $\sim$500 fb^{-1} over a decade of data-taking. This makes the risk of wear-out during Phase-2 too high to be taken, forcing an important revision of the readout system. Owing to this upgrade, precision timing measurements can also be envisioned, with a target precision of 30 ps on the arrival time of photons issued by H decays. The complete design of the Phase-2 CB upgrade is documented in the dedicated Technical Design Report (TDR) published in 2017 [9].

The Barrel Calorimeter Upgraded Design

The main upgrade foreseen for the CB is related to the ECAL Barrel (EB), and it encompasses a full replacement of both the front-end and back-end electronics. The design of this upgrade campaign is done abiding by the following guiding principles:

- adapt to the current on-detector mechanics,
- use ongoing common developments where possible,
- provide single crystal trigger primitives granularity,
- limit power dissipation,
- provide precise timing measurement.

The increased instantaneous and integrated luminosities foreseen for the HL-LHC do not pose any concern for the current led-tungstate crystals and the Avalanche Photo-Diodes (APDs). Therefore, to avoid the unnecessary risks posed by their replacement, the Phase-2 EB will operate the same Phase-1 components. Nevertheless, since the APD dark current scales logarithmically with the APD operating temperature, to meet the required performance, APDs will be operated at 9 °C, i.e. half of the Phase-1 temperature. To match the EB detector mechanics, the upgraded on-detector electronics are required to fit into the same physical space as the present one; hence, the individual EB passive boards will follow the same configuration as

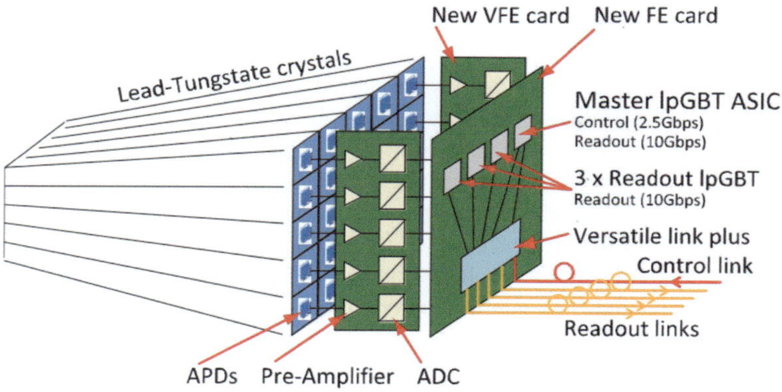

Fig. 4.2 Schematic of the upgrade ECAL barrel electronics architecture. Crystals and Avalanche Photo-Diodes (APDs) will not be replaced, while the new Very Front-End (VFE) and Front-End (FE) cards will follow the same Phase-1 configuration. Upgraded ASICs (Application-Specific Integrated Circuits) will be implemented based on industry standards featuring lpGBT (low-power GigaBit Transceiver) technology, ensuring single crystal readout [9]

the current ones. The Very Front-End (VFE) card, which provides the APDs pulse amplification, shaping, and digitization functions, will be replaced to improve the timing and noise performance that are subsequently used to discriminate anomalous signals (the so-called *spikes*). At the same time, to meet the required performance goals, the front-end card, the low-voltage distribution system, and the optical links will be replaced. Both the pre-amplification analogue ASICs (Application-Specific Integrated Circuits) and the digital converter ASICs will be upgraded to industry standards, implementing TSMC (Taiwan Semiconductor Manufacturing Company Limited) 130 nm CMOS (Complementary Metal-Oxide Semiconductor) and 65 nm lpGBT (low-power GigaBit Transceiver) technology. These components are naturally radiation tolerant and provide a 75% power consumption reduction compared to the current technology while lowering the material budget by 20%. Introducing these new improvements and components will allow individual crystal readout, mainly owing to the higher speed radiation tolerant lpGBT optical links. A schematic of the upgraded EB electronics architecture is given in Fig. 4.2. The higher transfer rates from the front-end electronics will need to be handled by an equally boosted back-end. For this reason, the off-detector electronics will be upgraded to the ATCA (Advanced Telecommunications Computing Architecture) industry standard [14], and the extensive use of state-of-the-art FPGAs will be made.

The second portion of the CB upgrade activity is represented by the HCAL Barrel (HB) upgrade. The main components of the HB detector consist of brass absorbers interspersed with active material tiles, front-end electronics with photo-sensors, and back-end electronics. Data from 2017 indicated that the expected radiation damage to the HB scintillators and fibres would have minimal impact on CMS physics performance during the full HL-LHC period. Hence, replacing the HB scintillators and fibres will not be necessary. The upgraded Phase-1 front-end electronics with Silicon Photo-Multipliers (SiPM) installed on the HB detector during the LS2 will continue to be used for the HL-LHC period, eliminating a significant cause of HB signal degradation. While scintillator signal reduction in the region of HB most exposed to radiation is expected, it will be compensated by the higher photo-detection efficiency of the SiPMs. Consequently, the impact of scintillator radiation damage on the CMS physics performance in the HB remains negligible for the entire HL-LHC operating period, rendering the replacement of HB scintillators unnecessary. However, the current HB back-end electronics, based on the μTCA (micro Telecommunications Computing Architecture) standard, cannot support the planned increase in the L1 trigger rate for Phase-2. To address this, the Phase-1 μTCA-based back-end electronics will undergo an upgrade to the ATCA standard using the same boards being developed for the EB and by other subdetectors. By adopting a homogeneous off-detector system, development and production resources can be optimized, spares can be shared, and long-term operations and maintenance are facilitated. Moreover, using industry standards will reduce upkeep costs compared to custom parts.

The Barrel Calorimeter Trigger Primitive Generation

In the Phase-1 CB architecture, the process of Trigger Primitives Generation (TPG; sometimes TPG can also stand for Trigger Primitives Generator) is performed in different ways for the EB and HB compartments: for ECAL, the TPG is performed on-detector while for HCAL, it is executed off-detector. In either case, the Phase-1 electronics can neither sustain the expected high L1 trigger rate foreseen for Phase-2 nor ensure the required latency. Therefore, the TPG will be performed in the upgraded Back-end Barrel Calorimeter Processors (BCPs) developed based on ATCA boards equipped with modern FPGAs and high-speed optical links for both EB and HB.

In Phase-2, the EB segmentation will follow the same Trigger Towers (TTs) mapping used in Phase-1, as detailed in Sect. 3.2.1. Nevertheless, the Trigger Primitives (TPs) will increase by a factor of 25 in granularity, with each crystal corresponding to one TP. At the TPG stage, the EB detector is divided into regions of $\eta \times \phi = 0.26 \times 0.35$ (i.e. $i\eta \times i\phi = 3 \times 4$ in discrete Cartesian coordinates) extension corresponding to 300 crystals composing 12 TTs. Each region is connected to one of the two FPGAs available per BCP. Therefore, to cover the whole 61200 crystals of the EB and account for overlaps and redundancies, 108 BCP cards and 216 FPGAs are needed. In each FPGA, spikes will be identified based on signal shape analysis and isolation; two options can then be envisioned: suppress spike crystals in the EB back-end or send them to the L1 trigger with a specific spike flag bit. Moreover, the TPG algorithm includes per-channel linearization (multiplication by gain ratios and channel inter-calibration constants) and conversion to transverse energy of the digitized signal. The data transfer between each FPGA and the L1 trigger will go through 25 Gb/s links, one every 20 crystals, at a rate of 40 MHz. Each crystal TP is encoded into a 16-bit digital variable: 10 bits are allocated to the E_T energy deposit with a foreseen Least Significant Bit (LSB) of 125 MeV, five bits are reserved to

Table 4.1 Barrel calorimeter trigger primitives data format used to send information to the Level-1 trigger

Trigger primitive	Quantities	Bits	Total bits	LSB
ECAL crystal	E_T	10		125 MeV
	Timing	5		$\mathcal{O}(60)$ ps
	Spike flag	1		
	Total		16	
HCAL trigger tower	E_T	10		0.25–0.5 GeV
	Feature bits	6		
	Total		16	

timing information with a yet undefined LSB of $\mathcal{O}(60)$ ps, the last bit is foreseen for the identification of the anomalous signals. In this encoding scheme, each crystal can carry up to 3.875 GeV of deposit and each TT $25 \times 3.875 = 96.875$ GeV at most. This value is lower than in Phase-1, but no loss of performance is expected by this reduction owing to the increased noise suppression capabilities of the upgraded EB.

For what concerns the HB, since no further upgrade of the front-end electronics is foreseen (besides the Phase-1 upgrade that took place in the LS2), the number of readout channels, the transverse $\eta \times \phi$ segmentation, and the number of longitudinal readout depths will remain analogous to Phase-1. Therefore, 2304 TTs will cover the whole HB, with an $i\eta \times i\phi = 16 \times 72$ segmentation, for every half barrel, each conveying the energy sum of the entire HCAL physical towers by summing four and three readout depths for respectively $|i\eta|$ regions 1 through 15 and $|i\eta|$ region 16. In each TT, individual depth measurements are first linearised into an E_T scale via a Look-Up Table (LUT) and then summed to the full TT energy deposit. Each HB TP is encoded into a 16-bit digital quantity: 10 bits are allocated to the E_T energy deposit with a foreseeable LSB of 0.25 or 0.5 GeV, while six bits are reserved as additional feature information. The definition of the algorithms for the latter six bits is still being studied, but the probable use will comprise the selection of isolated tracks for calibration purposes, trigger on muons or MIPs (Minimum Ionizing Particles), carry information on the shower shape, and aid lepton isolation. With no loss of generality, the best E_T encoding scheme is used in the following for the HB TTs: therefore, an LSB of 0.25 GeV is assumed, and each TT can carry up to 255.75 GeV of energy.

The overview of the CB TPs encoding into digital quantities is reported in Table 4.1. A summary of the geometry of the CB TTs is reported later on in Table 4.6 and further discussed in the text of Sect. 4.2.3.

4.1.2 *The High Granularity Calorimeter*

The CMS Phase-1 ECAL and HCAL endcap calorimeters have been designed to sustain an integrated luminosity up to $\sim$500 fb^{-1}. By the end of Run-3, they will have undergone an expected $\sim$400 fb^{-1}, making their use in Phase-2 impossible due to ECAL's lead-tungstate crystals and HCAL's plastic scintillators becoming opaque and effectively *blinding* this detector region. The new detector replacing the current endcaps must satisfy important criteria:

- withstand exceptional radiation levels,
- be highly dense to constrain the lateral compactness of showers,
- ensure high sensitivity in the busy forward region,
- be highly granular to disentangle the PU contributions,
- contribute to the Level-1 trigger decision.

An innovative detector has been designed to meet precisely these requirements: the High Granularity Calorimeter. Its complete design has been detailed in the HGCAL TDR published in 2018 [10], while subsequent developments are documented in the HGCAL public website [11].

The HGCAL design not only aims at addressing the technical necessities reported above but also strives for enhanced shower separation and particle identification. These two features are central in performing HL-LHC searches and measurements that include very forward and highly boosted objects. The main examples are the vector boson fusion and vector boson scattering mechanisms, for which it is paramount to trigger events based on the forward region without placing significant requirements on the rest of the event content.

The HGCAL Design

A cross-sectional view of the HGCAL detector is given in Fig. 4.3. In its latest design, the HGCAL detector features 47 layers divided into two compartments: 26 for the Electromagnetic Compartment (CE-E) and 21 for the Hadronic Compartment (CE-H). This number of longitudinal samplings is designed as a trade-off between the best shower reconstruction and the engineering requirements of the mechanical structure.

To meet the radiation hardness requirements, the active material used in the CE-E and the most forward part of the CE-H is silicon, alternated with layers of CuW, Cu, and Pb absorbers. Conversely, in the region of CE-H less subject to radiation, scintillator technology alternated with stainless steel absorbers is used. This configuration amounts to a total of 10 nuclear interaction lengths (λ_0), 1.3λ_0 for the CE-E and 8.5λ_0 for the CE-H. The CE-E alone will extend for a total of 27.7 radiation lengths (X_0). To further improve radiation resistance, the full system is cooled to $-30/35\,°C$ with liquid CO_2.

The choice of silicon as the main material of the HGCAL also ensures the high density of the detector, which is crucial to constrain the lateral spread of showers.

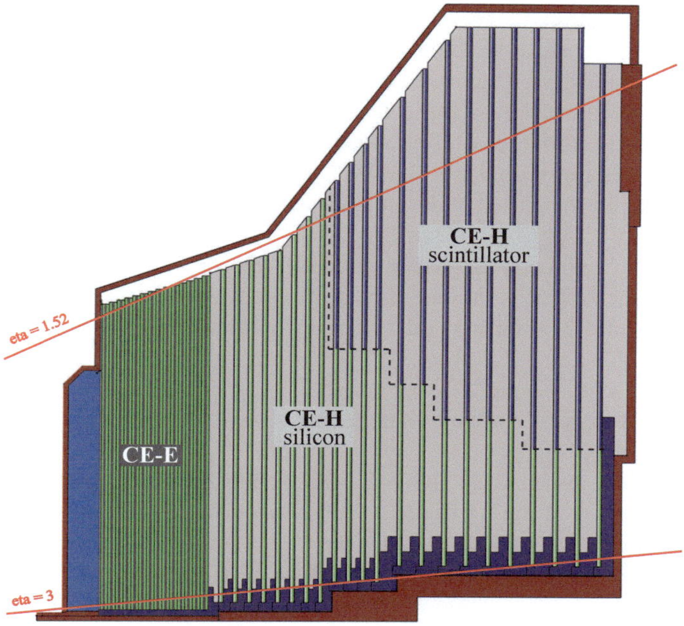

Fig. 4.3 Cross-sectional view of the HGCAL detector. In its latest design, it features 47 layers divided as follows: 26 silicon-based layers for the Electromagnetic Compartment (CE-E), 21 silicon- and scintillator-based layers for the Hadronic Compartment (CE-H), alternated with stainless steel absorber [10, 11]

To guarantee high sensitivity and outstanding PU rejection capabilities, the whole detector is laterally segmented to a significant degree. The silicon active material in the high occupancy sector is transversely segmented into hexagonal cells of surface 0.52–1.18 cm^2 and thickness 120–300 μm, depending on the detector region. The scintillating material, coupled to SiPMs for readout, in the low occupancy sector is transversely segmented with square shapes and sizes of 4–30 cm^2 depending on the pseudorapidity position. This geometrical configuration amounts to a total active area of 620 and 370 m^2 for the CE-E and CE-H compartments, respectively.

This design ensures the pseudorapidity coverage $1.5 \lesssim |\eta| \lesssim 3.0$ with a highly granular lateral and longitudinal segmentation. The enhanced lateral granularity, combined with the dense absorbers, yields effective individual shower discrimination in the detector. Moreover, the finely segmented longitudinal structure enhances PU rejection, particle identification, and energy resolution. These features of the design make the HGCAL a five-dimensional sampling calorimeter, the five dimensions corresponding to the three-dimensional position measured by the fine voxels of the detector, the energy deposit in each of the active material segments, and the timing information with an expected $\mathcal{O}(10\,\text{ps})$ resolution. For this reason, the HGCAL is sometimes referred to as an *imaging calorimeter*.

In this configuration, the HGCAL has a total of more than 6 M readout channels. The quantity of information streamed through them highly exceeds the available bandwidth of the trigger system; therefore, only about 1 M are dedicated to trigger purposes. This represents a significant advancement over Phase-1, amounting to more than a 500-fold increase. To contribute to the L1 trigger decision, raw data undergoes processing within the HGCAL readout electronics to construct highly granular TPs that can be used to develop advanced trigger algorithms, as elaborated in the following.

The HGCAL geometry described above is known as geometry `v16`; it was finalised in 2021, and it is the one used for the development of the L1 τ_h trigger described in Sect. 4.2. A further update is already foreseen for the coming years.

The HGCAL Trigger Primitive Generation

The raw input data stream from HGCAL is about 300 TB/s. This highly exceeds the available bandwidth of the L1 trigger system; therefore, an enormous and efficient data reduction needs to be performed via the trigger primitives generation. The TPG is performed in general-purpose Serenity ATCA platforms [15], a generic motherboard also shared by other Phase-2 subsystems, which is designed to withstand the unprecedented throughput of information from the Phase-2 detectors and efficiently manage it for several objectives.

The first tier in data reduction is the alternate readout of CE-E layers for triggering purposes; therefore, only 14 of the 27 layers of the electromagnetic compartment are exploited. Given the lower granularity of the CE-H compartment, all its layers are read out for TPG. A simplified schematic representation of the TPG process is reported in Fig. 4.4. The TPG is split into two main sections: one is implemented in the front-end electronics, and the other is executed in the back-end electronics. The second tier in data reduction is the condensation of the sensors' information into the so-called Trigger Cells (TCs). This is implemented in a custom ASIC designed for the HGCAL, the HGCROC (High Granularity Calorimeter Read-Out Chip). After the creation of the TCs, their information is sent to a second custom ASIC, the ECON-T (Endcap CONcentrator TPG), which selects TCs surpassing a pre-defined energy deposit threshold and produces the 8" modules sums. These first three steps are performed in the HGCAL front-end; their output is then sent to the HGCAL back-end, where the third tier of data reduction is performed in Xilinx Virtex UltraScale+ VU13P FPGAs [16]: the TCs are used to build the so-called 3D-clusters (CL^{3D}), and the energy sums are projected into an $\eta \times \phi$ binned energy deposit map. To achieve the maximum efficiency and meet the maximum latency requirement of $5\,\mu s$, the TPG procedure is performed regionally in a time multiplexed fashion whose detailed description is out of the scope of this Thesis.

The HGCROC receives the raw data from the silicon sensors and the scintillators' SiPMs and exploits it for building TCs. In the case of silicon sensors, each TC is obtained as the group of 3×3 adjacent silicon cells in the more granular region where the sensor area is $0.52\,cm^2$ or as the group of 2×2 adjacent silicon cells in the less

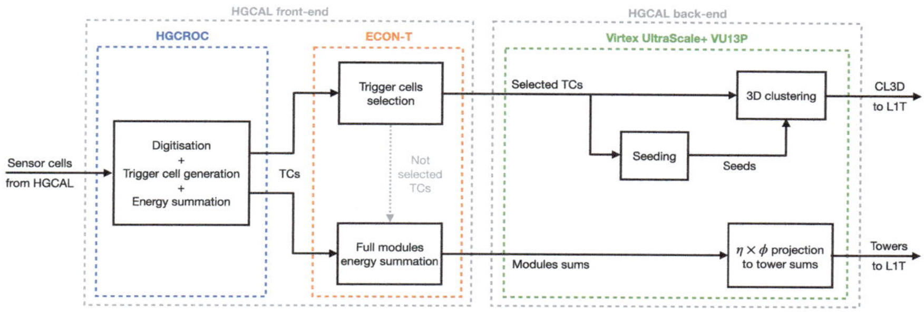

Fig. 4.4 Simplified schematic representation of the trigger primitives generation in the HGCAL detector. The information from the sensor cells of HGCAL is sent to the front-end where Trigger Cells (TCs) are built by the HGCROC chip. TCs are then sent to the ECON-T, where they are selected if above a threshold, and they are added together into module energy sums. The information is then passed to the back-end where in the Virtex UltraScale+ VU13P FPGAs the TCs are clustered into 3-dimensional trigger primitives, and the modules' sums are projected into trigger towers. The vertical dashed arrow in the ECON-T box represents the possibility of computing energy sums either from the full energy deposit or only from not selected TCs

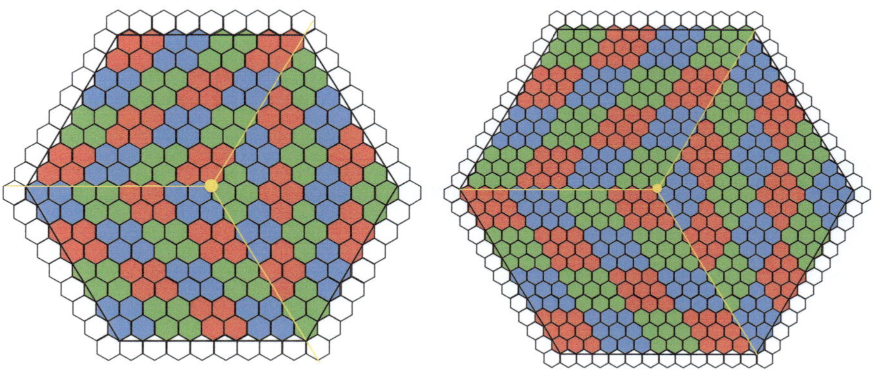

Fig. 4.5 Schematic illustration of the three-fold diamond configuration of sensor cells on hexagonal 8" silicon modules, showing the groups of sensor cells that get summed to form trigger cells, for the large 1.18 cm^2 sensor cells (left), and for the small 0.52 cm^2 cells (right) [10]

granular region where the sensor area is 1.18 cm^2. In both cases, the area covered by a TC is about 4 cm^2, and every 48 adjacent TCs are arranged in 8" modules. In the case of silicon sensors, given their simpler square geometry, TCs are formed by merging scintillator tiles over an azimuthal angle of 2.5°, corresponding to a similar η and ϕ extension ranging between 4 and 10 cm. For trigger purposes, the timing information is dropped to lower the information stream and the energy deposit in each TC is compressed into a digital variable of seven bits. The geometrical configuration of the TCs in one 8" silicon module, for the low-density and the high-density regions separately, is reported in Fig. 4.5.

The TCs built in the HGCROC are sent to the ECON-T, where their selection is performed. Multiple selection algorithms have been developed for the task, the one yielding the best compromise between performance and hardware resources usage being the so-called *mixed best choice and super trigger cell* approach. In this method, a fixed number of N TCs per module (the N *best choice*) is selected in the CE-E, with N being an adjustable parameter typically between 4 and 13 depending on the detector region; this selection leads to a fixed data size but variable readout channel content. In contrast, in the CE-H, TCs are further merged into so-called *super trigger cells*, which are all sent to the following stage of the TPG. In this approach, the grouping consists of 2×2 TCs in the higher-density modules and 4×4 TCs in the lower-density modules and the scintillator. This selection leads to both data size and readout channel content being fully fixed. The ECON-T also performs the summation of the full 8" modules energy deposit; this can be done under two different approaches: either all TCs or only those that are not selected as described above are used for the energy sum. The former method guarantees the full projection of the entire energy deposit onto the final $\eta \times \phi$ map for trigger usage, while the latter is designed only to recover the non-selected energy deposit for the computation of global variables like the missing transverse energy. In the following, the full sum of all TCs is always assumed for the energy projection map.

The heart of the TPG process runs in the Xilinx Virtex UltraScale+ VU13P FPGAs of the HGCAL back-end, where the TCs and module sums are converted into TPs. Two types of TPs are produced for each bunch crossing: the three-dimensional clusters CL^{3D} and the full depth $\eta \times \phi$ energy map binned in TTs. The TCs are clustered dynamically in both the transverse and longitudinal directions to form the CL^{3D} in two steps:

- **seeding**: TCs are projected onto a binned $(r/z, \phi)$ plane, and seeds are identified as local energy maxima in this two-dimensional histogram. The main parameter regulating the seeding is the energy threshold for a bin to be selected as a seed; a typical value is that of 5 MIP_T, where $MIP_T \equiv MIP/\sin\theta$.
- **clustering**: TCs are projected onto the $(x/z, y/z)$ plane and are clustered if they lie within a certain pre-determined distance. This parameter is adjustable, and typical values range from 0.01 to 0.05. The choice is optimized to maximize signal shower reconstruction and minimize PU contributions.

The high granularity information of the HGCAL is then exploited to derive the CL^{3D} properties: the η, ϕ, and z positions are defined as the energy-weighted barycentre, while several other variables related to the shape of the shower are computed based on the constituent TCs. Several algorithms and definitions are being explored for maximally PU discriminating shape variables that can be implemented in firmware; a list of the current set is reported in Table 4.2 alongside their detailed description. The second type of TP produced in the back-end is the traditional projective TTs map, which is obtained in two steps: first, a coarse map is produced per layer, and then the layers are combined into a single full-depth map. Given the complex geometry of the 8" modules, the matching from modules to TTs is performed via a LUT encoding

Table 4.2 List of CL^{3D} shape variables currently available in simulation and being explored as candidates for Phase-2 implementation

Variable	Description
Shower length	Length of the shower in number of layers, computed as (last layer – first layer) + 1
Core shower length	Length of the shower core in number of layers, computed as the maximum number of consecutive layers
$\sigma_{\eta\eta}^{\text{tot}}$	Energy weighted RMS of η-coordinates of TCs
$\sigma_{\eta\eta}^{\text{max}}$	Maximum of the per-layer energy weighted RMS of ϕ-coordinates of TCs
$\sigma_{\phi\phi}^{\text{tot}}$	Energy weighted RMS of ϕ-coordinates of TCs
$\sigma_{\phi\phi}^{\text{max}}$	Maximum of the per-layer energy weighted RMS of ϕ-coordinates of TCs
σ_{rr}^{tot}	Energy weighted RMS of r-coordinates of TCs
σ_{rr}^{max}	Maximum of the per-layer energy weighted RMS of r-coordinates of TCs
$\sigma_{rr}^{\text{mean}}$	Mean of the per-layer energy weighted RMS of r-coordinates of TCs
σ_{zz}^{tot}	Energy weighted RMS of z-coordinates of TCs
H/E	Ratio of energy deposits in the CE-H over CE-E

The energy weighted RMS is computed as $\sigma_{xx}^{*} = \sqrt{\frac{\sum_i^{\text{TC}^*} E_i \cdot (x_i - \langle x \rangle)^2}{\sum_i^{\text{TC}^*} E_i}}$

the detector's geometry. A summary of the geometry of the HGCAL TTs is reported later on in Table 4.6 and further discussed in the text of Sect. 4.2.3.

The TPs are finally sent to the L1 trigger, where the physics object reconstruction algorithms are implemented. The data format of the TPs is not yet fully defined, as a careful trade-off between physics performance and hardware resources needs to be found. In the configuration currently foreseen, the TTs are the first information to be sent to the L1 trigger. Each TT is encoded into a 16-bit digital quantity: 10 bits are allocated to the energy with an LSB of 0.25 GeV, three bits represent the hadronic over electromagnetic energy ratio, and three bits are reserved as possible feature bits; therefore, each TT can contain up to 255.75 GeV. After the TTs, the CL^{3D}s are sent ordered in decreasing p_T. Each CL^{3D} will have a fixed size of 256 bits, of which 14 bits are allocated to the E_T measurement with LSB 0.25 GeV; respectively nine, nine, and 12 bits are reserved for the η, ϕ, and z positions, with the former two having a $\pi/720$ LSB and the latter a 0.5 mm LSB; 16 bits are dedicated to each shape variable, with a maximum of eight variables being available. The full summary of the HGCAL TPs encoding into digital quantities is reported in Table 4.3.

The high granularity information of the HGCAL detector, encoded in all the variables of each CL^{3D} can then be maximally exploited for the definition and efficient identification of physics objects against PU. An exemplary reconstruction of 3D clusters is given in Fig. 4.6 as a purely pedagogical display to visualize what CL^{3D}s are. In this event, four 3D clusters are reconstructed and depicted in four different colours; a simplified envelope of the HGCAL detector is also shown. This display is obtained from a Monte Carlo (MC) sample simulating the production of τ candidates with no pile-up.

Table 4.3 HGCAL trigger primitives data format used to send information to the Level-1 trigger

Trigger primitive	Quantities	Bits	Total bits	LSB
Trigger tower	E_T	10		0.25 GeV
	H/E	3		
	Feature bits	3		
	Total		16	
3D cluster	E_T, $E_T^{e/\gamma}$, e/γ select bits	14, 14, 4	32	0.25 GeV
	$E_T^{\text{CE-E}}$, $E_T^{\text{core CE-E}}$, $E_T^{\text{CE-H}}$	8, 8, 8	24	0.25 GeV
	First layer z, spare	6, 2	8	
	η, ϕ, z, spare	9, 9, 12, 2	32	$\pi/720$, 0.5 mm
	TCs number, quality bits, spare	10, 10, 12	32	
	Shape information	8×16	128	
	Total		256	

4.1.3 The Phase-2 Level-1 Trigger System

Efficiently collecting datasets to be exploited in the ambitious HL-LHC physics program will be a challenging task. For this purpose, as part of the Phase-2 upgrade, the CMS Collaboration is fully redesigning its hardware-implemented Level-1 trigger. The Phase-2 L1 trigger builds on the other subdetectors' upgrades, exploiting improved tracking information and highly-granular calorimetric information in both barrel and endcap regions. Trigger data analysis will be performed through sophisticated algorithms such as PF reconstruction and the widespread use of ML techniques. The current conceptual system design is expected to take full advantage of state-of-the-art FPGAs and link technologies, providing a high-performance, low-latency computing platform for large throughput and refined data correlation across diverse sources. The architecture is developed to ensure the highest degree of flexibility in terms of triggering algorithms that are conceived to make the L1 trigger capable of performing close to real-time analysis in hardware platforms. The complete design and preliminary validation of the Phase-2 L1 trigger upgrade are documented in the dedicated TDR published in 2020 [8].

In the following, the design of the CMS Phase-2 L1 trigger is detailed, giving specific attention to the calorimeter trigger processing of TPs and highlighting the present status of the L1 τ_h trigger developments.

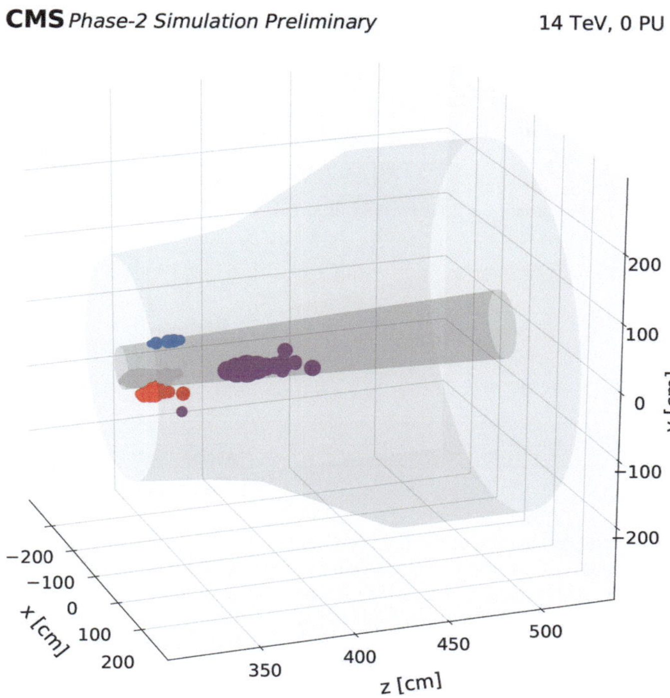

Fig. 4.6 Exemplary reconstruction of CL^{3D}s in the HGCAL detector for an event with no pile-up in which four 3D clusters are reconstructed and displayed in four different colours. The HGCAL geometrical outer envelope is shown in dim-grey, while the inner envelope where the beam pipe lies is shown in darker grey

Architecture

The foreseen architecture of the Phase-2 L1 trigger is reported in Fig. 4.7. This architecture aims at optimizing processing board numbers, interconnections, and latency while ensuring the needed flexibility and robustness of the system. By implementing intermediate global triggers, an optimal division of labour is achieved. Moreover, this system capitalizes on new hardware technologies to deliver computing power and high-speed data transfer for a global detector view. It takes inspiration from the Phase-1 upgrade choices, using generic stream-processing engines as data processing units instead of custom-designed processors. This option leaves ample room for further algorithm optimization, similar to the Phase-1 L1 trigger operation. This design enables both regional and time-multiplexed architecture options to be exploited. Finally, the system features large FPGAs like Xilinx Virtex UltraScale+ VU13P, equipped with up to 28 Gb/s transceivers, allowing the extensive development of advanced ML algorithms that can be implemented in FPGA firmware.

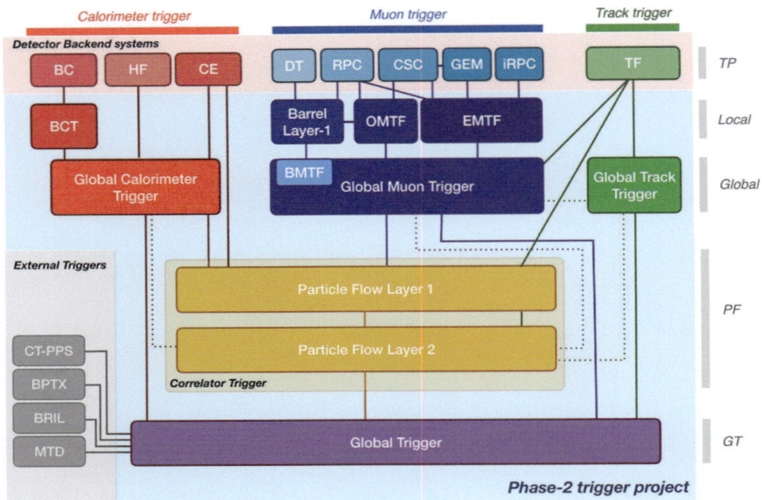

Fig. 4.7 Schematic representation of the foreseen Level-1 trigger upgraded system architecture in Phase-2. The information from the Barrel Calorimeters (BC), Hadron Forward Calorimeter (HF) and High Granularity Calorimeter (HGCAL, CE) detectors are used to reconstruct the e/γ, τ, and jet candidates, as well as global HT and p_T^{miss} quantities in the Global Calorimeter Trigger. The Drift Tube (DT), Resistive Plate Chamber (RPC), Cathode Strip Chamber (CSC), and Gaseous Electron Multiplier (GEM) detectors send their information to the Barrel, Overlap, and Endcap Muon Track Finders (BMTF, OMTF, and EMTF), which build tracks to be identified as μ candidates in the Global Muon Trigger. The Global Track Trigger uses the Track Finder (TF) track primitives to further increase the L1 performance. The Correlator Trigger exploits the high-level objects built by each global sub-triggers and employs a Particle Flow approach to event reconstruction. The Global Trigger collects the outputs from all previous steps and takes the event accept or reject decision [8]

The architecture of the Phase-2 L1 system is composed of four independent data processing paths converging in a single global trigger, complemented by the introduction of the innovative scouting system. This design choice, which includes several innovations compared to the Phase-1 system, reflects the need for the production of complementary trigger objects to achieve the best physics selectivity.

- **Calorimeter Trigger path**: it takes advantage of the upgrade of the barrel and endcap calorimeters, taking as input the highly granular calorimeter TPs over the entire pseudorapidity coverage of the detector. The crystal-based TPs from the EB are received by the Barrel Calorimeter Trigger (BCT) and they are processed as detailed in the following Section to build e/γ candidates. The crystal TPs are then merged into TTs and sent to the Global Calorimeter Trigger (GCT), where are implemented the reconstruction algorithms for τ_h, jet, and sum objects, which optimally exploit all calorimeter TPs and TTs. The L1 τ_h algorithm developed in this Thesis could be implemented in the GCT.
- **Track Trigger path**: this represents the first great innovation of the Phase-2 system; track TPs are produced in the tracker back-end from the information of the

outer tracker only and sent to the Global Track Trigger (GTT). In the GTT, the vertex and track-only object reconstruction is performed with the goal of subsequent muon and calorimeter deposit matching. This path will prove invaluable in identifying PU contributions and reducing the L1 trigger rate.

- **Muon Trigger path**: it profits from the redundant muon detectors infrastructure and improved η coverage from the GEM subdetector. The muon TPs are processed in a way analogous to the Phase-1 technique, in which separate algorithms analyse different areas of the detector, namely the Barrel Muon Track Finder (BMTF), Overlap Muon Track Finder (OMTF), and Endcap Muon Track Finder (EMTF). The Global Muon Trigger (GMT) receives both the muon track finders' and tracker track finders' output to build matched candidates.
- **Particle Flow Trigger path**: also referred to as Correlator Trigger (CT), is the second major innovation of the upgraded system. It receives as input the information from the previous three trigger paths and exploits it, as the name suggests, to implement a processing *a la* particle flow. It is subdivided into two layers, the first building the high-level candidates and the second implementing identification and isolation algorithms. This is the second possible implementation point for the L1 τ_h algorithm developed in this Thesis.
- **Global Trigger**: it gathers the output from the GCT, GMT, GTT, and CT to produce the event accept/reject determination based on a menu of algorithms. It optimally exploits correlation variables among different objects to achieve the best physics selectivity, often with algorithms designed for analysis-specific purposes. Owing to the largely improved computing power compared to Phase-1, sophisticated ML-based *topological* algorithms are being developed to target specific signatures of rare processes in order to enhance their selection efficiency.
- **Scouting system**: it is the third central innovation of the new L1 trigger; it guarantees the ability to perform trigger-less analysis of L1 data at the 40 MHz bunch-crossing rate by profiting from the spare optical links of the various processing boards. Moreover, its access to each L1 trigger subsystem makes it a great tool for real-time diagnostic of the whole system.

The presented architecture will have a total latency of 12.5 μs, corresponding to 500 bunch crossings and roughly three times longer than the one of Phase-1. This latency is dictated mainly by the track hardware reconstruction and matching time. The total output bandwidth is foreseen to be of 750 kHz, being 7.5 times larger than the Phase-1 output rate, thus allowing for energy thresholds comparable to Run-2 and Run-3 values to be retained in the much busier PU conditions of HL-LHC. Finally, exploiting state-of-the-art FPGAs and optical links will enable the selection of the events based on input data as high as $\sim$60 TB/s, corresponding to a 30-fold increase compared to the Phase-1 system. As is the case for the CB upgrade, wherever possible, the Phase-2 L1 trigger system will extensively use industry standards concerning boards, FPGAs, and optics, thus greatly facilitating long-term operations while reducing maintenance costs compared to custom parts.

Processing Trigger Primitives in the Barrel Calorimeter Trigger

This Section gives a brief overview of how the BCT processes the crystal TPs from the CB. This discussion is important to facilitate the understanding of the following Sections, where the inputs to the newly designed algorithm are discussed.

The EB crystal TPs will be received by the L1 trigger at the 40 MHz bunch crossing rate, where the BCT exploit them to reconstruct e/γ candidates before they are sent to the GCT. The L1 e/γ candidates are seeded by single TPs with $E_T > 1$ GeV, and the core cluster is defined as the 3×5 crystals surrounding the seed. To account for bremsstrahlung radiation, the 3×5 crystals above or below the core in the ϕ direction are added to the candidate if they have an energy deposit larger than $1/10$ that of the core. To reject misidentified jets, a shower shape variable is computed based on the ratio of energy deposits in different sub-regions of the e/γ candidate. Finally, an isolation variable is computed using the 27×27 crystals region around the seed. The use of crystal timing information is also being explored. The L1 e/γ candidates are then sent to the GCT. A detailed description of the algorithm can be found in the Phase-2 L1 trigger TDR [8].

After L1 e/γ candidates have been built, the crystals' information is used to build the EB TTs. As for the HB, the Phase-2 EB TTs will match the Phase-1 geometry amply discussed in Sect. 3.2.1. In this configuration, the 25 crystals pertaining to a TT are summed into a single quantity. A summary of the geometry of the CB TTs is reported later on in Table 4.6 and further discussed in the text of Sect. 4.2.3.

As a result of this processing, the information carried by the Phase-2 EB TT will be twofold: the energy deposit associated with the clustered e/γ candidates and the unclustered energy. These variables are separately accessible and are both encoded into 10-bit digital quantities with LSB 0.25 GeV.

Existent Level-1 τ_h Algorithms

The physics necessity of a dedicated L1 algorithm for τ_h leptons has already been discussed in Sect. 3.1. The wide range of physics analyses involving τ leptons necessitates high L1 selectivity over a wide p_T spectrum, therefore calling for diverse algorithmic approaches to be implemented simultaneously.

Multiple algorithms exploiting diverse subdetector inputs have been studied and precisely detailed in the 2020 Phase-2 L1 TDR [8]. The following will give a brief overview of only the two main algorithms which, over the three years separating the TDR and this Thesis, have been further optimized and proved to ensure the required performance.

- **CaloTau algorithm**: this algorithm relies solely on the TT information from the calorimeter subdetectors, i.e. CB, HGCAL, and HF. It implements a simplified version of the Phase-1 algorithm described in Sect. 3.2. This approach identifies τ_h candidates as fixed size $i\eta \times i\phi = 3 \times 5$ clusters seeded by a TT with energy deposit $E_T > 2.5$ GeV. Their calibration is performed by exploiting the crystal granularity of the EB TPs: three categories of calibration are built based on the number of crystal-based e/γ clusters (0, 1, or ≥ 2) found within the τ_h candidate. The isolation is performed similarly to the Phase-1 standard by defining a fixed $i\eta \times i\phi = 7 \times 7$ isolation region to calculate E_T^{iso}; the isolation requirements are optimized separately for the various regions of the CMS detector. This algorithm exploits calorimetric TT information only and is foreseen to be implemented in the GCT.

- **NNPUPPITAU algorithm**: this algorithm is based on a Neural Network (NN) that exploits particle- and event-level quantities based on a PF approach. L1 τ_h candidates are seeded iteratively by the charged PileUp Per Particle (PUPPI) candidate with the highest p_T. Within a cone of $\Delta R < 0.4$ from the seed, the 10 PUPPI candidates with the highest p_T are selected; their p_T, η, and ϕ distance from the seed, and PUPPI identification score are used as inputs to a dense NN. The output of the NN is used to determine if the candidate is a τ_h or not, while the p_T is evaluated as the sum of the transverse momenta of the PUPPI candidates within and angular distance $\Delta R < 0.1$ from the seed. This algorithm is built to optimally exploit track information and is foreseen to be implemented in the CT.

The most updated performance evaluation of the two algorithms detailed above is presented in Figs. 4.8 and 4.9, where matching efficiency and L1 rate are reported. For both algorithms, the matching efficiency is defined as the fraction of generated hadronically decaying τ leptons that are geometrically matched to an L1 τ_h candidate, and it is evaluated in a HH $\rightarrow$ bb$\tau\tau$ MC simulated sample with $\sqrt{s} = 14$ TeV and 200 interactions per bunch crossing. To assess the rate, a minimum-bias dataset produced with the same conditions is used; this dataset corresponds to single or

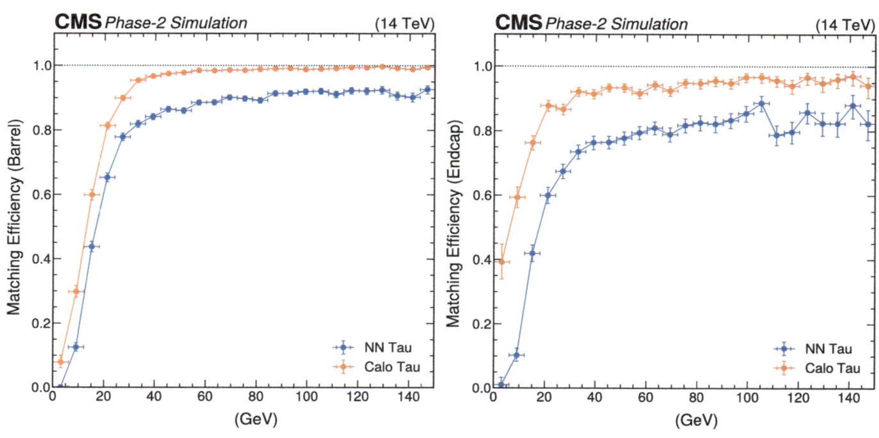

Fig. 4.8 Level-1 τ_h matching efficiency in the barrel (left) and in the endcap (right) for the CALOTAU (orange) and the NNPUPPITAU (blue) algorithms, as a function of generated p_T. The matching efficiency is computed as the fraction of generated τ_h that are geometrically matched to an Level-1 τ_h candidate; it is evaluated in HH $\rightarrow$ bb$\tau\tau$ Monte Carlo simulated sample with 200 pileup interactions

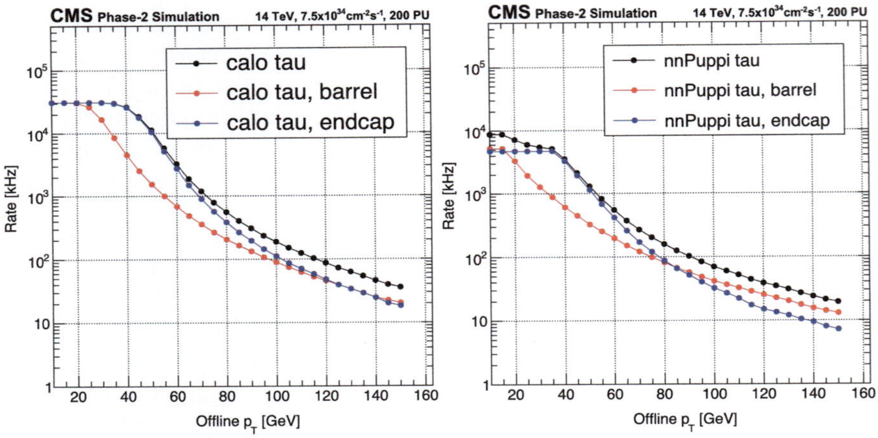

Fig. 4.9 Level-1 single-τ_h rate of the candidate for the CALOTAU (left) and the NNPUPPITAU (right) algorithms as a function of the offline p_T, defined as the generator p_T value at which the trigger turnon crosses the 90% efficiency point. The rate is evaluated in a Monte Carlo simulated minimum-bias dataset with 200 pileup interactions; the separate barrel and endcap contributions are reported

multiple diffractive inelastic interactions. In CMS data, the minimum-bias dataset is collected using the `MinBias` trigger, whose only requirement is the presence of one particle candidate with $p_T > 5$ GeV and $3 < |\eta| < 5$; in the MC simulated sample, the same requirement is enforced. The rate is presented as a function of the offline p_T, defined as the generator p_T value at which the trigger turnon crosses the 90% efficiency point.

As can be appreciated from the Figures, the NNPUPPITAU algorithm is more suited for triggers targeting lower thresholds as it keeps the rate under control, while CALOTAU is more fit for triggers with higher thresholds as it ensures 100% efficiency in the high-p_T regime. This difference resides in the optimal usage of a more advanced reconstruction technique by the former algorithm, which includes event-level information and can better reject fake candidates from the large HL-LHC PU in the low p_T regime. Conversely, the CALOTAU algorithm achieves better and more stable efficiency in the high p_T regime owing to a simpler and more robust approach to reconstruction. This complementarity can be exploited by simultaneously implementing both algorithms and identifying L1 τ_h candidates as the logical OR between the two. This approach achieves maximal reconstruction efficiency of high energy L1 τ_h candidates while still being able to lower the threshold at values comparable to Phase-1. A summary of the two algorithms' L1 trigger rate at fixed efficiency offline threshold is given in Table 4.4.

Table 4.4 Rates of the main Level-1 τ_h seeds in the current version of the Phase-2 L1 Menu. The offline p_T is defined as the generator p_T value at which the trigger turnon crosses the 90% efficiency point. The rate is estimated at $\langle PU \rangle = 200$

L1 τ_h seed	Offline threshold	Rate [kHz]
`CaloTau`	150	31
`DoubleCaloTau`	90/90	33
`DoubleNNPuppiTau`	52/52	28

4.2 The TauMinator Algorithm

In the HL-LHC era, the CMS Collaboration will strive for the precision measurement of the SM by tightening the current limits on its known parameters and extending the Run-3 physics analysis range. Primary importance will be given to the measurement of the SM λ_{HHH} coupling; moreover, many BSM models, some of which are discussed in Sects. 1.1.6 and 1.2.2, foresee the existence of heavy resonances that can decay to Higgs boson pairs. One of the most sensitive channels for this type of search is the HH $\rightarrow$ bb$\tau\tau$ owing to the τ lepton selection purity.

In this context, achieving the best signal selectivity on τ_h leptons at the L1 trigger is of crucial importance. As discussed above, some algorithms have already been developed using diverse approaches depending on the foreseen implementation subsystem of the Phase-2 L1 trigger. Nevertheless, large improvements are expected by exploiting the calorimetric information in a more refined algorithm that takes full advantage of the L1 trigger resources. As demonstrated by the NNPuppiTau algorithm, the extensive use of state-of-the-art FPGAs allows for implementing advanced ML algorithms in the L1 trigger hardware.

In this Section, the design, optimization, and firmware implementation of the innovative TauMinator[1] L1 τ_h trigger algorithm will be presented. This algorithm has been fully developed within this Thesis work, from conception to firmware testing, with the goal of profiting from the L1 trigger improvements to boost the performance of τ_h reconstruction from calorimeter information only. The TauMinator represents the first algorithm to ensure a unified treatment of all available TPs from the barrel and endcap of the Phase-2 CMS detector; at the same time, it is the first calorimeter-based algorithm that performs the identification and the calibration of τ_h candidates via a ML technique.

[1] The name TauMinator is intended as a pun between the well-known Terminator franchise and a misspelling of the word *miner*, as the algorithm would effectively do τ *mining*.

4.2.1 Algorithm Philosophy: How the Present Inspires the Future

As is customary with significant advancements, the TauMinator algorithm draws inspiration from past and present innovations. Namely, the work detailed in Chap. 3 inspired three considerations that led to this algorithm's design. These reflections will be presented in the following, and the philosophy of the TauMinator algorithm will be discussed.

The highly successful Run-2 and Run-3 L1 algorithms for reconstructing e/γ and τ_h candidates are based on one main principle: the dynamic recognition of TT patterns ascribable to an e/γ or τ_h energy deposit. This is achieved by dynamically clustering TTs around a seed according to predetermined rules based on the meticulous examination of the most likely shapes a cluster can assume. From this, the first consideration arises: can this method be boosted to avoid the definition of predetermined clustering rules and shape constraints?

The second reflection originates from the isolation procedure of L1 e/γ and τ_h candidates performed in Run-2 and Run-3. In this approach, the isolation energy E_T^{iso} is computed as the difference between the candidates' energy and the surrounding activity in an area extending over six and nine TTs in the η and ϕ directions, respectively. This choice is based mainly on physics considerations about the particle multiplicity in the hadronization of jets. Therefore, the second question arises: can this method be implemented in such a way that avoids the separation of clustering and isolation into two separate steps?

The novel Calibraton algorithm represents the third source of inspiration. Its design aims to calibrate individual TPs in the Run-3 calorimeter trigger Layer-1 by means of a NN. To achieve this, 9 $\times$ 9 groups of TPs are processed simultaneously to reconstruct the energy deposit of a target object; subsequently, the parameters of the NN are exploited to define single TP calibration factors. This approach profits from inherent correlations between the TTs in each 9 $\times$ 9 group. Hence the last reflection emerges: can this method be extended for deriving particle candidates' calibration that exploits the intrinsic correlation of TT energy deposits?

Two additional factors need to be pondered while trying to answer these three questions. The first is the exceptional amount of information that the HGCAL detector will provide to the L1 trigger; the TTs are just one of the two types of TP available: the CL^{3D} represents a unique possibility to exploit highly granular information for the rejection of PU contributions in the busy endcap region in 200 PU conditions. The second is the unprecedented availability of massive computing power in the L1 trigger; the use of state-of-the-art FPGAs grants the implementation of sophisticated algorithms unthinkable with the hardware available 10 years ago.

The philosophy of the TauMinator algorithm can then be summarized in the following five guidelines:

- boost the present τ_h shape recognition approach,
- avoid the need for a separate τ_h isolation,
- calibrate the τ_h candidate profiting of energy deposits correlations,

- exploit the highly granular information of HGCAL,
- maximally profit of the L1 FPGAs computing resources.

One solution exists that abides by all these five principles: the use of Convolutional Neural Networks (CNNs). This class of NNs is specifically designed to process pixel data and is generally used in image recognition and processing (a dedicated description of CNNs is given in Sect. 4.2.2). The TT map utilized within the L1 trigger system can be interpreted as a pixelated representation of the CMS calorimeters, making CNNs a natural approach. Following this idea, any τ_h candidate can be reconstructed as a fixed-size image of TTs, where each TT acts as a pixel, and a CNN can be trained to recognize patterns associated with τ_h lepton decays. This approach can perform both the rejection of background and the calibration of the τ_h candidate by exploiting the pattern recognition capabilities of a CNN embedded in FPGA firmware. Additionally, in the endcap region only, the CL^{3D} information can be included in the process owing to the intrinsic architecture of CNNs.

It is important to notice that the selection of CNNs as the preferred technical tool is also based on the necessity of their firmware implementation. From a pure performance point of view, the more general class of Graph Neural Networks (GNNs) would be a better choice; nevertheless, their embedding in FPGA firmware is much more hazardous, and it started being explored only at the time of writing this Thesis. This highlights how the entire development of the TauMinator algorithm is performed with the final FPGA firmware implementation as the main target.

4.2.2 Technical Interlude: CNNs and Datasets

In this Section, a brief overview of CNN technology is given alongside the summary of the samples used in the design of the TauMinator algorithm. Regarding the presentation of CNNs, this discussion is not intended as an in-depth course about ML techniques but rather as an intuitive exposition of CNNs and their power.

Convolutional Neural Networks

The main idea behind CNNs was inspired by studies on the striate cortex of cats [17]. This study showed that individual neurons in the visual cortex respond to stimuli from narrow portions of the entire visual field, the so-called *local receptive field*. A CNN is a software implementation of these biological findings, which Kunihiko Fukushima first introduced in 1980 [18] and which reached its modern image recognition conception in two publications by Yann LeCun in 1989 [19] and 1995 [20].

The fundamental building block of CNNs is the convolutional layer. Let's assume we have 2D image-like input data; this can be represented as the matrix $\boldsymbol{I}$ of dimension $F \times N \times M$, where F is the number of tiers of information, the so-called *filters*, and $N \times M$ is the pixel dimension of the input. In image processing, the filters are

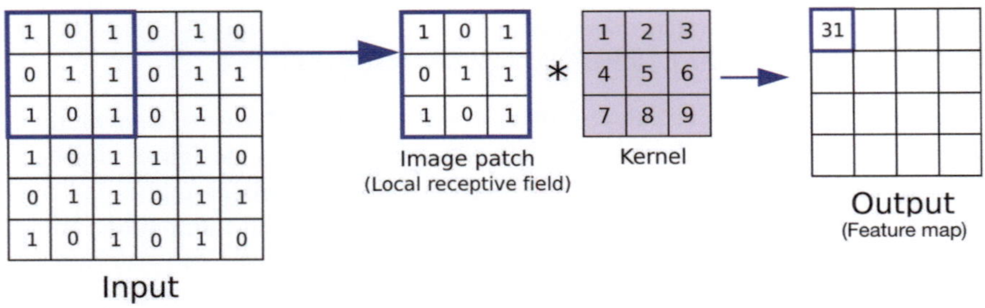

Fig. 4.10 Visual representation of the first step of a convolution. In this picture, the local receptive field of the network is a narrow input image patch of dimension 3 × 3; it undergoes the matrix product with the kernel of the convolutional layer to yield the first value of the output feature image

generally three and define the RGB colour components of the image. In a convolutional layer, we define the local receptive field of the network to be a subset of the input image of dimension $K \times L$ and define the *kernels* of the convolutional layer as the matrices $\boldsymbol{W}$ of dimension $K \times L$ containing the trainable parameters of the network, i.e. the weights. During training, the weights of the kernels are learnt through backpropagation, where the network adjusts the weights to minimize the error between the predicted feature maps and the true feature maps. We can then denote the convolution procedure as the following operation:

$$\boldsymbol{O}_{i,j} = (\boldsymbol{I} * \boldsymbol{W})_{i,j} = \sum_{k=0}^{K} \sum_{l=0}^{L} \sum_{m=0}^{F-1} \boldsymbol{I}_{i+k,j+l,m} \cdot \boldsymbol{W}_{k,l,m} \tag{4.1}$$

where $\boldsymbol{O}$ is the output matrix of the convolution. The matrix $\boldsymbol{O}$ is generally referred to as *feature map* because it represents the intermediate features that are built and learnt by the network during the convolution. Equation (4.1) represents one convolution of a single kernel; the complete set of convolutions performed by one layer is obtained by sliding the local receptive field of the network in steps of length S (the so-called *stride*) and iteratively performing the convolution operation until the whole input has been mapped. Therefore, the dimensions of $\boldsymbol{O}$ are fully determined by the shape of the input, the shape of the kernel, and the stride length; given the dimensions above, $\boldsymbol{O}$ would be of shape $\left(\left\lfloor \frac{N-K}{S} \right\rfloor + 1\right) \times \left(\left\lfloor \frac{M-L}{S} \right\rfloor + 1\right)$. A visual representation of the convolution operation is given in Fig. 4.10, where the matrix product between the input and the kernel is made explicit by numerical values. The local receptive field of the network is slid in strides over the input image, and different kernels are applied at each step; when the whole image has been covered, the output feature map will be fully filled.

In a CNN, multiple convolutional layers are stacked to learn different characteristics of the input data. In general, the first convolutional layer learns the presence of edges in the image, the second layer learns about basic shapes, and additional layers learn about ever more complex attributes of the data. Between each convolutional

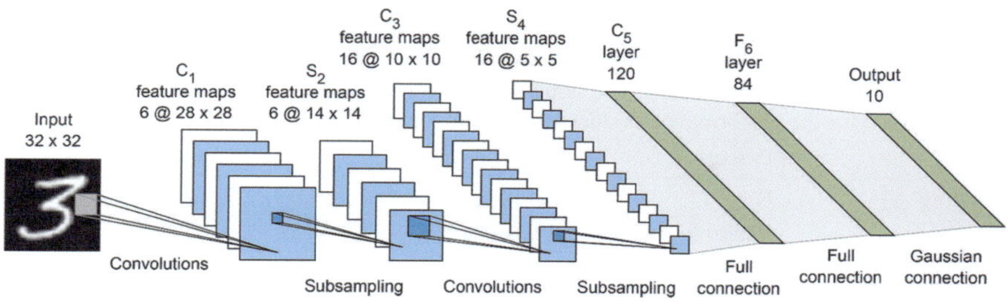

Fig. 4.11 Acrhitecture of the first modern Convolutional Neural Network (CNN), the LeNet-5 [20]. It presents all standard features of a CNN: two convolutional layers, each followed by a pooling layer, the typical feature map flattening, and three dense layers leading to the final output

layer, the *pooling* process is generally performed with dedicated layers. Pooling layers are used to downsample the feature maps and reduce their spatial dimensions, which helps to reduce the number of parameters in the network and improve computational efficiency. Max-pooling is a widely used technique where the maximum value from a local region of the feature map is taken (the region where pooling is performed does not need to have the same dimensions as the kernels). This procedure helps to retain the most important information while enhancing translational invariance.

In ordinary CNN architectures, after several convolutional and pooling layers, the final feature maps are usually flattened into a vector and passed through one or more fully connected layers. These dense layers use the extracted features to make high-level decisions and predictions, depending on the specific task the CNN is designed for. Figure 4.11 represents the original LeNet-5 structure designed by Yann LeCun in 1995. As just described, a series of convolutions and poolings are performed before feeding the flattened feature map into a fully connected network. This architecture represents the modern standard of CNNs.

Dataset Overview

The design and optimization of the TAUMINATOR algorithm presented in this Section are performed solely based on MC simulated samples. Various processes are used for different stages of the development. A mixture of different samples is used to achieve a statistical power large enough to train the CNNs of the algorithm. Two samples of SM $\mathrm{H} \rightarrow \tau\tau$ production via gluon fusion and vector boson fusion are used, both generated with POWHEG [21–23], respectively amounting to $1.0 \cdot 10^4$ and $5.0 \cdot 10^4$ events. These are complemented with two Drell-Yan $\mathbf{Z}/\gamma^* \rightarrow \ell\ell$ simulated samples produced in two exclusive regions of $m_{\ell\ell}$, both generated with MADGRAPH5_aMC@NLO [24], respectively amounting to $1.0 \cdot 10^4$ and $2.5 \cdot 10^4$ events. To obtain a fully unbiased performance evaluation, the datasets used in this step do not share any event with the training sample. For the τ_h reconstruction efficiency

Table 4.5 Simulated datasets used in the design of the TAUMINATOR algorithm, their corresponding generator, and total number of events available. Hadronization is simulated in all cases with PYTHIA, using the CP5 underlying event tune. The samples are separated between those used for the training and those used for the performance evaluation of the algorithm

Use	Process	MC generator	Events
Training	$H \to \tau\tau$	POWHEG	
	Vector boson fusion		$1.0 \cdot 10^4$
	Gluon fusion		$5.0 \cdot 10^4$
	$Z/\gamma^* \to \ell\ell$	MADGRAPH5_aMC@NLO	
	$10 < m_{\ell\ell} < 50\,\text{GeV}$		$1.0 \cdot 10^4$
	$m_{\ell\ell} > 50\,\text{GeV}$		$2.5 \cdot 10^4$
Performance	$HH \to bb\tau\tau$	MADGRAPH5_aMC@NLO	$5.0 \cdot 10^4$
	Minimum-bias	PYTHIA	$2.0 \cdot 10^6$

assessment, a SM $HH \to bb\tau\tau$ sample generated with MADGRAPH5_aMC@NLO and comprising $5.0 \cdot 10^4$ is used. To evaluate the TAUMINATOR rate, a minimum-bias dataset of $2.0 \cdot 10^6$ events is employed. The simulation of quark hadronization and fragmentation effects, underlying event, and multiple parton interactions leading to PU are performed with PYTHIA [25], with the CP5 [26] underlying event tune. A summary of the MC simulated samples used is given in Table 4.5.

To perform the supervised training of the CNNs of the TAUMINATOR algorithm, signal and background labels must be defined. We define an L1 object as being signal if found within an angular distance $\Delta R < 0.5$ from the hadronically decaying generated τ lepton. Following this scheme, any L1 object which does not satisfy this criterion is defined as background. No requirement is enforced on the τ_h decay mode to retain the highest possible inclusivity. This signal/background definition scheme will be intended for the remainder of the Chapter.

4.2.3 Trigger Tower Calorimeter Input

In Phase-2, as in Phase-1, the entire pseudorapidity coverage of the CMS detector calorimeters at the L1 trigger is organised in TTs. The following geometrical scheme is followed in the CB, where each TT corresponds to the extension $\eta \times \phi = 0.087 \times 0.087$, which covers a square of 5×5 ECAL crystals and one HCAL readout unit. Analogously to Phase-1, each half barrel is sectioned into 17 and 72 trigger primitives in the η and ϕ directions, with the discrete Cartesian coordinates of each TT are $i\phi \in [1, 72]$ and $i\eta \in [-17, 17] \setminus \{0\}$, making position 0 non-existent in both directions. In the endcap, the TTs are defined to obtain seamless coverage in the L1 trigger. In the current simulation, produced with geometry `v16`, the HGCAL TT projective map is segmented into 18 and 72 trigger primitives in the η and ϕ directions, respectively. In the endcap, the progressive numbering of the TTs is employed as an extension of

Table 4.6 Geometrical characteristics of the Phase-2 Level-1 calorimeter Trigger Towers (TTs). In the barrel ($|i\eta| \leq 17 \setminus \{0\}$), the same configuration of Phase-1 is maintained. In the endcap ($18 \leq |i\eta| \leq 35$), TTs are obtained from the projective map of the HGCAL 8" modules' energy sum

$\lvert i\eta \rvert$	$\lvert \eta_{\text{min}} \rvert$	$\Delta\eta$	$\Delta\phi$	EM calorimeters	HAD calorimeters
1–17	$(\text{n} - 1) \times 0.0870$	0.0873	0.0870	5×5 crystals	1 HB tower
18	1.4840	0.0810	0.0870	CE-E 8" module sums projection	CE-H module sums projection
19–35	$17 \cdot 0.0870 + (n - 18) \cdot 0.0850$	0.0850	0.0870	CE-E 8" module sums projection	CE-H modules sums projection

the barrel numbering. The first TT in HGCAL, i.e. $|i\eta| = 18$, has an extension of $\eta \times \phi = 0.081 \times 0.087$; all other TTs have an extension of $\eta \times \phi = 0.085 \times 0.087$. In this configuration, the discrete Cartesian coordinates of the HGCAL TTs are in the following ranges: $|i\eta| \in [18, 35]$ and $i\phi \in [1, 72]$. As is the case for the Phase-1 architecture, also in Phase-2, the position of the TTs does not need to be encoded into digital quantities as it is fully determined by the linking of the detector readout to the L1 trigger. A summary of the geometry of the TTs is reported in Table 4.6.

It is worth reminding at this point that each TT in the barrel carries three digital quantities of 16 bits each, for a total of 48 bits, 30 of which are reserved for the energy deposit as follows: 10 bits for the e/γ pre-clustered E_{T}, 10 bits for the ECAL unclustered E_{T}, and 10 bits for the HCAL E_{T}, all with LSB 0.25 GeV. In contrast, in the endcap, each TT is encoded into 16 bits, of which 10 are reserved for the E_{T} deposit with LSB 0.25 GeV.

The creation of L1 τ_{h} candidates is initiated by local energy maxima in exclusive regions extending five TTs in the η direction and nine TTs along the ϕ direction. Because of the nature of the τ_{h} decay, the energy deposit is expected both in the electromagnetic and hadronic calorimeters, so the total TT energy is used at this stage. The seeding is performed in exclusive regions, so no overlap between the clusters can be formed. Seeding trigger towers are required to satisfy an energy selection of $E_{\text{T}} \geq 2.5$ GeV; to ensure that not only the seed but entire clusters are contained in the HGCAL acceptance, seeds must have pseudorapidity $|i\eta| \leq 33$. All TTs within a distance of $|\Delta i\eta| \leq 2$ and $|\Delta i\phi| \leq 4$ from the seed are clustered in a single τ_{h} candidate if they have an energy deposit $E_{\text{T}} \geq 0.25$ GeV. Due to their characteristic dimensions, these clusters are referred to as $\text{CL}^{5\times9}$.

The selection of the clustering parameters is driven by the maximization of the algorithm's performance. Figure 4.12 reports the matching efficiency between the generated τ_{h} leptons and the $\text{CL}^{5\times9}$ as a function of the η and p_{T} of the generated τ_{h}. As it can be appreciated, enforcing tighter thresholds on the seeding TT especially reduces the reconstruction efficiency in the overlap region between the barrel and the

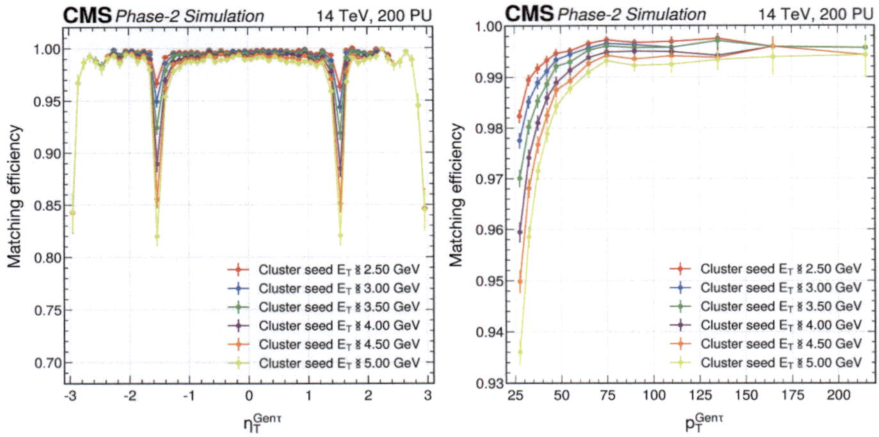

Fig. 4.12 Level-1 τ_h matching efficiency for the TAUMINATOR algorithm as a function of the of the η (left) and p_T (right) of the generated τ_h. Results are shown for the different E_T thresholds on the $CL^{5\times9}$ seeding TT; the lowest one is chosen to maximize reconstruction efficiency and to match the value used in the CALOTAU algorithm for comparability

endcap, and in the low p_T region. When the seeding threshold is highly increased, a significant loss of efficiency is also found in the high p_T regime. Following these considerations, the seeding threshold is fixed at the minimum $E_T \geq 2.5$ GeV to maximize the matching efficiency and, simultaneously, to be fully comparable with the CALOTAU algorithm, which uses the same value. The second parameter of the clustering is the extension of the clusters; several dimensions were tested ($i\eta \times i\phi \in \{9 \times 9, 7 \times 7, 5 \times 5, 5 \times 9, 5 \times 7, 3 \times 7, 3 \times 5\}$) and $i\eta \times i\phi = 5 \times 9$ was found to be the one yielding the best identification performance of the τ_h signal against the background. This finding is not surprising in light of the discussion presented in Sect. 3.2.2 when presenting the isolation variable used in the Run-3 algorithm. Moreover, this result is of major interest in light of the third guiding principle of the TAUMINATOR algorithm: *avoid the need for a separate* τ_h *isolation*. The conclusion that 5×9 is the best-performing dimension highlights how the CNN used for the identification can automatically enforce some isolation requirement on the τ_h candidates.

As discussed above, typical image inputs to CNNs are characterized by multiple tiers of information called filters. In both the barrel and the endcap, the presence of separate electromagnetic and hadronic contributions in each TT allows the definition of two filters for the $CL^{5\times9}$. Additionally, in the CB, the electromagnetic information is further split into two separate tiers: the e/γ clusters obtained from the crystal granularity TPs and the remaining unclustered energy deposit, resulting in a total of three possible filters. Figure 4.13 reports the contribution of the three filters in the barrel $CL^{5\times9}$ as a function of the generated τ_h transverse momentum. As it can be

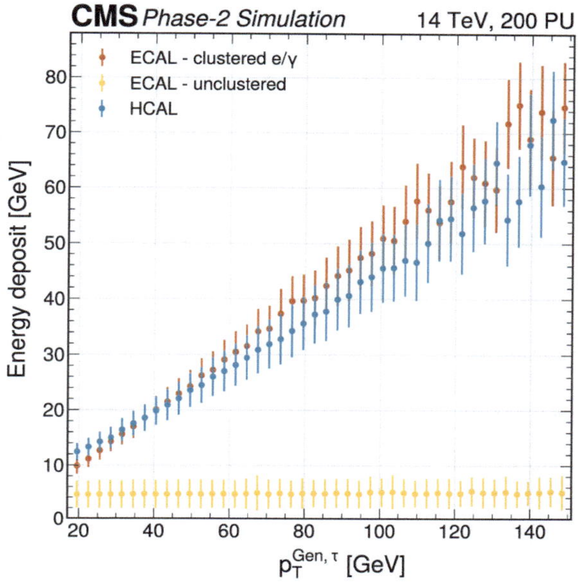

Fig. 4.13 Separate energy deposit contributions to the $CL^{5\times9}$ E_T in the barrel: pre-clustered energy deposit from the crystal-based e/γ algorithm (red), unclustered ECAL energy deposit (yellow), and HCAL energy deposit (blue)

appreciated, the electromagnetic contribution is concentrated almost solely in the clustered e/γ candidates, while the unclustered ECAL energy deposit has a constant behaviour over the entire p_T spectrum. Given this behaviour, the decision is taken to merge the electromagnetic contributions into one information filter. The choice to retain the unclustered energy deposit instead of dropping it is based on the possibility for the CNN to learn topological features related to it. The decision to retain only two filters, which is motivated here by physics reasoning, also represents an important point when considering the firmware implementation of the algorithm. Reducing the number of inputs to the CNN effectively lowers the amount of data to be handled in the FPGA. It simultaneously decreases the number of needed weights in the CNN, further reducing the need for hardware resources when implementing the algorithm in firmware. Moreover, as no e/γ crystal-based information is available in HGCAL, this makes the input from the barrel-endcap overlap region more consistent.

The position of the L1 τ_h candidates is defined as the centre of the seeding trigger tower $(\eta, \phi)_{seed}$, with no further adjustments, and it is inputted to the CNN encoded into a digital quantity of nine bits.

A representation of an exemplary $CL^{5\times9}$ is reported in Fig. 4.14, where the seeding trigger tower is highlighted in yellow, while the electromagnetic and hadronic energy deposit, respectively in red and blue, are shown with numerical values in GeV units with LSB 0.25 GeV, matching the firmware implementation.

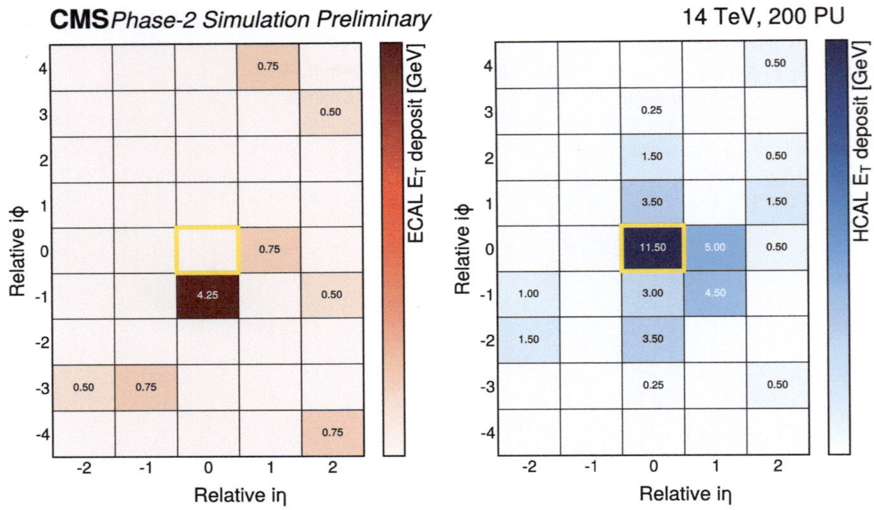

Fig. 4.14 Exemplary tower cluster of the TauMinator algorithm: the seeding trigger tower is highlighted in yellow; the electromagnetic energy deposit (red) and the hadronic energy deposit (blue) are shown with numerical values in GeV units; the cluster is arranged in units of η and ϕ position relative to the seeding tower

4.2.4 High Granularity Calorimeter Input

As discussed in Sect. 4.1.2, the HGCAL detector will provide two types of TPs to the L1 trigger: the TTs and the CL^{3D}s. The TTs from HGCAL are treated in a unified approach with those from the barrel as described above; however, special care needs to be taken for the CL^{3D}s.

Due to the high density of the HGCAL calorimeter and the TPG clustering procedure, single τ_h leptons produce single CL^{3D} clusters in $\sim$90% of the cases. Therefore, CL^{3D}-based L1 τ_h candidates can be selected as single clusters fulfilling $E_T > 4$ GeV. A preselection based on a BDT developed at the time of the TDR and trained for PU rejection is also applied [8]. The BDT exploits the shape variables to optimally reject PU-induced CL^{3D}s; its simple design allows for its implementation either in the GCT or the HGCAL back-end. The latter possibility would be more beneficial as it would save latency in the GCT, and its output could be encoded into the currently spare bits of the CL^{3D} digital variables. A minimal threshold of -0.10 is applied to the output score of the PU BDT. This value is optimized to strike the best balance between the L1 τ_h matching efficiency and the reduction of the fake rate from PU contributions. The dependency of the L1 τ_h matching efficiency and of the CL^{3D}s multiplicity on the selected PU BDT threshold is reported in Fig. 4.15. The left panel displays the efficiency as a function of the generated τ_h transverse momentum, while the right panel showcases the reduction in the number of CL^{3D} per endcap per event.

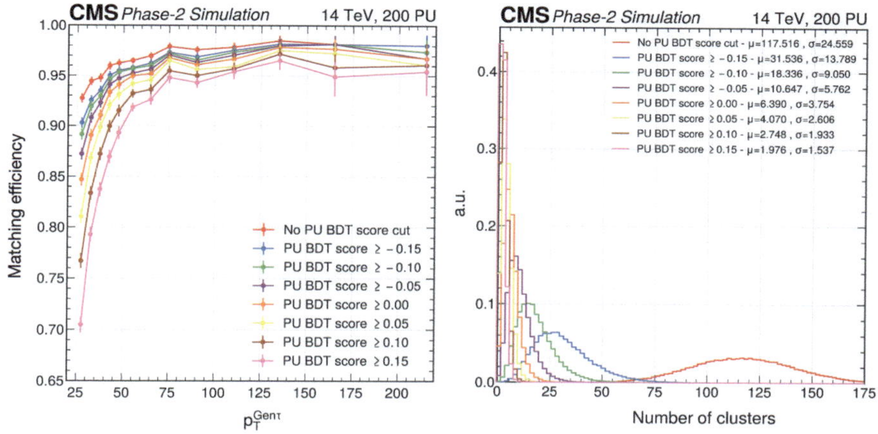

Fig. 4.15 L1 τ_h matching efficiency for the TAUMINATOR algorithm, restricted to the endcap region, as a function of the p_T of the generated τ_h (left). Distribution of the number of CL^{3D} per endcap per event (right). Results are shown for the different thresholds on the HGCAL PU BDT; the -0.10 threshold is chosen as the best trade-off between efficiency and CL^{3D} multiplicity reduction

The selected threshold value ensures an $\sim$84% reduction of the mean number of CL^{3D}, hence highly reducing the fake rate from PU contributions.

After CL^{3D} candidates are selected, the matching between $CL^{5\times9}$ and CL^{3D} is performed to ensure that they reconstruct the same τ_h lepton. Therefore, for $CL^{5\times9}$ satisfying $|\eta_{seed}| \geq 1.55 \sim |i\eta_{seed}| \geq 19$ the geometrical requirement $\Delta R(CL^{5\times9}, CL^{3D}) < 0.5$ is enforced. All $CL^{5\times9}$s and CL^{3D}s not satisfying this selection are rejected. While the physical transition point between barrel and endcap is at $|\eta| = 1.484$, the choice of the pseudorapidity value at which the geometrical matching between $CL^{5\times9}$ and CL^{3D} starts to be enforced is based on maximizing the L1 τ_h matching efficiency. The L1 matching efficiency as a function of η and p_T of the generated τ_h lepton is reported in Fig. 4.16 for three barrel/endcap demarcation points. As it can be appreciated, the use of the threshold $|\eta_{seed}| \geq 1.55$ maximizes the genuine τ_h selection efficiency over the whole p_T spectrum; as expected, the main contribution comes from the overlap region. Further increase of the threshold was also tested, but while the matching efficiency in the overlap region did not improve, the background rejection power of the algorithm decreased.

As discussed in Sect. 4.1.2, in addition to the standard variables of energy deposit and position, each CL^{3D} will be sent to the L1 trigger complemented by eight shape variables computed from the granular information of the detector. The list of variables currently studied is given in Table 4.2. This list exceeds the available number of bits allocated to shape variables; therefore, a selection needs to be made. To maximize the performance of the TAUMINATOR algorithm, a so-called *random backward skimming* approach is employed, as detailed in the following. The full set of $N = 17$ available variables is initially taken, and the algorithm is trained for the identification and

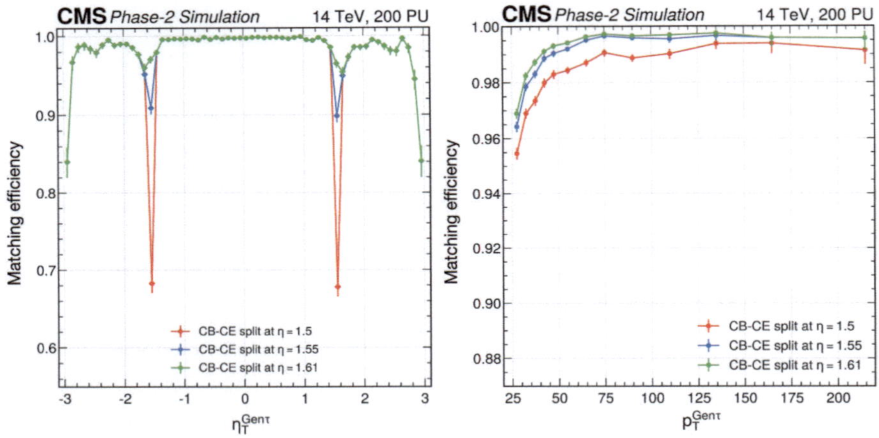

Fig. 4.16 L1 τ_h matching efficiency for the TAUMINATOR algorithm as a function of the of the η (left) and p_T (right) of the generated τ_h. Results are shown for the different $|\eta_{seed}|$ thresholds on the $CL^{5\times9}$ seeding TT after which geometrical matching with a CL^{3D} is required

calibration of L1 τ_h candidates; subsequently, all combinations of $N-1$ variables are formed, and separate trainings of the algorithm are performed for each combination set; the subset of variables yielding the highest performance is retained. This process is executed iteratively, preserving at each step the best-performing subset of variables until only one variable is left. At each step, the performance of the algorithm is defined as

$$\text{Loss} = \text{Loss}^{\text{Ident.}} + \text{Loss}^{\text{Calib.}}/100 \tag{4.2}$$

where $\text{Loss}^{\text{Ident.}}$ and $\text{Loss}^{\text{Calib.}}$ are the values of the τ_h identification and calibration loss function components of the optimized TAUMINATOR algorithm, respectively. These loss functions encode the error between the predicted and target quantity that the CNNs are learning, so the goal should be their minimization. At each step, both the identification and calibration of the TAUMINATOR algorithm are trained with the same set of variables: this choice is taken foreseeing the firmware implementation of the algorithm. Using different sets of variables, while possibly yielding better performance, would require using larger hardware resources for handling the data within the FPGAs. The evolution of the optimization metric is reported in Fig. 4.17, where the loss function used is shown both inclusively and split into its two components. As expected, the calibration component of the algorithm relies almost solely on the p_T and η of the CL^{3D}, while the identification needs a larger set of shape variables.

Following the backward skimming optimization of the CL^{3D} features, a final set of eight variables is chosen: p_T, η, shower length, core shower length, $\sigma_{\phi\phi}^{tot}$, σ_{zz}^{tot}, σ_{rr}^{tot}, and the z position. The differential distribution of these variables is given in Fig. 4.18,

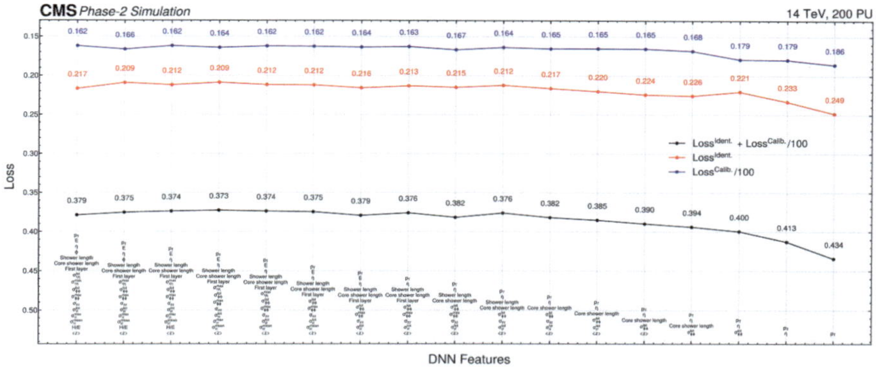

Fig. 4.17 Evolution of the optimization metric during the *random backward skimming* procedure employed to identify the CL^{3D} variables for input to the TauMinator endcap. At each step, the subset of $N - 1$ variables yielding the smallest loss is retained until one single variable remains. A set of eight variables is chosen in this manner: p_T, η, shower length, core shower length, $\sigma_{\phi\phi}^{tot}$, σ_{zz}^{tot}, σ_{rr}^{tot}, and the z position

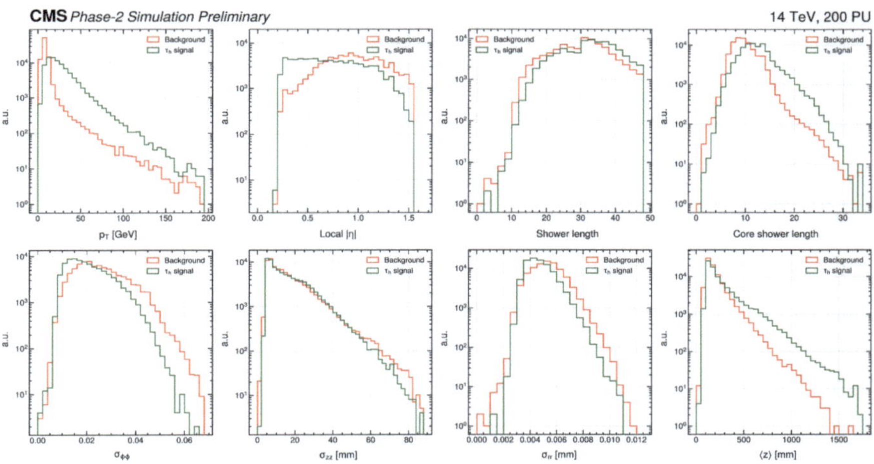

Fig. 4.18 Distributions of the CL^{3D} variables used in the TauMinator algorithm for the τ_h signal and the background samples. These variables are selected based on random backward skimming of all currently explored and available CL^{3D} variables

where their different behaviour under the signal and background hypothesis can be appreciated. Of the selected CL^{3D} feature, three are standard variables (i.e. p_T, η, and z position), while the other five are shape variables exploiting the HGCAL highly granular information.

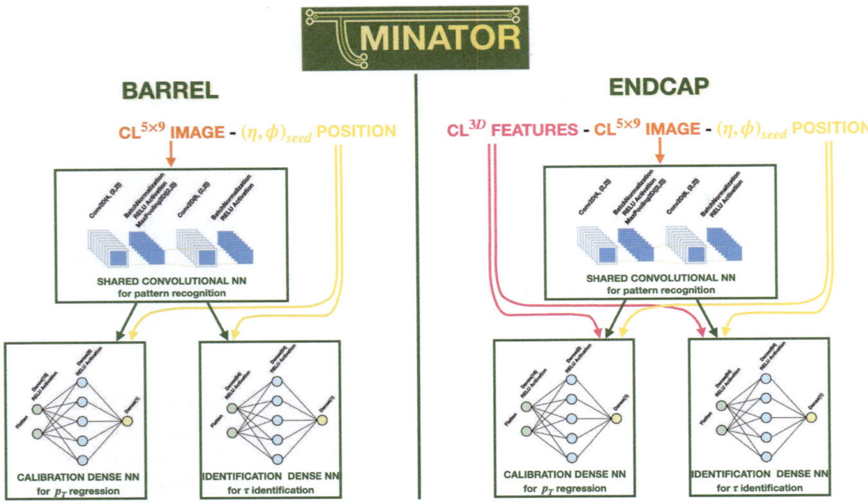

Fig. 4.19 Visual representation of the TAUMINATOR algorithm architecture. The TAUMINATOR comprises two sections: barrel and endcap with separation $|\eta| = 1.55 \sim |i\eta| = 19$. The CL$^{5\times9}$ identifies the input obtained from the trigger towers of the calorimeters, with $(\eta, \phi)_{\text{seed}}$ the seeding tower position, while CL^{3D} is the specific input from the HGCAL detector; the characteristics of both are fully detailed in the text. In each section of the algorithm, a standard convolutional network architecture is employed with the hyperparameters specified in the figure

4.2.5 Algorithm Architecture and Training

The architecture of the TAUMINATOR algorithm is reported in Fig. 4.19. Due to the different available TPs in the barrel and endcap areas, the algorithm is split into two independent compartments, one for each region; the optimization of the demarcation point between the two sections was detailed above. In the barrel section, corresponding to $|\eta| \leq 1.55 \sim |i\eta| \leq 18$, the input is represented by clusters of TTs built around a seed, the CL$^{5\times9}$. In the endcap section, corresponding to $|\eta| > 1.55 \sim |i\eta| \geq 19$, the input comprises CL$^{5\times9}$ built analogously to the barrel, associated with the HGCAL CL^{3D} according to their angular distance. In both partitions of the algorithm, the CL$^{5\times9}$ is processed by a CNN that performs the τ_h pattern recognition based on the TTs information; the additional information from the seeding TT and the CL^{3D} is concatenated to the CNN output and used as input to two dense NNs which perform the final identification and calibration of the τ_h candidate. A detailed description of the specific architecture of each NN component is given in the following, together with the training details of the TAUMINATOR algorithm.

The TAUMINATOR algorithm is implemented in Keras with a TensorFlow backend. The architecture of the CNNs in the barrel and endcap sections is identical, and it entails the following:

- Shared Convolutional NN
  - ◇ a 2D convolutional layer with kernel size 2×2, producing a four-dimensional feature map. Tensors dimensions: input $5 \times 9 \times 2$—output $4 \times 8 \times 4$,
  - ◇ a batch normalization layer followed by a Rectified Linear Unit (ReLU) activation function,
  - ◇ a max-pooling layer with subsampling window size 2×2. Tensors dimensions: input $4 \times 8 \times 4$—output $2 \times 4 \times 4$,
  - ◇ a 2D convolutional layer with kernel size 2×2, producing an eight-dimensional feature map. Tensors dimensions: input $2 \times 4 \times 4$—output $1 \times 3 \times 8$,
  - ◇ a batch normalization layer followed by a ReLU activation function,
  - ◇ a flattening layer. Tensors dimensions: input $1 \times 3 \times 8$—output 1×24.
- Identification Dense NN
  - ◇ a fully connected layer with 16 nodes and ReLU activation function,
  - ◇ a fully connected layer with eight nodes and ReLU activation function,
  - ◇ a fully connected layer with one node and sigmoid activation function.
- Calibration Dense NN
  - ◇ a fully connected layer with 64 nodes and ReLU activation function,
  - ◇ a fully connected layer with 64 nodes and ReLU activation function,
  - ◇ a fully connected layer with one node and linear activation function.

This structure is chosen foreseeing the firmware implementation of the algorithm. Convolutional layers are known to be particularly expensive in terms of hardware resources usage due to the type of operation they perform: the convolution defined in Eq. 4.1 necessitates the storage of several kernels, each being used multiple times. Therefore, when implementing a CNN in an FPGA, the use of resources can easily reach unfeasible levels. To prevent this, the TAUMINATOR exploits one single CNN, which is shared between the identification and calibration compartments. The choice of the hyperparameters of the architecture of the CNN is also made anticipating the firmware implementation of the algorithm. Thus, several combinations were tested, and a compromise was found between the dimensions of the network and the performance it yields. Smaller architectures were particularly difficult to achieve for the calibration component, which needs a substantially larger dense network compared to the identification; this limitation is further elaborated in the following.

In the particularly busy 200 PU environment of HL-LHC, the rejection of PU-induced contributions is of crucial importance; therefore, the training of the algorithm is performed giving precedence to the identification of τ_h candidates over their calibration. Therefore, the training procedure foresees two separate training loops (identification and calibration) to be performed for each section (barrel and endcap) of the TAUMINATOR algorithm. The first training is executed to obtain the highest possible L1 τ_h identification performance. The shared convolutional NN is then *frozen*, and its weights are fully set and will not be trained further. The calibration dense NN is then trained using the frozen version of the shared convolutional NN. Considering this procedure, it is now easier to understand why the fully connected calibration NN is considerably bigger in dimensions compared to the one used for identification: the calibration needs to be performed on a frozen version of the convolutional network. Moreover, the shared convolutional component entails max-pooling layers, which subsample the feature maps by selecting the largest features; this procedure is particularly well suited for the identification of the L1 τ_h candidates but, at the same time, it reduces the available information to be used for the calibration, effectively dropping part of the energy deposit in the intermediate feature maps.

The training in the barrel and endcap sections is identical, and it is executed with the following specifications:

- Shared Convolutional NN + Identification Dense NN
  - ◇ batch size $N = 1024$; testing sample = 10%,
  - ◇ binary cross entropy [27] loss function: $-\frac{1}{N}\sum_{i=0}^{N}[y_i \cdot \log \hat{y}_i + (1 - y_i) \cdot \log(1 - \hat{y}_i)]$,
  - ◇ Adam minimizer with initial learning rate 10^{-3},
  - ◇ learning rate reduction on the loss plateau, with reduction factor 1/2,
  - ◇ early stopping of the training if no improvement is made over the 10 previous epochs.
- Calibration Dense NN
  - ◇ convolutional NN weights frozen from the previous step,
  - ◇ sample weighting to achieve flattening of the τ_h generated p_T distribution,
  - ◇ batch size $N = 1024$; testing sample = 10%,
  - ◇ mean absolute percentage error loss function: $\frac{100}{N}\sum_{i=0}^{N} |\frac{y_i - \hat{y}_i}{y_i}|$,
  - ◇ Adam minimizer with initial learning rate 10^{-3},
  - ◇ learning rate reduction on the loss plateau, with reduction factor 1/2,
  - ◇ early stopping of the training if no improvement is made over the 10 previous epochs,

where we used the notation $\hat{y}$ and y to identify the predicted and target quantity that the NN is learning, respectively.

The last ingredient for performing the training of the CNNs is the definition of the training, testing, and validation datasets and their pre-processing. As introduced above, the sample used for the training is obtained from $H \rightarrow \tau\tau$ and Drell-Yan $Z/\gamma^* \rightarrow \ell\ell$ MC simulated datasets, in which the signal is identified as any τ_h lepton with $p_T > 18$ GeV matched to a TauMinator candidate, while all other TauMinator candidates constitute the background. This selection leads to a total of $\sim 3.5 \cdot 10^5$ L1 τ_h candidates, of which $\sim 2.5 \cdot 10^5$ are in the barrel and $\sim 1.1 \cdot 10^5$ in the endcap; these are complemented with $\sim 3.5 \cdot 10^5$ randomly selected background candidates, leading to a perfectly balanced dataset. Of the sample thus obtained, 80% is selected as the training dataset, and 20% is reserved as a validation set; of the training set, the CNN randomly selects 10% of the inputs as the testing sample. While the training and testing samples are both used at training time to keep overtraining under control, the validation sample is used only for additional control of overfitting after the training has been completed.

A pre-processing of the TT-based inputs is performed to maximize the performance. As detailed in Sect. 4.2.2, CNNs are inherently well suited for translation invariant problems; conversely, they are not naturally rotational invariant. By construction, the identification of τ_h leptons based on E_T deposits is a problem with both translational and mirror invariance along the ϕ axis of the $CL^{5\times9}$. Given a τ_h lepton with transverse momentum p_T^τ and position (η^τ, ϕ^τ), the associated $CL^{5\times9}$ is invariant under any translation $\phi^\tau + \varphi$ with $\varphi \in [0, 2\pi]$. Moreover, the $CL^{5\times9}$ is invariant if the energy deposit in it is mirrored along the ϕ axis in the following way: if we defined $E_T^{(i,j)}$ the energy deposit in the TT with $i\eta = i$ and $i\phi = j$, using the notation of Fig. 4.14 $i\eta \in [-2, 2]$ and $i\phi \in [-4, 4]$, the cluster can be mirrored applying the transformation $E_T^{(i,j)} \leftrightarrow E_T^{(i,-j)}$ (effectively this amounts to applying a charge conjugation $\tau^+ \leftrightarrow \tau^-$). While the former invariance will be learnt with ease by the CNN, the latter needs to be enforced; thus, the input dataset undergoes a so-called *augmentation*, in which all signal candidates are duplicated, and the $CL^{5\times9}$s associated with the duplicates are flipped over the ϕ direction while keeping all other input information unchanged.

Finally, the CL^{3D}-based inputs are pre-processed to facilitate the training and maximize performance. As can be appreciated in Fig. 4.18, the features associated with the HGCAL CL^{3D} assume largely different values. This sort of input is known not to be well suited for NNs, which tend to perform better when the inputs have comparable ranges. Therefore, a standard linear scaling of the kind $\frac{x-\langle x\rangle}{\sigma_x}$ is employed, where each input is scaled by the mean and standard deviation of the differential distribution of the input itself.

Identification Training Performance

The training of the shared convolutional NNs and the dense identification NNs is performed simultaneously with the procedure detailed above for the barrel and endcap independently. The performance of the barrel and endcap sections are reported separately in Fig. 4.20. The identification performance is evaluated with the Receiver Operating Characteristic (ROC) curve, which is a graphical illustration of the diagnostic ability of a binary classifier system obtained by portraying the true positive rate against the false positive rate as the discrimination threshold is varied. An important figure of merit is also the Area Under the ROC (AUC), which constitutes a single-valued compression of the ROC information; the AUC values are also reported in the Figures and amount to 0.97 and 0.96 for the barrel and endcap, respectively. Alongside the ROC curve, the output score of the CNNs is reported, showcasing a good separation of the τ_h signal from the background. In all cases, the results are reported for both training and validation samples, highlighting the robust training of the networks with no presence of overfitting.

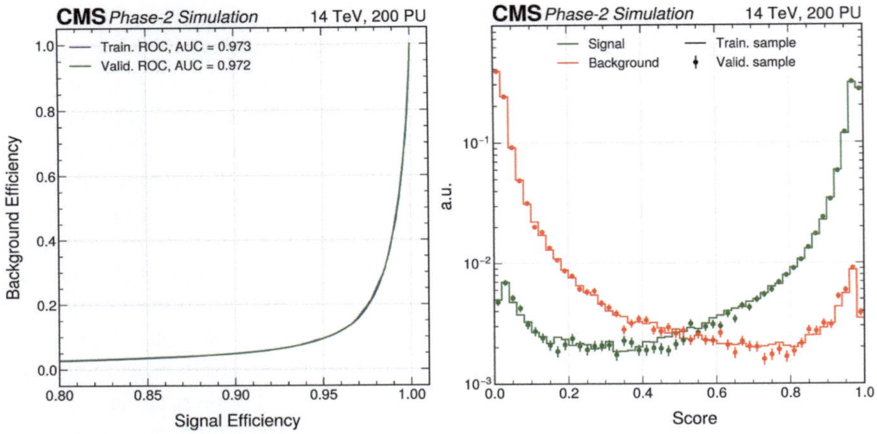

(a) TauMinator barrel section

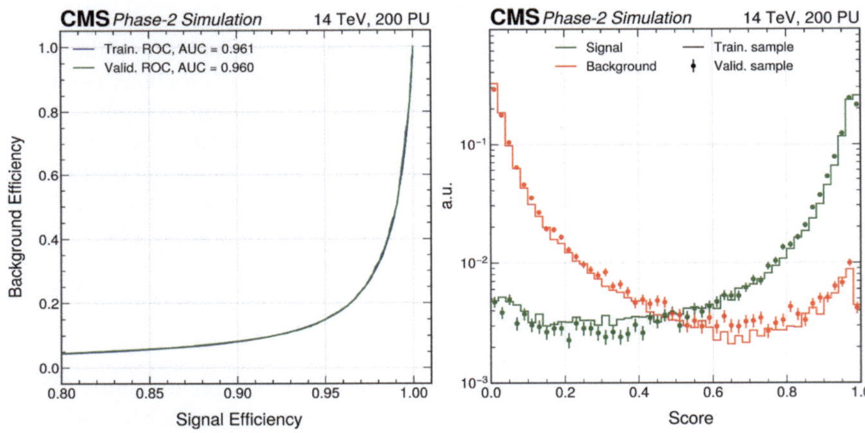

(b) TauMinator endcap section

Fig. 4.20 Receiver Operating Characteristic (ROC) curve (left) and identification score (right) for the barrel (**a**) and endcap (**b**) sections of the TauMinator algorithm. The results are shown for the training and validation datasets, showcasing no loss in performance and confirming the identification part of the CNN is not overfitting

Calibration Training Performance

The training of the dense calibration NNs is performed with the procedure detailed above, independently for the barrel and endcap, while keeping the convolutional NNs frozen. The performance of the barrel and endcap sections are reported separately in Fig. 4.21. The calibration performance is evaluated with the scale and response distributions. The former is shown in the form of the L1 calibrated p_T as a function of the generated τ_h transverse momentum; a very good scale of the L1 candidates is obtained

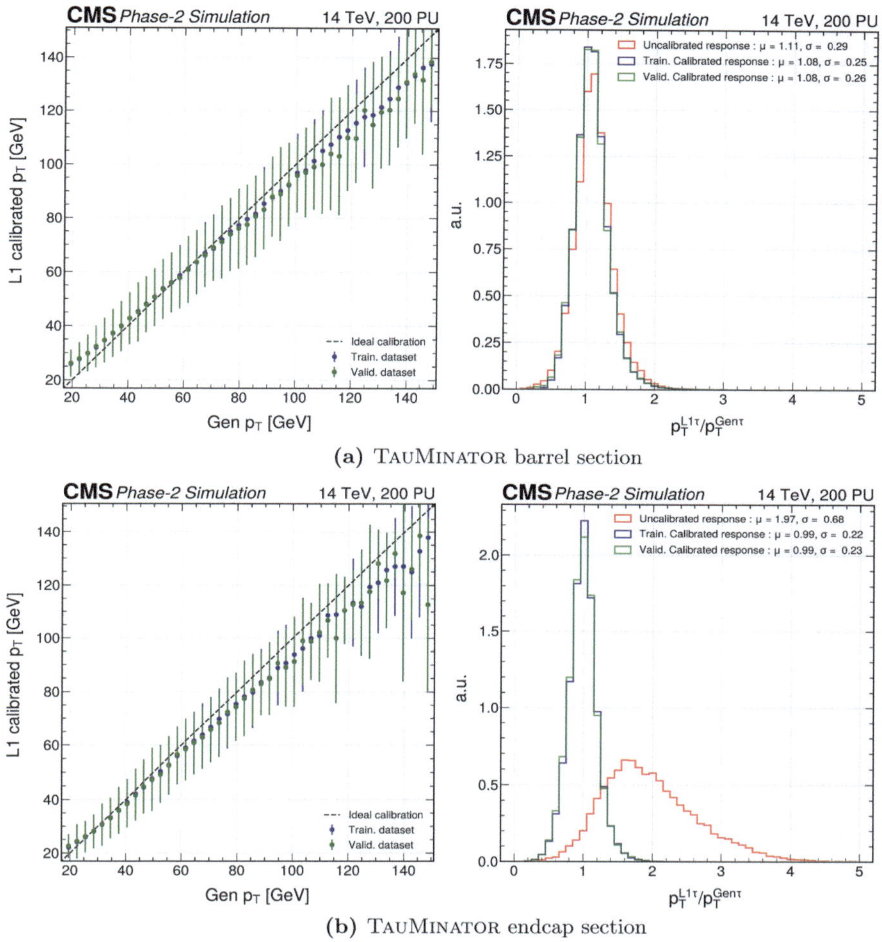

(a) TauMinator barrel section

(b) TauMinator endcap section

Fig. 4.21 Scale (left) and energy response (right) of the Level-1 τ_h with respect to the generated p_T for the barrel (**a**) and endcap (**b**) sections of the TauMinator algorithm. The results are shown for the training and validation datasets, showcasing no loss in performance and confirming the calibration part of the CNN is not overfitting

in the $30 \lesssim p_T \lesssim 60$ GeV region, while a small and consistent under-estimation at large p_T values is observed. Nevertheless, the energy response obtained as the ratio of the online estimated p_T over the generated p_T showcases the performance of the TauMinator algorithm, which reaches a reduction of the response RMS by 14% in the barrel and by an outstanding 74% in the endcap; the latter figure makes evident how fundamental the high granularity information is to mitigate the overwhelming PU contribution in the forward region. In all cases, the results are reported for both

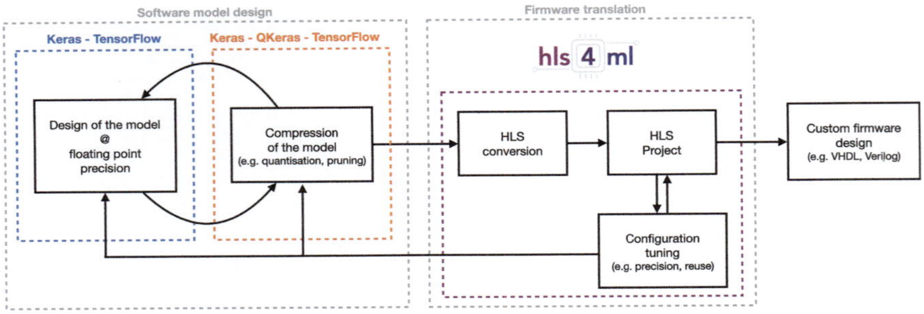

Fig. 4.22 Schematic representation of the workflow for the design of machine learning techniques for fast inference with the `hls4ml` package. A machine learning model designed with standard packages like Keras, and ideally compressed with tools like QKeras, can be translated into a custom firmware model that is implementable in FPGA. The firmware translation in `hls4ml` is achieved by exploiting the High-Level Synthesis (HLS) compiler package. The arrow identifies the typical progress of the design and feedback loops for more efficient architectures

training and validation samples, highlighting the robust training of the networks with no presence of overfitting.

4.2.6 *From Software to Firmware*

The TauMinator design outlined above is heavily influenced by the necessity to implement the CNNs into FPGA firmware; nevertheless, the architecture presented is built in Keras with TensorFlow backend to maximally exploit the interpretability of Python code when devising the algorithm. This results in an interpretable model, which uses a floating point precision architecture that is not easily implementable in FPGA firmware due to the enormous resources that floating point (32/64-bit) precision requires. Therefore, additional optimization steps need to be performed to achieve the final firmware-embedded model. A schematic representation of the workflow employed to reach this goal is reported in Fig. 4.22. The software design (blue box) was detailed in the previous Section, while this Section covers the compression of the model (orange box) as well as the firmware translation employing the `hls4ml` (High-Level Synthesis For Machine Learning) package [28] (purple box).

The compression of the TauMinator model aims at reducing the final firmware resources used by the CNN as much as possible while maintaining the desired performance. Two approaches are employed to reach this goal: the so-called *quantization* and *pruning* of the network. These two methods are exploited simultaneously to achieve maximal efficiency of the compression.

The quantization of a NN consists of the training of a network whose variables have been encoded into digital quantities of fixed precision; this is generally referred to as *quantization aware training*. In the case of the TauMinator this is achieved by

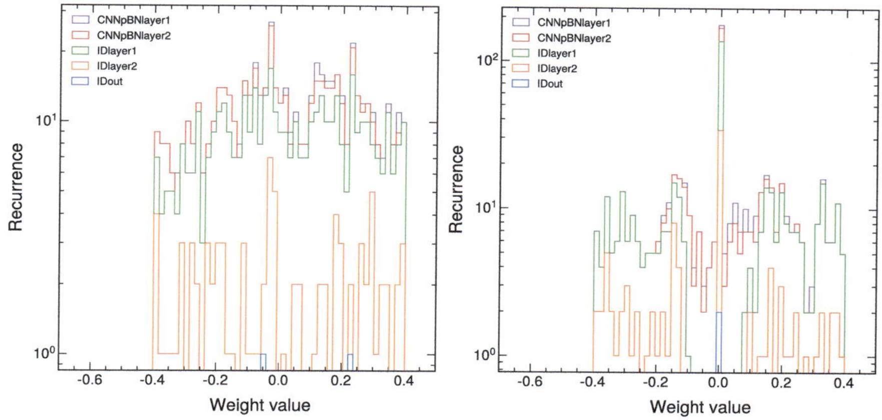

Fig. 4.23 Distribution of the TauMinator weights before (left) and after (right) the pruning procedure. Different colours represent different layers of the neural network. The large spike at 0.0 in the right panel corresponds to the pruned weights being removed; in this specific case, a pruning of 25% of the weights was enforced

exploiting the QKeras package with TensorFlow backend [29–31]. In this framework, all variables are encoded into `fixed<W,I>` (`ufixed<W,I>`) quantities which correspond to signed (unsigned) digital variable of `W` number of bits, of which `I` bits are reserved for the integer part. Following this approach, all weights and biases of the CNNs are encoded into `fixed<6,1>` quantities, while activation functions are encoded into several different digital quantities ranging from `fixed<8,1>` to `fixed<10,7>` depending on which layer they are associated to. Moreover, all inputs to the CNNs are encoded using the foreseen precision available in the Phase-2 L1 Trigger, as outlined in Tables 4.1 and 4.3 where the number of bits and the LSBs of the TPs are summarized.

Network pruning consists of systematically reducing the complexity of a NN by removing certain connections (i.e. weights) or entire neurons from the architecture, thus simplifying the NN structure. Pruning involves identifying and eliminating parameters that are deemed less important or redundant according to a pruning criterion, which assesses the relative contribution of each parameter to the network's overall performance. For the TauMinator algorithm, a magnitude-based pruning is adopted in association with the polynomial decay of the weights. In this approach, the magnitude of the weights is reduced in a polynomial manner, gradually encouraging their value to approach zero; when a weight reaches a low enough magnitude, it is set to a null value, thus effectively removing its impact on the network's computations. A visual representation of the effect of pruning on the identification network is reported in Fig. 4.23 where the distribution of the values of the weights of the NN is reported before and after pruning; after the pruning procedure is applied, a large spike appears at zero, highlighting the success of the strategy. In this specific case, a pruning of 25% of the weights was enforced.

The importance of network pruning is further understood by highlighting the relation between the FPGA resource usage and the number of arithmetical operations required for the application of the NN. The use of hardware resources is directly related to the volume of the data processed and the number of operations performed on this data. In the case of a NN, the total number of operations for its inference can easily be estimated to be

$$M_{\text{mult.}} = \sum_{l=0}^{L} N_{l-1} \times N_l \tag{4.3}$$

where L is the total number of layers and N_i is the number of neurons in the ith layer. This implies that the efficient removal of redundant weights and connections can reduce the number of multiplications and the resources used to perform them.

After quantization and pruning, the CNNs have been reduced to their simplest structure while using the smallest number of bits possible for weights encoding, thus ensuring an optimized firmware embedding of the algorithm. The final step toward FPGA implementation is the conversion of the software into a custom firmware design. This is achieved with the `hls4ml` package, which ensures the automatic translation of a trained NN into HLS (High-Level Synthesis) and VHDL (Very High-speed integrated circuits Hardware Description Language) code. Before detailing the use of `hls4ml`, it is first worth detailing some fundamental concepts about FPGAs and their available resources.

The most distinctive difference between software and firmware implementation of any algorithm is the capacity to execute independent operations in full parallelism. This owes to the intrinsic design of FPGAs, which can perform trillions of operations per second; nevertheless, this ability comes at the cost of the FPGA not being dynamically configurable during runtime. Therefore, crafting an optimal FPGA implementation necessitates a careful equilibrium between the consumption of FPGA resources and achieving the desired algorithm's throughput objectives. Several pivotal metrics govern the evaluation of a successful FPGA implementation:

- resources utilization, the most notable being:
  - ◇ Block RAM (BRAM): BRAMs are a memory resource embedded within FPGAs, organised as arrays of data storage elements. They allow for both reading and writing operations and are often used to implement data buffers and other memory-intensive tasks. Due to their speed and parallel access capabilities, they are particularly useful for applications requiring frequent data storage and retrieval.
  - ◇ Digital Signal Processing (DSP) blocks: DSPs are specialized hardware units designed for performing mathematical operations efficiently, especially those common in digital signal processing tasks. Each DSP block typically contains several multipliers, adders, and accumulators, allowing for fast execution of Multiply-Accumulate (MAC) operations. These blocks can be configured to perform various arithmetic operations like filtering and convolution.

- ◇ Flip-Flops (FFs): FFs are sequential logic elements used to store binary data. They can hold one bit of information and are edge-triggered, meaning they are synchronous to the FPGA clock (not to be confused with latches, which are identical in implementation to FFs but are asynchronous to the FPGA clock). FFs are the building blocks for registers and sequential logic circuits; they enable FPGAs to remember previous states and can create memory elements (buffers) within a digital design.
- ◇ Look-Up Tables (LUTs): LUTs are components to implement combinatorial logic functions. LUTs store predefined logic truth tables that define the output based on the input combinations, thus storing a small amount of input data and the corresponding outputs. By arranging and interconnecting multiple LUTs, complex logic functions can be built, putting LUTs at the heart of the FPGA's ability to perform various digital computations and implement different logic operations.

- Initiation Interval (II): II represents the minimum number of clock cycles required before the system can accept and start processing a new input, thus defining the time interval between the start of consecutive inferences. Achieving a lower initiation interval is desirable for real-time or high-throughput applications like the L1 trigger, as it allows the system to process data at a faster rate.
- Latency: latency refers to the total time delay or the number of clock cycles required for an operation to complete from the time the operation is initiated. It encompasses the time taken for data to travel through various stages of processing, including input buffering, computation, and output buffering.

This brief overview of the characteristics of an FPGA showcases how the direct design of an FPGA implementable algorithm is a highly non-trivial task, requiring the precise allocation of many different resources to specific components of an algorithm. This becomes even more complex when ML techniques are involved, with *hand-operated* implementation of a NN into FPGA firmware requiring several weeks of work and necessitating dedicated engineering support. To alleviate this highly demanding need when implementing ML algorithms in FPGA, the `hls4ml` package has been developed to ensure the automatic synthesization of a trained NN model into HLS and VHDL code. The `hls4ml` tool provides a large collection of adaptable parameters, facilitating users in exploring and tailoring the realm of trade-offs involving latency, II, and resource utilization pertinent to the specific application at hand. In practical terms, the time invested in executing the `hls4ml` automatic translation for a NN is notably shorter, spanning minutes to hours, compared to the considerable duration typically required to architect a dedicated NN configuration for FPGA compatibility. This capability facilitates the prototyping of ML algorithms, thus enhancing a physicist's capability to efficiently devise physics-oriented algorithms for the L1 trigger and largely reducing the time elapsing from algorithm conceptualization to practical application.

For the optimization of the TauMinator firmware implementation, several parameters provided by the `hls4ml` tool are exploited. The first is a high-level parameter that regulates how the Input/Output (IO) is handled; two strategies are available: the *parallel IO* and the *stream IO*, while the former consists of data being transmitted in parallel between the FPGA and the external system, the latter involves transmitting data as a continuous stream. Given the low-latency inference needed in the L1 trigger, the *parallel IO* is exploited. The second high-level parameter regulates the strategy used by the HLS compiler during the NN synthesization; two strategies are available: the *latency strategy*, which aims at reducing the total inference latency by exploiting more FPGA resources, and the *resources strategy*, which strives for lower resources usage at the cost of longer latency. Considering the well-known long latency issues that convolutional layers tend to create, the shared convolutional NNs of the TAUMINATOR are implemented using the *latency strategy*, while all other NNs use the *resources strategy*. The third high-level parameter is the *reuse factor* (R), which regulates the parallelization of the inference calculation; this effectively determines the trade-off between II, latency, throughput, and FPGA resource usage. All NNs of the TAUMINATOR algorithm are synthesized with $R = 1$, meaning that each multiplier is used only once in the computation of a layer's neuron output; this amounts to a full parallelization of the computations. This approach ensures the lowest inference latency at the cost of higher resources usage. The importance of the dedicated tuning of R is understood by highlighting the dependence of the latency on it. Consider a single layer of a NN; its inference latency can be expressed as

$$\mathcal{L} \sim \mathcal{L}_{\text{mult.}} + (R - 1) \cdot \text{II}_{\text{mult.}} + \mathcal{L}_{\text{activ.}} \tag{4.4}$$

where $\mathcal{L}_{\text{mult.}}$ is the latency of the multiplier, $\text{II}_{\text{mult.}}$ is the initiation interval of the multiplier, and $\mathcal{L}_{\text{activ.}}$ is the latency of the activation function computation; the approximation comes from possible additional latency for signal routing. Therefore, small R values are particularly suited for NN implementations that require large II.

Additionally, several low-level parameters are tuned to achieve the best firmware implementation. These parameters are all related to the precision of each computation performed inside the TAUMINATOR architecture. Owing to the high flexibility of `hls4ml`, a tailored precision is specified for each input, output and accumulation process in each layer of the CNNs; given their total number, they will not be specified here in the text. Nonetheless, one low-level parameter which is worth specifying is the *parallelization factor*, which sets the number of input instances that can be executed concurrently in a single clock cycle. This factor is exploited to reduce the II of the algorithm by parallelizing the input of the first layer of the CNN. A parallelization factor of 8 was chosen to reduce the II below the HL-LHC bunch crossing period, reaching an II of eight clock cycles (22.2 ns). This should be compared to an II of 55 clock cycles (148 ns) when a parallelization factor of 1 was used. This II reduction comes at the cost of the total number of resources used by the shared convolutional NN increasing by $\sim$15%.

The complete summary of the HLS implementation of the TAUMINATOR algorithm via the `hls4ml` tool is given in Figs. 4.24 and 4.25 for the barrel and endcap sections, respectively. All high- and low-level parameters exploited for the HLS compilation are explicitly reported in the flowcharts. The firmware synthesization is performed targeting a Xilinx Virtex UltraScale+ VU13P FPGA with part number

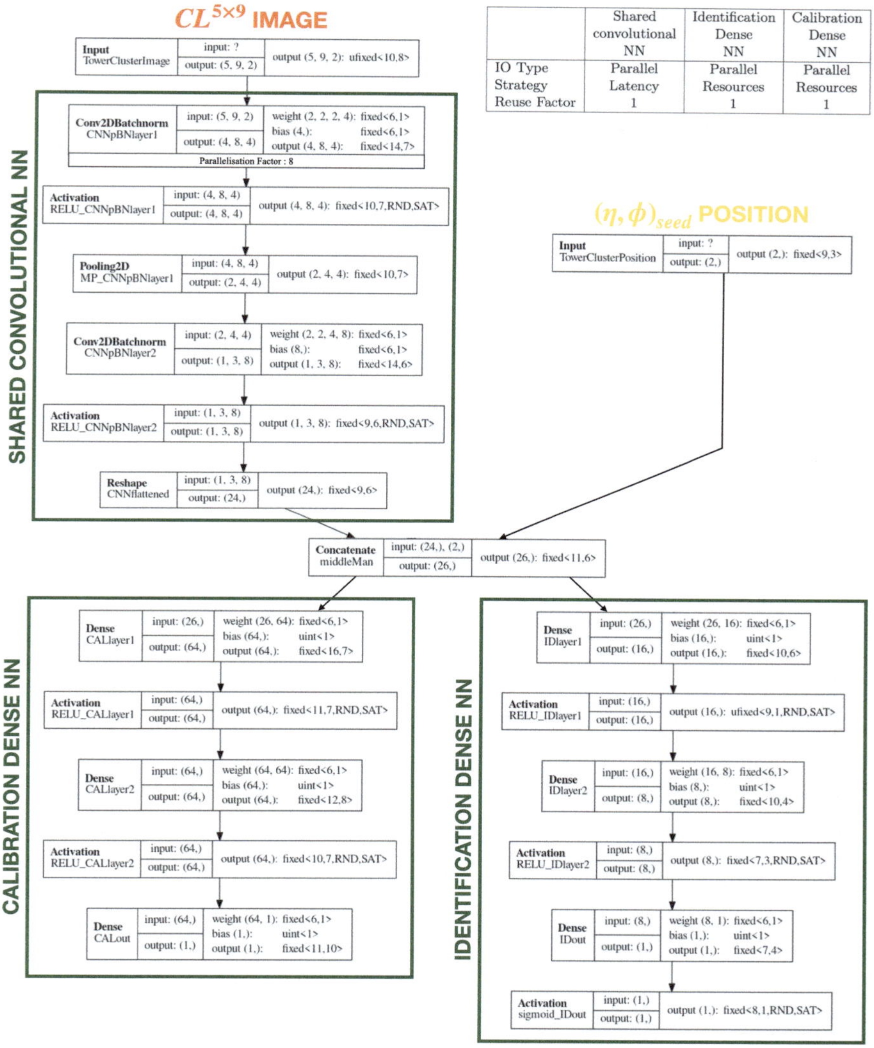

	Shared convolutional NN	Identification Dense NN	Calibration Dense NN
IO Type	Parallel	Parallel	Parallel
Strategy	Latency	Resources	Resources
Reuse Factor	1	1	1

Fig. 4.24 Flowchart summarizing the HLS implementation of the barrel section of the TauMinator algorithm via the `hls4ml` tool. The three green boxes highlight the three networks of the CNN architecture (the same naming of Fig. 4.19 is used); the inputs are highlighted in orange and yellow. Each black box in the diagram represents a layer of the neural network; the activation function employed can be inferred from the name of the activation layer. Each box specifies the input and output tensor dimensions and the bit encoding of the layer weights, biases, and output. The variable type `fixed<W,I>` (`ufixed<W,I>`) corresponds to a signed (unsigned) digital variable of `W` number of bits, of which `I` bits are allocated to the integer part. The acronyms `RND` and `SAT` stand for *rounding* and *saturation* of the variable, respectively. The table at the top summarises the input/output (IO) type, the synthesization strategy, and the reuse factor used for each neural network

	Shared convolutional NN	Identification Dense NN	Calibration Dense NN
IO Type	Parallel	Parallel	Parallel
Strategy	Latency	Resources	Resources
Reuse Factor	1	1	1

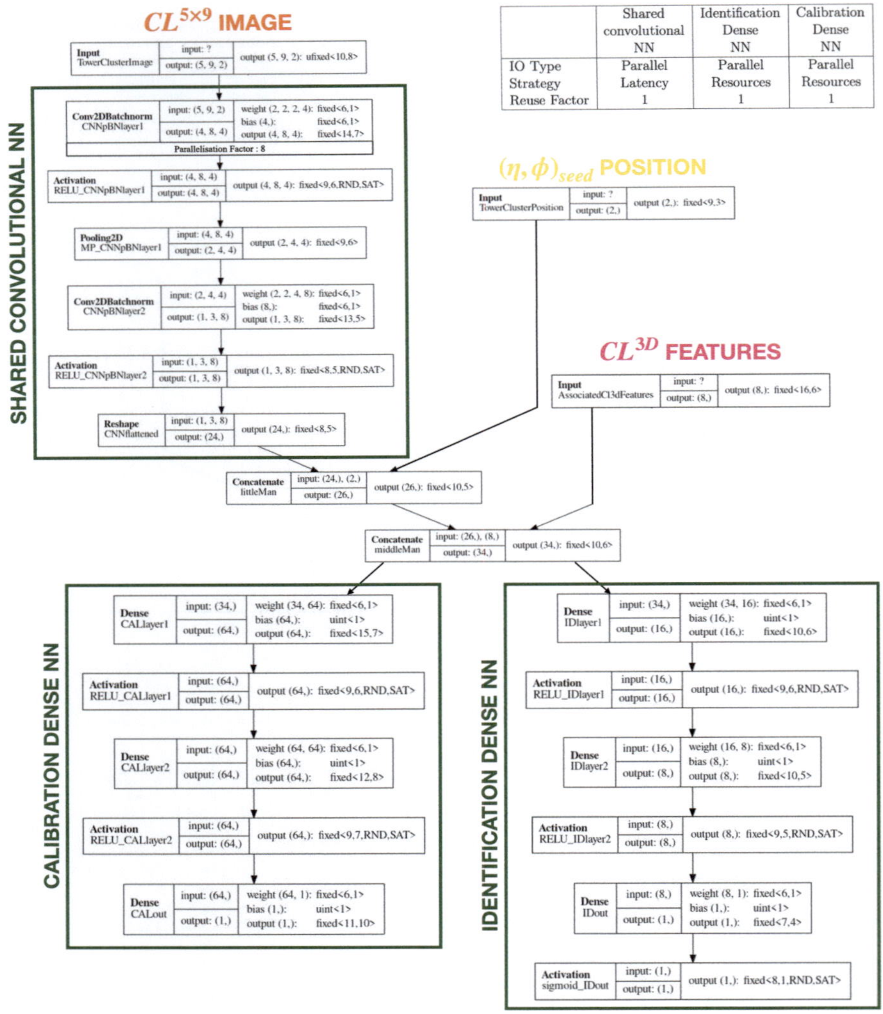

Fig. 4.25 Flowchart summarizing the HLS implementation of the endcap section of the TauMinator algorithm via the `hls4ml` tool. The three green boxes highlight the three networks of the CNN architecture (the same naming of Fig. 4.19 is used); the inputs are highlighted in orange, yellow, and pink. Each black box in the diagram represents a layer of the neural network; the activation function employed can be inferred from the name of the activation layer. Each box specifies the input and output tensor dimensions and the bit encoding of the layer weights, biases, and output. The variable type `fixed<W,I>` (`ufixed<W,I>`) corresponds to a signed (unsigned) digital variable of `W` number of bits, of which `I` bits are allocated to the integer part. The acronyms `RND` and `SAT` stand for *rounding* and *saturation* of the variable, respectively. The table at the top summarises the input/output (IO) type, the synthesization strategy, and the reuse factor used for each neural network

`xcvu13p-fhgb2104-2L-e`. All results reported in the following are obtained for this FPGA operated at a clock frequency of 360 MHz, which is typical for the first stages of the CMS L1 trigger.

The estimates of the main resources usage, the II, and the latency of each part of the TAUMINATOR algorithm are reported in Table 4.7 for the barrel and endcap sections, respectively. All components of the CNN architecture require a very small percentage of FPGA resources, generally remaining below 1% for all major resources, thus highlighting the great optimization power granted by the `hs4ml` package. It should be noted that the resources reported are for a single instance of the algorithm; therefore, the TAUMINATOR would be well suited for a time-multiplexed trigger architecture.

One important remark about resource usage is how the barrel and endcap sections of the TAUMINATOR algorithm require roughly the same amount of resources, despite the fact that for the HGCAL the CL^{3D} information is also exploited in the endcap. This is achieved owing to the optimal and independent compression and firmware optimization of the two sections of the algorithm. One exception is represented by the use of DPSs by the calibration NN, which are ~7.5 times higher in the barrel than in the endcap. This discrepancy is understood in light of the default treatment of variables of different bit widths in the HLS compiler: HLS automatically treats variables encoded into a number of bits ≤ 9 for LUT implementation, while it routes to DPS implementation variables encoded into a number of bits ≥ 10. By comparing the flowcharts in Figs. 4.24 and 4.25, we can notice that in the barrel, the two ReLU activation functions are encoded into `fixed<11,7>` and `fixed<10,7>` variables; conversely, in the endcap, they are encoded into `fixed<9,6>` and `fixed<9,7>`

Table 4.7 Summary of the main FPGA resources used by the barrel Table 4.7a and endcap Table 4.7b sections of the TAUMINATOR algorithm, alongside the Initialization Interval (II) and the Latency (Lat.) of each part of the algorithm. These results are obtained targeting a Xilinx Virtex UltraScale+ VU13P FPGA at a clock frequency of 360 MHz. The resources reported are for a single instance of the algorithm; therefore, the TAUMINATOR would be well suited for a time-multiplexed trigger architecture. The same naming of Fig. 4.19 is used for the networks

	LUT (%)	FF (%)	BRAM (%)	DSP (%)	II [ns]	Lat. [ns]
Shared convolutional NN	1.07	0.48	0.00	0.00	22.2	55.6
Identification dense NN	0.40	0.09	0.02	0.17	2.78	30.6
Calibration dense NN	1.68	0.39	0.00	3.28	2.78	38.9

(*a*) TAUMINATOR barrel section

	LUT (%)	FF (%)	BRAM (%)	DSP (%)	II [ns]	Lat. [ns]
Shared convolutional NN	1.06	0.40	0.00	0.00	22.2	55.6
Identification dense NN	0.53	0.03	0.02	0.16	2.78	27.8
Calibration dense NN	1.22	0.11	0.00	0.44	2.78	27.8

(*b*) TAUMINATOR endcap section

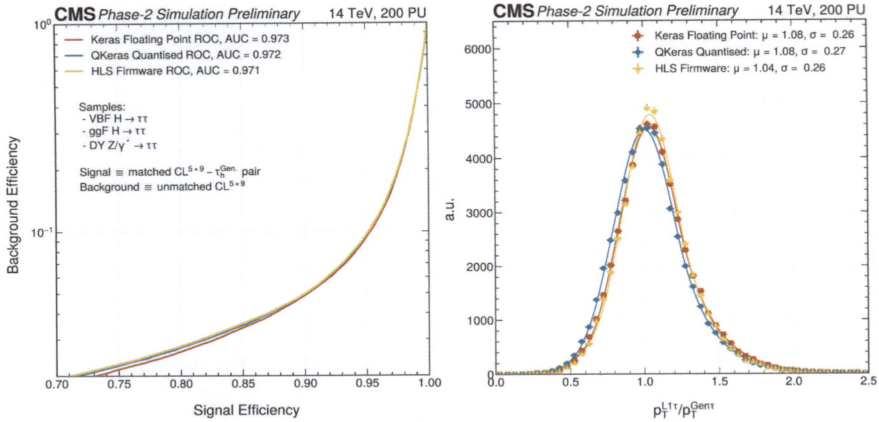

(a) TauMinator barrel section

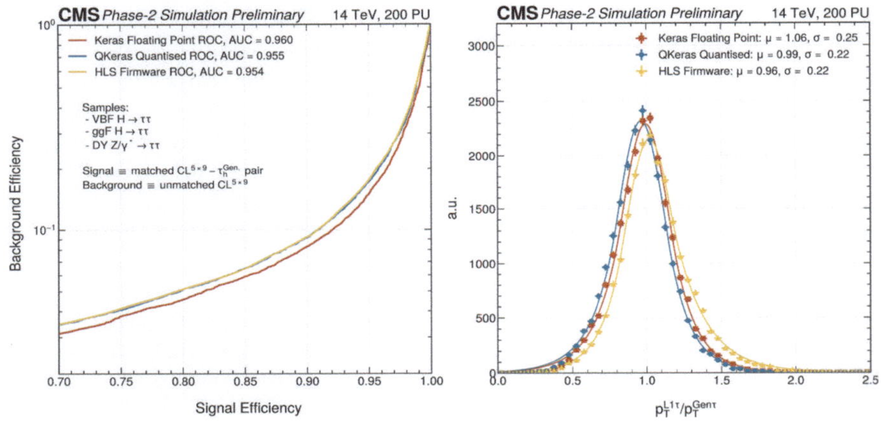

(b) TauMinator endcap section

Fig. 4.26 Receiver Operating Characteristic (ROC) curve (left) and energy response of the Level-1 τ_h with respect to the generated p_T (right) for the barrel (**a**) and endcap (**b**) sections of the TauMinator algorithm. The results are shown for the three steps of the design, i.e. Keras software (red), QKeras quantized and pruned software (blue), HLS firmware implementation (yellow), showcasing minimal loss of performance is achieved in all the steps

variables. This contrast plays at the edge of the HLS default behaviour, thus creating the understood resource usage difference.

When translating the TauMinator algorithm from software to firmware, concurrently to the optimization of the FPGA resource usage, it is imperative to preserve its physics performance. This is achieved by fine-tuning all the available variables for the network compression and all `hls4ml` parameters for the firmware synthesization while keeping track of the performance gain/loss at each change. The performance attained at each step of the design process (i.e. Keras software, QKeras quantized and pruned software, HLS firmware implementation) is reported in Fig. 4.26. As it can be appreciated, minimal loss in performance is achieved at each step, highlighting the successful adaptation of the TauMinator algorithm to the hardware constraints set by the L1 trigger FPGAs.

4.2.7 Firmware Deployment

The final step of the design of an FPGA implementable L1 trigger algorithm is the deployment of the firmware in a so-called *demonstrator board* and its validation to ensure that no differences exist with respect to the algorithm emulation in C++ and that the expected performance is achieved. This procedure is generally performed in several stages, from the test of the system in a controlled environment to the final installation and integration in the racks of the CMS service cavern at the LHC. Given the timeline of the TauMinator implementation at the HL-LHC, the only step that can be currently performed is the controlled environment test in a single FPGA board. A full implementation of the algorithm into firmware would require the definition of the complete pipeline of data from the detectors' back-end to the input of the TauMinator algorithm. Therefore, to be able to perform some hardware tests within the time constraints of this Thesis, the decision was taken only to perform a proof-of-concept test, as detailed below.

For this test, simulated data is used to build the inputs to the CNNs using C++ code. The choice of not implementing the clustering procedure into the FPGA is based on the knowledge that this step does not pose major problems for firmware implementation. The current Phase-1 implemented algorithms already perform non-trivial clustering processes, like the dynamical clustering of TTs for the Run-3 L1 τ_h candidates reconstruction, without significant challenges. Moreover, the firmware deployment is performed only for the shared convolutional NNs of the barrel and endcap section of the TauMinator, effectively the algorithm's most complex part. This choice is taken due to some hardware limitations of the board available for testing.

The TauMinator firmware deployment is performed in a Xilinx Alveo U200 Accelerator Platform [32] with part number `xilinx_u200_xdma_201830_2`. The firmware translation of the CNNs is achieved using the `hls4ml` tool, interfaced with the Vivado accelerator backend, using the same implementation parameters detailed above and reported in Figs. 4.24 and 4.25. The results obtained from operating the algorithm in the physical hardware are compared to the `hls4ml` emulated version of the networks. The results are reported in Fig. 4.27 for the barrel and endcap sections of the TauMinator algorithm; in both cases, the 24 outputs of the shared convolutional NN are compared between firmware and emulator prediction. In all cases, the bit-wise agreement is achieved.

This kind of test constitutes a preliminary proof-of-concept of the possibility of successfully implementing the TauMinator algorithm in FPGA firmware. The complete implementation of the algorithm is foreseen for the future years.

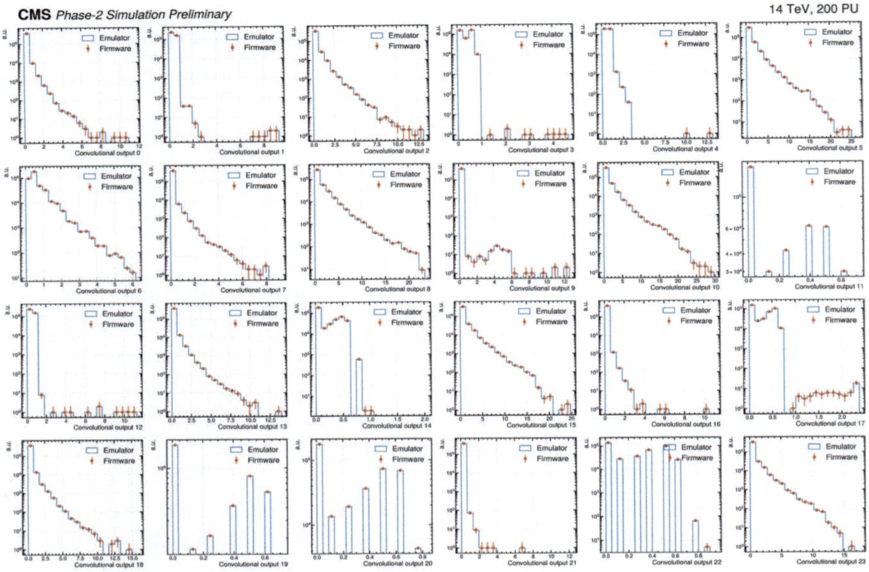

(a) TAUMINATOR barrel section

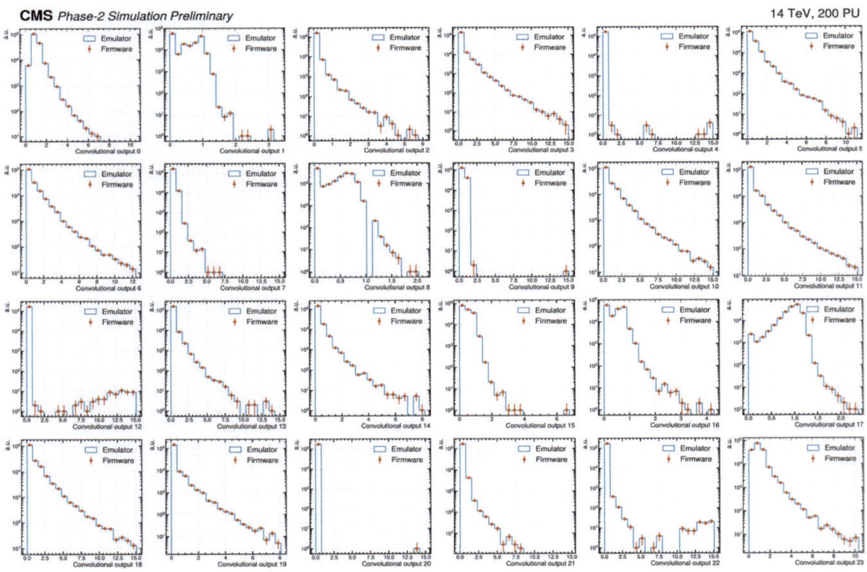

(b) TAUMINATOR endcap section

Fig. 4.27 Comparison between `hls4ml` emulator and firmware for the 24 output features of the convolutional part of the barrel (**a**) and endcap (**b**) sections of the TAUMINATOR algorithm. This comparison is obtained by deploying the algorithm on a Xilinx Alveo U200 board; the bit-wise agreement is reached on all variables

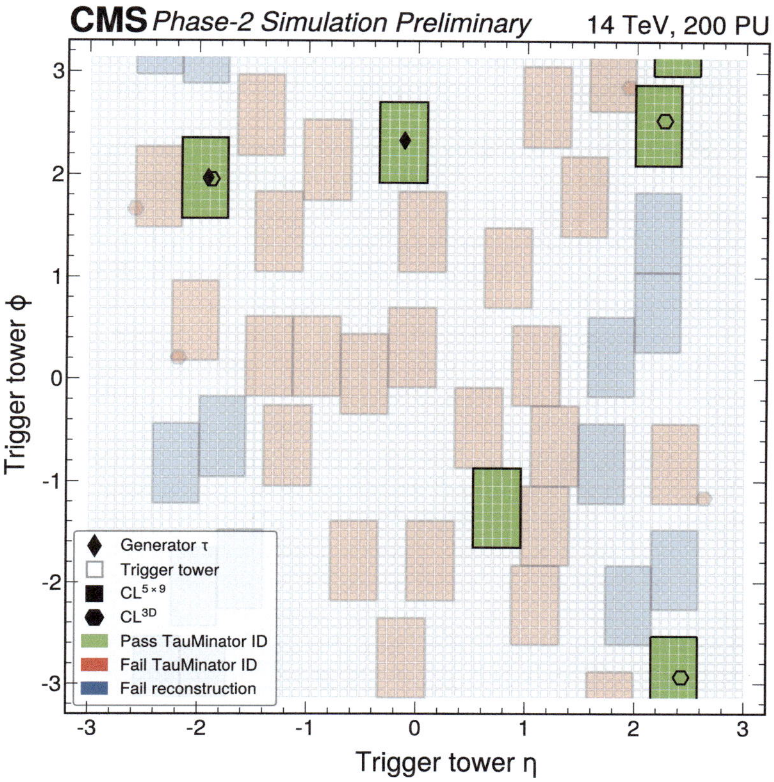

Fig. 4.28 A display of an HH → bb$\tau\tau$ event as the TauMinator algorithm reconstructs it: each dim-grey square represents a TT in input to the L1 Trigger; the coloured TTs represent the $CL^{5\times9}$ clusters reconstructed by the algorithm in the specific event. Solid black diamonds represent the event's two generator-level τ_h. The event is divided into two regions: the barrel for $|\eta| < 1.55$, and the endcap for $|\eta| > 1.55$. In the barrel, all $CL^{5\times9}$ clusters that are reconstructed are processed for identification and calibration using a series of neural networks designed for the purpose. Only the clusters satisfying the proximity requirement $\Delta R(CL^{5\times9}, CL^{3D}) < 0.5$ are processed for identification and calibration in the endcap. In the event display, the processed candidates passing the TauMinator ID are shown in green, while those failing it are shown in dim-red. The $CL^{5\times9}$ clusters in the endcap which do not pass the proximity requirement are shown in dim-blue

4.3 The TauMinator Performance

The Phase-2 of the LHC accelerator is scheduled to start in 2029 when the HL-LHC machine will begin operations. In this new era of experimental physics, collisions will be delivered at a centre-of-mass energy $\sqrt{s} = 14$ TeV and at an instantaneous luminosity of $5-7.5 \cdot 10^{34}\, \text{cm}^{-2}\, \text{s}^{-1}$. This will represent unprecedented conditions that will allow the CMS Collaboration to achieve a highly ambitious physics program. These running circumstances will also represent the most complex conditions ever experienced at particle colliders, with the number of simultaneous collisions per bunch crossing reaching as high as 200. In such a busy environment, efficiently collecting data will be of primary importance, and very high demands will be set for the Level-1 trigger.

In this context, a novel technique for the reconstruction, identification, and calibration of τ_h candidates in the L1 trigger has been developed within this Thesis work: the TAUMINATOR algorithm. This algorithm is the first of its kind, it exploits in a unified approach all TPs available in the barrel and endcap regions of the CMS detector and identifies L1 τ_h candidates exploiting a CNN architecture. The design and firmware implementation of the algorithm were discussed in the previous Sections; in the following, the official performance of the TAUMINATOR algorithm is presented. These results are obtained with the public version of the algorithm as it is implemented in the CMS Software (CMSSW); in all figures, the explicit comparison with the already existent CALOTAU algorithm is showcased.

To grasp further the workings of the algorithm, Fig. 4.28 reports the display of an HH $\rightarrow$ bb$\tau\tau$ event as the TAUMINATOR reconstructs it. The algorithm starts with the reconstruction of the CL$^{5\times9}$ clusters from TTs, which are represented by coloured rectangles of extension $\eta \times \phi = 5 \times 9$. In the barrel section of the algorithm, i.e. $|\eta| < 1.55 \sim |i\eta| \leq 18$, all CL$^{5\times9}$s are processed by the algorithm; in the endcap section, i.e. $|\eta| > 1.55 \sim |i\eta| \geq 19$, the CL^{3D} from the HGCAL detector, represented by hexagons, are also exploited and only clusters satisfying the proximity requirement $\Delta R(\text{CL}^{5\times9}, \text{CL}^{3D}) < 0.5$ are processed for identification and calibra-

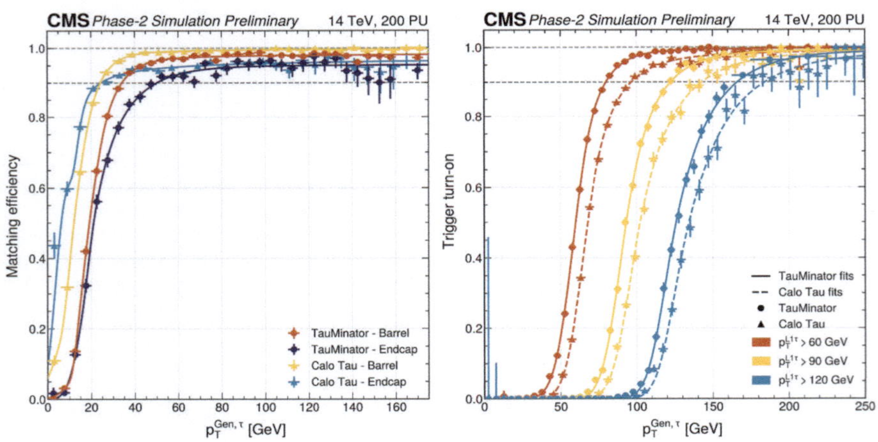

Fig. 4.29 Comparison of the matching efficiency (left) and the trigger turnons (right) of the TAUMINATOR algorithm and the CALOTAU algorithm, as a function of generated p_T. The matching efficiency is computed as the fraction of generated τ_h that are geometrically matched to an L1 τ_h candidate; the trigger turnon is defined as the fraction of matched L1 objects that pass a specific p_T threshold. The comparison is evaluated in MC simulated gluon fusion HH $\rightarrow$ bb$\tau\tau$, and the functional form of the fits consists of a cumulative Crystal Ball function convolved with an arc-tangent in the high p_T region

tion. The clusters failing the proximity requirement are shown in dim-blue colour, and those processed by the TauMinator algorithm are shown in green or dim-red if they pass or fail the identification, respectively. The two MC generated τ_h in the event are represented by solid black diamonds; in this specific event, both τ leptons are efficiently reconstructed by the TauMinator.

Figure 4.29 reports the matching efficiency and the trigger turnons as a function of generated p_T of the TauMinator algorithm compared to those of the CaloTau algorithm. The matching efficiency is computed as the fraction of generated τ_h that are geometrically matched to an L1 τ_h candidate; the trigger turnon is defined as the fraction of matched L1 objects that pass a specific p_T threshold. While the TauMinator matching efficiency is mostly comparable to the one of the CaloTau algorithm, showcasing a steep rise and a plateau approaching unity, the trigger turnons show a consistently better performance of the TauMinator algorithm owing to its better calibration of the L1 τ_h candidates. These results are evaluated in a MC simulated gluon fusion HH $\rightarrow$ bb$\tau\tau$ sample.

Figures 4.30 and 4.31, report the rate estimations obtained in a MC simulated minimum-bias dataset. All rates are presented as a function of the offline threshold, which is evaluated as the generator p_T value at which the trigger turnon

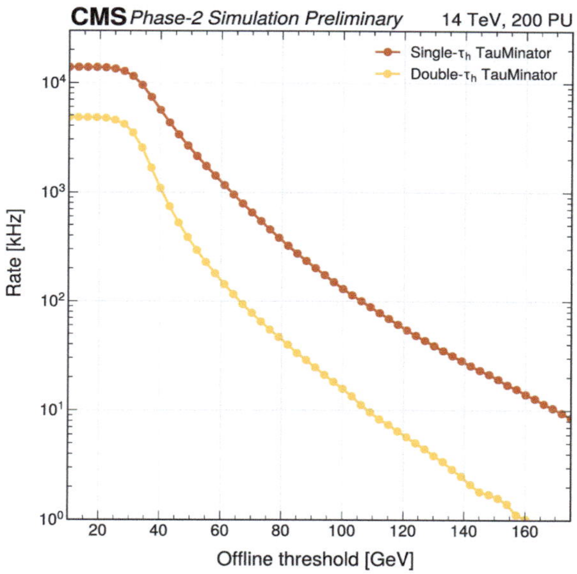

Fig. 4.30 Single-τ_h and Double-τ_h rates as a function of the offline p_T, defined as the generator p_T value at which the trigger turnon crosses the 90% efficiency point; for the Double-τ_h rate, the same threshold on both objects is applied. The TauMinator algorithm ensures the following improvements over the CaloTau algorithm: a reduction of the Double-τ_h rate by 21% (from 33 kHz to 26 kHz) at a threshold of 90 GeV on both objects; or conversely, a reduction of the threshold by 5 GeV on each object at a fixed rate of 33 kHz

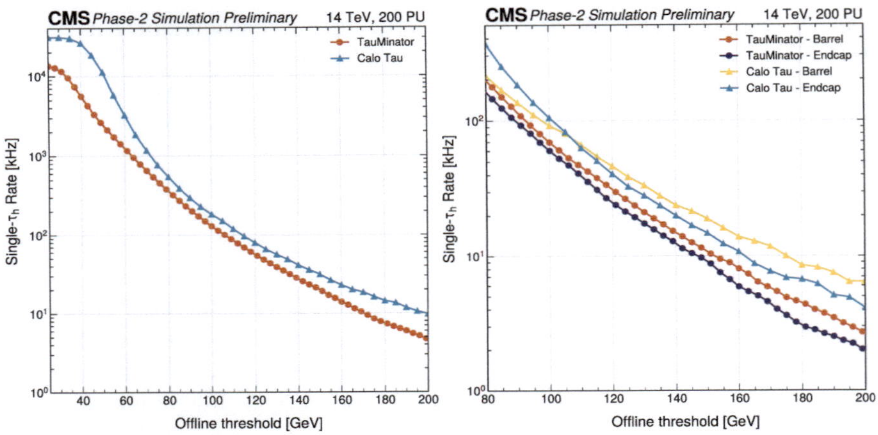

Fig. 4.31 Inclusive single-τ_h rate (left) and separate barrel and endcap single-τ_h rate contribution (right) as a function of the offline p_T, defined as the generator p_T value at which the trigger turnon crosses the 90% efficiency point, for the TauMinator algorithm and the CaloTau algorithm. The TauMinator algorithm ensures the following improvements over the CaloTau algorithm: a reduction of the inclusive rate by 37% (from 31.4 kHz to 19.8 kHz) at a threshold of 150 GeV; or conversely, a reduction of the threshold by 14 GeV at a fixed rate of 31.4 kHz; a reduction of the rate by roughly 40% (35%) in the barrel (endcap) at a threshold of 150 GeV

crosses the 90% efficiency point. The TauMinator consistently outperforms the CaloTau algorithm in both Double- and Single-τ_h rate estimations. In the case of Double-τ_h rates, the TauMinator algorithm ensures the following improvements: a reduction of the rate by 21% (from 33 kHz to 26 kHz) at a threshold of 90 GeV on both objects; or conversely, a reduction of the threshold by 5 GeV on each object at a fixed rate of 33 kHz. In the case of the Single-τ_h rate, the TauMinator algorithm guarantees the following advancements: a reduction of the inclusive rate by 37% (from 31.4 kHz to 19.8 kHz) at a threshold of 150 GeV; or conversely, a reduction of the threshold by 14 GeV at a fixed rate of 31.4 kHz. When considering the separate contributions to the rate from the barrel and endcap section, a reduction of the rate by roughly 40% (35%) in the barrel (endcap) at a threshold of 150 GeV is attained.

4.4 Triggering the Future with Machine Learning

In this Chapter, we discussed the development of a fully ML-based algorithm developed for the L1 trigger of CMS. This novel approach showcases important performance and exploits state-of-the-art techniques; nevertheless, it represents only a portion of the immense capabilities of ML techniques implemented in FPGAs. This Section serves as a brief perspective view of this fast-growing field.

The future of the L1 trigger is highly tied to the experience of the past. Run-1 demonstrated that a full field view of the detector was crucial to enhance selectivity; Run-2 and Run-3 proved that algorithms that treat TPs with dynamical association approaches are crucial to optimally reconstruct objects in high PU conditions and can be easily fitted to the changing running conditions; the planning of the Phase-2 upgrade is already showcasing that state-of-the-art hardware can be used to perform low latency inference based on ML techniques. Moreover, Run-3 demonstrates how the increased depth segmentation of the HCAL calorimeter can be exploited for partial tracking of the shower development. The HGCAL subdetector will bring this several steps further with the most granular calorimeter ever built. These few considerations bring forward one question: can this experience be condensed into one comprehensive approach?

One good candidate might be the use of three-dimensional (3D) ML vision algorithms with a global event view. Such an algorithm would be more complex than the current algorithms; while the most sophisticated algorithms for the Phase-2 L1 trigger now exploit CNNs, 3D ML vision would most probably require GNNs, which are technically an extension of CNNs, with additional invariance properties and increased performance. GNNs are particularly suited for sparse input like that of the energy deposits in the CMS deposit, as they process it by building information graphs and running convolutions on their nodes and edges. Important advancement is already being made in the implementation of GNNs in hardware within the L1 trigger community; therefore, it is not unlikely that they will become our next golden spear. The input to a 3D ML vision algorithm would be highly granular TPs, giving a full 3D mapping of the energy deposit in the CMS subdetectors. The network would then be trained to perform single- and multi-particle reconstruction by visualizing them in 3D. This would be a large improvement, especially for the calorimeter energy deposit, which would not be projected on the 2D plane of TTs anymore; this would fully exploit the 25-fold and 500-fold increase of readout channels in the barrel and endcap, respectively.

Would such an algorithm be flexible enough to adapt to the changing running conditions? Yes, if used in combination with Continual Learning (CL), an approach that is already being explored within the CMS L1 trigger team [33]. The idea is to perform a systematic retraining of ML techniques using the continuous stream of data as a training sample. Owing to largely improved hardware capabilities, it is not unlikely that CL could be implemented in online instrumentation, and ML technique could perform continuous unsupervised training based on the data that is being collected. It should be noted that this training would not happen in an FPGA but rather in a supporting GPU in which the algorithm is retrained, `hls4ml` is used for the automatic embedding of the updated structure into HDL, and the immediate implementation in the running hardware. This sort of procedure could even be performed at every inter-fill of the LHC.

The ultimate destination of the L1 trigger could finally be that of performing *online analysis* at 40 MHz, of which the Phase-2 L1 scouting system is the forebearer [8]. This would constitute the pinnacle of the L1 trigger capabilities, where an FPGA farm (analogous to the current HLT CPU+GPU farm) could perform a complete analysis of the data at the bunch crossing rate before taking the event accept or reject decision. Data buffering capabilities will eventually limit this design, as is already the case for the latency of the L1 trigger being limited by the buffering capacity of the tracker. Nevertheless, recent improvements in the field, allowing the storage of Tb of data with several seconds of buffer time [8], make this a viable possibility. In this context, ML techniques could be used to perform object reconstruction with offline-like performance, and simplified versions of offline analysis could be implemented to serve as seeds in the L1 menu. In this configuration, the event accept decision would be taken based on the event being eventually selected in a specific analysis signal region.

While the techniques defined above would still feature a standard triggering architecture, one idea could be redefining the concept of triggering in the first place. In the new configuration, the trigger would not serve only as the event selection tier but could serve as one section of a large reconstruction model. In the past few years, large language models the likes of GPT-4 [34] have redefined the boundaries of ML itself, with architectures including billions of weights. These enormous networks could also be envisaged in fundamental physics with the development of massive end-to-end networks. Such an architecture could take as input the raw information from the detectors (or even the more standard TPs) and be trained to perform the complete chain of trigger candidate building, online event selection, and offline object reconstruction. The architecture would be highly compartmentalised and *disassemblable* so that each piece could be implemented in specific hardware and software processors. In this configuration, the trigger would still serve as the data selection tier, but at the same time, it would perform a part of a much bigger scope end-to-end ML algorithm.

The ideas presented above correspond to a personal view of the direction the L1 trigger could take, owing to the great capabilities of ML implemented in FPGAs. These ideas stem from first-hand experience and from the knowledge that, at the time of writing this Thesis, physicists are developing ever more sophisticated algorithms, and engineers are designing ever more powerful hardware. Therefore, no matter the direction that will ultimately be taken, the certainty remains that the L1 Trigger is one of the most fundamental pieces in the architecture of a multipurpose detector like CMS. In the quest for new physics discoveries, the L1 trigger is our primary way to get *through the looking glass*.

References

1. Dainese A et al (2019) Report on the physics at the HL-LHC, and perspectives for the HE-LHC. Technical report. Geneva, Switzerland. https://doi.org/10.23731/CYRM-2019-007. https://cds.cern.ch/record/2703572
2. CMS Collaboration (2017) The phase-2 upgrade of the CMS muon detectors. Technical report. This is the final version, approved by the LHCC. Geneva, CERN. https://cds.cern.ch/record/2283189

3. CMS Collaboration (2018) Sensitivity projections for Higgs boson properties measurements at the HL-LHC. Technical report. Geneva, CERN (2018). https://cds.cern.ch/record/2647699
4. Contardo D et al (2015) Technical proposal for the Phase-II upgrade of the CMS detector. Technical report. Upgrade project leader deputies: Lucia Silvestris (INFN-Bari), Jeremy Mans (University of Minnesota) Additional contacts: Lucia.Silvestris@cern.ch, Jeremy.Mans@cern.ch. Geneva. https://doi.org/10.17181/CERN.VU8I.D59J. https://cds.cern.ch/record/2020886
5. Butler J et al (2015) CMS Phase II upgrade scope document. Technical report. Geneva, CERN. https://cds.cern.ch/record/2055167
6. Motta J (2024) Development and firmware implementation of a machine learning based hadronic Tau lepton Level-1 trigger algorithm in CMS for the HL-LHC. In: PoS EPS-HEP2023, p 590. https://doi.org/10.22323/1.449.0590
7. CMS Collaboration (2023) Hadronic Tau reconstruction in the CMS Phase-2 Level-1 trigger using NNs with calorimetric information. In: CMS detector performance summaries. https://cds.cern.ch/record/2868783
8. The Phase-2 upgrade of the CMS Level-1 trigger. Technical report. Final version. Geneva, CERN. https://cds.cern.ch/record/2714892
9. CMS Collaboration (2017) The Phase-2 upgrade of the CMS barrel calorimeters. Technical report. This is the final version, approved by the LHCC. Geneva, CERN. https://cds.cern.ch/record/2283187
10. CMS Collaboration (2017) The Phase-2 upgrade of the CMS endcap calorimeter. Technical report. Geneva, CERN. https://doi.org/10.17181/CERN.IV8M.1JY2. https://cds.cern.ch/record/2293646
11. CMS Collaboration (2023) HGCAL website. (Last visited: 2 Aug 2023). https://hgcal.web.cern.ch
12. CMS Collaboration (2017) The Phase-2 upgrade of the CMS tracker. Technical report. Geneva, CERN. https://doi.org/10.17181/CERN.QZ28.FLHW. https://cds.cern.ch/record/2272264
13. Collaboration CMS (2019) A MIP timing detector for the CMS Phase-2 upgrade. Technical report. Geneva, CERN. https://cds.cern.ch/record/2667167
14. PICMG (2023) PICMG - AdvancedTCA overview. (Last visited: 3 Aug 2023). https://www.picmg.org/openstandards/advancedtca/
15. Rose A et al (2019) Serenity: an ATCA prototyping platform for CMS Phase-2, p 115. https://doi.org/10.22323/1.343.0115
16. Xilinx AMD (2023) UltraScale+ FPGAs. (Last visited: 8 Aug 2023). https://docs.xilinx.com/v/u/en-US/ultrascale-plus-fpga-product-selection-guide
17. Hubel DH, Wiesel TN (1959) Receptive fields of single Neurones in the cat's striate cortex. J Physiol 148(3):574–591. https://doi.org/10.1113/jphysiol.1959.sp006308
18. Fukushima K (1980) Neocognitron: a self-organizing neural network model for a mechanism of pattern recognition unaffected by shift in position. Biol Cybern 36(4):193–202. https://doi.org/10.1007/bf00344251
19. LeCun Y et al (1989) Backpropagation applied to handwritten zip code recognition. In: Neural Comput 1(4):541–551. https://doi.org/10.1162/neco.1989.1.4.541
20. Lecun Y et al (1995) Learning algorithms for classification: a comparison on handwritten digit recognition. English (US). In: Oh JH, Kwon C, Cho S (eds) Neural networks. World Scientific, pp 261–276
21. Frixione S, Nason P, Oleari C (2007) Matching NLO QCD computations with parton shower simulations: the POWHEG method. JHEP 11:070. https://doi.org/10.1088/1126-6708/2007/11/070. arXiv:0709.2092 [hep-ph]
22. Re E (2011) Single-top Wt-channel production matched with Parton showers using the method. Eur Phys J C 71:1547. https://doi.org/10.1140/epjc/s10052-011-1547-z. arXiv:1009.2450 [hep-ph]
23. Campbell JM et al (2015) Top-pair production and decay at NLO matched with Parton showers. JHEP 04:114. https://doi.org/10.1007/jhep04(2015)114. arXiv:1412.1828 [hep-ph]
24. Alwall J et al (2014) The automated computation of tree-level and next-to-leading order differential cross sections, and their matching to Parton shower simulations. JHEP 07:79. https://doi.org/10.1007/jhep07(2014)079. arXiv:1405.0301 [hep-ph]

25. Sjöstrand T et al (2015) An introduction to 8.2. Comp Phys Commun 191:159. https://doi.org/10.1016/j.cpc.2015.01.024. arXiv:1410.3012 [hep-ph]
26. Sirunyan A et al (2020) Extraction and validation of a new set of CMS 8 tunes from underlying-event measurements. Eur Phys J C 80:4. https://doi.org/10.1140/epjc/s10052-019-7499-4. arXiv:1903.12179 [hep-ex]
27. Zhang Z, Sabuncu MR (2018) Generalized cross entropy loss for training deep neural networks with noisy labels. arXiv:1805.07836 [cs.LG]
28. Duarte et al J (2018) Fast inference of deep neural networks in FPGAs for particle physics. J Instrum 13(07):P07027–P07027. https://doi.org/10.1088/1748-0221/13/07/p07027
29. Moons et al B (2017) Minimum energy quantized neural networks. arXiv:1711.00215 [cs.NE]
30. Coelho CN et al (2021) Automatic heterogeneous quantization of deep neural networks for low-latency inference on the edge for particle detectors. Nat Mach Intell 3(8):675–686. https://doi.org/10.1038/s42256-021-00356-5
31. Wang E et al (2022) Enabling binary neural network training on the edge. arXiv:2102.04270 [cs.LG]
32. Xilinx AMD (2023) Alveo U200 data center accelerator card. (Last visited: 8 Aug 2023). https://www.xilinx.com/products/boards-and-kits/alveo/u200.html#overview
33. CMS Collaboration (2023) Continual Learning in the CMS Phase-2 Level-1 trigger. In: CMS detector performance summaries. https://cds.cern.ch/record/2859651
34. OpenAI (2023) GPT-4 technical report. arXiv:2303.08774 [cs.CL]

Chapter 5
The Search for $HH \to b\bar{b}\tau^+\tau^-$

The search for Higgs boson pair (HH) production in the $bb\tau\tau$ final state is one of the most challenging and yet most sensitive channels in our exploration of the Higgs boson (H) self-coupling (λ_{HHH}). Although initially designed as an explorative analysis meant to cover additional phase space in the study of HH production, the $HH \to bb\tau\tau$ is now the second most sensitive analysis in the Gluon Fusion (ggF) production channel and the most sensitive in the Vector Boson Fusion (VBF) production channel. Investigating this decay channel requires the experimental capability to identify and reconstruct several different types of final state objects and to use them to select signal-like events. This chapter is devoted to the presentation of the analysis strategy, the selection procedures, and the identification techniques, which are the foundation of the $HH \to bb\tau\tau$ search.

The τ lepton is an unstable particle whose decays contain neutrinos, thus rendering impossible the complete reconstruction of the event. At the same time, hadronic final states must be distinguished from instrumental backgrounds caused by the misidentification of quark and gluon jet. The complexity and heterogeneity of the final states demand the use of all the physics objects' information and require the exploitation of the excellent particle identification capabilities of the CMS detector. Furthermore, the presence of neutrinos in the final state, together with the tiny ggF cross section of 31.05 fb, hinders the possibility of a *simple bump hunt* but demands the use of novel and more complex analysis approaches to identify the signal on top of a huge background. Moreover, the need for the use of the event kinematic properties to reduce background contamination requires a thorough understanding of the process. For all these reasons, the $bb\tau\tau$ decay channel is probably one of the most challenging at the LHC. Figure 5.1 shows a candidate $HH \to bb\tau\tau$ event where one H decays to a bottom quark-antiquark pair (blue towers connected to the green jet cones), and the other decays to a $\tau^+\tau^-$ leptons pair (red and blue towers connected to the green jet cones).

The presented physics analysis is performed with proton-proton collision data recorded at a centre-of-mass energy $\sqrt{s} = 13$ TeV by the CMS experiment during

J. Motta, *Development of Machine Learning τ Trigger Algorithms and Search for Higgs Boson Pair Production*, Springer Theses,
https://doi.org/10.1007/978-3-031-96288-2_5

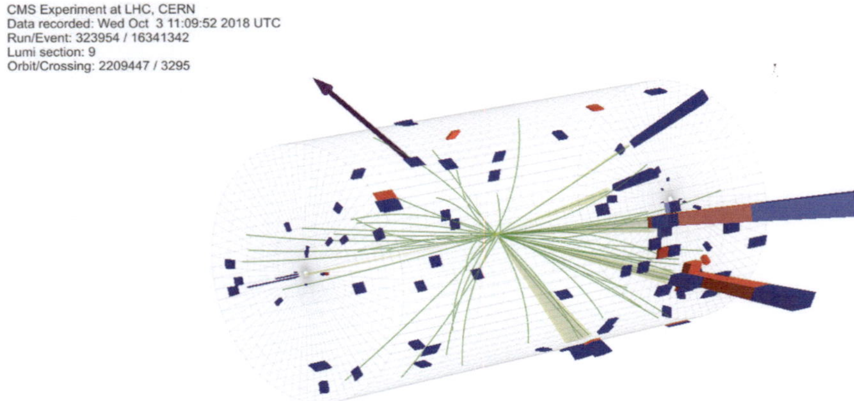

Fig. 5.1 Event recorded by the CMS detector in 2018 at a proton-proton centre of mass energy of 13 TeV. The event is a Higgs boson pair candidate produced via non-resonant gluon fusion production. One Higgs boson decays to a bottom quark-antiquark pair (blue towers connected to the green jet cones), and the other decays to a $\tau^+\tau^-$ lepton pair (red and blue towers connected to the green jet cones)

Run-2. The analyzed dataset has been collected during 2016, 2017, and 2018, corresponding to integrated luminosities of 35.9, 41.5, and 59.7 fb^{-1} respectively, yielding a total of 137 fb^{-1}, an unprecedented dataset size in high energy physics. This search is based on the previous publication of 2016 [1] and improves it by introducing several upgrades to the analysis flow, from better trigger selection to improved signal extraction, from enhanced signal modelling to strengthened limit setting.

The physics search presented in this Thesis is the result of the collective effort between scientists from several universities and institutes within the CMS Collaboration, which culminated in a publication in the Physics Letters B journal [2]. Within this team, I have had the chance to take part in most of the steps of the analysis. Most notably, I have been in charge of the evaluation of the event selection performance, the modelling of the top quark-antiquark pair ($t\bar{t}$) background and of the Effective Field Theory (EFT) benchmarks; I have contributed to the optimization of the discriminating variable used for the statistical interpretation and to the validation of the analysis statistical model, and I have been responsible for the production of the EFT interpretation of the results.

This chapter presents the core of the analysis strategy and is structured into five main parts. First, the $bb\tau\tau$ decay channel is presented, focusing on its experimental signature and associated challenges. These considerations are the bases for the analysis procedure described in the subsequent two Sections, covering the selection of physics objects based on the Particle Flow (PF) approach and the full event categorization. The discussion then moves to the modelling of signal and background processes with Monte Carlo (MC) techniques and their correction based on data-MC comparison. The last Section is finally devoted to the treatment of systematic uncertainties.

5.1 The bb$\tau\tau$ Decay Channel

In the selected final state, one H decays to a bottom quark-antiquark pair, while the second H decays to a pair of τ leptons. Nevertheless, neither of the four decay products is directly detectable in the CMS apparatus: the b quarks undergo hadronization and are detectable only in the form of b jets, while the τ leptons decay into lighter stable particles. As a consequence, the study of the HH $\rightarrow$ bb$\tau\tau$ process demands the investigation of several final states.

In the following, the notation τ_ℓ, where $\ell = \mathrm{e}, \mu$, is used for the leptonic decays of the τ, i.e. $\tau \rightarrow \ell \nu_\ell \nu_\tau$; we indicate by τ_h the hadronic decays in association with a ν_τ neutrino. The different decays of the τ lepton give rise to six possible final state configurations of the pair, summarized in Fig. 5.2, whose values are computed based on Table 2.2.

The search described in this Thesis only exploits a subset of the six available configurations: the $\tau_\mu\tau_\mathrm{h}$, $\tau_\mathrm{e}\tau_\mathrm{h}$, and $\tau_\mathrm{h}\tau_\mathrm{h}$ final states (also referred to as *channels*), which together account for 87.6% of all the decay modes. The fully leptonic decays (i.e. $\tau_\mathrm{e}\tau_\mathrm{e}$, $\tau_\mathrm{e}\tau_\mu$, and $\tau_\mu\tau_\mu$) are not considered because little sensitivity gain is expected, as not only do they have considerably smaller branching fractions, but they also suffer from large contamination from Drell-Yan lepton pair production (for $\tau_\mathrm{e}\tau_\mathrm{e}$ and $\tau_\mu\tau_\mu$) and from $\mathrm{t\bar{t}}$ (for $\tau_\mathrm{e}\tau_\mathrm{e}$, $\tau_\mathrm{e}\tau_\mu$ and $\tau_\mu\tau_\mu$).

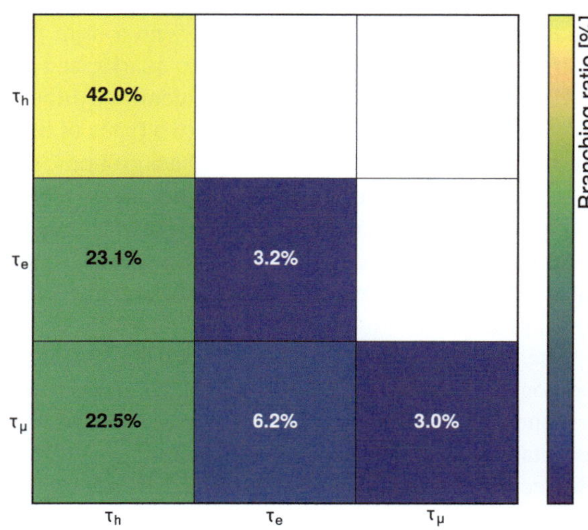

Fig. 5.2 Branching fractions of the six possible $\tau\tau$ decay final states

This search focuses on exploring non-resonant HH production in the two leading production mechanisms predicted by the Standard Model (SM): gluon fusion and vector boson fusion. High sensitivity to this process is of foremost importance to probe the nature of the Electro-Weak Symmetry Breaking (EWSB) and possible extensions of the SM as described in Chap. 1. For this reason, the non-resonant HH production is tested both under the SM hypothesis and under the hypothesis of an EFT Lagrangian parametrization where anomalous H couplings are accounted for.

The bb$\tau\tau$ search is affected by numerous sources of background, which we can qualitatively regroup into two categories: irreducible and reducible. By irreducible backgrounds, we mean those whose final state particle content is identical to the bb$\tau\tau$ signal; conversely, we categorize as reducible those whose final state is different from the signal but that, due to erroneous object identification, can be misidentified as such. The rates of misidentification at CMS are very low, but the production cross sections of many processes, like that of QCD multijet production, are several orders of magnitude larger than the signal, resulting in an overall sizeable contribution.

The main irreducible background is the t$\bar{\text{t}}$ production in the decay channel t$\bar{\text{t}}$ $\rightarrow$ b$\bar{\text{b}}$W$^{\pm}$W$^{\mp}$ $\rightarrow$ b$\bar{\text{b}}\ell^{\pm}\overset{(-)}{\nu}_{\ell}\tau^{\mp}\overset{(-)}{\nu}_{\tau}$ ($\ell = \text{e}, \mu, \tau$). The t$\bar{\text{t}}$ background affects all the channels considered and has a particular incidence on the cross-channels ($\tau_{\mu}\tau_{\text{h}}$, $\tau_{\text{e}}\tau_{\text{h}}$). The second most prominent irreducible background is Drell-Yan lepton pair production associated with jets in the specific process $\text{Z}/\gamma^* + \text{b}\bar{\text{b}} \rightarrow \ell\ell + \text{b}\bar{\text{b}}$. This constitutes only a fraction of the more general $\text{Z}/\gamma^* + \text{jj}$ case, with j a jet. In the following, this background will be referred to as *DY + jets* (or simply *DY*). Minor contributions from the rarer ZZ and ZH processes also affect this analysis.

The most important reducible background is the one of QCD multijet production. It is characterized by jets originating from quarks or gluons that are misidentified as τ or b candidates and especially affects the $\tau_{\text{h}}\tau_{\text{h}}$ channel. The contribution to the $\tau_{\text{e}}\tau_{\text{h}}$ and $\tau_{\mu}\tau_{\text{h}}$ channels is smaller because the selection requirements for the light leptons ensure a stringent rejection. Additional reducible backgrounds are the DY production of a τ pair in association with a light flavour quark pair, t$\bar{\text{t}}$ production with the decay of at least a W boson to quarks, and the W production in association with light jets when light jets are misidentified either as leptons or b jets.

To mitigate the influence of these two types of backgrounds, distinct approaches must be implemented. For reducible backgrounds, stringent criteria on object quality are enforced to minimize misidentification, albeit resulting in decreased signal efficiency. The optimal requirements are determined by striking a balance between these two factors. In contrast, irreducible background sources can only be statistically suppressed by leveraging their distinct kinematic characteristics relative to the signal processes.

The analysis strategy is designed to be sensitive to the many different signal topologies of the bb$\tau\tau$ final state; thus, it is effectively split into three main parts. The first one aims at reconstructing the decay products of the H with the highest possible acceptance for all final signatures by applying a series of general preselection criteria. The second part is designed to maximize the sensitivity to different signal topologies by applying dedicated selections and event categorizations. The final stage aims to further increase sensitivity by applying a dedicated discriminator based on a neural network to reject the residual background contamination. The complete workflow of the HH $\rightarrow$ bb$\tau\tau$ search is illustrated schematically in Fig. 5.3 where all steps of the analysis are reported, and the Sections of the text where they are described are specified. In the following, a summary of the entire workflow is given before treating each step in detail.

The preselection process starts with the application of trigger algorithms that target the presence of e, μ, τ_h, or combinations of them, as further discussed in Sect. 5.2.1. The application of the trigger strategy provides event candidates to exploit in the subsequent analysis. Hence, the leptons are required to pass baseline selection criteria that ensure the good identification of the H $\rightarrow$ $\tau\tau$ decay products. To target the H $\rightarrow$ bb decay, the presence of two jets is then required. The preselections' effect needs to be accounted for in the modelling of the MC simulation via the application of dedicated corrections. A comprehensive description of these aspects is given in Sects. 5.2.2, 5.2.3, and 5.4.1.

The application of the baseline selection criteria provides the analysis with object candidates that are exploited in the final event selection, which is based on the reconstruction of the H $\rightarrow$ $\tau\tau$ and H $\rightarrow$ bb as detailed in Sects. 5.3.1 and 5.3.2, respectively. In those events where the two H candidates are found, a specific categorization strategy is used to enhance the signal sensitivity of the analysis. The categorization process first probes the presence of the two jets typical of the VBF topology to maximize the statistical power to target this production mechanism; subsequently, a neural network multi-classifier is employed to further enhance the sensitivity to VBF HH production. The events not satisfying the VBF selection criteria are categorized based on the spatial overlap and quality of the two jets from the selected H $\rightarrow$ bb candidate. Additionally, the invariant mass of the jet and lepton pairs are exploited to enhance the rejection of background events that do not align with the decay hypothesis of a Higgs boson pair. The categorization strategy is fully detailed in Sect. 5.3.3.

Once the signal regions are defined via a precise categorization of events, the signal versus background discrimination process is finally performed by applying a deep neural network classifier trained to target the rejection of $t\bar{t}$ events, the main residual background contaminant. The same classifier is employed in all categories of the analysis; its architecture and training are detailed in Sect. 5.3.4.

A combination of MC simulation and data-driven approaches is employed to model the expected signal and background contributions as outlined in Sect. 5.4. The obtained differential distributions are first compared to data in signal-depleted regions to validate their agreement with data; then, they are used for statistical analysis to obtain the results of the search. All the systematic uncertainties arising from the procedure summarised above and affecting the search presented in this Thesis are detailed in Sect. 5.5.

The statistical analysis of the data is finally performed to obtain the results of the HH $\rightarrow$ bb$\tau\tau$ search, which is interpreted both within the SM and in the context of an EFT Lagrangian parametrization. This last part of the workflow, alongside the results, is described in detail in Chap. 6.

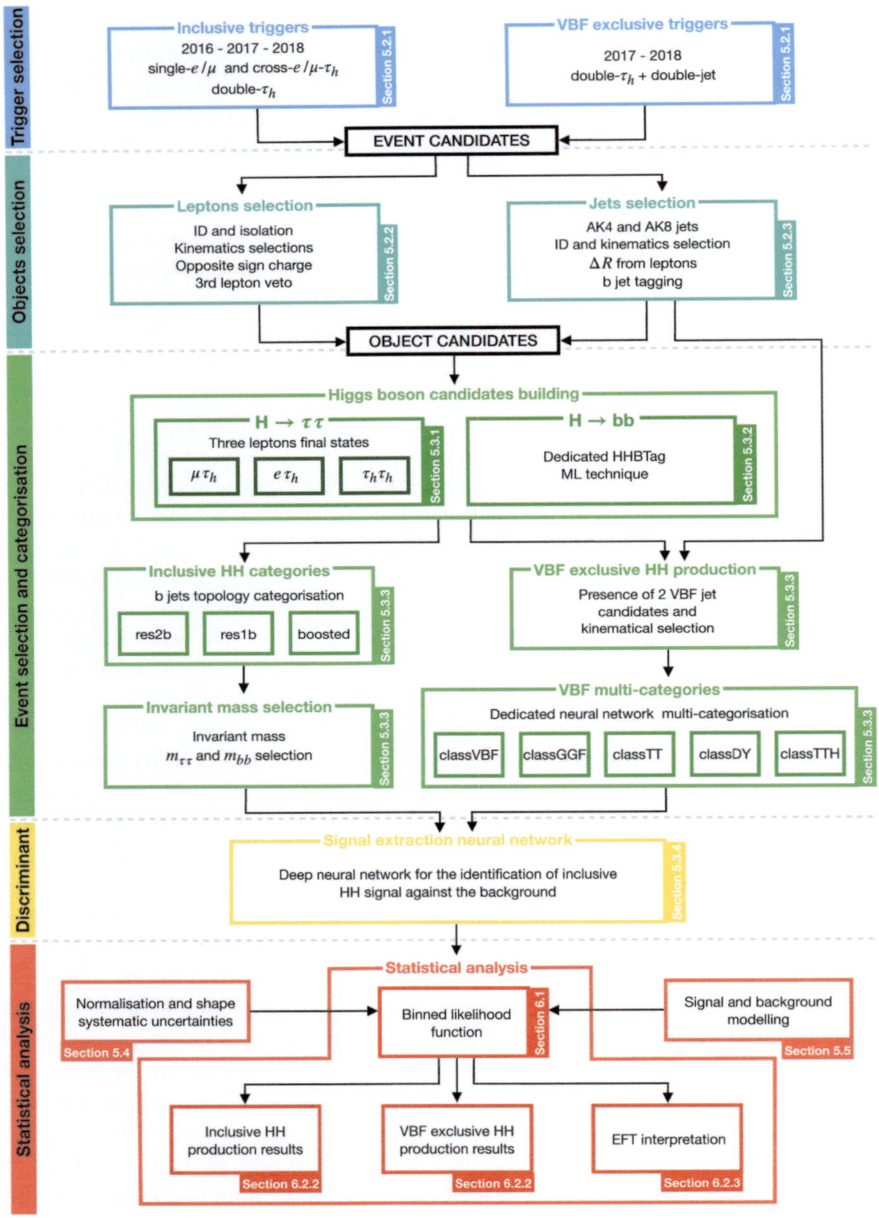

Fig. 5.3 Illustration of the HH → $bb\tau\tau$ analysis workflow: events are selected based on a dedicated trigger strategy; object selection imposes general requirements to reconstruct H candidates with high efficiency; event selection and categorization are based on the kinematical properties of the H candidates to yield eight categories per each of the three lepton final states; a deep neural network is employed for signal extraction and its output serves as discriminating variable; statistical analysis of the data based on a likelihood function is employed to obtain the search results. Each step of the workflow is fully detailed in the text, in the specific Sections reported in this figure

5.2 Physics Objects Preselection

The HH $\rightarrow$ bb$\tau\tau$ search starts with the application of a series of general preselection criteria aimed at reconstructing with the highest possible acceptance two H candidates in a proton-proton collision event. First, a specific trigger strategy is employed to select data online, exploiting the excellent selection capabilities of the Level-1 (L1) trigger and High-Level Trigger (HLT); both generic and analysis-oriented trigger paths are employed. Secondly, the selection of physics objects focuses on the reconstruction of final states particles compatible with the H $\rightarrow \tau\tau$ and H $\rightarrow$ bb decays. The application of this baseline preselection ensures that all signal topologies of the bb$\tau\tau$ final state are addressed simultaneously.

5.2.1 *Trigger Requirements*

The data events recorded offline for this analysis are selected with a set of L1+HLT triggers that require the presence of electrons, muons, τ leptons, or a combination of them in the event; no requirement on the presence of jets is made at this stage. This choice is taken to fully exploit the excellent reconstruction performance of leptons in the CMS detector and maximize the sensitivity to the bb$\tau\tau$ final state. All the trigger paths containing a τ lepton are seeded by L1 τ_h candidates reconstructed with the trigger algorithm described in detail in Chap. 3, which has been optimized in the context of this Thesis.

The trigger strategy follows in the steps of the one used in the previously published result and expands it with the introduction of new cross-object and VBF trigger paths. The former target the semi-leptonic decay channels, i.e. $\tau_e\tau_h$ and $\tau_\mu\tau_h$, to increase the overall signal acceptance by lowering the p_T selection on the τ_h candidate at the price of requiring the presence of an additional well-identified lepton in the event. The latter aims at increasing the acceptance of exclusive VBF HH production by targeting its specific final state topology. The detailed trigger object requirements of each HLT path used for this analysis are reported in Tables 5.1, 5.2, and 5.3, for all the data-taking periods. Whenever two or more paths are used at the same time, the logical OR is always intended to maximize efficiency.

Table 5.1 reports the trigger paths targeting the $\tau_\mu\tau_h$ final state; therefore, all paths contain one well-identified μ lepton. The reconstruction of muons at HLT is initiated by the corresponding candidate at L1; it uses first the muon system information only and then includes that from the tracking subdetectors via extrapolation of the trajectories. Two different types of isolation can be enforced on the muon candidate: the first one is a calorimetric criterion based on the Electromagnetic Calorimeter (ECAL) and Hadronic Calorimeter (HCAL) subdetectors' energy deposits; the second is a tracking criterion based on the tracks reconstructed around the muon at HLT. To cope with the increasing instantaneous luminosity delivered by the LHC during the three years of data-taking, the additional pseudorapidity requirement of $|\eta| < 2.1$

Table 5.1 Trigger paths used in the $\tau_\mu \tau_h$ channel for the three years of data-taking and the corresponding integrated luminosity recorded with each. Within each year, the logical OR of all the paths is always intended. The paths requiring one τ_h have changed during the 2018 data-taking after the implementation of the HPS algorithm. The Isolation (Iso) criteria reported in the table are detailed in the text

Year	Trigger requirement	Int. lumi [fb^{-1}]
2016	One μ, $p_T > 22$ GeV, Calo. Iso	2.8
	One μ, $p_T > 22$ GeV, $\|\eta\| < 2.1$, Calo. Iso	2.8
	One μ, $p_T > 22$ GeV, Tracker Iso	33.5
	One μ, $p_T > 22$ GeV, $\|\eta\| < 2.1$, Tracker Iso	33.5
	One μ, $p_T > 19$ GeV, $\|\eta\| < 2.1$ + One τ_h, $p_T > 20$ GeV, Loose PF Iso	36.3
2017	One μ, $p_T > 24$ GeV, Calo. Iso	3.6
	One μ, $p_T > 27$ GeV, Calo. Iso	41.5
	One μ, $p_T > 20$ GeV, $\|\eta\| < 2.1$, Calo. Iso + One τ_h, $p_T > 27$ GeV, $\|\eta\| < 2.1$, Loose PF Iso	41.5
2018	One μ, $p_T > 24$ GeV, Calo. Iso	59.7
	One μ, $p_T > 27$ GeV, Calo. Iso	59.7
	One μ, $p_T > 20$ GeV, $\|\eta\| < 2.1$, Calo. Iso + One τ_h, $p_T > 27$ GeV, $\|\eta\| < 2.1$, Loose PF Iso	17.7
	One μ, $p_T > 20$ GeV, $\|\eta\| < 2.1$, Calo. Iso + One HPS τ_h, $p_T > 27$ GeV, $\|\eta\| < 2.1$, Loose PF Iso	42.0

is applied to some trigger paths. The difference introduced by this requirement is accounted for in the subsequent data analysis strategy by applying the same selection offline, as discussed in the following. The HLT muon can then be used either alone or in association with an HLT τ_h for the single-μ paths or for the cross paths, respectively; different p_T and isolation requirements are applied on the muon depending on data-taking conditions and trigger bandwidth availability.

Table 5.2 reports the trigger paths targeting the $\tau_e \tau_h$ final state; therefore, all paths contain one well-identified electron. The reconstruction of electrons at HLT is initiated by the corresponding candidate at L1; it uses first the ECAL energy deposit around the L1 candidate and then matches it with a track, with an approach similar to the one used offline. A single isolation criterion is computed using the scalar p_T sum of the PF clusters and HLT tracks that lay in a cone of $\Delta R < 0.3$ around the candidate; several selections are defined on the discriminator value, generally referred to as Working Points (WPs), and correspond to different signal efficiencies and misidentification probabilities. The selected working point for the electron isolation in this analysis is always the most stringent, referred to as *tight*. The HLT electron can then be used either alone or in association with an HLT τ_h for the single-e paths or for the cross paths, respectively; different p_T and isolation requirements are applied on the electron depending on data-taking conditions and trigger bandwidth availability.

Table 5.2 Trigger paths used in the $\tau_e\tau_h$ channel for the three years of data-taking and the corresponding integrated luminosity recorded with each. Within each year, the logical OR of all the paths is always intended. The paths requiring one τ_h have changed during the 2018 data-taking after the implementation of the HPS algorithm. The Isolation (Iso) criteria reported in the table are detailed in the text

Year	Trigger requirement	Int. lumi [fb^{-1}]
2016	One e, $p_T > 25$ GeV, $\|\eta\| < 2.1$, Tight Iso	36.3
2017	One e, $p_T > 32$ GeV, Tight Iso, L1 Double-e/γ	41.5
	One e, $p_T > 32$ GeV, Tight Iso	41.5
	One e, $p_T > 35$ GeV, Tight Iso	41.5
	One e, $p_T > 24$ GeV, $\|\eta\| < 2.1$, Tight Iso + One τ_h, $p_T > 30$ GeV, $\|\eta\| < 2.1$, Loose PF Iso	41.5
2018	One e, $p_T > 32$ GeV, Tight Iso	59.7
	One e, $p_T > 35$ GeV, Tight Iso	59.7
	One e, $p_T > 24$ GeV, $\|\eta\| < 2.1$, Tight Iso + One τ_h, $p_T > 30$ GeV, $\|\eta\| < 2.1$, Loose PF Iso	17.7
	One e, $p_T > 24$ GeV, $\|\eta\| < 2.1$, Tight Iso + One HPS τ_h, $p_T > 30$ GeV, $\|\eta\| < 2.1$, Loose PF Iso	42.0

Table 5.3 reports the trigger paths targeting the $\tau_h\tau_h$ final state. The reconstruction of τ_h at HLT is initiated by the corresponding candidate at L1, and two different approaches have been used during Run-2. The first one, used for 2016, 2017, and the first 17.7 fb^{-1} of 2018, is a simple cone-shaped reconstruction that builds candidates starting from PF jets and clusters PF charged hadrons, photons, and neutral hadrons within an angular distance of $\Delta R = 0.08 - 0.12$ depending on the p_T of the object. The second method, deployed in 2018 and used for the collection of 42 fb^{-1}, uses a method similar to the offline Hadrons-Plus-Strips (HPS) algorithm, where multiple combinations of charged hadrons and photons within the signal cone are computed and ranked based on their consistency with a genuine τ_h. The highest ranking combination is selected as the reconstructed τ_h candidate. For the τ_h candidate, the isolation is defined based on the tracks within a cone of radius $\Delta R = 0.4$; in 2016, the isolation requirement changed from using only tracks with at least five hits in the tracking subdetector to using only tracks with at least three hits; at the same time an additional criterion on the scalar p_T sum of neutral candidates reconstructed was also applied (combined isolation). In the triggers used in this analysis, the loose, medium, and tight WPs are all used depending on the trigger path considered. The HLT τ_h are always used either in pairs for the double-τ_h trigger paths or in association with an e/μ for the cross paths; different p_T and isolation requirements are applied on the τ_h depending on data-taking conditions and trigger bandwidth availability.

Table 5.3 Trigger paths used in the $\tau_h\tau_h$ channel for the three years of data-taking and the corresponding integrated luminosity recorded with each. Within each year, the logical OR of all the paths is always intended. The paths requiring one τ_h have changed during the 2018 data-taking after the implementation of the HPS algorithm. The isolation (Iso) criteria reported in the table are detailed in the text

Year	Trigger requirement	Int. lumi [fb^{-1}]
2016	Two τ_h, $p_T > 35$ GeV, $\|\eta\| < 2.1$, Medium Iso	27.6
	Two τ_h, $p_T > 35$ GeV, $\|\eta\| < 2.1$, Medium comb. Iso	8.7
2017	Two τ_h, $p_T > 35$ GeV, $\|\eta\| < 2.1$, Tight-Tight PF Iso	41.5
	Two τ_h, $p_T > 40$ GeV, $\|\eta\| < 2.1$, Medium-Tight PF Iso	41.5
	Two τ_h, $p_T > 40$ GeV, $\|\eta\| < 2.1$, Tight-None PF Iso	41.5
	Two τ_h, $p_T > 20$ GeV, $\|\eta\| < 2.1$, Loose-None PF Iso + Two jets, $p_T^1 > 115$ GeV, $p_T^2 > 40$ GeV, $m_{jj} > 620$ GeV	27.1
2018	Two τ_h, $p_T > 35$ GeV, $\|\eta\| < 2.1$, Tight-Tight PF Iso	17.7
	Two τ_h, $p_T > 40$ GeV, $\|\eta\| < 2.1$, Medium-Tight PF Iso	17.7
	Two τ_h, $p_T > 40$ GeV, $\|\eta\| < 2.1$, Tight-None PF Iso	17.7
	Two HPS τ_h, $p_T > 35$ GeV, $\|\eta\| < 2.1$, Medium-None Iso	42.0
	Two τ_h, $p_T > 20$ GeV, $\|\eta\| < 2.1$, Loose-None PF Iso + Two jets, $p_T^1 > 115$ GeV, $p_T^2 > 40$ GeV, $m_{jj} > 620$ GeV	17.7
	Two HPS τ_h, $p_T > 20$ GeV, $\|\eta\| < 2.1$, Loose-None PF Iso + Two jets, $p_T^1 > 115$ GeV, $p_T^2 > 40$ GeV, $m_{jj} > 620$ GeV	42.0

The effect of the introduction of the cross-object trigger paths in this analysis can be appreciated in Fig. 5.4, where the 2D distributions of simulated SM HH events produced via the ggF mechanism are presented as a function of the generator level p_T of the muon(electron) and of the τ_h. The offline thresholds allowed by the single-lepton and cross-object triggers are superimposed in red and green, respectively. The use of single-lepton triggers offers the advantage of not imposing a specific threshold on the τ candidate; therefore, the offline selection for the τ_h is only dictated by the capabilities of the HPS offline reconstruction algorithm and can be lowered to $p_T > 20$ GeV. Conversely, sacrificing some coverage in the τ_h low p_T region, already accounted for by events triggering the single-lepton paths, reduces the lepton p_T threshold, thereby enabling access to additional phase space. In the case of the single-μ triggers, their sole use ensures the reconstruction of roughly 30% of the HH signal events where one τ decays to a muon and the other one hadronically; in contrast, the sole use of the single-e triggers guarantees the reconstruction of approximately 23% of the HH signal events where one τ decays to an electron and the other one hadronically. By comparison, the exclusive use of the cross-$\mu\tau$ or cross-$e\tau$ triggers would yield a rough 26 and 21% efficiency in reconstructing HH signal events where one τ decays to a muon or an electron and the other one to hadrons, respectively. Therefore, the exploitation of their combination ensures an increase in acceptance to 34 and 28% for the $\mu\tau$ and $e\tau$ decay modes, respectively.

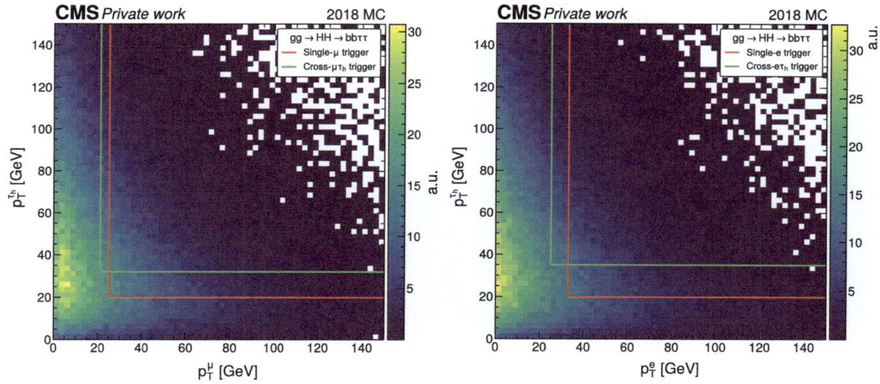

Fig. 5.4 2D distributions of $(p_T^{\mu}, p_T^{\tau_h})$ (left) and $(p_T^{e}, p_T^{\tau_h})$ (right) for simulated standard model gluon fusion Higgs boson pair production events. The offline thresholds allowed by the use of the single-lepton triggers are indicated in red, while those corresponding to the cross triggers are indicated in green

In 2017, an additional trigger path was designed and deployed to target the VBF HH production mechanism by exploiting its distinctive kinematics and topology. This trigger uses the presence of additional jets with high p_T and high invariant mass of their system (m_{jj}) as a handle to target VBF events; moreover, these additional requirements on the jets guarantee the possibility to reduce the p_T threshold on the τ_h candidates, which is curtailed to the lowest possible reconstruction value of 20 GeV. Introducing this new trigger path secures a 17% increase in acceptance of VBF events, which is a major achievement for a signal as rare as the one under consideration. The offline thresholds allowed by the new VBF H $\rightarrow \tau\tau$ trigger are shown in Fig. 5.5 in comparison with the standard double-τ triggers.

The final efficiency of the trigger strategy adopted in this analysis is reported in Fig. 5.6; the result is shown separately for the gluon fusion and vector boson fusion production mechanisms as a function of the HH invariant mass (m_{HH}). In both figures, the lower panel reports the efficiency of the trigger strategy computed as the ratio between the two distributions in the upper panel; as expected, the efficiency increases at larger values of m_{HH}.

All the selections detailed above are identically applied to both data and MC simulated events. Given the impossibility of a perfect reproduction of the behaviour of data in simulation, any systematic difference arising in the trigger efficiency needs to be accounted for and corrected. To achieve this goal, the so-called *trigger Scale Factors* (SFs) are computed event by event as $SF = \epsilon_{data}/\epsilon_{MC}$, where ϵ_{data} and ϵ_{MC} are the efficiency of the trigger requirements in data and in MC simulation, which both need to be evaluated.

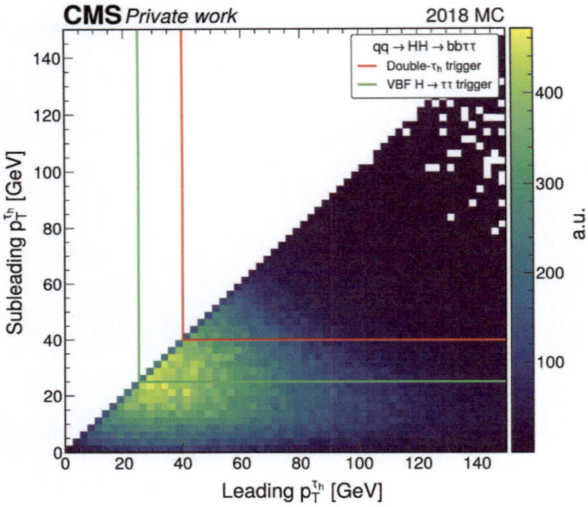

Fig. 5.5 2D distributions of simulated standard model vector boson fusion Higgs boson pair production events as a function of the p_T of the two reconstructed τ leptons. The offline thresholds allowed using the standard double-τ triggers are indicated in red, while those corresponding to the new vector boson fusion $H \rightarrow \tau\tau$ trigger are indicated in green

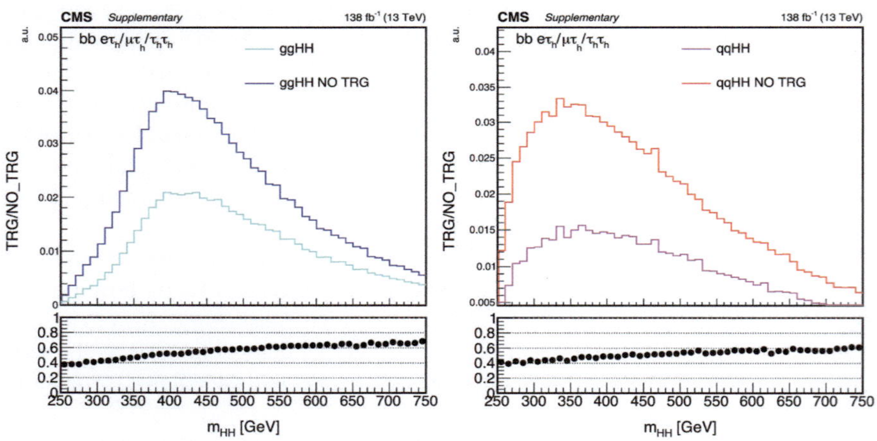

Fig. 5.6 Final efficiency of the trigger strategy adopted in the HH $\rightarrow bb\tau\tau$ search as a function of the Higgs bosons pair invariant mass (m_{HH}). The result is shown separately for the gluon fusion (left) and the vector boson fusion (right) production mechanisms. The lower panel reports the efficiency of the trigger strategy computed as the ratio between the two distributions in the upper panel

In the semi-leptonic channels, the effect of the use of the logical OR between single and cross-object triggers needs to be carefully considered when computing the efficiency for the calculation of the trigger SFs. A reasonable assumption in this context is that of the full decorrelation of the efficiency of the two objects in the cross-triggers; under this hypothesis, the efficiency of the logical OR between single and cross-object triggers is fully factorizable and can be computed as follows:

$$\epsilon^{\mathrm{S}\vee\mathrm{C}} = \epsilon_{\ell}^{\mathrm{S}} \cdot (1 - \epsilon_{\tau}^{\mathrm{C}}) + \epsilon_{\ell}^{\mathrm{C}} \epsilon_{\tau}^{\mathrm{C}} \tag{5.1}$$

where the superscripts S and C represent the single- and cross-triggers, respectively, while the subscripts represent which object of the trigger path is considered. The approach to the derivation of such efficiencies for each specific object and trigger is detailed in the following.

For both the single-μ and cross-$\mu\tau$ triggers, the muon efficiency is evaluated using a tag-and-probe technique on $Z \to \mu\mu$ events analogous to the one described in Sect. 3.4.1 for the evaluation of the L1 τ_h trigger performance. This strategy secures the collection of an unbiased sample of muon candidates with respect to trigger requirements, whose efficiency can therefore be estimated. This approach requires one *tag*-μ with stringent trigger, reconstruction, and identification criteria, and then uses criteria based on the kinematics of the $Z \to \mu\mu$, namely the reconstructed mass $m_{\mu\mu}$ to be in a neighbourhood of $m_Z = 91.2$ GeV [3], to identify a second *probe*-μ without directly applying trigger criteria on it. The trigger efficiency is evaluated as the fraction of *probe*-μ passing the trigger under consideration. The SFs are henceforth derived as a function of reconstructed muon p_T and η. A similar strategy is followed to derive the electron SFs in the single-e and cross-eτ triggers, exploiting $Z \to ee$ events. Finally, the efficiency and SFs of the τ candidates of the triggers are provided centrally by the CMS τ Performance Object Group (POG) using a similar tag-and-probe technique on $Z \to \tau_\mu \tau_h$ data events collected with single-μ triggers and $Z \to \tau\tau$ MC simulated events; the SFs are provided as a function of the reconstructed τ lepton p_T and decay mode.

5.2.2 *Pre-selection of H → ττ Objects*

The exploration of the $HH \to bb\tau\tau$ process initiates with an in-depth analysis of the $H \to \tau\tau$ decay. This entails the reconstruction and selection of electrons, muons, τ_h objects, and the calculation of MET resulting from neutrinos present in the event. The primary objective of this initial stage is to reconstruct the decay products of $H \to \tau\tau$ with optimal acceptance. To accomplish this, a set of quality criteria specifically optimized for this investigation is employed on each object reconstructed using the standard algorithms detailed in Sect. 2.4. Moreover, as previously outlined for the trigger selections, precise corrections are derived and applied to address any disagreement between data and MC simulated events. A description of all the criteria used for the $\tau\tau$ candidates' identification and the related correction are detailed in this Section.

Muons

As described in Sect. 2.4.2, muons can be reconstructed in three different ways: exclusively from the hits in the tracker subdetector (tracker muons), only from the signature in the muon chambers detectors (standalone muons), or with a combination of both to further improve the reconstruction (global muons). For each track, the number of hits in it and its quality are the basis of muon identification. Three WPs are defined accordingly and are referred to as *loose*, *medium*, and *tight*. The selected signal muons in this search are required to satisfy the tight WP, whose main criterion is the reconstruction by the global muon algorithm. Additionally, it applies stringent requirements on the χ^2 of the track fit, and it applies minimal requirements on the number of hits in the muon chambers, strip tracker, and pixel detectors. These rigorous selections all aim at suppressing the erroneous identification of hadrons escaping the calorimeter volume, cosmic rays, and in-flight decays of other particles to the per cent level. The tight WP is not the only one exploited in this search; the medium WP is also employed when defining veto muons, as detailed below. To be selected, muon candidates must be compatible with the primary vertex; thus, each candidate muon track must have a distance from the primary vertex fulfilling $\Delta_{xy} < 0.045$ cm in the transverse plane and $\Delta_z < 0.2$ cm on the longitudinal axis.

A non-negligible fraction of the background to reconstructed muons in CMS comes from jets. This is primarily due to two factors: the misidentification of charged particles within jets and the presence of genuine muons originating from semileptonic weak decays of b or c quarks within the jets. To reduce such contamination, an isolation requirement can be defined on the muon candidate as follows:

$$\mathcal{I}^{\ell}_{rel} = \frac{\sum p_{\mathrm{T}}^{\mathrm{ch.}} + \max\left(0, \sum p_{\mathrm{T}}^{\mathrm{nt.}} + \sum p_{\mathrm{T}}^{\gamma} - \frac{1}{2}\sum p_{\mathrm{T}}^{\mathrm{PU}}\right)}{p_{\mathrm{T}}^{\ell}}, \qquad \text{with} \quad \ell = \mu, \mathrm{e} \tag{5.2}$$

where $\sum p_{\mathrm{T}}^{\mathrm{ch.}}$, $\sum p_{\mathrm{T}}^{\mathrm{nt.}}$, and $\sum p_{\mathrm{T}}^{\gamma}$ are the scalar sums of the transverse momenta of charged hadrons, neutral hadrons, and photons originating from the primary vertex, while $\sum p_{\mathrm{T}}^{\mathrm{PU}}$ is the scalar sum of transverse momenta of charged hadrons not originating from the primary vertex, i.e. pileup (PU). This isolation is computed for muons using PF candidates within an angular distance $\Delta R < 0.4$ of the direction of the candidate, and two different thresholds are applied on it: signal muons are required to pass the tighter $\mathcal{I}^{\mu}_{rel} < 0.15$, while the veto muons are only required to satisfy the looser $\mathcal{I}^{\mu}_{rel} < 0.3$ selection.

Finally, kinematical selections are applied to each muon candidate as follows. Each muon is required to be within the pseudorapidity region $|\eta| < 2.1$; for the muons passing the single-μ trigger, thresholds of $p_T > 23$ GeV and $p_T > 25$ GeV are applied to candidates of the 2016 and 2017–2018 data-taking periods, respectively; for the muons passing the cross-$\mu\tau$ trigger, thresholds of $p_T > 20$ GeV and $p_T > 21$ GeV are applied to candidates of the 2016 and 2017–2018 data-taking periods, respectively.

Correction factors obtained, analogously to the trigger SFs, with a tag-and-probe technique on $Z \rightarrow \mu\mu$ are derived to account for differences between the data and the MC simulation in the muon identification and isolation efficiencies. The agreement of the MC simulation with the observed data after the application of said corrections can be appreciated in Fig. 5.7 for the muon p_T and η distributions for events selected in the $\tau_\mu\tau_h$ final state. The two bottom panels display the ratio between the data and the background hypothesis and render explicit the good data-MC agreement.

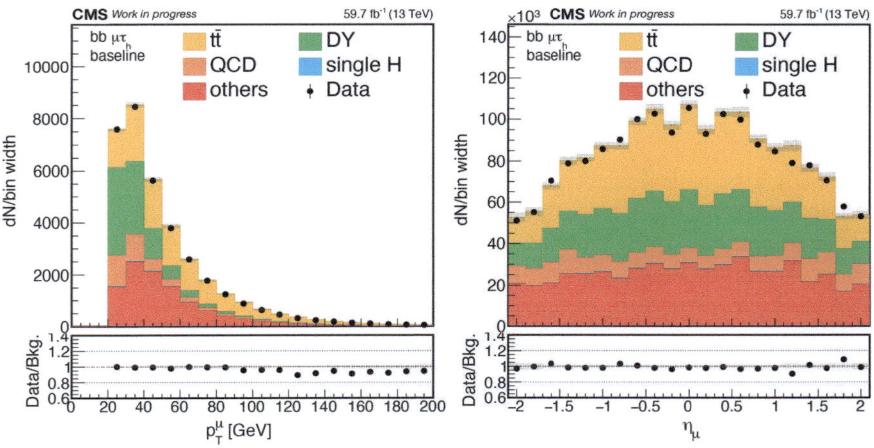

Fig. 5.7 Muon p_T (left) and η (right) distributions for events selected in the $\tau_\mu\tau_h$ final state. The black points and error bars represent the data; the background histograms are stacked and organised to highlight the largest and most interesting contributions. The bottom panel reports the ratio between the data and the background hypothesis. The shaded band in all panels represents the statistical uncertainty only

Electrons

The reconstruction of electrons is initiated by clusters of energy deposits in the ECAL detector, which undergo matching to tracks in the inner silicon tracker, which are in turn refitted with a Gaussian-Sum Filter (GSF), as described in Sect. 2.4.3. In contrast to muons, the identification of electrons does not rely on the quality of track fits or the number of hits but is based on a multivariate approach consisting of a Boosted Decision Tree (BDT) classifier [4, 5]. This BDT combines the information from observables sensitive to the bremsstrahlung magnitude along the electron trajectory, the momentum and geometrical matching between the clusters and their associated trajectories, shower-shape observables, electron conversion variables, and the PF isolation components of Eq. 5.2. The BDT discriminant is designed to be of general purpose, therefore it is trained on all electrons with $p_T > 10$ GeV in three independent pseudorapidity intervals, two for the barrel and one for the endcaps. As for any discriminant, several WPs are defined on the discriminator value. In this search, electron candidates must fulfil the tight identification WP, corresponding to a signal efficiency of about 80%. The tight WP is not the only one exploited in this search; the loose WP is also employed when defining veto electrons, as detailed below.

As for muons, an isolation criterion is imposed to further suppress the contamination from the hadron jet background. For electrons, the relative isolation $\mathcal{I}^{e}_{rel}$ is computed based on PF candidates reconstructed within a distance $\Delta R < 0.3$ and it is required to be under a threshold of $\mathcal{I}^{e}_{rel} < 0.10$, while the veto electrons are only required to satisfy the looser $\mathcal{I}^{e}_{rel} < 0.3$ selection. Moreover, consistency of the GSF track with the primary vertex is enforced by requiring a relative distance between the two of $\Delta_{xy} < 0.045$ cm in the transverse plane and $\Delta_z < 0.2$ cm in the longitudinal direction.

Finally, kinematical selections are applied to each electron candidate as follows: all electrons are required to be within the pseudorapidity region $|\eta| < 2.1$; for the 2016 data-taking period, a threshold of $p_T > 26$ GeV is applied; for the 2017–2018 data-taking period, thresholds of $p_T > 33$ GeV and $p_T > 25$ GeV are applied to candidates passing single-e trigger or the cross-eτ trigger, respectively.

Correction factors obtained, analogously to the trigger SFs, with a tag-and-probe technique on Z → ee are derived to account for differences between the data and the MC simulation in the electron identification and isolation efficiencies. The agreement of the MC simulation with the observed data after the application of said corrections can be appreciated in Fig. 5.8 for the electron p_T and η distributions for events selected in the $\tau_e\tau_h$ final state. The two bottom panels display the ratio between the data and the background hypothesis and render explicit the good data-MC agreement.

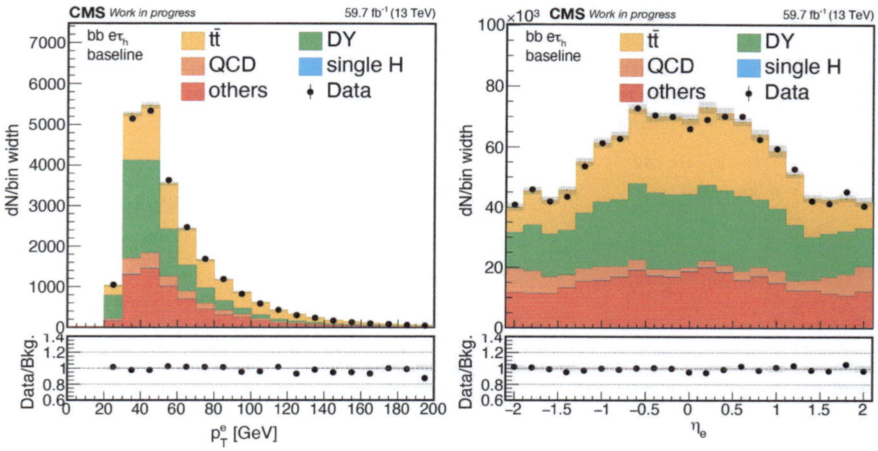

Fig. 5.8 Electron p_T (left) and η (right) distributions for events selected in the $\tau_e\tau_h$ final state. The black points and error bars represent the data; the background histograms are stacked and organised to highlight the largest and most interesting contributions. The bottom panel reports the ratio between the data and the background hypothesis. The shaded band in all panels represents the statistical uncertainty only

Taus

Hadronic decays of τ leptons are reconstructed with the HPS algorithm detailed in Sect. 2.4.4. It is fruitful here to recall that the HPS algorithm is seeded by PF jets, it builds τ candidates by considering all possible combinations of PF constituents and strips within a signal cone, and it chooses the highest-ranking combination as the final τ_h candidate. This method gives the possibility to target seven different Decay Modes (DMs) of the τ: $h^\pm$, $h^\pm\pi^0$, $h^\pm\pi^0\pi^0$, $h^\pm h^\mp h^\pm$, $h^\pm h^\mp h^\pm\pi^0$, $h^\pm h^{\mp/\pm}\pi^0$, and $h^\pm h^{\mp/\pm}\pi^0\pi^0$, where the categories with two charged particles are designed to target the decay modes with three charged particles of which one is lost, due either to geometrical acceptance or to misidentification. In the analysis presented here, only four of the seven available decay mode categories are used: a maximum of one π^0 is allowed and no decay modes with two charged particles are considered.

The main background for τ_h identification is represented by quark and gluon jets, the second one is constituted by misidentified electrons and muons. In the previous $HH \rightarrow bb\tau\tau$ analysis to distinguish genuine τ_h from the background, a combination of isolation and three different multivariate identifiers, each trained on a specific background source, were used. In the scope of the analysis presented here, a new method is used for the identification of hadronically decaying τ leptons: the DeepTau algorithm [6]. This newly developed algorithm is based on a deep convolutional neural network with an architecture specifically designed for the task at hand. This algorithm mainly exploits low-level information in the form of 188 input features (e.g. PF charged hadron tracks, electron cluster shapes, muon track hits and quality,

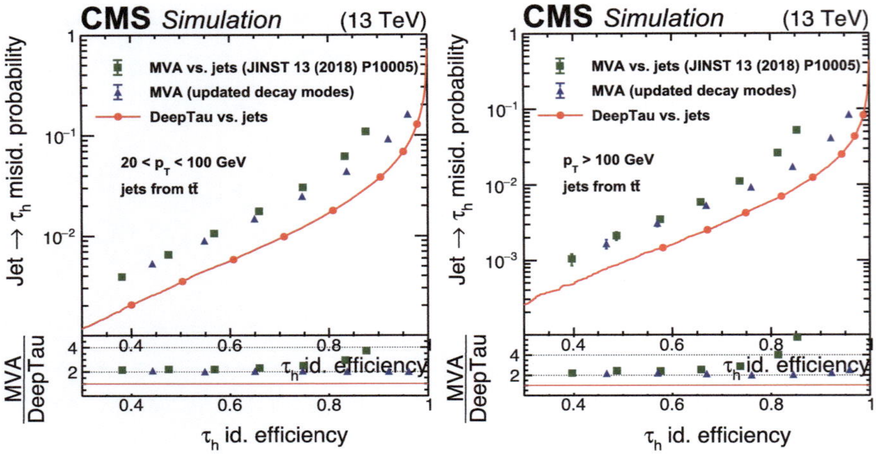

Fig. 5.9 Probability for quark and gluon jets to be misidentified as τ_h versus the efficiency for genuine τ_h identification, for low-p_T (left) and high p_T (right) candidates. On each plot three different discriminators are reported: the old multivariate method before (green) and after (blue) the update of the decay modes definition, and the new DeepTau algorithm (red) [6]

HCAL energy deposits, etc.), and combines them with 47 high-level variables (e.g. τ_h four-momentum, isolation, primary vertex compatibility, etc.) to further improves the performance and the stability of the network. The DeepTau algorithm, compared to the previous methods, has the great advantage of being a multi-classifier discriminant, namely three different identification outputs are produced at the same time by the network: DeepTauVsJet, DeepTauVsMu, and DeepTauVsEle, which target the rejection of jets, muons, and electrons, respectively; several WPs are defined for each of the three discriminants. Figure 5.9 reports the efficiency for quark and gluon jets to be misidentified as τ_h versus the efficiency for genuine τ_h identification in two different regions of p_T of the candidate, for different versions of the τ_h identification algorithm. The old multivariate approach is reported both before and after the update of the decay modes definition, in green and blue respectively, while the new DeepTau algorithm is reported in red. The DeepTau WPs are indicated as full circles. This comparison shows an improvement of at least a factor two across the entire p_T spectrum, with improvements up to a factor four and above when targeting particularly high τ_h identification efficiency. The WPs selected for this analysis are the following: medium DeepTauVsJet, tight DeepTauVsMu, and very-loose DeepTauVsEle. The full description of the DeepTau algorithm design, input variables, training procedure, and validation can be found in [6].

As for muons and electrons, consistency between the primary vertex and the τ_h candidates is enforced by requiring a relative distance between the two of $\Delta_{xy} < 0.045$ cm in the transverse plane and $\Delta_z < 0.2$ cm in the longitudinal direction. Contrary to the other objects, no explicit isolation condition is applied to τ leptons as this information is already fully exploited in the identification algorithm.

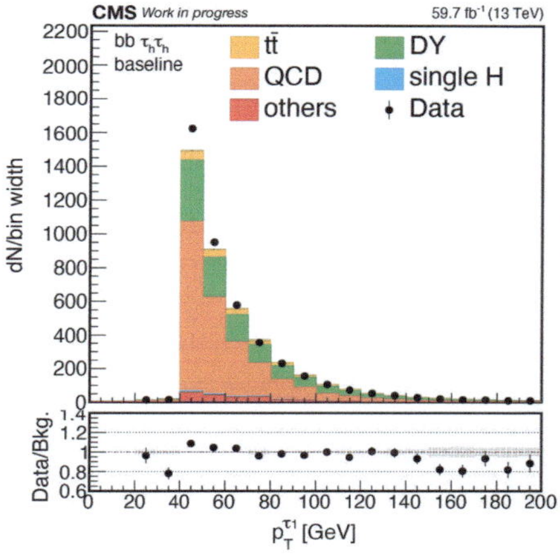

Fig. 5.10 Most isolated τ lepton p_T (left) and η (right) distributions for events selected in the $\tau_h\tau_h$ final state. The black points and error bars represent the data; the background histograms are stacked and organised to highlight the largest and most interesting contributions. The bottom panel reports the ratio between the data and the background hypothesis. The shaded band in all panels represents the statistical uncertainty only

Finally, kinematical selections are applied to each τ candidate as follows. The τ candidates in events that pass the single-object triggers are required to be within the pseudorapidity region $|\eta| < 2.3$, and the transverse momentum threshold is as low as the HPS reconstruction algorithm allows, i.e. $p_T > 20\,\text{GeV}$, for all years of data-taking. Conversely, τ candidates in events that pass the cross-object triggers are required to be in the pseudorapidity region $|\eta| < 2.1$, and the transverse momentum thresholds are: $p_T > 25\,\text{GeV}$ for cross-$\mu\tau$ in 2016, $p_T > 32\,\text{GeV}$ for cross-$\mu\tau$ in 2017–2018, $p_T > 35\,\text{GeV}$ for cross-eτ in 2017–2018. Lastly, τ candidates in events that pass the double-τ trigger and VBF H $\rightarrow \tau\tau$ are required to be in the pseudorapidity region $|\eta| < 2.1$, while a threshold of $p_T > 40\,\text{GeV}$ is enforced for the former and of $p_T > 25\,\text{GeV}$ for the latter.

Correction factors obtained, analogously to the trigger SFs, with a tag-and-probe technique on Z $\rightarrow \tau\tau$ are derived to account for differences between the data and the MC simulation in the electron identification and isolation efficiencies. The agreement of the MC simulation with the observed data after the application of said corrections can be appreciated in Fig. 5.10 for the leading τ lepton p_T and $|\eta|$ distributions for events selected in the $\tau_h\tau_h$ final state. The two bottom panels display the ratio between the data and the background hypothesis and render explicit the good data-MC agreement.

Missing Transverse Momentum

As discussed in depth in Sect. 2.4.6, the transverse momentum imbalance vector ($\vec{p}_T^{miss}$) vector is reconstructed using the PF algorithm; despite not being directly used for the selection of bb$\tau\tau$ signal event, both the magnitude and the direction of $\vec{p}_T^{miss}$ are exploited in combination with other observables to efficiently reject the $t\bar{t}$ background, as it is further discussed in the following. The dominant source of $\vec{p}_T^{miss}$ in the bb$\tau\tau$ final state stems from the neutrinos produced in the decays of the two τ leptons. A smaller fraction is also due to neutrinos originating from the decays of B hadrons resulting from the hadronization of the two b quarks. However, this contribution constitutes only a minor portion of the overall magnitude of $\vec{p}_T^{miss}$. Although no explicit requirement is applied on $\vec{p}_T^{miss}$, so-called MET filters are used to reject events where a massive, unphysical, or uninteresting $\vec{p}_T^{miss}$ is reconstructed. These filters use timing, pulse shape, and topology of the signals from the subdetectors to perform the selection.

During the 2017 data-taking, a combination of factors caused a steep deterioration of the $\vec{p}_T^{miss}$ agreement between data and MC simulation. The increase in PU due to routine LHC operations, combined with ECAL crystals ageing at large pseudorapidity values and inadequate mitigation of out-of-time PU due to unforeseen LHC bunch filling scheme, contributed to substantial noise in the forward regions of the detector, generating an artificial energy imbalance in the transverse plane. This resulted in large disagreements with the MC simulation. Therefore, in 2017, a corrected version of $\vec{p}_T^{miss}$ is defined excluding PF candidates with $p_T < 50$ GeV and $2.65 < |\eta| < 3.14$, and specific corrections are applied to the MC to account for it. The agreement of $\vec{p}_T^{miss}$ magnitude between the MC simulation and the observed data can be appreciated in Fig. 5.11 for the $\tau_\mu\tau_h$ final states in the 2017 and 2018 data-taking. The two bottom panels display the ratio between the data and the background hypothesis and render explicit the good data-MC agreement.

Other Selections

To reject those cases in which the same PF candidate is used to reconstruct and identify two different objects, the two reconstructed leptons in each of the final states are always required to be separated by an angular distance of $\Delta R > 0.5$.

Moreover, to account for charge conservation in the H $\rightarrow \tau\tau$ decay, the two candidates are required to have opposite electric charges. This requirement is very efficient owing to the very low charge misidentification in CMS; for the decay τ_h decay modes considered in this analysis (i.e. those without missing charged hadrons), the charge misassignment is evaluated in a Z $\rightarrow \tau\tau$ sample and estimated to be 1% on average, 2% for τ_h with $p_T \approx 200$ GeV, and 8% for τ_h with $p_T \approx 1$ TeV [6]; the

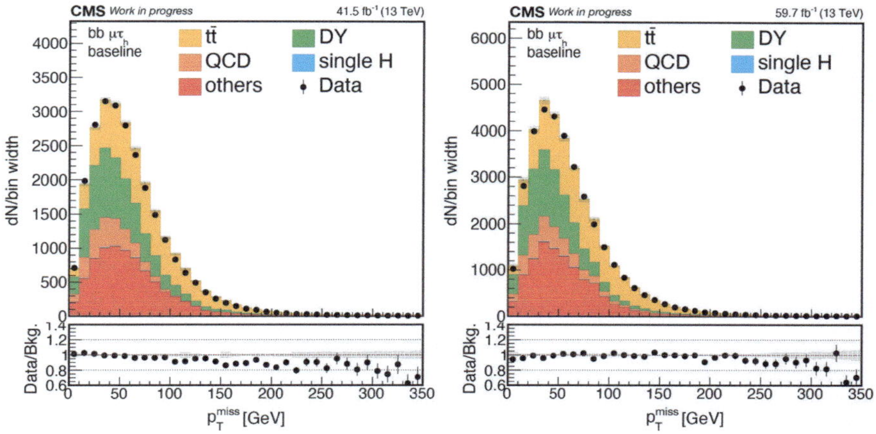

Fig. 5.11 Distribution of the $\vec{p}_{\mathrm{T}}^{\mathrm{miss}}$ vector magnitude in events selected in the $\tau_\mu \tau_\mathrm{h}$ final state for the 2017 (left) and 2018 (right) data-taking periods. The black points and error bars represent the data; the background histograms are stacked and organised to highlight the largest and most interesting contributions. The bottom panel reports the ratio between the data and the background hypothesis. The shaded band in all panels represents the statistical uncertainty only

charge misidentification for muons and electrons is smaller still and it is of the order $\mathcal{O}(10^{-3})$ [7, 8].

Additionally, each reconstructed lepton must correspond to the HLT candidates on which the trigger decision is taken by requiring a $\Delta R < 0.5$ geometrical distance between the two; this selection is enforced for all muons, electrons, and taus for both single and cross-object paths. The offline-to-HLT match ensures that no bias is introduced when applying to MC simulation events the trigger SFs previously described.

Finally, to suppress background contribution from Drell-Yan production, events containing additional isolated electrons or muons are rejected; this requirement is highly efficient for signal events, as no additional leptons are expected in the case of the HH $\rightarrow$ bb$\tau\tau$ signal process. Electrons used for this veto must have $p_\mathrm{T} > 10\,\mathrm{GeV}$, $|\eta| < 2.5$, $\mathcal{I}^{\mathrm{e}}_{rel} < 0.3$ or pass the loose WP of the identification BDT; muons used for this veto must have $p_\mathrm{T} > 10\,\mathrm{GeV}$, $|\eta| < 2.4$, $\mathcal{I}^{\mu}_{rel} < 0.3$. For both electrons and muons, consistency with the primary vertex is enforced with the requirements $\Delta_{xy} < 0.045\,\mathrm{cm}$ in the transverse plane and $\Delta_z < 0.2\,\mathrm{cm}$ on the longitudinal axis.

The object selections for each candidate that were detailed in the text above are summarized in Table 5.4 for the $\tau_\mu \tau_\mathrm{h}$ final state, in Table 5.5 for the $\tau_\mathrm{e} \tau_\mathrm{h}$ final state, and in Table 5.6 for the $\tau_\mathrm{h} \tau_\mathrm{h}$ final state. In all tables, the selections are organized per object and per data-taking period; for simplicity, only the transverse momentum selections that for 2016 differ from those of 2017–2018 are written explicitly.

Table 5.4 Offline selections for the $\tau_\mu\tau_h$ channel. The selections for the three data-taking years are reported; for simplicity, only the transverse momentum selections that for 2016 can differ from those of 2017–2018 are written explicitly

	HLT paths	Selections—2017–2018	Selections—2016
μ	All	$\|\Delta_{xy}\| < 0.045$ cm, $\|\Delta_z\| < 0.2$ cm $\mathcal{I}^{\mu}_{rel} < 0.15$ Tight ID	
	Single-μ	$p_T > 25$ GeV, $\|\eta\| < 2.1$	$p_T > 23$ GeV, $\|\eta\| < 2.1$
	Cross-μτ	$p_T > 21$ GeV, $\|\eta\| < 2.1$	$p_T > 20$ GeV, $\|\eta\| < 2.1$
τ_h	All	DM: $h^\pm$, $h^\pm\pi^0$, $h^\pm h^\mp h^\pm$, $h^\pm h^\mp h^\pm \pi^0$ $\|\Delta_{xy}\| < 0.045$ cm, $\|\Delta_z\| < 0.2$ cm Medium DEEPTAUVSJET Tight DEEPTAUVSMU Very-loose DEEPTAUVSELE	
	Single-μ	$p_T > 20$ GeV, $\|\eta\| < 2.3$	$p_T > 20$ GeV, $\|\eta\| < 2.1$
	Cross-μτ	$p_T > 32$ GeV, $\|\eta\| < 2.1$	$p_T > 25$ GeV, $\|\eta\| < 2.1$
Pair		$\Delta R(\mu, \tau_h) > 0.5$ Opposite charge HLT-RECO matching 3rd lepton veto and di-μ veto	

5.2.3 *Pre-selection of H → bb and VBF Objects*

After the pre-selection of the candidates for the reconstruction of the H → $\tau\tau$ decay, the exploration of the HH → bb$\tau\tau$ process continues with the study of the H → bb decay products, which are detected as hadron jets in the CMS detector. The main goal of this stage is to reconstruct the decay products of H → bb with optimal acceptance and preserve the jets characteristic of the VBF topology. A set of quality criteria specifically optimized for this investigation are employed, and precise corrections are applied to address any disparities observed between data and MC simulated events. A description of all the criteria used for the bb and VBF candidates' identification and the related correction are detailed in this Section.

Jets Selection

As detailed in Sect. 2.4.5, jets are reconstructed from PF candidates using the anti-k_T algorithm with two different distance parameters: $R = 0.4$ (AK4 jets) and $R = 0.8$

Table 5.5 Offline selections for the $\tau_e\tau_h$ channel. The selections for the three data-taking years are reported; for simplicity, only the transverse momentum selections that for 2016 can differ from those of 2017–2018 are written explicitly

	HLT paths	Selections—2017–2018	Selections—2016
e	All	$\|\Delta_{xy}\| < 0.045$ cm, $\|\Delta_z\| < 0.2$ cm $\mathcal{I}^{e}_{rel} < 0.10$ Tight BDT ID	
	Single-e	$p_T > 33$ GeV, $\|\eta\| < 2.1$	$p_T > 26$ GeV, $\|\eta\| < 2.1$
	Cross-$\tau_e\tau_h$	$p_T > 25$ GeV, $\|\eta\| < 2.1$	*Absent HLT path*
τ_h	All	DM: $h^{\pm}$, $h^{\pm}\pi^0$, $h^{\pm}h^{\mp}h^{\pm}$, $h^{\pm}h^{\mp}h^{\pm}\pi^0$ $\|\Delta_{xy}\| < 0.045$ cm, $\|\Delta_z\| < 0.2$ cm medium DEEPTAUVSJET tight DEEPTAUVSMU very-loose DEEPTAUVSELE	
	Single-e	$p_T > 20$ GeV, $\|\eta\| < 2.3$	$p_T > 20$ GeV, $\|\eta\| < 2.3$
	Cross-$\tau_e\tau_h$	$p_T > 35$ GeV, $\|\eta\| < 2.1$	*Absent HLT path*
Pair		$\Delta R(e, \tau_h) > 0.5$ Opposite charge HLT-RECO matching 3rd lepton veto and di-e veto	

(AK8 jets), the former is what we could define as *standard jets* while the latter is devised to better cover boosted topologies where two or more jets can have large overlap areas. In what follows, unless otherwise specified, we will refer to AK4 jets simply as jets.

To veto jets that are poorly reconstructed or that originate from instrumental noise, a PF jet identification criterion is defined based on several jet-related observables: the charged hadron multiplicity, the fraction of charged and neutral hadrons clustered within the jet, and the fraction of energy deposited in the ECAL detector by the neutral and charged hadron candidates. In this analysis, the tight WP of such discriminant is employed, and its efficiency is evaluated in data with events where at least a pair of well-separated jets with large invariant mass are found. The assessment is conducted using a tag-and-probe technique: the tag jet is required to pass the tight WP, while the efficiency of the WP is defined as the fraction of probe jets in the di-jet events that pass the same tight WP.

To reject jets originating from PU, an additional discriminator is applied to any jet with $p_T < 50$ GeV. The so-called *pileup jets*, do not originate from the primary vertex and are generally reconstructed from the overlap of several low energetic

Table 5.6 Offline selections for the $\tau_h\tau_h$ channel. The selections for the three data-taking years are reported; for simplicity, only the transverse momentum selections that for 2016 can differ from those of 2017–2018 are written explicitly

	HLT paths	Selections—2017–2018	Selections—2016
Both τ_h	All	DM: $h^\pm$, $h^\pm\pi^0$, $h^\pm h^\mp h^\pm$, $h^\pm h^\mp h^\pm \pi^0$	
		$\lvert\Delta_{xy}\rvert < 0.045$ cm, $\lvert\Delta_z\rvert < 0.2$ cm	
		Medium DEEPTAUVSJET	
		Very-loose DEEPTAUVSMU	
		Very-very-loose DEEPTAUVSELE	
	di-τ_h	$p_T > 40$ GeV, $\lvert\eta\rvert < 2.1$	
	VBF H → $\tau_h\tau_h$	$p_T > 25$ GeV, $\lvert\eta\rvert < 2.1$	*Absent HLT path*
Pair		$\Delta R(\tau_h, \tau_h) > 0.5$	
		Opposite charge	
		HLT-RECO matching	
		3rd lepton veto	

jets; for this reason, they tend to be broader than genuine jets and their constituents are usually not compatible with the primary vertex. Several WPs are defined for a dedicated discriminator based on a BDT, exploiting jet shape and tracking variables; in this analysis, the loose WP is employed for the rejection of pileup jets.

AK8 jets are employed in this search to better cover the boosted topologies of the H → bb decay where the hadronization products of the two jets partially overlap. To identify the substructure of AK8 jets, the soft drop grooming algorithm [9] is used. The algorithm iteratively decomposes an AK8 jet into sub-jets by recursively removing the soft, wide-angle radiation. This technique is able to efficiently identify the two quarks in the H → bb decays and, at the same time, mitigate the contribution from initial state radiation, PU, and underlying event, that have a sizeable contribution to large radius jets like the AK8. This mitigation results in an additional improvement of the jet invariant mass resolution. Furthermore, for AK8 jets the PileUp Per Particle Identification (PUPPI) algorithm is employed to further improve PU mitigation; this algorithm exploits both global information from the event, as well as local information to identify PU at the particle level.

Finally, basic kinematical requirements are applied on all jets: each PF jet must satisfy $p_T > 20$ GeV and $|\eta| < 4.7$. As previously discussed, for data collected in 2017 a series of factors contributed to the arising of a considerable data-MC disagreement at large pseudorapidity values. Therefore, as for the definition of $\vec{p}_T^{\,miss}$, all jets with $p_T < 50$ GeV and $2.65 < |\eta| < 3.14$ are removed.

In the Run-2 data-taking, the jet energy resolution measured in data was found to be substantially worse than in the MC simulation. For this reason, a specific procedure, referred to as *smearing*, has been put in place to correct the MC simulation and recover the good agreement with data. As the name of the method suggests, it consists of introducing additional smearing to the MC simulated jets; two different approaches have been defined, in this analysis both are used depending on the particle-level matching of the jet.

If a particle-level jet is found to satisfy the following matching criteria to the reconstructed jet:

$$\Delta R < \frac{R_{\text{jet}}}{2} \qquad \text{and} \qquad |p_{\text{T}} - p_{\text{T}}^{\text{prtcl.}}| < 3 p_{\text{T}} \sigma_{\text{JER}} \tag{5.3}$$

where R_{jet} is the jet cone size parameter (e.g. 0.4 for AK4 jets), p_{T} is the jet transverse momentum, $p_{\text{T}}^{\text{prtcl.}}$ is the transverse momentum of the corresponding jet clustered from generator-level particles, and σ_{JER} is the relative resolution measured in simulation; then the four-momentum of the reconstructed jet is scaled by a factor

$$c_{\text{JER}} \equiv 1 + (s_{\text{JER}} - 1) \cdot \frac{p_{\text{T}} - p_{\text{T}}^{\text{prtcl.}}}{p_{\text{T}}} \tag{5.4}$$

where s_{JER} is the data-to-simulation core resolution scale factor.

Otherwise, if Eq. 5.3 are not satisfied, the reconstructed jet four-momentum is scaled by a factor

$$c_{\text{JER}} \equiv 1 + \mathcal{N}(0, \sigma_{\text{JER}}) \sqrt{\max(s_{\text{JER}}^2 - 1, 0)} \tag{5.5}$$

where $\mathcal{N}(0, \sigma_{\text{JER}})$ is a number sampled from a normal distribution with mean zero and variance the square of the relative resolution measured in simulation.

Identification of b Jets

As for any quark, b quarks are detected in the CMS detector as hadron jets. Originating from b quarks, b jets have the distinctive feature of containing long-lived B mesons and hadrons, whose typical lifetime is of the order of the picosecond and can be exploited for identification purposes. At the expected energies of production at the LHC, the B hadrons typically travel a few millimetres away from the primary vertex and decay to light leptons with a distinctive signature containing what is referred to as *secondary vertex*. The necessity of probing the presence of a secondary vertex and the study of its properties requires the restriction of the pseudorapidity range to the tracker acceptance; therefore, only jets with $|\eta| < 2.4$ are considered for b-tagging. Several multivariate techniques have been developed in CMS for the efficient tagging of b jets; in the analysis presented here, the DEEPJET algorithm [10] is employed. This method is based on a convolutional recursive deep neural network architecture that exploits 650 input variables, divided into four categories: global variables (e.g. jet kinematics, number of vertices), charged PF candidate features (e.g.

tracks quality and kinematics), neutral PF candidate features (e.g. energy deposit, candidate-jet relative position), and secondary vertex features associated with the jet (e.g. position, p_T). As for all other discriminants introduced in this analysis, several WPs can be defined and used depending on the different trade-offs between the efficiency of genuine b jet identification and the misidentification probability for light flavour jets. In this search, both the predefined WPs and the DEEPJET discriminator output are simultaneously used as detailed in Sects. 5.3.2 and 5.3.4. Depending on the final category considered, the loose and medium working points are used, which correspond respectively to an efficiency (misidentification rate for light-flavour and gluon jets) of 94 (10) and 84 (1)% [10].

To correct the discrepancy between MC simulation and data, multiple reweighting techniques are designed by the POG in charge of the b-tagging tools. In this analysis, considering the direct use of the DEEPJET discriminator output, the scale factors are computed per event to modify the shape of the discriminator score. The objective is to accurately predict the distributions of the b-tagging discriminant; this is achieved solely by modifying the weights of the selected MC events, eliminating the need to reintroduce MC events that failed the selection process. It is crucial to emphasize that this method does not involve any event migration between different b-tagging bins. The computation of the event SFs is as follows:

$$\mathrm{SF}_{\text{b-tag}} = \prod_{i=0}^{\mathrm{N_{jets}}} \omega(D^{(i)}, p_T^{(i)}, \eta^{(i)}) \tag{5.6}$$

where the weights ω are provided by the POG as a function of the discriminator score D, the transverse momentum p_T of the jet, and its pseudorapidity η. The ω weights are computed using the following tag-and-probe technique in two different regions: one denoted as Heavy Flavour (HF) region, enriched in $t\bar{t}$ di-lepton events, and one denoted as Light Flavour (LF) region, enriched in Z+jets di-lepton events.

$$\mathrm{HF}: \omega(D, p_T, \eta) = \frac{\mathrm{DATA} - \mathrm{MC_{udsg+c}}}{\mathrm{MC_b}} \tag{5.7}$$

$$\mathrm{LF}: \omega(D, p_T, \eta) = \frac{\mathrm{DATA} - \mathrm{MC_{b+c}}}{\mathrm{MC_{udsg}}} \tag{5.8}$$

In the HF (LF) region, a specific jet is required to be (anti-)tagged. For each lepton channel (i.e. ee, eμ, μμ), the expected yield in the MC simulation is normalized to the corresponding data in order to correct only the shape. Subsequently, the MC simulation is once again normalized to the data per (p_T, η) bin, and histograms are constructed for the b-tagging discriminant D of the probed jets. To ensure accuracy, contamination arising from light flavour quark or gluon (b quarks), determined in the MC simulation, is subtracted from the data. The resulting distribution is then divided by the MC distribution of b quark (light flavour quark or gluon) probe jets. The values of the measured SF as a function of the leading jet p_T in the event are

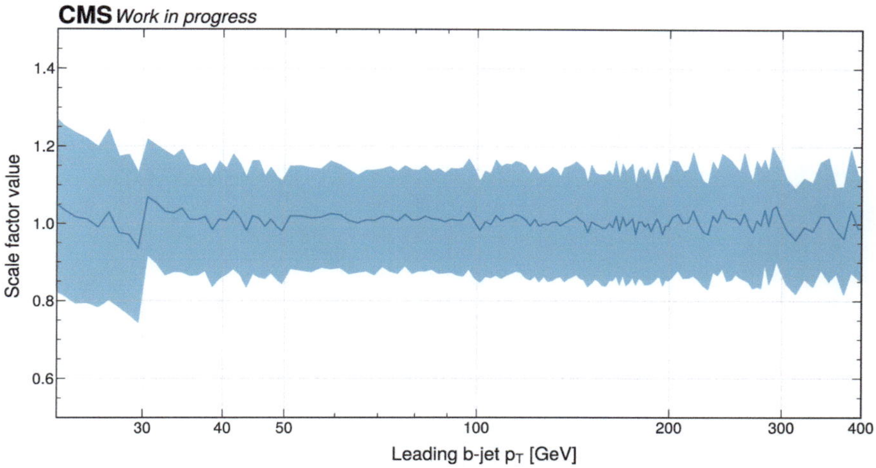

Fig. 5.12 Measured b tagging scale factor as a function of the leading b-jet p_T in the event. The shaded bands denote the systematic uncertainty associated with the jet energy scale and jet energy resolution

Table 5.7 Values of the r factors used to correct the b-tag event weights and preserve the normalization of the MC samples

Year	Final state	r factor
2016	$\tau_\mu \tau_h$	1.0081
	$\tau_e \tau_h$	1.0068
	$\tau_h \tau_h$	1.0103
2017	$\tau_\mu \tau_h$	0.9993
	$\tau_e \tau_h$	0.9949
	$\tau_h \tau_h$	0.9547
2018	$\tau_\mu \tau_h$	1.0039
	$\tau_e \tau_h$	1.0040
	$\tau_h \tau_h$	0.9795

shown in Fig. 5.12. The SFs assume weights between 0.9 and 1.1 over the whole p_T spectrum; this is not surprising as they are employed for a minor morphing of the discriminator shape rather than accounting for the efficiency of a hard selection criterion. The shaded bands reported denote the systematic uncertainty associated with the jet energy scale and jet energy resolution.

The primary purpose of the event weights obtained through this method is to modify the shape of the b-tagging discriminant. At the same time, it is crucial to preserve the expected event yields before applying any b-tag selection criteria. This entails ensuring that the number of events remains unchanged before and after incorporating b-tag weights. Given that analysis phase spaces can be arbitrary and

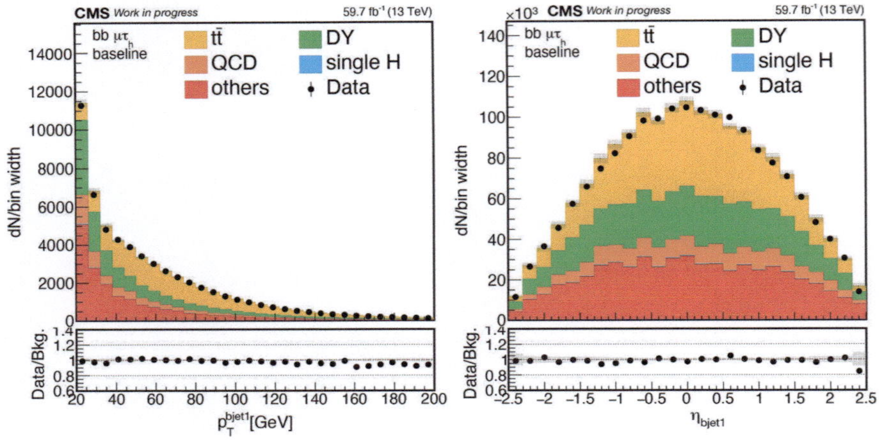

Fig. 5.13 Leading b jet p_{T} (left) and η (right) distributions for events selected in the $\tau_\mu\tau_{\mathrm{h}}$ final state. The black points and error bars represent the data; the background histograms are stacked and organised to highlight the largest and most interesting contributions. The bottom panel reports the ratio between the data and the background hypothesis. The shaded band in all panels represents the statistical uncertainty only

significantly differ from the scale factor measurement phase space (which employs the tag-and-probe selection detailed above with precisely two jets), the average b-tag event weight doesn't need to be precisely 1.0. Thus, to correct for residual yield modifications, we measure the sum of event weights both before and after applying the b-tag event weights without requiring any b-tag selection in either case. The resulting ratio, denoted as $r = \sum \omega_{\text{before}} / \sum \omega_{\text{after}}$, represents a phase space extrapolation factor that is multiplied by the b-tag event weight. The r factor values are reported in Table 5.7 for each year and channel.

The final agreement of the MC simulation with the observed data after the application of said corrections can be appreciated in Fig. 5.13 for the leading b jet p_{T} and $|\eta|$ distributions for events selected in the $\tau_\mu\tau_{\mathrm{h}}$ final state. The two bottom panels display the ratio between the data and the background hypothesis and render explicit the good data-MC agreement.

5.3 Event Selection and Categorization

The objects preselected according to the criteria outlined in the previous Section are used to reconstruct the two Higgs bosons in their respective H → $\tau\tau$ and H → bb decays. The events are first categorized in three channels based on the $\tau\tau$ final state; then, two b jets are coupled based on their likelihood to be the decay products of the second H boson; subsequently, the two jets characteristic of the VBF topology are probed. The events are then further classified based on the bb quality and topology, as well as based on a Machine Learning (ML) multi-classifier, into eight categories. This yields a total of 72 signal regions (i.e. 3 years $\times$ 3 channels $\times$ 8 categories) that are simultaneously explored.

5.3.1 $H \rightarrow \tau\tau$ Final State Assignment and Selection

The determination of the $\tau\tau$ final state in an event is based solely on offline information and probes the presence of electrons, muons, and τ leptons that satisfy the preselection criteria outlined in Sect. 5.2.2. In each event, the presence of at least one τ_h candidate is required; the type of the pair is then based on the selection of the other candidate to be a τ_μ, a τ_e, or a second τ_h. If at least one muon candidate is identified, the event is classified as $\tau_\mu\tau_h$. Conversely, if the event contains at least one electron candidate but no muons, it is classified as $\tau_e\tau_h$. Finally, in cases where neither electrons nor muons are detected but at least one other τ_h is found, the event is classified as $\tau_h\tau_h$. If no pair is identified that meets the selection criteria, the event is discarded.

Once the pair type is determined, all potential pairs of the same kind within the event are examined. At this stage, the isolation and pair electric charge requirements are not applied, as these conditions are reserved for subsequent stages, where they are employed to define background-enriched regions for data-driven estimations. Therefore, due to the absence of the isolation and charge criteria resulting in limited separation between genuine candidates and jets, multiple pairs of the same kind can be found in each channel; in these cases, ambiguities are resolved using a dedicated choice procedure. This decision-making process is particularly delicate in the $\tau_h\tau_h$ channel due to the substantial number of jets observed in an event, many of which can potentially be misidentified as τ_h.

In the $\tau_\mu\tau_h$ and $\tau_e\tau_h$ final states, the muon or electron are always respectively assigned as the leading candidate. In the case of classification as $\tau_h\tau_h$, both permutations of the candidates are considered. Pairs are initially sorted based on the isolation of their leading lepton. In rare cases where the isolation of the first candidates is identical, the pair with the higher transverse momentum of the leading candidate is given priority in the sorting process. If the p_T values are also equal (indicating a shared leading candidate among the pairs), preference is given to the pair with the more isolated trailing candidate. If ambiguities persist due to the equivalent isolation of the second candidate, priority is granted to the pair with the higher p_T of the sub-leading lepton. Once all pairs have been sorted, the first pair that satisfies the baseline selections is selected. This selection strategy is employed to maximize event purity and eliminate any potential overlap between the three distinct final states.

Finally, to ensure the complete orthogonality of the three channels, events with multiple electron and muon candidates are removed by applying an additional lepton veto. As a result, only one muon or electron candidate per event is selected, ensuring the mutual exclusivity of the three final states and univocally determining the $\tau\tau$ final state.

5.3.2 H → bb Selection and VBF Jets Assignment

The selection of the two jets forming the H → bb candidate is made among those jets that adhere to the preselection criteria outlined in Sect. 5.2.3. Therefore, two jets with $p_T > 20$ GeV and $|\eta| < 2.4$ need to be selected; at this stage, the additional requirement is made for the b jet candidates to have an angular distance $\Delta R > 0.5$ from each of the $\tau\tau$ candidates to prevent them from being reconstructed by the same PF objects. After these basic geometrical selections, the b-tagging of the jets needs to be executed.

In the previous HH → bb$\tau\tau$ analysis, the choice was made to strictly enforce the working points of the b-tagging discriminator and, in case more than two jets were found to be compatible with the H → bb decay, those with the largest output of the b-tagging discriminant were chosen. Figure 5.14 reports the selection of the b jets (b tag) and their association as decay products of the Higgs boson (bb candidate) in green and teal, respectively. As it can be appreciated, this procedure made the H → bb selection step (b tag + bb candidate) the second least efficient in the analysis strategy devised in 2016.

To remedy this inefficiency, a new, more sophisticated approach has been developed based on a Recursive Neural Network (RNN) architecture: the HHBTAG algorithm, whose condensed design overview is detailed in the following. RNNs are a class of powerful Neural Networks (NNs) used for modelling sequential data, such as time series or natural language. Unlike the simpler feed-forward NNs that process data in one direction without feedback, RNNs employ a feedback mechanism where

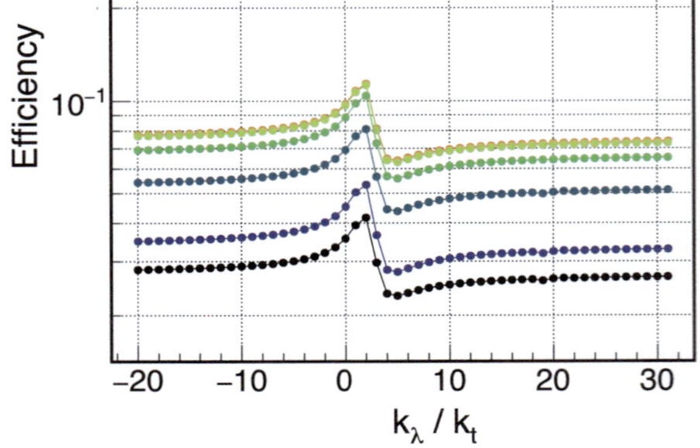

Fig. 5.14 Cumulative efficiency of the selection steps as a function of the anomalous κ_λ/κ_t hypothesis in the previous public result [1, 11]. The selection of the b jets (b tag) and their association as decay products of the Higgs boson (bb candidate) jointly constitute the second least efficient step of the analysis strategy devised in 2016

the output from a layer is saved and fed back to the input. This allows the network to build a memory of previous terms in the sequence, enabling predictions of the next term while considering past information. By sharing parameters and weights between neurons, RNNs incorporate feedback loops or *back-propagation through time*, introducing an additional time variable. This capability enables RNNs to leverage both current and earlier inputs, making adjustments during back-propagation to gradually improve predictions. While RNNs have shown remarkable results in various applications, these network architectures are generally more challenging to train; due to the presence of the time component, the traditional layer-by-layer back-propagation approach is not feasible. Moreover, the vanishing and exploding gradient problems [12] become amplified, necessitating additional considerations. However, one solution to address this challenge is the use of Long Short Term Memory (LSTM) [13], an innovative recurrent network architecture. When combined with an appropriate gradient-based learning algorithm, LSTM proves capable of learning long-term dependencies and mitigating the issues associated with training RNNs. The HHBTAG discriminant is constructed in Keras with a TensorFlow backend. The architecture of the model entails the following components:

- five LSTM layers employing a sigmoid activation function,
- 10 time distributed dense layers utilizing a sigmoid activation function,
- one time distributed dense layer featuring a single unit and using a sigmoid activation function,
- following each layer of the two first points, a batch normalization layer is applied,
- binary cross entropy is used as the loss function, minimized with the Adam algorithm.

For the training and the evaluation of the HHBTAG performance, the entire MC dataset is divided into two distinct samples based on the event number: the even and the odd event number sample; separate training, testing, and performance assessment processes are conducted for each sample. In one case, the *odd sample* is used for training and the *even sample* for testing, while in the other case, the *even sample* is employed for training and the *odd sample* for testing. This approach ensures the use of the full dataset for training and, at the same time, guarantees a comprehensive evaluation of the model's performance across different event subsets. The RNN input variables are chosen to maximize the performance of the algorithm while keeping their number constrained. These features are summarized in Table 5.8, ordered by their permutation importance, alongside their description. The inputs can be split into four different categories: the b jet candidate score and kinematic variables, the kinematic variables of the other objects in the event, the angular distances of the b jet candidate from the other objects in the event, and global event variables.

Contrary to all other discriminants presented hitherto, for the HHBTAG algorithm, there is no need for the definition of working points as through the application of this method, we aim not at selecting a class of objects but rather two specific b jets that are compatible with the $H \to bb$ decay. Therefore, after the application of the trained RNN to all jet candidates that satisfy the baseline b jet selections, the two

Table 5.8 Lists of input variables used by the HHBTAG algorithm for the H → bb selection, ordered by their permutation importance

Variable	Description
DEEPJET score	Score of the DEEPJET discriminant applied to the b jet candidate
η(jet)	Pseudorapidity of the b jet candidate
$\Delta\phi$(jet,$H_{\tau\tau}$)	Azimuthal angle distance between the b jet candidate and the H → $\tau\tau$ system defined as the sum of the visible four-momenta of the two τ candidates
p_T(jet)	Transverse momentum of the b jet candidate
$\frac{E}{p_T}$(jet)	Energy of the b jet candidate, normalized to its transverse momentum
$\frac{M}{p_T}$(jet)	Mass of the b jet candidate, normalized to its transverse momentum
$\Delta\eta$(jet,$H_{\tau\tau}$)	Pseudorapidity distance between the b jet candidate and the H → $\tau\tau$ system defined as the sum of the visible four-momenta of the two τ candidates
$p_T(H_{\tau\tau})$	Transverse momentum of the H → $\tau\tau$ system defined as the sum of the visible four-momenta of the two τ candidates
$p_T(\tau_1) + p_T(\tau_2)$	Scalar sum of the transverse momenta of the two τ candidates from the H → $\tau\tau$ decay
$\eta(H_{\tau\tau})$	Pseudorapidity of the H → $\tau\tau$ system defined as the sum of the visible four-momenta of the two τ candidates
$\Delta\phi(\phi_{p_T^{miss}}, H_{\tau\tau})$	Azimuthal angle distance between the missing transverse energy and the H → $\tau\tau$ system defined as the sum of the visible four-momenta of the two τ candidates
$\frac{p_T^{miss}}{p_T(H_{\tau\tau})}$	Missing transverse energy magnitude normalized to the scalar sum of the transverse momenta of the τ candidates
$\tau\tau$ final state	Final state reconstructed for the H → $\tau\tau$ decay
Year	Year of data-taking

candidates with the highest HHBTAG score are chosen as the two decay products of the H boson.

The efficacy of this selection process is evaluated in MC simulation of non-resonant HH production and compared to two available standard algorithms: DEEPJET and DEEPCSV [14]. The former has already been introduced, and the latter, whose description is out of the scope of this Thesis, is an additional b-tagger that is available within CMS and represents the standard before the introduction of the DEEPJET algorithm. The performance comparison is presented in Fig. 5.15, where for all three algorithms, the two candidates with the highest discriminant score are chosen to reconstruct the H → bb decay. The results are presented for all $\tau\tau$ pair final states to highlight the good performance of the HHBTAG algorithm in all regions of the analysis. In this Figure, the efficacy of the objects' identification is evaluated in terms of the purity of the selected b jet pair candidate:

$$\mathcal{P} = \frac{N^{true}}{N} \tag{5.9}$$

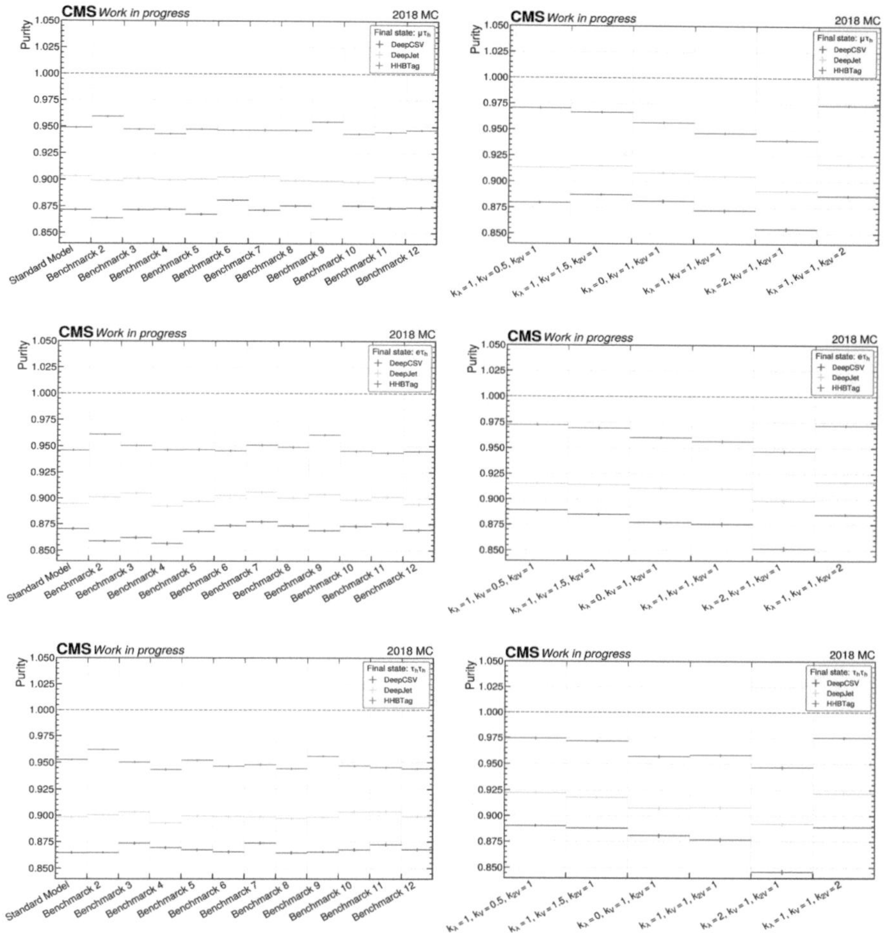

Fig. 5.15 Performance of the HHBTag algorithm, compared to the standard DeepJet and DeepCSV discriminators. The performance is evaluated in terms of purity, as defined in Eq. 5.9, computed in Monte Carlo simulation for the gluon fusion (left column) and vector boson fusion (right column) production mechanisms, in the $\tau_\mu \tau_h$ (top row), $\tau_e \tau_h$ (central row), and $\tau_h \tau_h$ (bottom row) channels. For the gluon fusion production mechanism, the purity is shown for the standard model prediction as well as for some of the effective field theory benchmarks detailed in Sect. 5.4.3. For the vector boson fusion production mechanism, the purity is shown for the standard model prediction as well as for the beyond the standard model scenarios described in Sect. 1.2.2, identified by the value of the coupling strength modifiers

where the purity $\mathcal{P}$ granted by a classifier is defined as the ratio between the number of events in which the candidate selection matches the ground truth (N^{true}) and the total number of events in which the H → bb candidate is reconstructed with the classifier under consideration (N). The comparison focuses only on non-resonant simulated samples detailed later, in Sects. 5.4.2 and 5.4.3. For all production mechanisms, signal scenarios, and years, the HHBTAG algorithm achieves purity ranging from 92 to 98%, surpassing the previously best-performing DEEPJET discriminator, which attains purity spanning from 86 to 92%. Therefore, selecting the H → bb candidate as the two b jet candidates with the highest HHBTAG score consistently and considerably enhances the performance for non-resonant HH production scenarios for all $\tau\tau$ pair final states.

After the assessment of the H → bb candidate, the two jets typical of the VBF signature are probed. From all remaining jets at this stage, all possible pairs are formed from those with $p_{\text{T}} > 30\,\text{GeV}$ and $|\eta| < 4.7$; the pair with the highest invariant mass (M_{jj}) is then designated as the candidate VBF jets pair. This choice is based on the characteristic kinematics of the VBF process, for which two jets with large pseudorapidity distance are expected, therefore having an invariant mass $M_{jj} = 2\sqrt{p_{\text{T}}^{(1)} \cdot p_{\text{T}}^{(2)} \cdot \cosh(\Delta\eta_{jj})}$, where $\Delta\eta_{jj} = \eta^{(1)} - \eta^{(2)}$. Additionally, to reduce background contamination, the following two criteria are imposed: $M_{jj} > 500\,\text{GeV}$ and $\Delta\eta_{jj} > 3$.

Moreover, for events selected within the exclusive VBF trigger phase-space, i.e. those belonging to the $\tau_{\text{h}}\tau_{\text{h}}$ channel where both τ_{h} satisfy the threshold $p_{\text{T}} > 25\,\text{GeV}$ and at least one of them has $p_{\text{T}} < 40\,\text{GeV}$ as described in Sect. 5.2.1, an additional set of selection criteria is imposed on the VBF jet candidates to ensure trigger efficiency plateau. In these events, the VBF jet candidates must satisfy the condition $M_{jj} > 800\,\text{GeV}$, where the leading VBF jet should fulfil $p_{\text{T}} > 140\,\text{GeV}$ and the subleading one should satisfy $p_{\text{T}} > 60\,\text{GeV}$.

5.3.3 Signal Regions Definition and Event Categorization

The preceding steps of the analysis aimed to establish the presence of viable H → $\tau\tau$ and H → bb candidates within each event and, if so, identify the most probable H candidates. However, due to the HH signal cross section being orders of magnitude smaller than that of other processes, the events that pass the selections remain predominantly contaminated by background contributions. Therefore, a categorization strategy needs to be devised to maximize the HH signal fraction in the three $\tau\tau$ final state channels.

An important development of the analysis presented in this Thesis, over the previous HH → bb$\tau\tau$ public results, is the detailed investigation of the VBF production mechanism. However, the production cross section for the VBF mechanism is 1.726 fb, 18 times smaller than the already tiny ggF production cross section. To better cope with this disparity, the choice is taken to give precedence to VBF candidates in the categorization. Therefore, if two VBF jet candidates are found as detailed above, the event is classified as originating from VBF production. To further improve the sensitivity, VBF events are subsequently categorised using a dedicated NN, as presented below.

If an event is not categorised as VBF, it is automatically identified as originating from the ggF mechanism for which three orthogonal categories are defined to enhance the sensitivity. Moreover, only for the ggF class, a selection is performed on the invariant mass of the two H candidates to further enhance the signal fraction in each category.

The categorization strategy is discussed in detail in the following and results in a total of 72 signal regions (i.e. 3 years $\times$ 3 channels $\times$ 8 categories).

Multi-categorization for VBF Topology

A criticality of the small VBF production cross section value is that even applying tight event selection, this category is still highly contaminated by background events as well as misclassified ggF HH events. Therefore, a multi-categorization approach based on an advanced ML technique was devised to overcome this problem and increase the power of the statistical analysis of the data. In this approach, a tailored NN is trained to assign probability estimates for an event to belong to categories associated with the different signal and background processes that contribute to the event yield in the VBF phase space. A brief description of the design, architecture, input variables, and performance of this new method is given in the following.

The multi-classification strategy aims to correctly identify the signal and background process in the VBF signal region using the 74 input variables summarized in Table 5.9 alongside their description. These features can be classified into two main categories: Low-Level (LL) four-vector components of particles and High-Level (HL) variables.

To fully exploit the discriminative power of the variables of category LL, they are first processed by a Lorentz Boost Network (LBN) [15]. LBNs are a recent development in the ML domain; they combine input four-vectors to create combined particles and corresponding rest frames according to weighted sums using trainable weights. Each of these *artificial* particles can be boosted to any of the *artificial* rest frames through Lorentz transformations. This process can extract a set of physics-inspired higher-level variables based on the low-level inputs. In the model described here, the LBN is configured to find up to 30 particle combinations and boost them into 30 independent rest frames. The extracted high-level features include the energy, transverse momentum, pseudorapidity, and mass of each Lorentz-boosted particle, as well as the angular separation between pairs of boosted particles. The set of distilled features is fed to a standard dense NN jointly with the already available high-level observables of the HL set. The architecture of this model entails the following components:

Table 5.9 Lists of input variables used by the VBF multi-classification algorithm

Objects	Variables	Description
Leptons	E, p_T, η, ϕ	Final state leptons' four-vector components
b jets	E, p_T, η, ϕ DEEPJET, HHBTAG scores	b jets' four-vector components and discriminators' scores
VBF jets	E, p_T, η, ϕ DEEPJET, HHBTAG scores	VBF jets' four-vector components and discriminators' scores
Central jets	E, p_T, η, ϕ DEEPJET, HHBTAG scores	Additional central jets' four-vector components and discriminators' scores
Forward jets	E, p_T, η, ϕ DEEPJET, HHBTAG scores	Forward jets' four-vector components and discriminators' scores
MET	p_T^{miss}, $\phi(\vec{p}_T^{miss})$	Missing transverse energy magnitude and azimuthal direction
H → $\tau\tau$	E, p_T, η, ϕ	H → $\tau\tau$ candidate's four-vector components
H → bb	E, p_T, η, ϕ	H → bb candidate's four-vector components
Global event	$\tau\tau$ final state	Final state of the $\tau\tau$ pair
	Year	Year of data-taking

- four hidden layers of 128 nodes, each utilizing a hyperbolic tangent activation function
- one output layer with nine units using a softmax activation function
- following each hidden layer, a batch normalization layer is applied
- a random unit dropout with a probability of 5% per unit per forward pass is applied
- the custom loss function

$$L(\hat{y}(x), \hat{y}_G(x), y, y_G) = \frac{1}{N}\sum_{n}^{N} -y_n \log \hat{y}_n(x) \quad + \quad \frac{1}{N}\sum_{n}^{N} -y_{G,n} \log \hat{y}_{G,n}(x) \quad + \quad L_2(\lambda)$$

minimized with the Adam optimizer. The loss function comprises three terms: the first two represent two standard Categorical Cross Entropy (CCE) functions, whereas the third term denotes the typical L_2 normalization of the Adam optimizer. The first CCE is applied to all nine output classes and compares the predicted output $\hat{y}_n(x)$ of an event n given the inputs x with its corresponding true value y_n. The second CCE is applied to the two groups of signal (i.e. all HH processes) and background (i.e. all background processes). This effectively increases the relative importance of signal events and, more importantly, introduces a balancing for false classifications.

The sum of the nine outputs of the NN is normalized to unity; in this way, each output represents the probability of an event pertaining to a specific signal or background process. The targeted processes are: gluon fusion HH, vector boson fusion

HH, $t\bar{t}$, DY, and $t\bar{t}H$; where the VBF signal is split into the two hypotheses of $\kappa_{2V} = 1$ or $\kappa_{2V} = 0.2$; the $t\bar{t}$ background is divided into the three cases of its final state being fully-leptonic, semi-leptonic, or fully-hadronic; the $t\bar{t}H$ process is branched into events with either $H \rightarrow \tau\tau$ or $H \rightarrow bb$ decays. This choice is made to fully exploit the kinematical and topological differences between each sub-process. The contribution from the QCD multijet background is not considered, as it is determined based on a data-driven approach that prevents the simple definition of truth labels.

The ultimate VBF multi-categorization is performed by first summing together the outputs of the sub-processes to yield five final categories. Two categories target the HH signal and are denoted as `classGGF` and `classVBF`; three categories target background processes and are denoted as `classttH`, `classTT`, and `classDY`. Each event is classified into the category with the highest output probability, ensuring mutual exclusivity between classes. The performance of this classification algorithm is reported in Fig. 5.16 where the row-normalized process classification accuracy and confusion rates in simulated events are reported. The performance is shown for the three Run-2 years separately to highlight the stability of the VBF multi-classifier over different data-taking conditions. Regardless of the QCD background not being targeted by the NN, its contribution is reported for completeness; it should thus be noted that QCD events are desirably assigned mostly to the category for the $t\bar{t}$ with which it shares a similar topology.

Categorization of ggF Candidate Events

Events not pertaining to the VBF class discussed above are automatically assigned to the ggF class. To further enhance the sensitivity of the analysis in this phase space region, events are categorized based on the topology of the two b jet candidates.

The angular separation of the two b quarks produced in the H decay can be computed to be

$$\Delta R(\mathrm{b}, \mathrm{b}) = \frac{m_{\mathrm{H}}}{p_{\mathrm{T}}^{(\mathrm{H})}} \cdot \frac{1}{\sqrt{z(1-z)}} \geq \frac{250\,\mathrm{GeV}}{p_{\mathrm{T}}^{\mathrm{H}}} \tag{5.10}$$

where m_{H} and $p_{\mathrm{T}}^{(\mathrm{H})}$ are respectively the mass and transverse momentum of the SM H, while z is the fraction of mass carried by one of the two b quarks. Based on this parametrization, three topologies can be observed experimentally. For low values of transverse momentum, i.e. $p_{\mathrm{T}}^{(\mathrm{H})} \lesssim 300\,\mathrm{GeV}$, the two jets have a large enough angular distance $\Delta R(\mathrm{b}, \mathrm{b}) \gtrsim 0.8$ for them to be efficiently reconstructed by the AK4 anti-k_{T} algorithm. For intermediate values of transverse momentum, i.e. $300 \lesssim p_{\mathrm{T}}^{(\mathrm{H})} \lesssim 600\,\mathrm{GeV}$, the separation of the objects is $0.4 \lesssim \Delta R(\mathrm{b}, \mathrm{b}) \lesssim 0.8$, so that the two jets partially overlap and can be reconstructed either individually by the AK4 algorithm or together as a single jet by the AK8 algorithm. In the extreme case of high values of transverse momentum, i.e. $p_{\mathrm{T}}^{(\mathrm{H})} \gtrsim 600\,\mathrm{GeV}$, the separation is $\Delta R(\mathrm{b}, \mathrm{b}) \lesssim 0.4$ and the AK4 algorithm is unable to reconstruct two distinct jets.

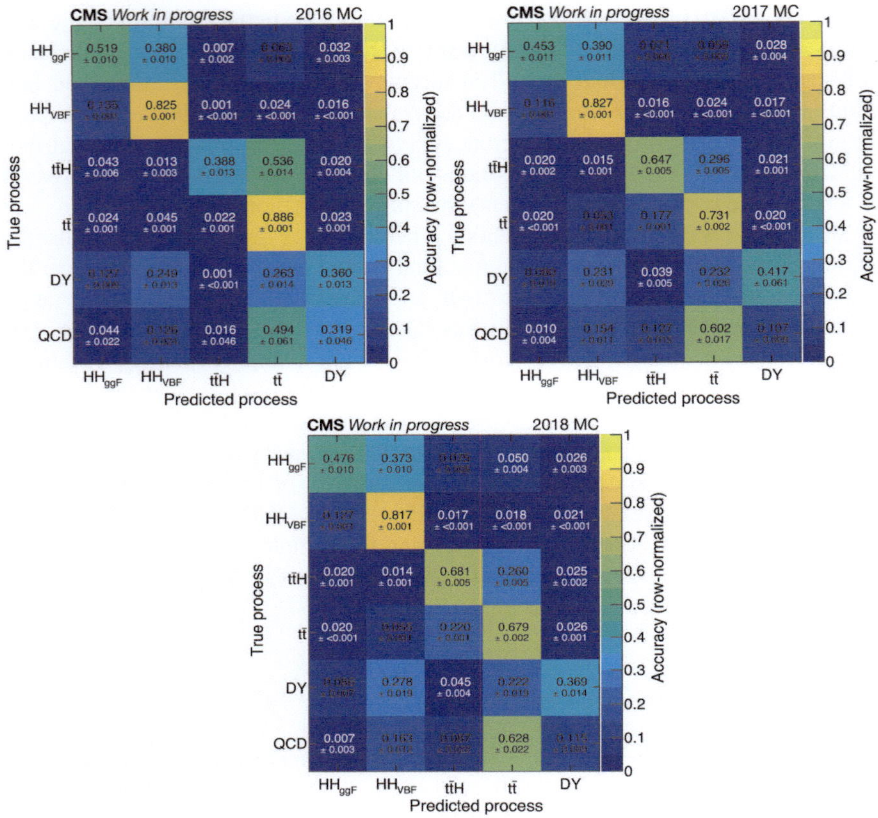

Fig. 5.16 Row-normalized process classification accuracy and confusion rates in simulated events for the vector boson fusion multi-categorization neural network. The result is presented separately for the 2016 (top left), 2017 (top right) and 2018 (bottom) data-taking periods. Quoted uncertainties encompass the effect of the Monte Carlo simulation being statistically limited

In this analysis, only the non-resonant production mechanisms presented in Sects. 1.2.1 and 1.2.2 are considered. Under this assumption, the probability of producing an extremely boosted H is highly attenuated; therefore, no specific category is dedicated to this regime. In contrast, the three following categories are devised to maximize the sensitivity in the case of a low and moderate boost of the H → bb candidate:

- `boosted`: events where an AK8 jet with invariant mass $m > 30$ GeV and passing the loose DeepJet b-tagging WP is matched to two AK4 subjets with distance $\Delta R < 0.4$
- `res2b`: events where the two AK4 jets with the highest HHBTag score both pass the medium DeepJet b-tagging WP

- `res1b`: events where only one of the two AK4 jets with the highest HHBTAG score passes the medium DEEPJET b-tagging WP, while the second passes the loose WP.

Invariant Mass Selection

The kinematics inherent to the HH $\rightarrow$ bb$\tau\tau$ decay process can be leveraged to further enhance the sensitivity of the analysis in the ggF categories. To this end, the τ and b candidates can be exploited to reconstruct the invariant masses of the H $\rightarrow \tau\tau$ ($m_{\mathrm{H}\tau\tau}$) and H $\rightarrow$ bb (m_{Hbb}) candidates, which provide a clear HH signal signature.

The reconstruction of $m_{\mathrm{H}\tau\tau}$ is performed using the SVFIT algorithm [16], which is based on a likelihood technique quantifying the level of compatibility between a Higgs boson mass hypothesis and the measured momenta of the τ decay products complemented by the missing transverse energy reconstructed in the event. The kinematics of the $\tau \rightarrow \tau_{\mathrm{h}}\nu_{\tau}$ and $\tau \rightarrow \ell\nu_{\ell}\nu_{\tau}$ ($\ell = \mu$, e) decays are parametrized by only two and three variables, respectively: the fraction of τ lepton energy in the laboratory frame carried by the visible decay products, one angle specifying the orientation of the τ lepton momentum vector with respect to the momentum vector of the visible decay products, and for the leptonic decays the invariant mass of the neutrino system. Given the presence of two τ leptons in the decay of the H boson, in the fully hadronic and semi-leptonic final states, four and five observables are needed to fully constrain the energy and momentum of both τ lepton in the laboratory system, respectively. In the case not considered in this search where both τ leptons decay to leptons, six observables would be needed. This sets against the only two observables available to constrain the momenta of the neutrinos, i.e. the magnitude and the azimuthal angle of the $\vec{p}_{\mathrm{T}}^{\,\mathrm{miss}}$ vector. Moreover, the use of the missing transverse momentum to this end posits the hypothesis that the missing transverse energy in the event owes mainly to these neutrinos. In the SVFIT algorithm, the fact that the τ lepton decay kinematics is underconstrained by measurable observables is handled via a likelihood approach to computing the conditional probability $P(m_{\mathrm{H}\tau\tau}|\boldsymbol{\theta}, \mathbf{p})$ under the marginalization of the unconstrained parameters $\boldsymbol{\theta}$, using the measure leptons' momenta $\mathbf{p}$.

An improvement of the resolution on $m_{\mathrm{H}\tau\tau}$ by about 30% over the visible invariant mass of the selected τ lepton decay products is achieved by the use of the SVFIT algorithm. Nonetheless, a residual shift of the H boson mass peak below the expected 125 GeV value endures, owing to the assumption of the missing transverse energy originating exclusively from the neutrinos in the τ decay. Indeed, a small fraction of the $\vec{p}_{\mathrm{T}}^{\,\mathrm{miss}}$ vector is due to the B hadrons and mesons decays from the b quarks hadronization. Figure 5.17 demonstrates the favourable agreement observed between the MC simulation and data for the distribution of the $m_{\mathrm{H}\tau\tau}$ variable across all three $\tau\tau$ final states in the 2018 data-taking period. The three bottom panels display the ratio between the data and the background hypothesis and render explicit the good data-MC agreement.

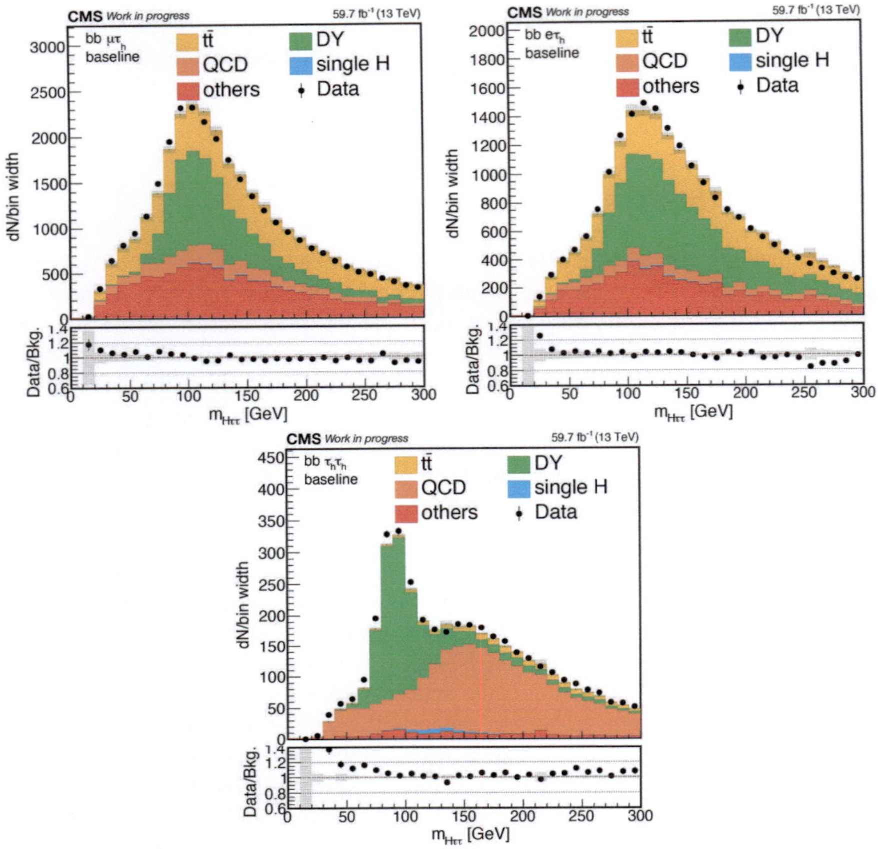

Fig. 5.17 Distribution of the $m_{\mathrm{H}\tau\tau}$ invariant mass reconstructed with the SVFIT algorithm in the $\tau_\mu\tau_\mathrm{h}$ (left), $\tau_\mathrm{e}\tau_\mathrm{h}$ (centre), and $\tau_\mathrm{h}\tau_\mathrm{h}$ (right) final states for the baseline selection region. The black points and error bars represent the data; the background histograms are stacked and organised to highlight the largest and most interesting contributions. The bottom panel reports the ratio between the data and the background hypothesis. The shaded band in all panels represents the statistical uncertainty only

The reconstruction of m_{Hbb} is performed solely using the visible observables in the event. In the cases where two AK4 b jets are reconstructed, the invariant mass of the jets' system is computed to be m_{Hbb}, In the cases where one single AK8 b jet is reconstructed, m_{Hbb} equals the mass of the AK8 jet. Also in this case, a residual shift of the H boson mass peak below the expected 125 GeV value is due to the fraction of missing transverse energy from the B hadrons and mesons decays in the b quarks hadronization. The good agreement of the m_{Hbb} modelling between MC simulation and data can be appreciated in Fig. 5.18 where the invariant mass is reported for the 2018 data-taking period in the three $\tau\tau$ final states. The three bottom panels display

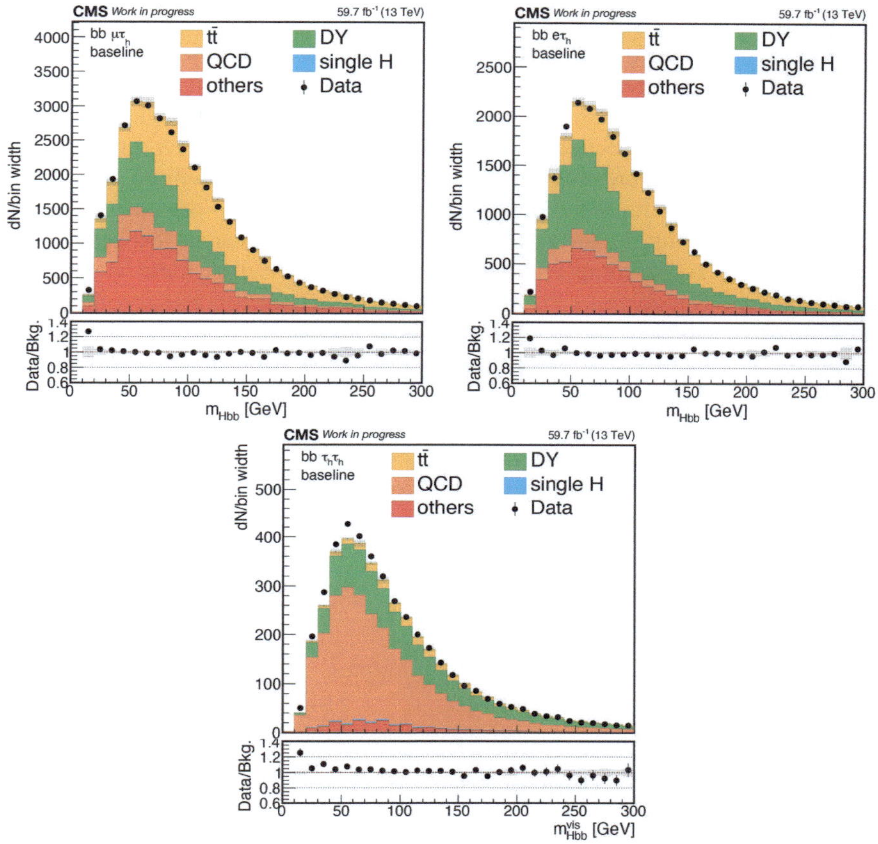

Fig. 5.18 Distribution of the m_{Hbb} invariant mass reconstructed from b jets in the $\tau_{\mu}\tau_{\mathrm{h}}$ (left), $\tau_{\mathrm{e}}\tau_{\mathrm{h}}$ (centre), and $\tau_{\mathrm{h}}\tau_{\mathrm{h}}$ (right) final states for the baseline selection region. The black points and error bars represent the data; the background histograms are stacked and organised to highlight the largest and most interesting contributions. The bottom panel reports the ratio between the data and the background hypothesis. The shaded band in all panels represents the statistical uncertainty only

the ratio between the data and the background hypothesis and render explicit the good data-MC agreement.

The invariant masses of the two H candidates can then be combined into a single invariant mass selection encoding the HH kinematics. In earlier iterations of the HH $\rightarrow$ bb$\tau\tau$ analysis, a rectangular selection was employed [17, 18], later replaced by an enhanced elliptical selection [1]. Moreover, previous to the analysis presented here, the invariant mass selection criterion was optimized by employing the signal-to-background ratio (S/B) as a metric. However, with the introduction of the new Deep Neural Network (DNN) (cf. Sect. 5.3.4) for signal event identification, the HH signal extraction now relies entirely on the discriminatory power of the DNN. In this new context, the objective of the invariant mass criterion shifts from improving the S/B ratio to enhancing the DNN performance, which in turn is now charged with the maximization of the sensitivity. To assess the impact of the invariant mass selection on the analysis sensitivity, studies were conducted by varying the signal efficiency of the invariant mass selection and retraining the DNN under each scenario. The results revealed that a more discriminating mass selection, despite improving the S/B ratio, led to poorer limits on the HH $\rightarrow$ bb$\tau\tau$ cross section. This outcome highlights the superior capability of the DNNs in effectively distinguishing between signal and background events compared to the classical mass selection. Consequently, the primary objective of applying the mass selection is no longer focused on enhancing the S/B ratio but rather on effectively removing background events that significantly deviate in regions where no signal overlap is expected while maintaining a high signal efficiency. The choice to retain an elliptically shaped selection is motivated by its ability to better encapsulate the signal shape compared to rectangular selections, while still retaining a manageable level of complexity. The search for the optimal invariant mass selection is performed via random sampling of the free parameters of the ellipsis; two different criteria are obtained depending on the H $\rightarrow$ bb being reconstructed as two resolved AK4 jets (`res2b` and `res1b` categories) or one boosted AK8 jet (`boosted` category).

$$\text{Resolved AK4}: \frac{(\mathrm{m}_{\mathrm{H}\tau\tau} - 129\,\mathrm{GeV})^2}{(53\,\mathrm{GeV})^2} + \frac{(\mathrm{m}_{\mathrm{Hbb}} - 169\,\mathrm{GeV})^2}{(145\,\mathrm{GeV})^2} < 1 \tag{5.11}$$

$$\text{Boosted AK8}: \frac{(\mathrm{m}_{\mathrm{H}\tau\tau} - 128\,\mathrm{GeV})^2}{(60\,\mathrm{GeV})^2} + \frac{(\mathrm{m}_{\mathrm{Hbb}} - 159\,\mathrm{GeV})^2}{(94\,\mathrm{GeV})^2} < 1 \tag{5.12}$$

The invariant mass criterion of Eq. 5.11 is presented in Fig. 5.19, superimposed to the distribution of SM HH signal and inclusive background events in the $(m_{\mathrm{Hbb}}, m_{\mathrm{H}\tau\tau})$ plane for the 2018 data-taking period. From this Figure, the shape of the elliptical mass selection can be explained: the larger extension of the ellipsis in the m_{Hbb} direction is due to the fact that the background distribution spreads to a greater extent in the $m_{\mathrm{H}\tau\tau}$, therefore a better trade-off between signal efficiency and background rejection is obtained in this manner.

At this point, it is worth reminding that the eight categories just defined (i.e. `res2b`, `res1b`, `boosted`, `classGGF`, `classVBF`, `classttH`, `classTT`, `classDY`) are defined for each of the three final states of the $\tau\tau$ pair, i.e. $\tau_\mu\tau_h$, $\tau_e\tau_h$, and $\tau_h\tau_h$. The same strategy is replicated for the three years of data-taking, yielding a total of 72 signal regions (i.e. 3 years $\times$ 3 channels $\times$ 8 categories) that are simultaneously explored in the statistical analysis detailed in Sect. 6.2.

5.3.4 Deep Neural Network for Signal Extraction

The candidate selection process and the definition of the signal region exhibits excellent capability in effectively suppressing reducible backgrounds and, to a certain extent, irreducible backgrounds. However, despite all the selection criteria, the signal region remains subject to large contamination from events originating from background processes. Namely, in the $\tau_\mu \tau_h$ and $\tau_e \tau_h$ final states, a large contribution from the $t\bar{t}$ and Drell-Yan backgrounds is present due to the direct production of a prompt muon or electron and the lower p_T thresholds applied on them. Conversely, the $\tau_h \tau_h$ channel is mainly affected by the QCD multijet and Drell-Yan processes due to the large number of jets that can be misidentified as τ_h, and genuine $\tau\tau$ pair production, respectively. These observations motivate the development of a dedicated technique for the identification of the signal, with specific attention to its discrimination against the major background contributions. A novel multivariate algorithm, employing a DNN, has been devised for this search to effectively eliminate the remaining background contamination in all categories of the analysis by exploiting the kinematic characteristics inherent to the HH $\rightarrow$ bb$\tau\tau$ processes under investigation.

Similarly to what is done for the HHBTAG detailed in Sect. 5.3.2, in order to mitigate the statistical uncertainty associated with the density distributions of the MC data sample used for the training of the DNN, a strategy of data division is employed. The MC data is evenly split into two distinct sets based on the event number, and two separate discriminators are trained, each using a different subset of the data. During the inference phase, the trained discriminators are applied to make predictions on the respective halves of the data that were not utilized for their training. This methodology ensures that all MC simulated events are used for inference, thereby achieving the desired outcome. It should be noted, however, that this approach does not entail the provision of an explicit validation sample, thereby

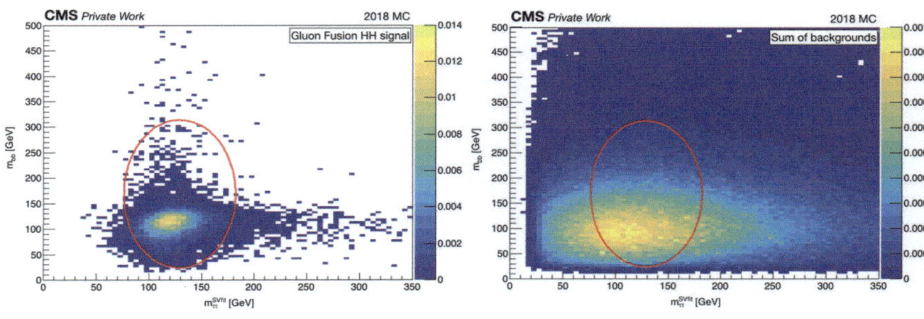

Fig. 5.19 2D distributions of $(m_{\text{Hbb}}, m_{\text{H}\tau\tau})$ for the HH SM signal (left) and for the sum of MC backgrounds (right). The normalized process yields after the $\tau\tau$ and bb candidates' selections and before the invariant mass requirements are shown for the 2018 data-taking period. The presence of two b-tagged jets passing the medium identifier working point is required; the three $\tau\tau$ decay channels considered in the analysis are merged. The red ellipses show the region selected by the selection in formula (5.11)

precluding any fine-tuning of the hyperparameters, which may introduce potential biases to the model. Nevertheless, the investigation conducted in Reference [19] revealed that the choice of architecture parameters and training scheme exhibits minimal influence on the overall performance, thereby indicating that the omission of extensive tuning is unlikely to compromise significantly the efficacy of this approach.

Over 100 data features are computed, encompassing components of final-state four-momenta, relative angles between final states in different rest-frames, reconstructed masses, transverse masses, as well as features generated by reconstruction and tagging algorithms. Given the expectation of redundant information within this extensive feature set, a selection process is employed to identify the most relevant and indispensable features specifically suited for our classification task. The first removal of unnecessary inputs targets groups of strongly monotonically correlated features. Hierarchical clustering is employed to identify groups of features exhibiting monotonic relationships, using Spearman's Rank-Order Correlation Coefficient (SROCC) [20]. Features with an SROCC value below 0.2 are considered correlated within a group. Subsequently, random forest classifiers [21] are used to assess the impact of removing features from each correlated group on the classifier's performance. Initially, models are trained using the full set of remaining features. Then, a correlated feature is temporarily removed, and new models are trained. If the average performance does not decrease, the feature is deemed safe for removal. If the removal of two or more features from a correlated cluster does not result in any performance loss, the single feature that leads to the largest performance improvement is selected for removal. This process is repeated on the remaining features within the cluster until only one feature remains or no further features can be safely removed. The second but most important approach for feature selection involves evaluating their permutation importance, which measures the degradation in model performance when a specific feature is randomly shuffled while others remain unchanged. If a feature is important, shuffling it should diminish prediction accuracy by destroying the information it carries. However, it should be noted that feature importance can vary depending on the model training process and the specific training dataset; to address this, multiple models are trained on subsamples of the original dataset, and features with an average permutation importance greater than or equal to a predefined threshold are deemed important. This process is repeated on new subsamples, and features are selected based on the fraction of times they are identified as important. The final selection tier aims at removing variables that showcase strong non-monotonic correlations among each other. The interdependence between features can be assessed by employing regression techniques to predict the values of one feature based on the remaining features as inputs. The importance of the input features in the regression model can then be examined to identify which features play a crucial role in predicting the values of the target feature. After this procedure is applied, a total of 26 features are retained as input to the DNN; they are listed in Table 5.10 alongside their detailed description.

Each of the two discriminators is an ensemble of 10 NNs trained using 10-fold stratified cross-validation on the training data, with weights assigned based on their validation performance. The procedure is as follows:

Table 5.10 Lists of input variables used by the deep neural network for signal extraction. The first part of the table reports input variables in descending order of importance evaluated random forest models. The second part of the table reports the additional six categorical input features with no specific ordering criterion

Variable	Description
DEEPJET CvsB b jet 1	Charm versus bottom DEEPJET discriminator score binned in WPs for the first b jet
DEEPJET CvsL b jet 2	Charm versus light flavour DEEPJET discriminator score binned in WPs for the second b jet
χ^2_{KinFit}	χ^2 quality of the HH system invariant mass HHKINFIT fit
$m_{\mathrm{HH}}^{\mathrm{KinFit}}$	HH system invariant mass computed with HHKINFIT
$m_{\mathrm{H}\tau\tau}$	Invariant mass of the $\mathrm{H}\tau\tau$ candidate
$\Delta R(\tau, \tau) \cdot p_{\mathrm{T}}^{(\mathrm{H}\tau\tau)}$	Angular distance between the leptons on the $\tau\tau$ pair scaled by the transverse momentum of the $\mathrm{H}\tau\tau$ candidate
$m_{\mathrm{T}}(\ell)$ ($\ell = \mathrm{e}, \mu, \tau_{\mathrm{h}}$)	Transverse mass of the first lepton in the $\tau\tau$ pair
$p_{\mathrm{T}}(\tau_{\mathrm{h}})$	Transverse momentum of the light-lepton, or softest τ_h selected
$\Delta R(\tau, \tau)$	Angular distance between the leptons on the $\tau\tau$ pair
$\Delta\phi(\mathrm{H}\tau\tau, \vec{p}_{\mathrm{T}}^{\mathrm{miss}})$	Azimuthal angle distance between the $\mathrm{H}\tau\tau$ candidate and the missing transverse energy
$m_{(\mathrm{Hbb})}$	Mass of the Hbb candidate
HHBTAG b jet 2	HHBTAG score of the second b jet
m_{HH}^{vis}	HH system invariant mass computed from visible components
$\Delta\phi(\mathrm{H}\tau\tau, \mathrm{Hbb})$	Azimuthal angle distance between the two H bosons
$\Delta R'(\tau, \tau)$	Angular distance between the leptons on the $\tau\tau$ pair, in the rest frame of the $\mathrm{H}\tau\tau$ candidate plus missing transverse energy system
$p_{\mathrm{T}}(\mathrm{Hbb})$	Transverse momentum of the H reconstructed in the bb pair
$p_{\mathrm{T}}(\ell)$ ($\ell = \mathrm{e}, \mu, \tau_{\mathrm{h}}$)	Transverse momentum of the first lepton in the $\tau\tau$ pair
p_{T} b jet 1	Transverse momentum of the first b jet
$\xi''(\tau\tau, \mathrm{bb})$	Angle between the decay plane of the taus and the decay plane of the b jets, in the rest frame of the HH system
$\cos\theta'(\tau_{\mathrm{h}}, \mathrm{H}\tau\tau + \vec{p}_{\mathrm{T}}^{\mathrm{miss}})$	Cosine of the angle between the second τ lepton and the direction of flight of the $\mathrm{H}\tau\tau$ candidate plus missing transverse energy system, in the rest frame of the latter
Categorical variable	**Description**
ISBOOSTED	Boolean for events pertaining to the `boosted` category
$\tau\tau$ final state	Reconstructed final state of the $\mathrm{H} \to \tau\tau$ decay
ISVBF	Boolean for events pertaining to one of the VBF multi-categories
DEEPJET WP b jet 1	DEEPJET working point of the first b jet
DEEPJET WP b jet 2	DEEPJET working point of the second b jet
Year	Year of data-taking

1. The training data is divided into 10 folds, ensuring that the fractions of samples, channels, and jet categories in each fold closely match the overall fractions in that half of the training data.
2. A network is trained on nine folds, using the tenth fold as validation data to monitor the network's generalization during training.
3. Once the training cycle is complete, the model state corresponding to the best performance on the validation data is saved, and the model is assigned an importance weight based on the inverse of its loss on the validation data.
4. Subsequent networks are trained, each using a different fold of the data as the validation sample until all 10 folds of the training data have been utilized exactly once.
5. The importance weights of the ensemble are normalized to ensure their sum equals 1.0.
6. The ensemble's prediction for a given data point is calculated as the sum of each model's prediction for that data point multiplied by the respective model's importance weight: $p_{\text{ensemble}}(x) = \sum_i^N \omega_i p_i(x)$.

In this process, the architecture of each of the networks entails the following components:

- entity embedding of categorical features, with embedding size $M = (1 + N)/2$ for feature cardinality $N(12, 13)$
- six densely connected hidden layers of widths equal to the number of input features plus the sum of each embedding size for categorical input features, each utilizing a swish-1 activation function [22]
- single output neuron, with sigmoid activation function
- a random unit dropout with a probability of 10% per unit per forward pass is applied
- cyclical schedule for learning rate and β_1 over 15 epochs with cosine interpolation [23]
- event-weighted binary cross entropy is used as a loss function, minimized with the Adam optimizer.

5.3.5 *Final Selection Efficiency*

In this section, the event selection criteria are summarized, and their effectiveness in capturing the collision events is assessed, particularly focusing on their efficiency in identifying the signal processes. These criteria are the trigger requirements described in Sect. 5.2.1, the preselections of H $\rightarrow \tau\tau$ and H $\rightarrow$ bb object described in Sects. 5.2.2 and 5.2.3, the final state selection detailed in Sects. 5.3.1 and 5.3.2, and the invariant mass selection defining the signal region as reported in Sect. 5.3.3. Rather than reporting each selection's efficiency separately, for the sake of simplicity, the choice is made to group selections into five macro-categories. The first is the

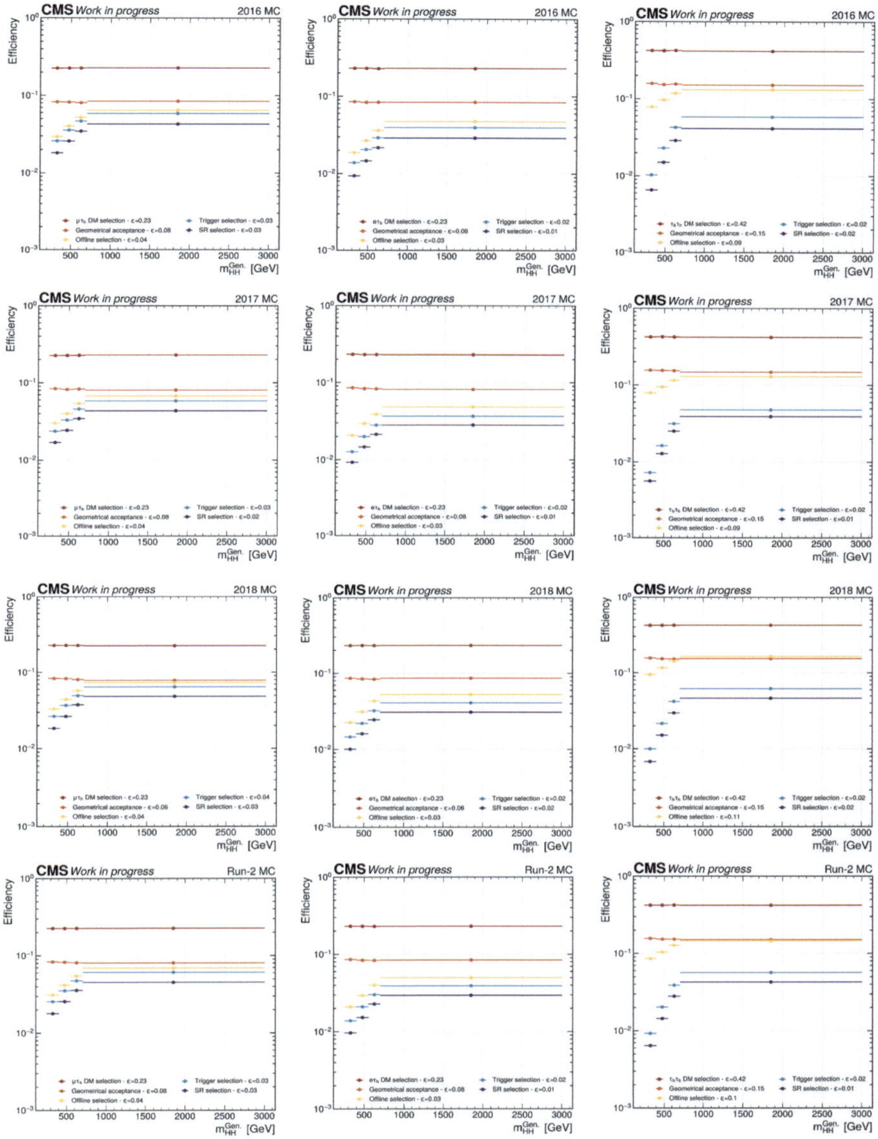

Fig. 5.20 Efficiency of the event selections as a function of the reconstructed mass of the Higgs bosons pair for the ggF production mechanism evaluated in MC simulation for the 2016 (first row), 2017 (second row), 2018 (third row), and Run2 (fourth row) data-taking periods. The result is shown per $\tau\tau$ final state: $\tau_\mu\tau_h$ (left column), $\tau_e\tau_h$ (centre column), and $\tau_h\tau_h$ (right column)

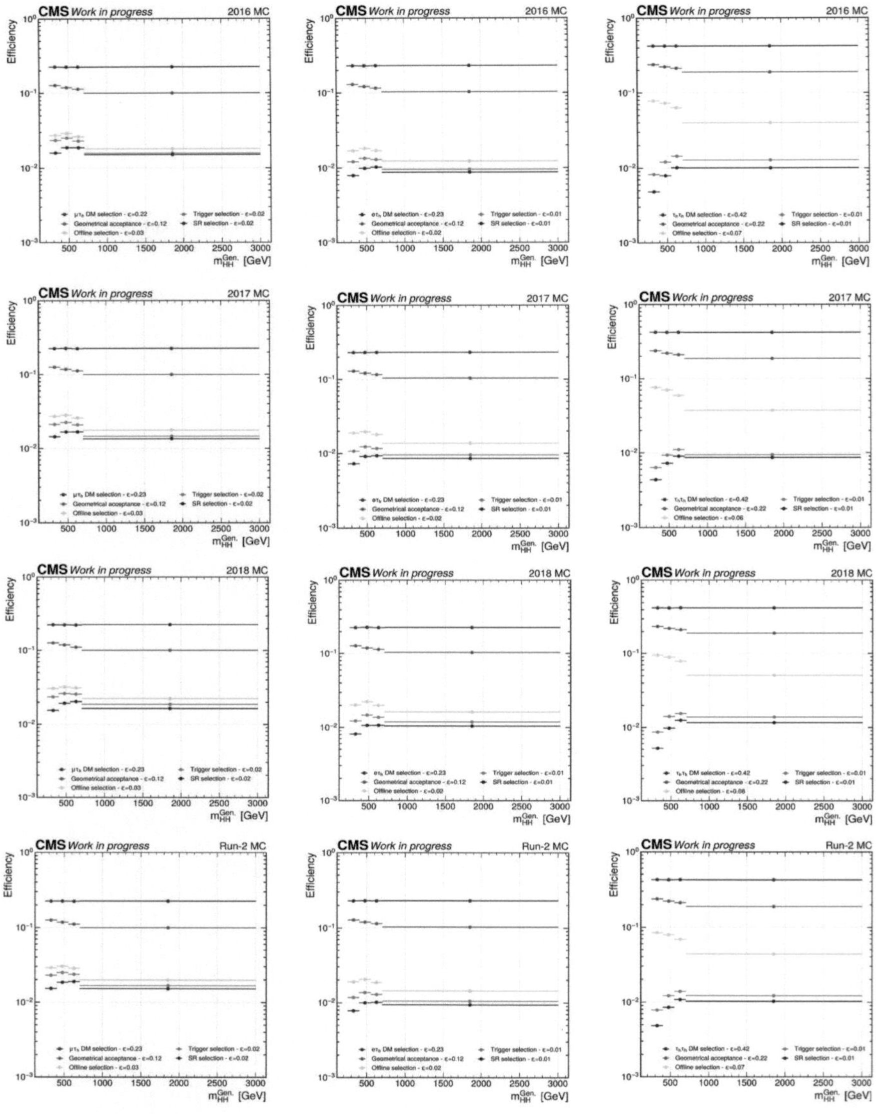

Fig. 5.21 Efficiency of the event selections as a function of the reconstructed mass of the Higgs bosons pair for the VBF production mechanism evaluated in MC simulation for the 2016 (top row), 2017 (centre row), and 2018 (bottom row) data-taking periods. The result is shown per $\tau\tau$ final state: $\tau_{\mu}\tau_{h}$ (left column), $\tau_{e}\tau_{h}$ (centre column), and $\tau_{h}\tau_{h}$ (right column)

selection of the final state of the $\tau\tau$ pair and referred to as *decay mode selection* (DM selection); the second encompasses all criteria reducing the pseudorapidity acceptance of the analysis and is therefore referred to as *geometrical acceptance*; the third entails all the offline selections that are summarized in Tables 5.4, 5.5, and 5.6; the fourth category regroups all trigger requirements that are applied on each event and summarized in Tables 5.1, 5.2, and 5.3; the last selection reported is the invariant mass criterion defined in Sect. 5.3.3 and referred to as *signal region (SR) selection.*

The efficiency of the selections is studied as a function of the MC generator simulated mass of the HH pair, denoted as $m_{\mathrm{HH}}^{Gen.}$, and it is reported in Figs. 5.20, and 5.21 for the ggF and VBF production mechanisms, respectively. The efficiency of each group of selections is reported per final state of the $\tau\tau$ pair and per year of data-taking, as well as for the inclusive Run-2 MC simulation. The choice is made to report all channels and all years separately to showcase how the analysis strategy detailed above is resilient to the different data-taking conditions encountered during Run-2 and ensures a consistent selection performance notwithstanding the modifications that the CMS detector has undergone in the 2016 and 2017 year End Technical Stop (YETS).

As can be appreciated from the figures, the biggest limiting factor, affecting all final states and all years, to efficiently reconstructing HH $\rightarrow$ bb$\tau\tau$ events is the geometrical acceptance of the analysis, which is itself driven by the need to identify two b jet candidates within the tracker acceptance of $|\eta| < 2.4$, and the requirement of the τ candidates to be within $|\eta| < 2.1$. Then, for the $\tau_{\mu}\tau_{\mathrm{h}}$ and $\tau_{\mathrm{e}}\tau_{\mathrm{h}}$, the second important limit is represented by the offline selection, which needs to be particularly tight to guarantee the reconstruction of genuine candidates in the harsh collision environment. Finally, specifically in the $\tau\tau$ channel, the larger restriction is represented by the trigger strategy, which entails the application of p_{T} thresholds much higher than those of the other two channels.

5.4 Modelling of Physics Processes

The precise modelling of the phenomenology of the physical processes studied and their interaction with the CMS detector is central to the definition of the analysis blueprint and its optimization. Signal and background processes are simulated via MC techniques based on the most precise generators and state-of-the-art calculations. Residual imperfections in the modelling, stemming from inaccurate reproduction of the interaction with the detector or from the perturbative order at which theoretical calculations were performed, can then be corrected by the application of dedicated weights to the MC events. Where the MC simulation is not accurate enough, and its reweighting does not correct for the residual discrepancies with data, approaches consisting of the entire estimation or partial correction of these processes in signal-free data regions, usually referred to as *data-driven methods*, are adopted to improve their description in the signal regions. In the following, these methods are discussed for the processes relevant to the bb$\tau\tau$ analysis.

5.4.1 Properties of Monte Carlo Simulation and Reweighting

The generation of MC simulated samples involves a comprehensive consideration of both physics and experimental effects, which is carried out in several steps. The initial step involves generating the hard scatter interaction, wherein various event generators are utilized depending on the specific process under investigation. For the search presented in this Thesis, two generators are used: the MADGRAPH5_aMC@NLO generator, of which different tags are used (2.2.2, 2.3.2.2, 2.3.3, 2.4.2, and 2.6.0), or the POWHEG v2.0 generator, depending on the process and year. The second stage is the simulation of quark hadronization and fragmentation effects, underlying event, and multiple parton interactions leading to PU; this is achieved using PYTHIA v8.226 or v8.230 depending on the year considered. For this second step, depending on process and year, the underlying event tunes CUETP8M1 [24] and CP5 [25] are used, while the NNPDF 3.0 [26] and NNPDF 3.1 [27] parton distribution functions are employed. Finally, the CMS detector response is modelled using the GEANT4 toolkit [28] and the same reconstruction algorithms used for data that are described in Sect. 2.4 are applied. Nevertheless, MC generators do not always reproduce exactly the state-of-the-art theoretical predictions, and the simulation of the detector can contain effects not observed in data and vice-versa. To cope with these residual discrepancies, dedicated reweighting of the simulated events is needed to achieve the best replication of data behaviour. This is done by means of SFs that are computed for each possible source of discrepancy; some of these SFs have already been detailed in Sects. 5.2 and 5.3, where they have been introduced alongside the selections that make them necessary. In the following, the reweighting procedure is reported for the additional effects accounted for in this search.

Pileup Reweighting

The process of producing simulated datasets for analysis is computationally demanding and typically spans several weeks. Consequently, the MC samples are generated before the precise PU distribution of the actual data is known. Since PU influences the detector response and the reconstruction of physics objects, the simulated events are adjusted by applying weights based on the ratio between the PU distribution in the data and that in the MC. Following this approach, a precise morphing of the MC to data PU distribution is achieved.

τ_h Energy Scale

A correction is applied to account for the difference in the τ_h candidates' energy scale between MC and data due to electrons being misidentified as hadronic τ leptons; it is measured with a tag-and-probe technique in $Z/\gamma^* \rightarrow \tau\tau \rightarrow \mu\nu_\mu\nu_\tau\tau_h\nu_\tau$ events separately for the different decay modes, and split into endcap and barrel regions.

No energy scale correction is needed for muons misidentified as taus. The τ_h energy scale SFs are computed by the POG and provided to all analyses in CMS.

Level-1 ECAL Trigger Prefiring

During the 2016 and 2017 detector operations, a gradual timing shift of the ECAL crystal signal response in the endcap region was observed due to their ageing. This timing shift was propagated to the ECAL trigger primitives, resulting in their wrong association with earlier bunch crossings, an effect referred to as *prefiring* [29]. The fraction of ECAL trigger primitives produced in earlier bunch crossing strongly depends on the pseudorapidity and reaches significant values for the very forward regions of the calorimeter. As a consequence of one of the L1 trigger rules forbidding two consecutive bunch crossings to be selected [30], the prefiring induced by the ECAL mistiming resulted in small fractions of events being lost. The effect was accounted for and corrected before the 2018 operation began; nonetheless, its impact is sizeable for the previous two years of data-taking. The prefiring behaviour cannot be accounted for by the simulation, so appropriate weights are applied to encode the probability of an event to prefire according to the p_T and η of the forward jets and photons in the event.

Level-1 Trigger Prescale

In cases where the instantaneous luminosity at the start of an LHC fill surpasses a specific threshold, certain trigger paths might be *prescaled*. Any HLT path is assigned an adjustable prescale factor P that reduces the trigger rate of $1/P$ by retaining only one event accept decision every P occurrences in order to manage the rate of accepted events. To account for this, events triggered by prescaled paths are adjusted using the weights provided in Table 5.11. Additionally, the VBF trigger was operational only during a fraction of the 2017 data-taking runs; therefore, a correction factor must also be applied to the events that triggered only the VBF path. It should be noted that for the single lepton triggers the prescale weights are already included in the trigger efficiency SFs described in Sect. 5.2.1, and they are reported here only for completeness.

2017 Fully Hadronic Channel Dedicated Correction

After applying all the recommended corrections, a satisfactory agreement between data and MC simulation is observed for all years of data-taking in all channels, with the exception of the 2017 $\tau_h\tau_h$ final state. In this channel, there remains a discrepancy of approximately 20% in the region where the Drell-Yan background contribution is dominant, as it can be appreciated in the top row of Fig. 5.22. To address this disagreement, a region enriched with Drell-Yan events is defined by

Table 5.11 Level-1 trigger prescale scale factors applied to Monte Carlo simulated events. The single lepton triggers the prescale weights are already included in the trigger efficiency scale factors described in Sect. 5.2.1, and they are reported here only for completeness

Year	Trigger requirement	Prescale weight
2016	One μ, $p_T > 22$ GeV, Calo. Iso	0.795221971
	One μ, $p_T > 22$ GeV, $\|\eta\| < 2.1$, Calo. Iso	0.923782353
	One μ, $p_T > 22$ GeV, Tracker Iso	0.795221971
	One μ, $p_T > 22$ GeV, $\|\eta\| < 2.1$, Tracker Iso	0.923782353
2017	One μ, $p_T > 24$ GeV, Calo. Iso	0.91613901
	Two τ_h, $p_T > 20$ GeV, $\|\eta\| < 2.1$, Loose-None PF Iso + Two jets, $p_T^1 > 115$ GeV, $p_T^2 > 40$ GeV, $m_{jj} > 620$ GeV	0.65308574
2018	Two HPS τ_h, $p_T > 20$ GeV, $\|\eta\| < 2.1$, Loose-None PF Iso + Two jets, $p_T^1 > 115$ GeV, $p_T^2 > 40$ GeV, $m_{jj} > 620$ GeV	0.990342

imposing the regular $\tau_h\tau_h$ channel selections reported in Sects. 5.2.2 and 5.3.1, along with an additional criterion on the angular separation between the two taus, i.e. $\Delta R(\tau_h, \tau_h) < 2$, while the requirement for at least two b jet candidates is relaxed to increase the statistics. This region exhibits relatively pure genuine τ_h candidates, and it is utilized to derive additional scale factors. Four different categories are established based on the requirement that both τ candidates have the same decay mode; within each category, a correction factor (SF_{DM}) is extrapolated for each decay mode of the τ leptons considered in this analysis. The values of the scale factors thus obtained, alongside the related uncertainties obtained from the fit, are listed in Table 5.12. The effect of this additional weighting of MC events can be appreciated in the bottom row of Fig. 5.22; after applying the SF_{DM} correction, the discrepancy between data and MC simulation is significantly mitigated, leaving only a minor disagreement in the very low-mass region. Nevertheless, this particular region is excluded from the analysis through the application of the invariant mass selections defined in Sect. 5.3.3; therefore, no further adjustments are required.

5.4.2 SM HH Signal Modelling

The main target of the analysis presented in this Thesis is the evaluation of observed upper limits on the non-resonant HH production cross section as predicted by the SM. To achieve this, a careful and precise simulation of the two production mechanisms accessible at the LHC given the Run-2 integrated luminosity of 138 fb^{-1}, i.e. ggF and VBF, needs to be performed. To this end, a similar modelling strategy is used for the two production mechanisms in order to model the HH production at any arbitrary value of the κ_λ and κ_{2V} modifiers for the ggF and VBF mechanisms, respectively. These techniques are detailed in the following.

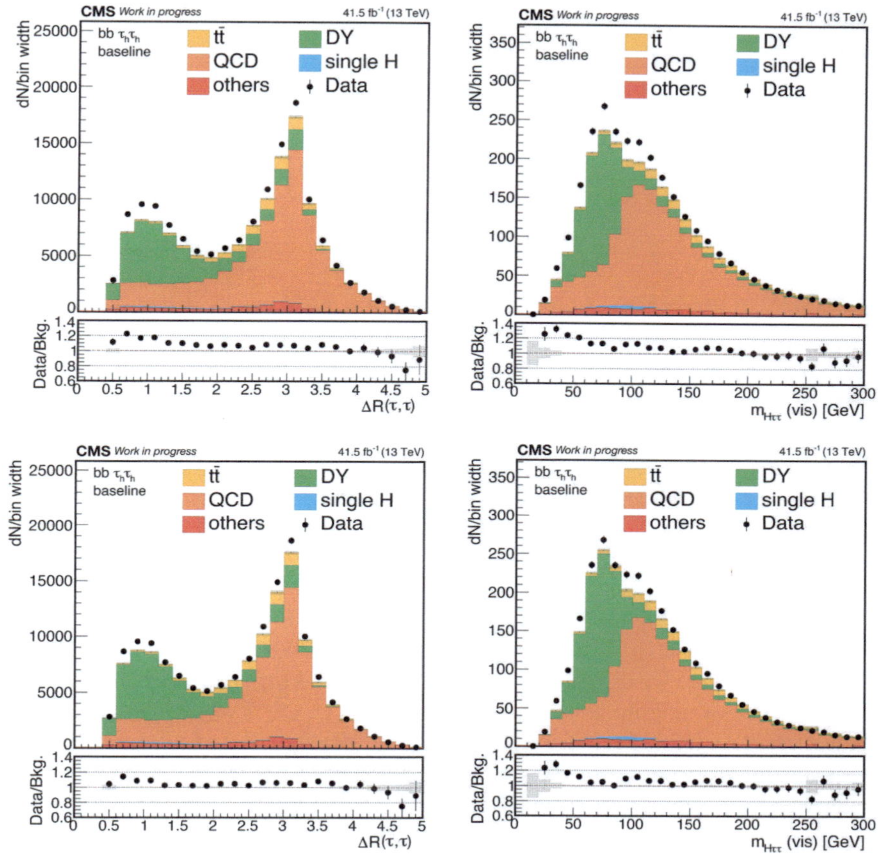

Fig. 5.22 Distribution of the spatial separation ΔR between the two τ candidates (left column) and of the invariant mass of the visible decay products of the $\tau\tau$ pair (right columns) in the $\tau_h\tau_h$ final state for the 2017 data-taking period. The distributions are reported before (top row) and after (bottom row) the application of the dedicated scale factor defined in this analysis. The black points and error bars represent the data; the background histograms are stacked and organised to highlight the largest and most interesting contributions. The bottom panel reports the ratio between the data and the background hypothesis. The shaded band in all panels represents the statistical uncertainty only

NLO Gluon Fusion Modelling

At Leading Order (LO) approximation, the amplitude of the ggF HH production can be split into two components as

$$\mathcal{A} = \kappa_\lambda \kappa_t \mathcal{A}_\triangle + \kappa_t^2 \mathcal{A}_\Box \tag{5.13}$$

Table 5.12 Dedicated scale factors derived for the $\tau_h\tau_h$ final state in 2017 data-taking period as a function of the decay mode of the τ_h candidate

Decay mode	SF_{DM}
$h^\pm$	$1.078^{+0.034}_{-0.036}$
$h^\pm\pi^0$	$1.112^{+0.023}_{-0.023}$
$h^\pm h^\mp h^\pm$	$0.984^{+0.063}_{-0.067}$
$h^\pm h^\mp h^\pm\pi^0$	$0.759^{+0.178}_{-0.259}$

where $\mathcal{A}_\triangle$ ad $\mathcal{A}_\square$ respectively identify the contribution from the triangle and box Feynman diagrams reported in Sect. 1.2.1. The production cross section of Higgs boson pairs via a gluon fusion loop is then calculated as the square of the amplitude. Using the same naming convention, the cross section reads:

$$\begin{aligned}\sigma(\kappa_\lambda, \kappa_t) \sim |\mathcal{A}|^2 &= \kappa_\lambda^2\kappa_t^2|\mathcal{A}_\triangle|^2 + \kappa_t^4|\mathcal{A}_\square|^2 + \kappa_\lambda\kappa_t^3\mathcal{I}_{\triangle\square} = \\ &= \boldsymbol{k}(\kappa_\lambda, \kappa_t) \cdot \boldsymbol{A}\end{aligned} \tag{5.14}$$

where $\mathcal{I}_{\triangle\square} = |\mathcal{A}_\triangle^*\mathcal{A}_\square + \mathcal{A}_\square^*\mathcal{A}_\triangle|$ is the interference term between the two Feynman diagrams and where we defined the vector of couplings $\boldsymbol{k}(\kappa_\lambda, \kappa_t) = (\kappa_\lambda^2\kappa_t^2, \kappa_t^4, \kappa_\lambda\kappa_t^3)$ and the vector of amplitude components $\boldsymbol{A} = (|\mathcal{A}_\triangle|^2, |\mathcal{A}_\square|^2, \mathcal{I}_{\triangle\square})$. This makes it apparent that, for a fixed set of $(\kappa_\lambda, \kappa_t)$ values, the HH signal can be described as the linear sum of only three components, making possible the definition of the following equation:

$$\begin{aligned}\boldsymbol{\Sigma} &= \boldsymbol{K} \cdot \boldsymbol{A} = \\ &= \begin{pmatrix} k_1^1 & k_1^2 & k_1^3 \\ k_2^1 & k_2^2 & k_2^3 \\ k_3^1 & k_3^2 & k_3^3 \end{pmatrix} \begin{pmatrix} |\mathcal{A}_\triangle|^2 \\ |\mathcal{A}_\square|^2 \\ \mathcal{I}_{\triangle\square} \end{pmatrix}\end{aligned} \tag{5.15}$$

where we defined the vector of cross sections $\boldsymbol{\Sigma}(\kappa_\lambda, \kappa_t) = (\sigma_1(\kappa_\lambda, \kappa_t), \sigma_2(\kappa_\lambda, \kappa_t), \sigma_3(\kappa_\lambda, \kappa_t))$, and where we denote with (k_i^1, k_i^2, k_i^3) the three elements of the coupling vector for a specific choice of κ_λ and κ_t. Finally, by simple inversion of (5.15) and substitution in (5.14), we get:

$$\sigma(\kappa_\lambda, \kappa_t) = \boldsymbol{k}^{\mathrm{T}}(\kappa_\lambda, \kappa_t) \cdot \boldsymbol{K}^{-1} \cdot \boldsymbol{\Sigma} \tag{5.16}$$

which shows that the modelling of the cross section at any arbitrary value of the κ_λ modifier can be obtained by a linear combination of three know inputs. This can be applied to build the differential $d\sigma/dx$ distribution as well.

In the HH → $bb\tau\tau$ search presented in this Thesis, this approach is used to model the ggF HH signal at the time of the maximum likelihood fit of the final discriminating

Table 5.13 Combinations of the κ_λ and κ_t modifiers used as inputs to solve Eq. 5.16 and model Higgs boson pair production via the gluon fusion production mechanisms. The cross section is computed for a Higgs boson mass $m_H = 125.09$ GeV [31]

κ_λ	κ_t	$\sigma(\kappa_\lambda, \kappa_t)$ [fb]
1	1	$31.05^{+0.68}_{-1.55}(\text{scale})^{+1.24}_{-5.59}(m_t)$
2.45	1	$13.12^{+0.30}_{-0.67}(\text{scale})^{+0.52}_{-2.89}(m_t)$
5	1	$94.82^{+4.65}_{-8.34}(\text{scale})^{+12.33}_{-3.79}(m_t)$

variable described in Sect. 6.1. To do so, three MC simulated samples, produced with the POWHEG v2.0 generator at Next-to-Leading Order (NLO) precision, are used simultaneously to solve Eq. 5.16 and obtain the desired modelling of the gluon fusion signal differential distribution. The values of the κ_λ and κ_t modifiers of the three samples used are reported in Table 5.13; the cross section is computed for a Higgs boson mass $m_H = 125.09$ GeV.

LO Vector Boson Fusion Modelling

A similar procedure to what was described above for the ggF HH production is used for the modelling of the vector boson fusion mechanism. At LO approximation, the cross section of the VBF HH production can be split into six components:

$$\begin{aligned}\sigma(\kappa_\lambda, \kappa_V, \kappa_{2V}) \sim |\mathcal{A}|^2 &= |\kappa_{2V}\mathcal{A}_X + \kappa_V^2\mathcal{A}_W + \kappa_V\kappa_\lambda\mathcal{A}_Y|^2 = \\ &= \kappa_{2V}^2|\mathcal{A}_X|^2 + \kappa_V^4|\mathcal{A}_W|^2 + \kappa_V^2\kappa_\lambda^2|\mathcal{A}_Y|^2 + \\ &\quad + \kappa_{2V}\kappa_V^2\mathcal{I}_{XW} + \kappa_{2V}\kappa_V\kappa_\lambda\mathcal{I}_{XY} + \kappa_V^3\kappa_\lambda\mathcal{I}_{WY}\end{aligned} \tag{5.17}$$

where $\mathcal{A}_X$, $\mathcal{A}_W$, and $\mathcal{A}_Y$ are the contributions from the three (ordered) VBF Feynman diagrams reported in Sect. 1.2.1, and we defined $\mathcal{I}_{XW}$, $\mathcal{I}_{XY}$, and $\mathcal{I}_{WY}$ as their respective interferences. Applying the same procedure described above, the following relation can thus be demonstrated to hold:

$$\sigma(\kappa_\lambda, \kappa_V, \kappa_{2V}) = \boldsymbol{k}^T(\kappa_\lambda, \kappa_V, \kappa_{2V}) \cdot \boldsymbol{K}^{-1} \cdot \boldsymbol{\Sigma} \tag{5.18}$$

where the same naming convention is used as for the ggF HH modelling, with the only difference that, given the presence of an additional Feynman diagram, the vectors $\boldsymbol{k}$ and $\boldsymbol{\Sigma}$ have six entries each, while the matrix $\boldsymbol{K}$ has a 6×6 shape.

Analogously to what is done for the ggF HH signal modelling, this approach is used to model the VBF HH signal at the time of the maximum likelihood fit of the final discriminating variable described in Sect. 6.1. To accomplish this, six MC simulated samples, produced with the MADGRAPH5_aMC@NLO 2.3.2 (2.6.0) generator at LO precision for the 2016 (2017–2018) campaign, are used simultaneously to solve

Table 5.14 Combinations of the κ_λ, κ_V, and κ_{2V} modifiers used as inputs to solve Eq. 5.18 and model Higgs boson pair production via the vector boson fusion production mechanisms. The cross section is evaluated from the Monte Carlo simulation at leading order approximation

κ_λ	κ_V	κ_{2V}	$\sigma(\kappa_\lambda, \kappa_t)$ [fb]
1	1	1	1.726
1	0.5	1	10.824
1	1.5	1	66.018
0	1	1	4.609
2	1	1	1.423
1	1	2	14.218

Eq. 5.18 and obtained the desired modelling of the vector boson fusion signal differential distribution. The values of the κ_λ, κ_V, and κ_{2V} modifiers of the six samples used are reported in Table 5.14; the cross section is evaluated from the MC simulation at LO.

5.4.3 *BSM HH Signal Modelling*

In order to explore various Beyond the SM (BSM) scenarios, it becomes necessary to model the signal for multiple sets of couplings. Due to the resource intensity of the MC generation, only a limited number of fully simulated BSM signal samples are available; therefore, sophisticated weighting techniques are employed to overcome this limit. These techniques allow us to effectively model additional BSM scenarios based on a small set of fully simulated samples.

In the context of non-resonant HH production, extensive modelling of diverse HH signals becomes a crucial undertaking. BSM scenarios, in particular, are of significant interest, as they stem from the EFT Lagrangian parametrization detailed in Sect. 1.1.5. This parametrization defines a five-dimensional space of parameters that significantly influence the signal behaviour; these parameters are the five H couplings λ_{HHH}, y_t, c_2, c_g, and c_{2g}. The H self-coupling and the top Yukawa coupling are generally traded for their modifiers $\kappa_\lambda = \lambda_{HHH}/\lambda_{HHH}^{SM}$ and $\kappa_t = y_t/y_t^{SM}$. To comprehensively explore the non-resonant HH production mechanism, it is imperative to investigate both the SM prediction and the various BSM scenarios originating from points in this five-dimensional parameter space.

For the ggF HH samples, a dedicated reweighting technique can be exploited at two different levels of precision for the modelling of a large number of BSM scenarios. This method involves merging all the simulated ggF datasets into a single sample and its reweighting to the desired BSM scenario following the five-dimensional parametrization mentioned in the previous paragraph, as precisely detailed below. This approach can use input samples either generated at LO or at NLO precision. The former choice guarantees the availability of a substantial number of simulated signal events, but it does not account for parton emission at the matrix element level due to the limited precision of the simulation; in contrast, the latter choice ensures the better modelling of higher order effects at the price of a smaller number of available events. In this analysis, both reweighting techniques are adopted to tackle different problems: for the training of the ML techniques used in the search, the ggF LO modelling is employed to produce events; in contrast, the ggF NLO modelling is used for the creation of the EFT benchmarks used to probe BSM physics scenarios.

The event weighting technique is based on the fundamental observation that HH production via gluon fusion constitutes a $2 \rightarrow 2$ scattering process. At LO approximation and prior to the hadronization effects, the two H bosons are emitted back-to-back in the azimuthal direction with identical transverse momentum. Furthermore, the assumption of isotropic production in the azimuthal angle of the HH allows us to neglect its influence. The effects stemming from the parton distribution functions and higher order corrections at NLO precision can be both accounted for by applying a Lorentz boost to the HH pair. Hence, in the centre-of-mass frame of the collision, the HH production is fully parametrized by two variables that can be chosen to be the invariant mass of the Higgs boson pair and the absolute value of the cosine of the polar angle of one Higgs boson with respect to the beam axis ($|\cos\theta^*|$).

When applying this reweighting technique at NLO precision, the simulated samples used as input are the same as reported in Sect. 5.4.2. Conversely, when applying this procedure at LO approximation, the input samples are produced with the MADGRAPH5_aMC@NLO 2.4.2 generator for the campaigns of all three years considered in this analysis. In either case, the datasets are combined and the event differential distribution in the (m_{HH}, $|\cos\theta^*|$) plane is encoded in a bi-dimensional histogram normalized to unity with 36 and four bins in the first and second variable, respectively. Such distributions are reported in Fig. 5.23 for the three years of data-taking considered in this analysis, both at LO and NLO approximation. If we denote by f^i_{SUM} the fraction of events in each bin i of the histograms, a weight associated with each event can be computed based on the m_{HH} and $|\cos\theta^*|$ values using the ratio of the HH differential cross sections in the corresponding bin for the target BSM coupling combination to the predicted SM yield in the same bin.

As discussed in Sect. 1.2.2, the ratio of the total HH cross section to the SM prediction, i.e. R_{HH}, is parametrized with the formulae in Eqs. 1.62 and 1.63 for the LO approximation and the NLO precision correction, respectively. These equations are a generic expression for the interference of the diagrams stemming from the EFT Lagrangian parametrization; therefore, the same relations hold in each bin of the (m_{HH}, $|\cos\theta^*|$) histograms, provided that the coefficients $\mathrm{A}_j|_{j=1,\ldots,23}$ become a function of the bin number, i.e. $\mathrm{A}^i_j|^{i\in 2\mathrm{D}(m_{\mathrm{HH}},|\cos\theta^*|)}_{j=1,\ldots,23}$. In this context, Eqs. 1.62 and 1.63 can be written for each bin to be:

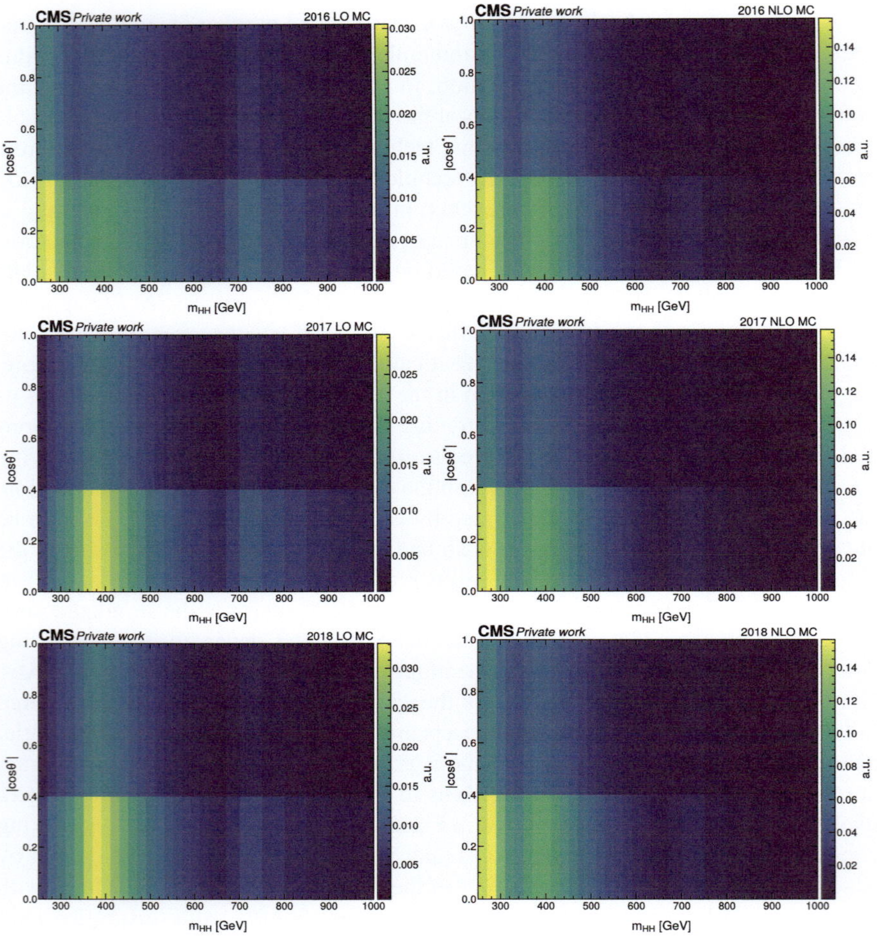

Fig. 5.23 2D distributions of $(m_{\mathrm{HH}}, |\cos\theta^*|)$ used in the reweighting to BSM scenarios. The normalized yield is shown for the samples generated at LO (left column) and NLO (right column) approximation. The three years of data-taking are reported: 2016 (top row), 2017 (centre row), and 2018 (bottom row). In each panel, the distributions are obtained by merging in a single dataset all available MC simulated HH samples for different combinations of couplings; at LO precision, twelve samples generated with MADGRAPH5_aMC@NLO 2.4.2 are used; at NLO precision, six samples simulated with POWHEG 2.0 are employed. For visualization simplicity, only the region of phase space with $m_{\mathrm{HH}} < 1000$ GeV is reported, while the full histograms extend up to $m_{\mathrm{HH}} = 5000$ GeV

$$R^i_{\mathrm{HH}} = \frac{\sigma^i_{\mathrm{HH}}}{\sigma^{i,\mathrm{SM}}_{\mathrm{HH}}} \overset{\mathrm{LO}}{=} \mathrm{A}^i_1\kappa_t^4 + \mathrm{A}^i_2 c_2^2 + \mathrm{A}^i_3\kappa_t^2\kappa_\lambda^2 + \mathrm{A}^i_4 c_g^2\kappa_\lambda^2 + \mathrm{A}^i_5 c_{2g}^2$$
$$+ \mathrm{A}^i_6 c_2\kappa_t^2 + \mathrm{A}^i_7\kappa_\lambda\kappa_t^3 + \mathrm{A}^i_8\kappa_t\kappa_\lambda c_2 + \mathrm{A}^i_9 c_g\kappa_\lambda c_2 \tag{5.19}$$
$$+ \mathrm{A}^i_{10} c_2 c_{2g} + \mathrm{A}^i_{11} c_g\kappa_\lambda\kappa_t^2 + \mathrm{A}^i_{12} c_2\kappa_t^2$$
$$+ \mathrm{A}^i_{13}\kappa_\lambda^2 c_g\kappa_t + \mathrm{A}^i_{14} c_{2g}\kappa_t\kappa_\lambda + \mathrm{A}^i_{15} c_g c_{2g}\kappa_\lambda$$

in the LO approximation, while at NLO precision the coefficient A^i_j are modified and new terms appear, yielding in each bin the additional correction:

$$\Delta R^i_{\mathrm{HH}} = \frac{\Delta\sigma^i_{\mathrm{HH}}}{\sigma^{i,\mathrm{SM}}_{\mathrm{HH}}} \overset{\mathrm{NLO}}{=} \mathrm{A}^i_{16}\kappa_t^3 c_g + \mathrm{A}^i_{17}\kappa_t c_2 c_g + \mathrm{A}^i_{18}\kappa_t c_g^2\kappa_\lambda + \mathrm{A}^i_{19} c_g\kappa_t c_{2g}$$
$$+ \mathrm{A}^i_{20}\kappa_t^2 c_g^2 + \mathrm{A}^i_{21} c_2 c_g^2 + \mathrm{A}^i_{22} c_g^3\kappa_\lambda + \mathrm{A}^i_{23} c_g^2 c_{2g} \tag{5.20}$$

To derive the A^i_j coefficients, the R^i_{HH} ratio is computed relying on MC simulated events for different combinations of the five EFT couplings on which it depends and interpolated as a function of them. The details on the fit procedure and the verification of its quality, as well as the values of the A^i_j coefficients, which are not reported here given their extremely large number, can be found in References [32–34].

Therefore, the event weight ω used to morph the merged dataset to any generic BSM combination of the five EFT couplings is fully determined by the parametrization in Eqs. 5.19 and 5.20, and the bi-dimensional histograms of normalized event counts described above. The intermediate value Γ is defined as:

$$\Gamma(\kappa_\lambda, \kappa_t, c_2, c_g, c_{2g}|i) \overset{\mathrm{DEF}}{=} \frac{1}{f^i_{\mathrm{SUM}}} \times \frac{\hat{\sigma}^i_{\mathrm{HH}}(\kappa_\lambda, \kappa_t, c_2, c_g, c_{2g})}{\sigma_{\mathrm{HH}}(\kappa_\lambda, \kappa_t, c_2, c_g, c_{2g})}$$
$$= \frac{\hat{f}_{\mathrm{SM}^i}}{f^i_{\mathrm{SUM}}} \times \frac{R^i_{\mathrm{HH}}(\kappa_\lambda, \kappa_t, c_2, c_g, c_{2g})}{R_{\mathrm{HH}}(\kappa_\lambda, \kappa_t, c_2, c_g, c_{2g})} \tag{5.21}$$

where the use of the hat denotes the evaluation of the specific value from direct calculation rather than from estimating it from MC simulation. In this way, the MC simulation of the SM HH production can be used to further increase the event count of the SUM dataset rather than employing it for the evaluation of f_{SM^i}. The event weight ω is defined as follows, normalized to the sum over the N simulated MC events, to modify the differential distribution of the events but not their normalization:

$$\omega = \frac{\Gamma}{\sum_N \Gamma} \tag{5.22}$$

The validity of the procedure is evaluated with a closure test where the distribution obtained with the application of the weight from Eq. 5.22 is compared to the generated samples. This is illustrated in Fig. 5.24, where the good agreement of the two methods is shown for three hypotheses of the κ_λ and κ_t modifiers. The bottom panel displays

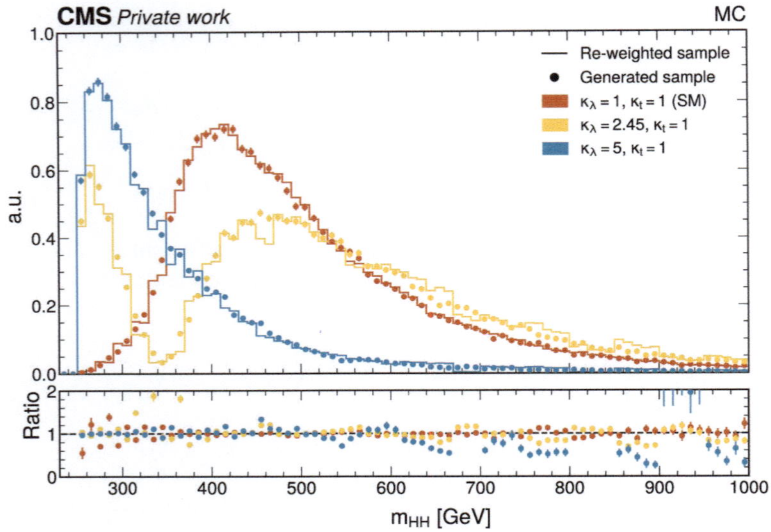

Fig. 5.24 Comparison of the generator level $m_{\rm HH}$ distribution obtained in a Monte Carlo sample generated with a specific set of Higgs boson couplings (points) and with the event weighting procedure (solid lines). The bottom panel reports the ratio between the differential distributions obtained with the weighting procedure and those from the Monte Carlo simulation

the ratio between the MC simulated and reweighted sample, rendering explicit their agreement and proving the validity of the technique.

5.4.4 $t\bar{t}$ *Background Modelling*

The major source of background contaminating the HH → $bb\tau\tau$ signal regions is the $t\bar{t}$ production. This process is modelled relying on MC simulation at NLO precision with the POWHEG v2.0 generator. The multiparton, parton shower, and hadronization effects are simulated with PYTHIA v8.226 and NNPDF 3.0 parton distributions functions for the 2016 dataset, while PYTHIA v8.230 and NNPDF 3.1 parton distributions functions are used for the 2017 and 2018 datasets. For all three years, the underlying event tune CP5 is employed.

While the shape of the process is well-modelled by the MC simulation, the normalization of the background shows a disagreement with respect to the observed data, as can be appreciated in Fig. 5.25. Analogous disagreements have been observed by other analyses [35, 36] within the CMS Collaboration, their origin being the complexity of $t\bar{t}$ MC simulation.

In order to address this normalization discrepancy, a new dedicated approach for the estimation of the normalization of the $t\bar{t}$ background is introduced in this analysis.

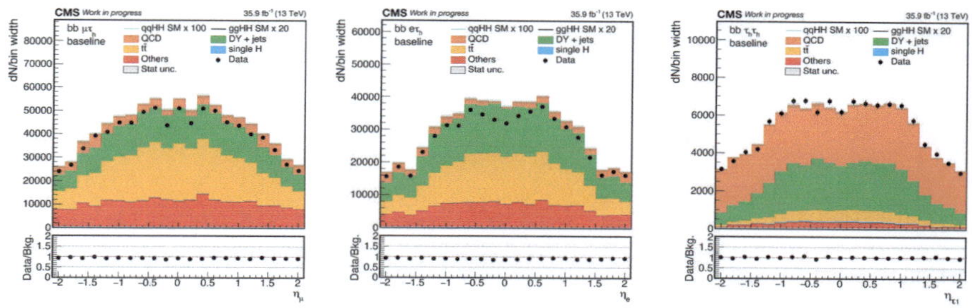

Fig. 5.25 Distributions of the η of the first lepton in the baseline selection for the three channels in 2016: $\tau_{\mu}\tau_{h}$ (left), $\tau_{e}\tau_{h}$ (centre), and $\tau_{h}\tau_{h}$ (right). The black points and error bars represent the data; the background histograms are stacked and organised to highlight the largest and most interesting contributions. The bottom panel reports the ratio between the data and the background hypothesis. The shaded band in all panels represents the statistical uncertainty only. The two channels with the biggest $t\bar{t}$ contamination show data over MC ratio disagreement. As expected, in the channel where the $t\bar{t}$ contamination is the smallest, the discrepancy is almost absent

To this end, a specific $t\bar{t}$ Control Region (CR) is defined where a set of scale factors are fitted to correct the discrepancy. In the following, these scale factors will be referred to as *ttSF*, and their definition and validation will be outlined.

Definition of the $t\bar{t}$ Control Region

The definition of the CR follows four main considerations:

- it must be enriched in $t\bar{t}$ events
- it must be as depleted as possible of the other backgrounds
- it must be as similar as possible to the SR while remaining orthogonal to it
- it must not depend on the τ pair final state

To make the CR independent of the τ pair categorization, it is defined with the following procedure: events are first categorized following the strategy outlined in Sect. 5.3.1, then the $\tau_{\mu}\tau_{h}$, $\tau_{e}\tau_{h}$, and $\tau_{h}\tau_{h}$ events are merged in a single CR for each year. The decision to use three different CRs for the three years is based on two considerations. Firstly, the CMS experimental apparatus has undergone several changes throughout Run-2; secondly, the disagreement observed is different in the three years both in our analysis and in the other analyses that report the discrepancy. Nevertheless, it should be noted that the disagreement for the 2017 and 2018 data-taking periods is comparable and smaller with respect to 2016. This is a reflection of the differences in the simulation of $t\bar{t}$ events for 2016 with respect to 2017 and 2018, and the upgrade of some CMS subdetectors in the 2016 YETS.

To make the CR as similar as possible to the SR, and at the same time enriched in $t\bar{t}$ events, the same selections outlined in Sect. 5.3.2 for the `res2b` category are applied. This ensures that a genuine $b\bar{b}$ pair, conceivably coming from the $t\bar{t}$ decay,

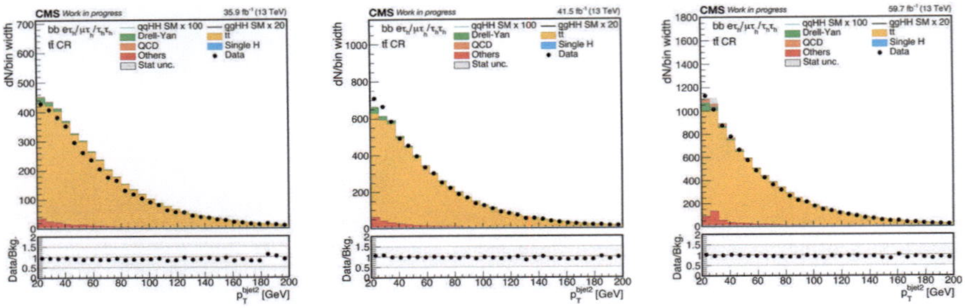

Fig. 5.26 Distribution of the trailing b jet transverse momentum in the $t\bar{t}$ control region for the three years of data-taking: 2016 (left), 2017 (centre), and 2018 (right). The black points and error bars represent the data; the background histograms are stacked and organised to highlight the largest and most interesting contributions. The bottom panel reports the ratio between the data and the background hypothesis. The shaded band in all panels represents the statistical uncertainty only

is found. To make the CR orthogonal to the SR, the elliptical mass selection defined in Eq. 5.11 is inverted. Therefore, the mass selection used for the $t\bar{t}$ CR is:

$$\frac{(m_{\mathrm{H}\tau\tau} - 129\,\mathrm{GeV})^2}{(53\,\mathrm{GeV})^2} + \frac{(m_{\mathrm{Hbb}} - 169\,\mathrm{GeV})^2}{(145\,\mathrm{GeV})^2} > 1, \tag{5.23}$$

This mass selection not only guarantees orthogonality with the SR but also ensures that the CR is enriched in $t\bar{t}$ events and, at the same time, depleted of HH signal events.

The distribution of the trailing b-jet transverse momentum in the $t\bar{t}$ CR for the three years is reported in Fig. 5.26. As can be appreciated, the data over MC ratio panel highlights the normalization issues of the $t\bar{t}$ process. Moreover, the disagreement is more pronounced for 2016, where it spans beyond 10%, than for 2017 and 2018, where it is contained below 10%. As mentioned above, no shape dependence is observed in the disagreement. Therefore, the decision to keep the MC shape modelling was taken.

As it can be grasped from Fig. 5.26, the contamination of the $t\bar{t}$ CR from other backgrounds is very small, being less than 7% for all three years. The $t\bar{t}$ CR defined is thus pure enough and safe to be used for the computation of the ttSF.

The study of the $t\bar{t}$ CR and the fit of the ttSF is done in two steps: firstly, a series of CR-only fits are performed; secondly, a fit of the CR together with a subset of the SR is performed. While the first step has the goal of confirming that the disagreement has no shape dependence, the second step has the target of validating the approach. Since no shape trend is observed in the disagreement, the variable to be used for the fits can be arbitrarily chosen. For simplicity, the score of the DNN in the $t\bar{t}$ CR is used. This choice additionally ensures that any further effect coming from the application of the DNN is consistently taken into account between the CR and the SR.

Table 5.15 Results of the fits of the control regions under the three binning paradigms for the three years. The ttSF fitted values and their errors are reported

	2016	2017	2018
CR (1bin)	0.908 ± 0.006	0.988 ± 0.006	0.966 ± 0.009
CR (3bin)	0.908 ± 0.006	0.988 ± 0.006	0.967 ± 0.009
CR (20bin)	0.908 ± 0.006	0.988 ± 0.006	0.974 ± 0.008

Control Region Fit of ttSF

The CR-only fits are performed under three different binning paradigms of the DNN score: one single bin, three variable width bins (i.e. $[0, 0.05, 0.3, 1]$), and twenty constant width bins. In all the cases, the fit is performed applying the Barlow-Beeston approach [37], which ensures the correct treatment of the systematic uncertainties arising from the finite number of MC events. This method represents such systematic uncertainties by introducing a separate nuisance parameter multiplying the expected number of events from each background source in each bin. Though this method adds a very large number of new free parameters in the likelihood, the problem is tractable in a profile likelihood approach since the values of the parameters that maximize the likelihood within a bin can be found independently of those in all the other bins.

Apart from those coming from the Barlow-Beeston approach, the only other nuisance parameter introduced at this stage is the rate parameter ttSF, whose value will be extracted via the fit. The decision not to introduce any other nuisance parameter at this stage comes from the fact that this fit does not have the goal of constraining the systematic uncertainties of the analysis but only evaluating the ttSF. Moreover, introducing other nuisances at this stage could lead to a poor fitting of the ttSF rate parameter as the modelling of the $t\bar{t}$ normalization could be encoded in the modification of nuisance parameters that are, in reality, not related to it.

The results of the CR-only fit are reported in Table 5.15 for the three years and the three binning paradigms described above. As can be seen, no shape dependence of the fit can be appreciated. The central value of the fit remains almost constant at the variation of the binning paradigm, and when it changes, the variations are well-compatible within one standard deviation. This gives the possibility to pick the single binning paradigm with no loss of modelling quality. The single bin CR approach will be intended in the following.

Control Region Plus Signal Region Fit of ttSF

Having performed the CR-only fits and having assessed that no shape dependence is present, the second step is the validation of the method in a subset of the SR. The two categories chosen for the validation are the `res2b` and the `res1b` since they present the highest $t\bar{t}$ contribution. Moreover, given the current exclusion limit on κ_λ and the

Table 5.16 Results of the control region plus signal region for the three years; the control region is always shaped in a single bin. The ttSF fitted value and its error are reported

	2016	2017	2018
CR+SR	0.908 ± 0.006	0.988 ± 0.006	0.968 ± 0.008

small signal yield expected at low values of the DNN score, the region of events scoring below 0.5 is unblinded. Moreover, by construction, this region contains the majority of the $t\bar{t}$ events contaminating the SR.

The CR-plus-SR fits are performed under a single binning paradigm: the CR will have only one bin, while the SR will have the same variable width binning, from 0 to 0.5, applied during limit setting. The Barlow-Beeston approach is applied both to the CR and to the SR. In addition to this, the ttSF rate parameter is applied to both regions. For the same reasons explained above, no further nuisances are applied to the CR. Nevertheless, all the normalization nuisance parameters reported in Sect. 5.5 are included for the SR. This is because, even if the goal here is not to constrain the other nuisance parameters, their inclusion is bound to the fact that we are effectively unblinding part of the SR.

The values of the ttSF from the CR-plus-SR fit are reported in Table 5.16 for the three years separately. As can be appreciated, the result of the CR-plus-SR fit is compatible with the results reported above from the CR-only fits. This agreement validates the procedure and highlights that the ttSF correctly propagates from CR to SR.

Application of ttSF

Given the good agreement between the CR and CR-plus-SR fits outlined above, no further fit of CR-plus-SR is required, and the ttSF value can be taken from the CR-only fit. According to the QCD estimation procedure explained in Sect. 5.4.5, the ttSF is used to scale the $t\bar{t}$ MC yield before the QCD evaluation in order for it to be different from zero.

The plots reported in Figs. 5.27a, b, and 5.28 show the improvement obtained by the application of the ttSF for the three years. The numerical values of the ttSF used are those reported in the first row of Table 5.15. In all cases, the distributions before and after the ttSF application are reported in the upper and lower rows, respectively. As it can be appreciated, the improvement is substantial and consistent throughout the three years and the two SR categories. In all Figures, the bottom panels display the ratio between the data and the background hypothesis and render explicit the good data-MC agreement achieved after the correction of the $t\bar{t}$ normalization when employing the method described above.

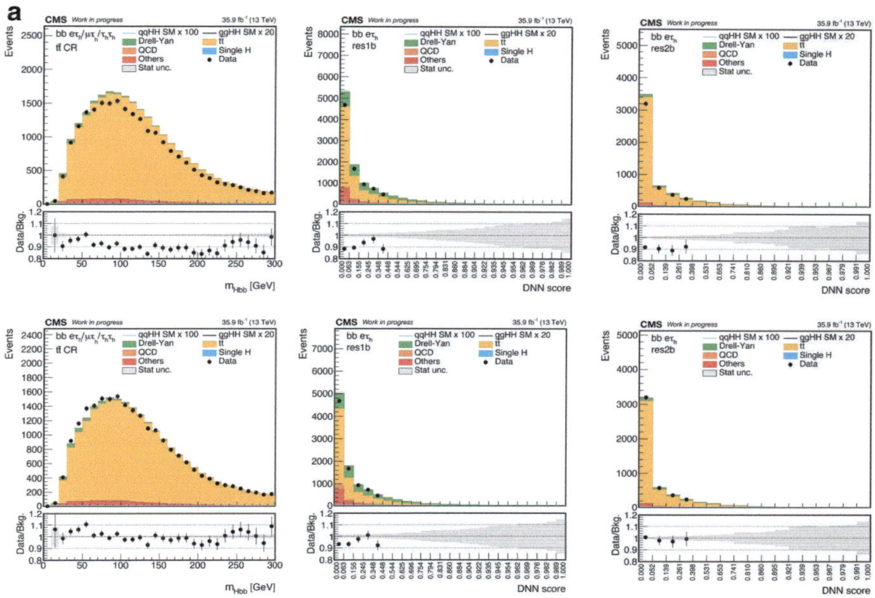

Fig. 5.27 Distributions of m_{Hbb} in the $t\bar{t}$ control region (left column), and of the deep neural network score in the `res1b` (central column) and `res2b` (right column) categories for the 2016 (**a**) and 2017 (**b**) data-taking periods. The distributions are reported before (top row) and after (bottom row) the application of the ttSF. The black points and error bars represent the data; the background histograms are stacked and organised to highlight the largest and most interesting contributions. The bottom panel reports the ratio between the data and the background hypothesis. The shaded band in all panels represents the statistical uncertainty only

5.4.5 *Multijet Background Modelling*

The modelling and estimation of multijet QCD background, where one or two jets are misidentified as a τ_{h} candidate, present significant challenges. There are two key factors that hinder the use of MC samples with accurate modelling. Firstly, the low probability of a quark or gluon jet being identified as a τ_{h} candidate (ranging between 10^{-2} and 10^{-3}), combined with the requirement for two additional jets in the event to satisfy the b-tagging criteria, necessitates an exceedingly large sample to ensure an adequate number of simulated events in the signal regions. Secondly, the imperfect modelling of the misidentification of quark and gluon jets as τ_{h} by the MC simulation, which is mainly due to its dependency on detector effects. To address these challenges, a data-driven approach is adopted, where the multijet background contribution is estimated from jet-enriched regions containing events independent from those in the signal region. This strategy overcomes the limitations of the MC simulation, allowing for a more reliable estimation of the multijet background.

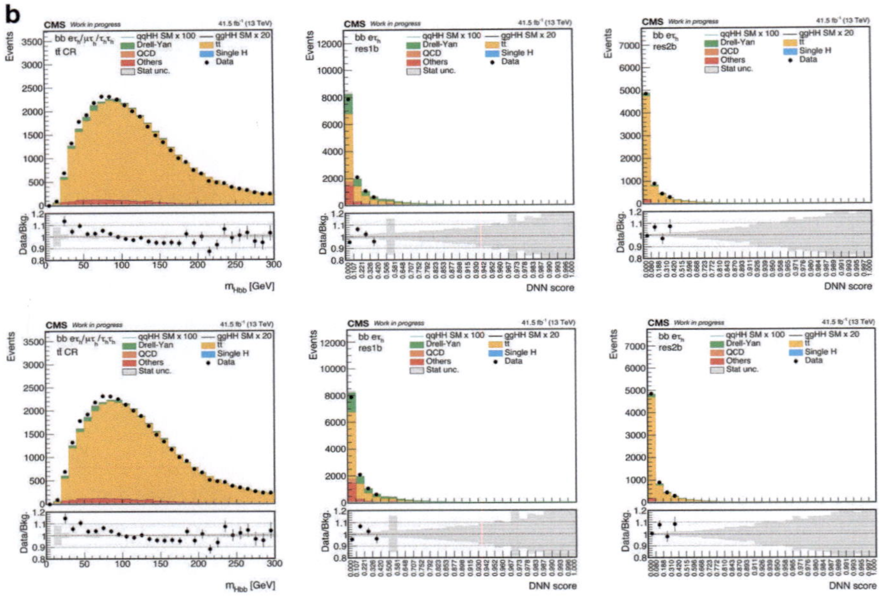

Fig. 5.27 (continued)

The data-driven estimation is performed with the so-called *ABCD method*, consisting of the use of a signal region (A) plus three sidebands (B, C, D). Region A corresponds to the signal region as defined by the selection in Sects. 5.2 and 5.3, where for each category, a τ candidate pair in the $\tau_\mu\tau_h$, $\tau_e\tau_h$, and $\tau_h\tau_h$ final state with Opposite-Sign (OS) electric charge is found, and where the τ_h candidates satisfy the medium WP of the DEEPTAUVSJET discriminator. In the three sideband regions, these selections are alternatively lifted. In region B, the pair charge requirement is inverted (Same-Sign, SS) while the DEEPTAUVSJET criterion is maintained. In region C, it is the DEEPTAUVSJET discriminator selection to be loosened from medium to VVVLose (Very-Very-Very-Loose) WP while the charge requirement is maintained. Finally, in region C, the pair charge requirement is inverted and at the same time the DEEPTAUVSJET discriminator selection is loosened from medium to VVVLose. For consistency with the channels containing a muon or an electron, in the $\tau_h\tau_h$ final state, the DEEPTAUVSJET selection is inverted only for the τ lepton with subleading p_T. Due to the limited statistics, in the `boosted` category only, the b-tagging requirements are consistently loosed to use the DEEPTAUVSJET loose WP instead of the medium WP in all four regions; for the same limitation, at this stage, the five VBF categories defined in Sect. 5.3.2 are considered as a single macro-category. A schematic representation of the regions' definition is reported in Fig. 5.29.

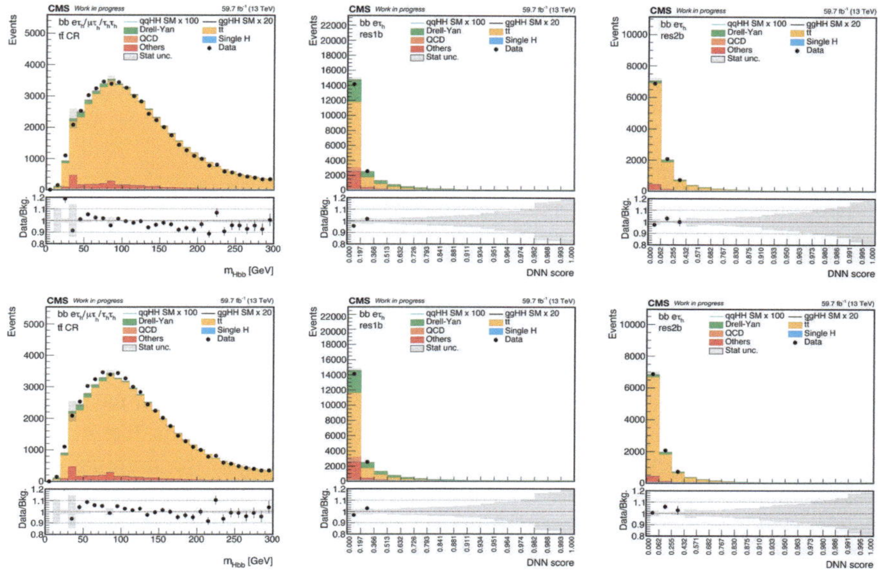

Fig. 5.28 Distributions of m_{Hbb} in the $\mathrm{t\bar{t}}$ control region (left column), and of the deep neural network score in the `res1b` (central column) and `res2b` (right column) categories for the 2018 data-taking periods. The distributions are reported before (top row) and after (bottom row) the application of the ttSF. The black points and error bars represent the data; the background histograms are stacked and organised to highlight the largest and most interesting contributions. The bottom panel reports the ratio between the data and the background hypothesis. The shaded band in all panels represents the statistical uncertainty only

Fig. 5.29 Schematic representation of the regions used for the QCD background data-driven estimation. OS and SS stand for opposite and same sign charge of the lepton pair, respectively. Medium and VVVLose (Very-Very-Very-Loose) working points are the two working points used for the DEEPTAUVSJET discriminator. Therefore, region A represents the signal region defined in Sects. 5.2 and 5.3

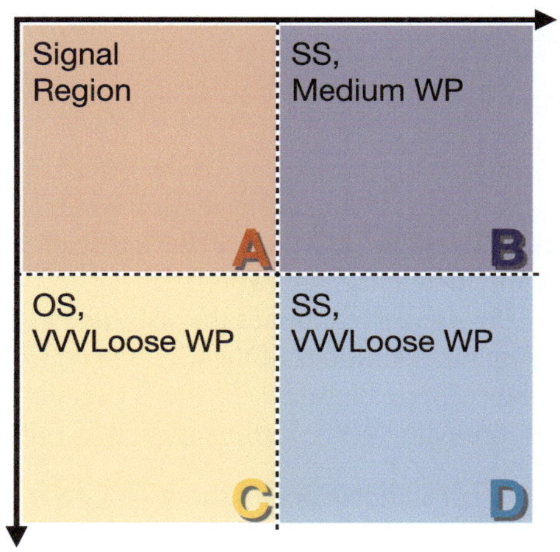

In the B, C, and D regions, the number of QCD background events (N_i where $i =$ B, C, D) is estimated by subtracting from the total number of observed data events (N_i^{data}) the yields of the other backgrounds estimated using the MC simulation (N_i^{bkg}):

$$N_i = N_i^{\text{data}} - N_i^{\text{bkg}}$$

In this configuration, the differential distribution of the multijet background, also referred to as *shape*, can be arbitrarily taken from region B or C without modifying the good modelling of data with MC. Nevertheless, it is important to note that this selection has strong implications when evaluating systematic uncertainties. Indeed, in order to include in the analysis an uncertainty on the shape of the QCD background, the necessity arises to define an alternative template of the multijet differential distribution. One option could be to choose either one of regions B or C for the nominal shape, and the other as the alternative shape used in the maximum likelihood fit described in Sect. 6.1 to encode one standard deviation from the nominal shape value. This choice presents one criticality: the uncertainty is by construction completely asymmetrical, having both the up and down variations modelled by the same shape. This results in the problem that the fitting process can use the corresponding nuisance parameter to account for potential discrepancies between data and MC in one direction by slightly adjusting it in that same direction. However, differences in the opposite direction require modifications to other nuisance parameters, resulting in an inherently asymmetric behaviour of the statistical model.

To solve this issue, the decision is taken to model the differential distribution of the QCD multijet background as the mean of the two shapes in regions C and B, leaving the two original shapes to be used as the variations used by the statistical model as uncertainty. Finally, the yield of the QCD background in the SR (N_{A}) is estimated as:

$$N_{\text{A}} = \frac{N_{\text{C}} \times N_{\text{B}}}{N_{\text{D}}} \tag{5.24}$$

Several assessments have been conducted to validate the reliability of the ABCD method for the QCD multijet background estimation presented here. A brief overview is provided in the following.

The first test examines the stability of the estimated QCD normalization by modifying the definitions of the C and D regions using four different DeepTau working points. These alternative C and D regions are defined based on the DeepTauVSjet discriminator WP selection applied as follows:

- the τ_{h} candidate passes the VVVLoose WP but fails the VVLoose WP,
- the τ_{h} candidate passes the VVLoose WP but fails the VLoose WP,
- the τ_{h} candidate passes the VLoose WP but fails the Loose WP,
- the τ_{h} candidate passes the Loose WP but fails the Medium WP.

In these four cases, the ratio of yields in regions C and D is computed and compared with the value obtained using the standard definition of C and D quarters. The measurements demonstrated a satisfactory level of compatibility.

In the second test, not only the C/D ratio is estimated, but the full ABCD evaluation of the QCD background is performed using the four alternative categories defined above. This approach permits the estimation of the QCD contribution in regions close to the analysis signal regions but with a poor signal population (due to the use of a very loose working point) and enhanced QCD contribution. These tests also resulted in good agreement.

The third test involves comparing the QCD estimation obtained from the ABCD method with direct data minus MC subtraction in a sideband region with negligible HH signal presence. This sideband region is defined by inverting the elliptic mass selection from the `res1b` category. Once again, the QCD estimation obtained from the ABCD method showed good agreement with the direct data minus MC subtraction in this sideband, thus validating the QCD estimations used in the analysis.

5.4.6 Drell-Yan Background Modelling

The estimation of the Drell-Yan background originating from $Z/\gamma^* \rightarrow \ell\ell$, where $\ell = e, \mu, \tau$, is performed relying on a mixed MC and data-driven approach. The MC simulation of the DY background is produced with the MADGRAPH5_aMC@NLO 2.3.3 (2.4.2) generator for the 2016 (2017–2018) campaigns. The process is simulated at LO, where up to four jets produced at Matrix Element (ME) are included. In order to enhance the selected event count, an inclusive simulated sample is augmented with exclusive samples, each requiring the emission of one, two, three, or four jets at the ME level. Additionally, two samples demanding the emission of one or two b quarks at the ME level are simulated. To ensure consistency between the jets generated at the ME level and those from the parton shower simulation of the inclusive sample, the MLM merging scheme [38] is employed for matching. To further optimize the generation process, the knowledge of the signal region invariant mass selection defined in Eqs. 5.11 and 5.12 allows the restriction of generated events to those fulfilling $m_{\ell\ell} > 50\,\text{GeV}$, as otherwise they would not enter the signal categories used in this analysis. The MC samples described above are combined and normalized to the theoretical cross section $\sigma(Z/\gamma^* \rightarrow \ell\ell) = 6077.22 \pm 1.49(\text{integration}) \pm 14.78(\text{pdf}) \pm 121.54(\text{scale})$ computed with the FEWZ simulation code [39] at Next-to-Next-to Leading Order (NNLO) precision using the NNPDF 3.1 parton distribution functions.

Table 5.17 Trigger paths used in the μμ sideband for the three years of data-taking and the corresponding integrated luminosity recorded with each. Within each year, the logical OR of all the paths is always intended

Year	Trigger requirement	Int. lumi [fb^{-1}]
2016	One μ, $p_T > 22$ GeV, Calo. Iso	28.6
2017	One μ, $p_T > 24$ GeV, Calo. Iso	38.1
	One μ, $p_T > 27$ GeV, Calo. Iso	41.5
2018	One μ, $p_T > 24$ GeV, Calo. Iso	59.7
	One μ, $p_T > 27$ GeV, Calo. Iso	

The necessity of relying on LO MC generation of DY events stems from the very narrow phase space explored by this analysis. Indeed, the number of DY events satisfying the requirement on the presence of two additional jets in the event, both fulfilling the b-tagging criteria, is two to three orders of magnitude smaller than the actual number of generated events. Therefore, the use of the LO approximation helps overcome this by allowing the generation of much larger event samples. Nevertheless, it leads to a known mismodelling of the emission of jets of different flavours. To cope with this effect, a dedicated sideband enriched in $Z/\gamma^* \rightarrow \mu\mu$ events is used to derive corrective factors.

To be selected in the $Z/\gamma^* \rightarrow \mu\mu$ sideband, the trigger requirements detailed in Table 5.17 need to be met; within each year, if multiple trigger paths are used, the logical OR of all the paths is always intended. Additionally, the following offline selections are applied: the muon candidates must be compatible with the primary vertex, thus each candidate track must have a distance from the primary vertex fulfilling $\Delta_{xy} < 0.045$ cm in the transverse plane and $\Delta_z < 0.2$ cm on the longitudinal axis; basic kinematical selections of $p_T > 20$ GeV and $|\eta| < 2.4$ are imposed on each candidate; muons must be tightly isolated satisfying the $\mathcal{I}^{\mu}_{rel} < 0.15$ criterion calculated in a cone of $\Delta R < 0.4$; finally, the muon pair should exhibit opposite sign charges and have an invariant mass $m_{\ell\ell} > 50$ GeV. Concurrently, a veto is applied on additional lepton, and to reduce the QCD and $t\bar{t}$ contributions, the missing transverse energy is bound to satisfy $p_T^{miss} < 45$ GeV. All jet selection criteria are kept unchanged with respect to the signal categories.

The correction scale factors for DY normalization are then estimated in 18 control regions within the μμ sideband phase space. The CRs are defined to be orthogonal as follows:

- Events are split into three macro-categories based on the number of b-tagged jets:
  - ◇ Z + **light flavour jets**: in this category targeting the reducible background contribution from jet mistagging, no jet candidate is found to pass the b-tagging requirements
  - ◇ Z + **one b-tagged jet**: in this category targeting the irreducible background to the `res1b` category, only one jet candidate satisfies the b-tagging criteria
  - ◇ Z + **two b-tagged jets**: in this category targeting the irreducible background to the `res2b` category, two jet candidate fulfil the b-tagging selection
- Each macro-category is divided into six CRs based on the transverse momentum of the reconstructed Z candidate:
  - ◇ **very-low p_T**: where $0 < p_T^{(Z)} < 10$ GeV for all data-taking years
  - ◇ **low p_T**: where $10 < p_T^{(Z)} < 50$ GeV and $10 < p_T^{(Z)} < 30$ GeV, for the 2016 and 2017–2018 data-taking periods, respectively
  - ◇ **medium-low p_T**: where $50 < p_T^{(Z)} < 80$ GeV and $30 < p_T^{(Z)} < 50$ GeV, for the 2016 and 2017–2018 data-taking periods, respectively
  - ◇ **medium-high p_T**: where $80 < p_T^{(Z)} < 110$ GeV and $50 < p_T^{(Z)} < 100$ GeV, for the 2016 and 2017–2018 data-taking periods, respectively
  - ◇ **high p_T**: where $110 < p_T^{(Z)} < 190$ GeV and $100 < p_T^{(Z)} < 200$ GeV, for the 2016 and 2017–2018 data-taking periods, respectively
  - ◇ **very-high p_T**: where $p_T^{(Z)} > 190$ GeV and $p_T^{(Z)} > 200$ GeV, for the 2016 and 2017–2018 data-taking periods, respectively

To match the 18 CRs thus defined, the MC simulated events are split into the same categories based on the number of B partons in the generated process and the generator-level transverse momentum of the Z boson. The invariant mass distribution of the muon pair is fitted simultaneously across all categories, incorporating the 18 pre-defined MC templates described above, whose normalizations are the free parameters of the fitting process. As a result, 18 correction factors are derived for each of the categories, resulting in a correction of both the normalization and the differential distribution of the DY background contribution. The uncertainties associated with the scale factors and their covariance matrix are also derived from the fit and included as sources of systematic errors. The good agreement of the di-muon angular distance $\Delta R(\mu, \mu)$ modelling between MC simulation and data can be appreciated in Fig. 5.30 where the invariant mass is reported for the 2018 data-taking period. The two bottom panels display the ratio between the data and the background hypothesis and render explicit the good data-MC agreement achieved after the application of the correction detailed above.

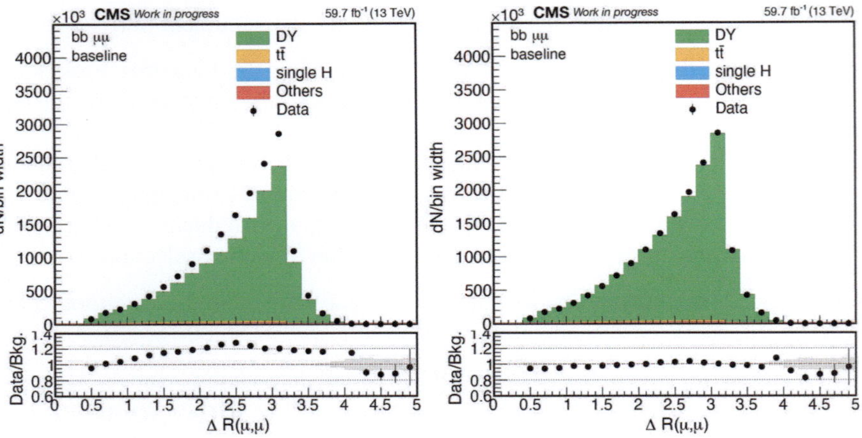

Fig. 5.30 Distributions of the di-muon angular distance $\Delta R(\mu, \mu)$ before (left) and after (right) the application of the Drell-Yan correction factors for the 2018 data-taking period. The black points and error bars represent the data; the background histograms are stacked and organised to highlight the largest and most interesting contributions. The bottom panel reports the ratio between the data and the background hypothesis. The shaded band in all panels represents the statistical uncertainty only

5.4.7 Other Backgrounds Modelling

Other background processes that affect the $bb\tau\tau$ decay channel are W boson production in association with jets, single t quarks production, electroweak production of a vector boson in association with jets, $t\bar{t}$ production in association with one or two vector bosons, and production of vector bosons pairs and triplets. Additionally, SM single Higgs boson production via ggF, VBF, Higgs-strahlung, and $t\bar{t}$ associated production are all added as background sources in this analysis. Considering their mild contribution to the total event yield in the signal region, all of these processes are modelled relying solely on MC simulation for both their differential distribution and normalization.

The generation of $W \to \ell\nu_\ell$, where $\ell = e, \mu, \tau$, is performed with MadGraph5_aMC@NLO 2.2.2 (2.4.2) for the 2016 (2017–2018) campaign. In all years, the samples are generated at LO precision in exclusive regions of the transverse momentum scalar sum of the partons emitted at the ME level, and the parton shower merging is performed in the MLM scheme. The sample is normalized to the theoretical cross section computed at NNLO precision with the FEWZ simulation code to be $\sigma(W \to \ell\nu_\ell) = 61.53^{+2.81}_{-2.58} \times 10^3$ pb.

The simulation of single t quark production is performed for all years with the POWHEG v2.0 generator at NLO precision. Both the s- and t-channel processes are normalized to the theoretical cross section at NNLO precision computed to be $\sigma(t_{s\text{-channel}}) = 71.7 \pm 1.4(\text{scale}) \pm 3.4(\text{PDF})$ pb and $\sigma(t_{t\text{-channel}}) = 216.99^{+6.62}_{-4.64}$ (scale) $\pm\, 6.16(\text{PDF} + \alpha_S)$ pb, respectively.

Electroweak production of W^+, W^-, and Z boson decaying to final state leptons in association with two jets is simulated with MadGraph5_aMC@NLO 2.3.2 (2.4.2) for the 2016 (2017–2018) campaigns at LO precision. The three processes are normalized to the cross sections at LO obtained from the MC generator, corresponding to $\sigma(W^+) = 25.69 \pm 0.51$ pb, $\sigma(W^-) = 20.25 \pm 0.41$ pb, and $\sigma(Z) = 3.987 \pm 0.080$ pb.

The production of a $t\bar{t}$ pair in association with one vector boson is simulated with MadGraph5_aMC@NLO 2.2.2 and 2.4.2, both with and without FxFx merging [40], depending on the specific subprocess. The included final states of the vector bosons are: $W/Z \to qq$, $W \to \ell\nu_\ell$, and $Z \to \ell\ell$. The samples are normalized to their theoretical cross sections computed at NLO precision be $\sigma(t\bar{t}Z|Z \to qq) = 0.5297 \pm 0.0008$ pb, $\sigma(t\bar{t}Z|Z \to \ell\ell) = 0.2529 \pm 0.0004$ pb, $\sigma(t\bar{t}W|W \to \ell\nu_\ell) = 0.2043 \pm 0.0020$ pb, $\sigma(t\bar{t}W|W \to qq) = 0.4062 \pm 0.0021$ pb. The production of a $t\bar{t}$ pair in association with two vector bosons is included for completeness; it is simulated with MadGraph5_aMC@NLO 2.2.2 and 2.4.2 for the 2016 and 2017–2018 campaigns, respectively; it has an inclusive cross section of $\mathcal{O}(10^{-3})$ pb.

The production of two and three vector bosons is simulated with the POWHEG 2.0 and MadGraph5_aMC@NLO 2.2.2 generators, respectively. Given the elevated number of combinations of initial and final states, the included MC samples are only reported in the summary table below alongside their cross section.

Finally, the simulation of the SM H production via the ggF, VBF, ZH, WH, and $t\bar{t}H$ mechanisms is simulated with the POWHEG v2.0 generator for all years. The ggF and VBF productions are included with the $H \to \tau\tau$ final state and they are normalized to their theoretical cross section; the former is calculated with the iHixs program [41, 42] at Next-to-NNLO (N^3LO) QCD precision and NLO Electro-Weak (EW) accuracy to be $\sigma(H_{ggF}) = 48.61^{+2.08}_{-3.15}(\text{theory}) \pm 0.90(\text{PDF})^{+1.26}_{-1.27}(\alpha_S)$ pb, the latter is calculated with the HAWK program [43] at NNLO QCD precision and NLO EW accuracy to be $\sigma(H_{VBF}) = 3.766^{+0.016}_{-0.012}(\text{scale}) \pm 0.079(\text{PDF} + \alpha_S)$ pb. The ZH production is included in the $Z \to \ell\ell$+$H \to bb$, $Z \to qq$+$H \to bb$, and $Z \to$ any+$H \to \tau\tau$ final states and it is normalized to the theoretical cross section computed to be $\sigma(ZH) = 0.880^{+0.030}_{-0.024}(\text{scale}) \pm 0.015(\text{PDF} + \alpha_S)$ pb. The WH production mechanism is included in the $W \to$ any+$H \to \tau\tau$ final state and it is normalized to the theoretical cross section split into its W^+ and W^- contributions, to be $\sigma(W^+H) = 0.831 \pm 0.006(\text{scale}) \pm 0.015(\text{PDF} + \alpha_S)$ pb and $\sigma(W^-H) = 0.527 \pm 0.003(\text{scale}) \pm 0.010(\text{PDF} + \alpha_S)$ pb, respectively. For both the Higgs-strahlung production mechanisms, the cross section is computed at NNLO QCD precision and NLO EW corrections with the HAWK program. Finally, the $t\bar{t}H$ production is included for $H \to bb$ and $H \to \tau\tau$ final states and is normalized to its theoretical cross section computed at NLO accuracy for both QCD and EW corrections to be $\sigma(t\bar{t}H) = 0.5071^{+0.0294}_{-0.0467}(\text{scale}) \pm 0.0183(\text{PDF} + \alpha_S)$ pb. All the SM H boson production cross sections are quoted as computed in References [44, 45], and all assume $m_H = 125.09$ GeV.

The complete list of the simulated processes described in this Section, of the generator used for the simulation, and of the corresponding theoretical cross section, is reported in Table 5.18.

Table 5.18 Summary of the simulated processes, the generator used, and the normalization cross section. In the table, MG5 and POW2 denote MADGRAPH5_aMC@NLO and POWHEG 2.0, respectively

Process	Generator	σ[pb]
$t\bar{t}$	POW2	$\sigma(t\bar{t}) = 833.9^{+20.5}_{-30.0}$(scale)±
		$\pm 21.0(\text{PDF} + \alpha_S)^{+23.2}_{-22.5}$(mass)
Fully hadronic		$45.7 \cdot 10^{-2} \times \sigma(t\bar{t})$
Fully leptonic		$10.5 \cdot 10^{-2} \times \sigma(t\bar{t})$
Semi leptonic		$21.9 \cdot 10^{-2} \times \sigma(t\bar{t})$
$Z/\gamma^* \to \ell\ell$ + jets	MG5	6077.22 ± 1.49(integ.)±
		±14.78(PDF) ± 121.54(scale)
$W \to \ell\nu_\ell$ + jets	MG5	$\sigma(W) = 61.53^{+2.81}_{-2.58} \times 10^3$
$0 \leq H_T < 100$		$96.5 \cdot 10^{-2} \times \sigma(W)$
$100 \leq H_T < 200$		$2.7 \cdot 10^{-2} \times \sigma(W)$
$200 \leq H_T < 400$		$7.1 \cdot 10^{-3} \times \sigma(W)$
$400 \leq H_T < 600$		$1.0 \cdot 10^{-3} \times \sigma(W)$
$600 \leq H_T < 800$		$2.4 \cdot 10^{-4} \times \sigma(W)$
$800 \leq H_T < 1200$		$1.1 \cdot 10^{-4} \times \sigma(W)$
$1200 \leq H_T < 2500$		$2.6 \cdot 10^{-5} \times \sigma(W)$
$2500 \geq H_T$		$6.3 \cdot 10^{-5} \times \sigma(W)$
Single t	POW2	
s-channel		71.7 ± 1.4(scale) ± 3.4(PDF)
t-channel		$216.99^{+6.62}_{-4.64}$(scale) ± 6.16(PDF + α_S)
EWK	MG5	
W^+ + jets		25.69 ± 0.51
W^- + jets		20.25 ± 0.41
Z + jets		3.987 ± 0.080
$t\bar{t}$ + V	MG5	
V= Z → qq		0.5297 ± 0.0008
V= Z → $\ell\ell$		0.2529 ± 0.0004
V= W → $\ell\nu_\ell$		0.2043 ± 0.0020
V= W → qq		0.4062 ± 0.0021
$t\bar{t}$ + VV (V = Z,W)	MG5	$\mathcal{O}(10^{-3})$
Single H	POW2	
ggF		$48.61^{+2.08}_{-3.15}$(theory) ± 0.90(PDF)$^{+1.26}_{-1.27}(\alpha_S)$
VBF		$3.766^{+0.016}_{-0.012}$(scale) ± 0.079(PDF + α_S)
ZH		$0.880^{+0.030}_{-0.024}$(scale) ± 0.015(PDF + α_S)
W^+H		0.831 ± 0.006(scale) ± 0.015(PDF + α_S)
W^-H		0.527 ± 0.003(scale) ± 0.010(PDF + α_S)
$t\bar{t}H$		$0.5071^{+0.0294}_{-0.0467}$(scale) ± 0.0183(PDF + α_S)

(continued)

Table 5.18 (continued)

Di-boson	POW2	
$ZZ \to \ell\ell\ell\ell$		1.26
$ZZ \to \ell\ell\nu\nu$		0.564
$ZZ \to \ell\ell qq$		5.52
$ZZ \to qq\nu\nu$		4.07
$WZ \to \ell\ell\ell\nu$		4.43
$WZ \to \ell\nu qq$		10.71
$WZ \to \ell\nu\nu\nu$		3.06
$WZ \to \ell\ell qq$		5.595
$WW \to \ell\ell\nu\nu$		12.18
$WW \to \ell\nu qq$		50.00
$WW \to qqqq$		51.72
Tri-boson	MG5	
ZZZ		0.0147
WZZ		0.057
WWZ		0.168
WWW		0.209

5.5 Systematic Uncertainties

Various sources of systematic uncertainty influence the modelling of signal and background processes that were described above. They originate from limited knowledge of the background and signal processes, discrepancies between simulation and data, and imperfect knowledge of the detector response. They are categorized as *normalization* and *shape* uncertainties: while the first affect only the total event yield of the processes, the latter affect also the differential distribution of the events. The uncertainties described in this Section are introduced as nuisance parameters in the statistical analysis of the data described in Sect. 6.1.2, with log-normal priors.

All systematic uncertainties considered in this analysis are presented in the following Sects. 5.5.1 and 5.5.2, and they are summarized in Table 5.19.

5.5.1 Normalization Uncertainties

Normalization uncertainties impact signal and background modelling differently for simulated processes and data-driven estimates. For simulated processes, uncertainties encompass trigger, reconstruction, and identification efficiency of the final state objects, as well as integrated luminosity and theoretical cross section uncertainties. In contrast, data-driven estimates are mainly subject to uncertainties arising from the limited number of events used for the estimation process. The following normalization uncertainties are considered in this search.

- Different sources of uncertainty in the integrated luminosity measurement are considered and treated as correlated. Their values are obtained from dedicated Van-der-Meer scans and the stability of detector response during data-taking. The integrated luminosities for the 2016, 2017, and 2018 data-taking years have 1.2–2.5% individual uncertainties [46–48], while the overall uncertainty for the 2016–2018 period is 1.6%. These uncertainties are applied only to the signals and to background processes estimated from simulation. Since the normalizations of the $t\bar{t}$, $Z/\gamma^* \rightarrow \ell\ell$, and multijet backgrounds are obtained from data, they are not subject to the integrated luminosity uncertainties.
- Electron and muon reconstruction, isolation, and identification uncertainties are determined from the simulation-to-data scale factors; a value of 1% for both electrons and muons is obtained [8, 49]. An additional uncertainty of 3 (15)% for τ leptons with $p_T < 100$ GeV (>100 GeV) is added in the $\tau_h\tau_h$ channel [6].
- During the 2016–2017 data-taking, a gradual shift in the timing of the inputs of the ECAL L1 trigger in the region at $|\eta| > 2.0$ caused a trigger inefficiency. For events containing an electron (a jet) with $p_T > 50$ (100) GeV, in the region $2.5 < |\eta| < 3.0$ the efficiency loss is approximately 10–20%, depending on p_T, η, and time. Correction factors were computed from data and applied to the acceptance evaluated by simulation. An uncertainty of 2% is assigned to this effect.
- The uncertainty in the pileup reweighting technique is estimated by varying the values of the applied pileup weights by their uncertainty. The resulting systematic uncertainty is estimated to have a value of 1% and it is correlated among all channels and categories in each year.
- The normalization of the $t\bar{t}$ background is taken from a fit to a CR per year, as described in Sect. 5.4.4. The uncertainty in the scale factors obtained in these CRs is purely statistical and is always below 1%.
- The normalization of the $Z/\gamma^* \rightarrow \ell\ell$ background is taken from a fit to 18 CRs per year, as described in Sect. 5.4.6. The uncertainties in the scale factors obtained in these CRs are propagated to the SR taking into account their correlation and range from 0.1 to 60% depending on the year and CR considered.
- The multijet background contribution is determined from data in jet-enriched regions, as described in Sect. 5.4.5. Two normalization uncertainties are derived in order to take into account the event yield statistical fluctuations in these CRs and their dependence on the τ_h isolation requirement used to define them.
- The uncertainties in the normalizations of the backgrounds modelled relying solely on the simulated events range from 2 to 10%.

- The theoretical uncertainty in the cross section of HH production via ggF: $^{+6\%}_{-23\%}$ (scale + m_t), $\pm 3\%$ (PDF + α_S) fb [50]; and via VBF: $^{+0.03}_{-0.04}\%$ (scale) and $\pm 2.1\%$ (PDF + α_S) [51]. These uncertainties are only considered when upper limits are quoted with respect to the SM and for the likelihood scans, they are not included for the upper limits on the cross section.
- The theoretical uncertainty in the H branching fractions [45] amount: for the decay to $b\bar{b}$, to $\pm 0.65\%$ (theory), $^{+0.72}_{-0.74}\%$ (m_q), $^{+0.78}_{-0.80}\%$ (α_S), where m_q is the quark mass; and for the decay to τ leptons to $^{+1.16}_{-1.17}\%$ (theory), $^{+0.98}_{-0.99}\%$ (m_q), 0.62% (α_S).
- The uncertainty in the modelling of the VBF signal in PYTHIA8 is estimated through samples generated setting the dipole recoil option, which affects the initial-state parton showers, to ON [52]. The ratio of integrated yields in the dipole recoil ON/OFF samples is taken as the uncertainty: it varies from 10% for largely populated categories, such as classVBF, to 70% for categories with poor VBF signal contribution, such as classttH.

5.5.2 Shape Uncertainties

Shape uncertainties are incorporated to account for the inherent limitations in characterizing background processes, stemming from imperfect simulation and the scarcity of events in either the simulation or the data sidebands. These uncertainties answer to the constrained knowledge regarding the accurate representation of background shapes. The following shape uncertainties are considered in this analysis.

- The uncertainty in the measurement of the energy of τ_h leptons [6]. The uncertainties are derived by combining low- and high-p_T measurements in $Z \to \tau\tau$ and in the off-shell $W^* \to \tau\nu$ events. Four different uncertainties are included to take into account the different τ_h decay modes considered in this analysis. When considering the uncertainty for a particular decay mode, the shift is applied only to the τ_h candidates that are reconstructed with that particular decay mode, while all other τ_h candidates are left unchanged.
- Uncertainties related to the calibration of Jet Energy Scale (JES) and Resolution (JER). For the JES, 11 separate sources of uncertainty are included per year: those appearing in multiple years are treated as fully correlated, while those appearing only in one year are treated as uncorrelated. For the JER, alternative templates are produced by shifting all the jet-related features. These shifts stem from the use of scale factors to smear the energy of simulated jets to match the observed energy resolution in data.
- Separate uncertainties in the energy scale of electrons misidentified as τ_h candidates are provided to take into account two different decay modes, $h^{\pm}$ and $h^{\pm}\pi^0$. The uncertainty in the energy scale of muons misidentified as τ_h candidates is 1%, uncorrelated across decay modes.

- The uncertainties arising from the application of the DeepTau identification scale factors are determined using a tag-and-probe procedure as a function of the τ_h candidate p_T. Five uncertainties are computed for the identification of taus against jets and are calculated in p_T bins of the τ_h candidate. All these uncertainties are used in the $\tau_\mu\tau_h$ and $\tau_e\tau_h$ channels, while, since in the $\tau_h\tau_h$ channel both leptons are required to have p_T above a threshold, only the highest p_T bin uncertainty can be applied. Two uncertainties are present in the identification of τ leptons against electrons, one for the barrel and one for the endcap, and are treated as uncorrelated across η bins of the τ_h candidate.
- The shape uncertainty in the multijet contribution is determined from the two alternative templates described in Sect. 5.4.5.
- The uncertainties arising from the misidentification of jets as τ_h candidates are determined from $\tau_\mu\tau_h$ CRs defined inverting the charge sign requirement on the τ leptons and imposing that neither of the b jet candidates passes the medium working point of the DeepJet algorithm. Six uncertainties, one for the barrel and one for the endcap for each year are derived and treated as uncorrelated across years and detector regions.
- Trigger efficiencies and scale factors are measured in a $Z \rightarrow \tau\tau \rightarrow \mu\nu_\mu\nu_\tau\tau_h\nu_\tau$ enriched region with a tag-and-probe procedure and fitted separately for data and simulation. The uncertainty, determined as a function of the reconstructed τ_h candidate p_T, is taken from the fit. The scale factors are obtained from the data-to-simulation ratio and their uncertainties are propagated accordingly. Four uncertainties are included to take into account the different τ_h decay modes considered in this analysis. They are applied to the τ_h candidate of each channel. Two additional trigger uncertainties are used to cover the cases where the τ lepton decays to an electron or a muon. Finally, one uncertainty is added for the $\tau_h\tau_h$ final state in 2017–2018 to take into account the jet candidates scale factors of the VBF trigger.
- The uncertainty in the b-tagging efficiency. This takes into account the contamination from udscg (cb) jets in heavy- (light-) flavour regions, as well as statistical fluctuations in both data and simulation samples used for the computation of the b-tagging efficiency.
- The uncertainties in the pileup jet identification scale factor as functions of p_T and η.
- The uncertainties due to the limited number of simulated events entering individual bins of the distributions. They are introduced in a bin-by-bin fashion using the Barlow-Beeston approach, which introduces a set of nuisance parameters multiplying the expected number of events in each bin of each background source.

Table 5.19 Systematic uncertainties affecting the normalization and the differential distribution of the different processes considered. The last column reports the CMS-specific name used in the statistical interpretation; * denotes any combination of symbols (e.g. year, process, channel); these names can be used for better interpretation of Fig. 6.10

	Systematic	Process	CMS name
Normalization uncertainty	$\mathcal{B}(\mathrm{H} \to \tau\tau)$	Signals	BR_hbb
	$\mathcal{B}(\mathrm{H} \to \mathrm{bb})$	Signals	BR_htt
	Luminosity	All	lumi_13TeV_*
	PU reweigthing	All	CMS_PUreweight_*
	e reco.	All	CMS_bbtt_effEle_*
	μ reco.	All	CMS_bbtt_effMu_*
	ECAL trg. prefiring	All (2016–2017)	CMS_l1_ecal_prefiring_*
	$\mathrm{t\bar{t}}$ norm.	$\mathrm{t\bar{t}}$	CMS_bbtt_ttCRnorm_*
	Multijet norm.	Multijet	CMS_bbtt_*_*_*_QCDshape
	Sgn/Bkg. cross section	All involving H	alpha_s
	Sgn./Bkg. cross section	All involving H	pdf_*
	Sgn/Bkg. cross section	All involving H and V	QCDscale_*
	Sgn. cross section	VBF signal	qqHH_pythiaDipoleOn
Shape uncertainty	DY norm.	DY	CMS_bbtt_*_DYSFunc*
	Trg. SFs	All	CMS_bbtt_*_trigSF*
	Jet-τ_{h} fake rate	All	CMS_bbtt_*_jetTauFakes_*
	Custom τ_{h} SF 2017	All	CMS_bbtt_customSF2017DM*
	b-tag	all	CMS_btag_*_*
	PU jet id.	All	CMS_eff_j_PUJET_id_*
	Jet resol.	All	CMS_res_j_*
	Jet scale	All	CMS_scale_j_*_*
	τ_{h} scale	All	CMS_scale_t_*_*
	DeepTau eff.	All	CMS_eff_t_id_*_*
	Multijet shape	Multijet	CMS_bbtt_*_*_*_QCDshape
	Bin-by-bin	All (low stats.)	autoMCStats

References

1. CMS Collaboration (2018) Search for Higgs boson pair production in events with two bottom quarks and two tau leptons in proton-proton collisions at $\sqrt{s} = 13\,\mathrm{TeV}$. Phys Lett B 778:101–127. https://doi.org/10.1016/j.physletb.2018.01.001. https://doi.org/10.1016%5C%2Fj.physletb.2018.01.001
2. CMS Collaboration (2023) Search for nonresonant Higgs boson pair production in final state with two bottom quarks and two tau leptons in proton-proton collisions at $\sqrt{s} = 13\,\mathrm{TeV}$. Phys Lett B 842:137531. ISSN: 0370-2693. https://doi.org/10.1016/j.physletb.2022.137531. https://www.sciencedirect.com/science/article/pii/S0370269322006657
3. Tanabashi M, and others (Particle Data Group) (2018) Review of particle physics. Phys Rev D 98:030001. https://doi.org/10.1103/PhysRevD.98.030001. https://link.aps.org/doi/10.1103/PhysRevD.98.030001

4. Friedman JH (2001) Greedy function approximation: a gradient boosting machine. Ann Stat 29(5). https://doi.org/10.1214/aos/1013203451, https://doi.org/10.1214%2Faos%2F1013203451
5. Speckmayer P et al (2010) The toolkit for multivariate data analysis, TMVA 4. J Phys: Conf Ser 219(3):032057. https://doi.org/10.1088/1742-6596/219/3/032057, https://doi.org/10.1088%2F1742-6596%2F219%2F3%2F032057
6. CMS Colaboration (2022) Identification of hadronic tau lepton decays using a deep neural network. J Instrum 17(07):P07023. https://doi.org/10.1088/1748-0221/17/07/p07023, https://doi.org/10.1088%2F1748-0221%2F17%2F07%2Fp07023
7. CMS Collaboration (2021) Performance of the CMS muon trigger system in proton-proton collisions at $\sqrt{s} = 13$ TeV. J Instrum 16(07):P07001. https://doi.org/10.1088/1748-0221/16/07/p07001, https://doi.org/10.1088%2F1748-0221%2F16%2F07%2Fp07001
8. CMS Collaboration (2021) Electron and photon reconstruction and identification with the CMS experiment at the CERN LHC. J Instrum 16(05):P05014. https://doi.org/10.1088/1748-0221/16/05/p05014, https://doi.org/10.1088%2F1748-0221%2F16%2F05%2Fp05014
9. Larkoski AJ et al (2014) Soft drop. J High Energy Phys 2014(5). https://doi.org/10.1007/jhep05(2014)146, https://doi.org/10.1007%2Fjhep05%282014%29146
10. Bols E et al (2020) Jet flavour classification using DeepJet. J Instrum 15(12):P12012–P12012. https://doi.org/10.1088/1748-0221/15/12/p12012, https://doi.org/10.1088%2F1748-0221%2F15%2F12%2Fp12012
11. Cadamuro L (2018) Search for Higgs boson pair production in the $b\bar{b}\tau^+\tau^-$ decay channel with the CMS detector at the LHC. Springer. ISBN: 978-3-030-04055-0. https://doi.org/10.1007/978-3-030-04055-0
12. Hochreiter S (1998) The vanishing gradient problem during learning recurrent neural nets and problem solutions. Int J Uncertain, Fuzziness Knowl-Based Syst 06(02):107–116. https://doi.org/10.1142/s0218488598000094, https://doi.org/10.1142%2Fs0218488598000094
13. Hochreiter S, Schmidhuber J (1997) Long short-term memory. Neural Comput 9(8):1735–1780. https://doi.org/10.1162/neco.1997.9.8.1735, https://doi.org/10.1162%2Fneco.1997.9.8.1735
14. CMS Collaboration (2018) Identification of heavy-flavour jets with the CMS detector in pp collisions at 13 TeV. J Instrum 13(05):P05011–P05011. https://doi.org/10.1088/1748-0221/13/05/p05011, https://doi.org/10.1088%2F1748-0221%2F13%2F05%2Fp05011
15. Erdmann M et al (2019) Lorentz Boost Networks: autonomous physics-inspired feature engineering. J Instrum 14(06):P06006–P06006. https://doi.org/10.1088/1748-0221/14/06/p06006, https://doi.org/10.1088%2F1748-0221%2F14%2F06%2Fp06006
16. Bianchini L et al (2014) Reconstruction of the Higgs mass in H → $\tau\tau$ Events by Dynamical Likelihood techniques. J Phys: Conf Ser 513(2):022035. https://doi.org/10.1088/1742-6596/513/2/022035, https://doi.org/10.1088%2F1742-6596%2F513%2F2%2F022035
17. CMS Collaboration (2016) Search for non-resonant Higgs boson pair production in the $bb\tau\tau$ final state using 2016 data. Tech. rep. CMS-PAS-HIG-16-028. CERN. https://cms-results.web.cern.ch/cms-results/public-results/preliminary-results/HIG-16-028/index.html
18. CMS Collaboration (2016) Search for resonant Higgs boson pair production in the $bb\tau\tau$ final state using 2016 data. Tech. rep. CMS-PAS-HIG-16-029. CERN. https://cms-results.web.cern.ch/cms-results/public-results/preliminary-results/HIG-16-029/index.html
19. Strong GC (2020) On the impact of selected modern deep-learning techniques to the performance and celerity of classification models in an experimental high-energy physics use case. Mach Learn: Sci Technol 1(4):045006. https://doi.org/10.1088/2632-2153/ab983a, https://doi.org/10.1088%2F2632-2153%2Fab983a
20. Spearman Rank Correlation Coefficient. In: The concise encyclopedia of statistics. Springer, New York, pp 502–505. https://doi.org/10.1007/978-0-387-32833-1_379, https://doi.org/10.1007%2F978-0-387-32833-1_379
21. Breiman L (2001) Random forests. Mach Learn 45(1):5–32. https://doi.org/10.1023/a:1010933404324, https://doi.org/10.1023%2Fa%3A1010933404324
22. Ramachandran P, Zoph B, Le QV (2017) Searching for activation functions. arXiv: 1710.05941 [cs.NE]

23. Smith LN, Topin N (2019) Super-convergence: very fast training of neural networks using large learning rates. In: Pham T (ed) Artificial intelligence and machine learning for multi-domain operations applications. SPIE, May 2019. https://doi.org/10.1117/12.2520589, https://doi.org/10.1117%2F12.2520589
24. Khachatryan V et al (2016) Event generator tunes obtained from underlying event and multiparton scattering measurements. Eur Phys J C 76:155. https://doi.org/10.1140/epjc/s10052-016-3988-x, arXiv: 1512.00815 [hep-ex]
25. Sirunyan A et al (2020) Extraction and validation of a new set of CMS 8 tunes from underlying-event measurements. Eur Phys J C 80:4. https://doi.org/10.1140/epjc/s10052-019-7499-4, arXiv: 1903.12179 [hep-ex]
26. Ball RD et al (2015) Parton distributions for the LHC run II. JHEP 04:40. https://doi.org/10.1007/JHEP04(2015)040. arXiv: 1410.8849 [hep-ph]
27. Ball RD et al (2017) Parton distributions from high-precision collider data. Eur Phys J C 77:663. https://doi.org/10.1140/epjc/s10052-017-5199-5, arXiv: 1706.00428 [hep-ph]
28. Agostinelli S et al (2003) GEANT4—a simulation toolkit. Nucl Instrum Meth A 506:250. https://doi.org/10.1016/S0168-9002(03)01368-8
29. CMS Collaboration (2020) Performance of the CMS Level-1 trigger in proton-proton collisions at $\sqrt{s} = 13$ TeV. J Instrum 15(10):P10017–P10017. https://doi.org/10.1088/1748-0221/15/10/p10017, https://doi.org/10.1088%2F1748-0221%2F15%2F10%2Fp10017
30. CMS Collaboration. CMS TriDAS project: Technical Design Report, Volume 1: The Trigger Systems. Technical design report. CMS. http://cds.cern.ch/record/706847
31. Amoroso S et al (2020) Les Houches 2019: Physics at TeV Colliders: Standard Model Working Group Report. arXiv: 2003.01700 [hep-ph]
32. Carvalho A et al (2016) Higgs pair production: choosing benchmarks with cluster analysis. J High Energy Phys 2016(4):1–28. https://doi.org/10.1007/jhep04(2016)126, https://doi.org/10.1007%2Fjhep04%282016%29126
33. Buchalla G et al (2018) Higgs boson pair production in non-linear Effective Field Theory with full mt-dependence at NLO QCD. JHEP 09:57. https://doi.org/10.1007/JHEP09(2018)057, arXiv: 1806.05162
34. Azatov A et al (2015) Effective field theory analysis of double Higgs boson production via gluon fusion. Phys Rev D 92(3). https://doi.org/10.1103/physrevd.92.035001, https://doi.org/10.1103%2Fphysrevd.92.035001
35. CMS Collaboration (2021) Measurement of differential $t\bar{t}$ production cross sections in the full kinematic range using lepton + jets events from proton-proton collisions at $\sqrt{s} = 13$ TeV. Phys Rev D 104(9). https://doi.org/10.1103/physrevd.104.092013, https://doi.org/10.1103%2Fphysrevd.104.092013
36. Aldaya M et al. Measurement of multi-differential cross sections for top quark pair production in the dilepton final state at $\sqrt{s} = 13$ TeV with the full Run2 dataset. https://cms.cern.ch/iCMS/jsp/db_notes/noteInfo.jsp?cmsnoteid=CMS%20AN-2019/228
37. Barlow R, Beeston C (1993) Fitting using finite Monte Carlo samples. Comput Phys Commun 77(2):219–228. ISSN: 0010-4655. https://doi.org/10.1016/0010-4655(93)90005-W
38. Alwall J et al (2007) Comparative study of various algorithms for the merging of parton showers and matrix elements in hadronic collisions. Eur Phys J C 53(3):473–500. https://doi.org/10.1140/epjc/s10052-007-0490-5, https://doi.org/10.1140%2Fepjc%2Fs10052-007-0490-5
39. Li Y, Petriello F (2012) Combining QCD and electroweak corrections to dilepton production in the framework of the FEWZ simulation code. Phys Rev D 86(9). https://doi.org/10.1103/physrevd.86.094034, https://doi.org/10.1103%2Fphysrevd.86.094034
40. Frederix R, Frixione S (2012) Merging meets matching in MC@NLO. J High Energy Phys 2012(12). https://doi.org/10.1007/jhep12(2012)061, https://doi.org/10.1007%2Fjhep12%282012%29061
41. Buehler S (2013) Precise inclusive Higgs predictions using iHixs. In: Proceedings of 10th international symposium on radiative corrections (applications of quantum field theory to phenomenology) — PoS(RADCOR2011). Sissa Medialab, Jan 2013. https://doi.org/10.22323/1.145.0021, https://doi.org/10.22323%2F1.145.0021

42. Dulat F, Lazopoulos A, Mistlberger B (2018) iHixs 2 — Inclusive Higgs cross sections. Comput Phys Commun 233:243–260. https://doi.org/10.1016/j.cpc.2018.06.025, https://doi.org/10.1016%2Fj.cpc.2018.06.025
43. Denner A et al (2015) HAWK 2.0: a Monte Carlo program for Higgs production in vector-boson fusion and Higgs strahlung at hadron colliders. Comput Phys Commun 195:161–171. https://doi.org/10.1016/j.cpc.2015.04.021, https://doi.org/10.1016%2Fj.cpc.2015.04.021
44. Dainese A et al (2019) Report on the physics at the HL-LHC, and perspectives for the HE-LHC. Tech. rep. Geneva, Switzerland. https://doi.org/10.23731/CYRM-2019-007, https://cds.cern.ch/record/2703572
45. de Florian D et al (2017) Handbook of LHC Higgs cross sections: 4. Deciphering the nature of the Higgs sector. CERN yellow reports: monographs. Geneva, CERN. https://doi.org/10.23731/CYRM-2017-002, https://cds.cern.ch/record/2227475
46. Sirunyan AM et al (2021) Precision luminosity measurement in proton-proton collisions at $\sqrt{s} = 13$ TeV in 2015 and 2016 at CMS. Eur Phys J C 81:800. https://doi.org/10.1140/epjc/s10052-021-09538-2, arXiv: 2104.01927 [hep-ex]
47. CMS Collaboration (2017) CMS luminosity measurement for the 2017 data-taking period at $\sqrt{s} = 13$ TeV. Tech. rep. CMS-PAS-LUM-17-004. CERN. https://cds.cern.ch/record/2621960
48. CMS Collaboration (2019) CMS luminosity measurement for the 2018 data-taking period at $\sqrt{s} = 13$ TeV. Tech. rep. CMS-PAS-LUM-18-002. CERN. https://cds.cern.ch/record/2676164
49. Sirunyan AM et al (2018) Performance of the CMS muon detector and muon reconstruction with proton-proton collisions at $\sqrt{s} = 13$ TeV. JINST 13:P06015. https://doi.org/10.1088/1748-0221/13/06/P06015, arXiv: 1804.04528 [physics.ins-det]
50. Grazzini M et al (2018) Higgs boson pair production at NNLO with top quark mass effects. J High Energy Phys 2018(5). ISSN: 1029-8479. https://doi.org/10.1007/jhep05(2018)059, http://dx.doi.org/10.1007/JHEP05(2018)059
51. Dreyer FA, Karlberg A (2018) Vector-boson fusion Higgs pair production at N3LO. Phys Rev D 98(11). ISSN: 2470-0029. https://doi.org/10.1103/physrevd.98.114016, http://dx.doi.org/10.1103/PhysRevD.98.114016
52. Cabouat B, Sjöstrand T (2018) Some dipole shower studies. Eur Phys J C 78:226. https://doi.org/10.1140/epjc/s10052-018-5645-z, arXiv: 1710.00391v2

Chapter 6
The Results on HH → $b\bar{b}\tau^+\tau^-$

Abstract The first part of this chapter is devoted to the description of the effort to identify a discriminating observable that, being maximally sensible to the presence of the investigated signal, can be used as input to a statistical method. The second part of this chapter will then cover the results obtained within the Standard Model (SM) framework as well as their interpretation in various Beyond the SM (BSM) scenarios, with special regard for Effective Field Theories (EFT). The bb$\tau\tau$ analysis, as any other analysis in CMS, starts from the information deposited in each subdetector of the CMS experiment, which is optimally exploited and combined by Level-1 (L1) trigger and High-Level Trigger (HLT) to select the most interesting events to be recorded for storage (Sect. 2.3) based on a trigger menu targeting multiple compelling signatures. The journey continues with the offline analysis of the data, whose first step is the reconstruction of physics objects via the Particle Flow (PF) algorithm (Sect. 2.4). In the bb$\tau\tau$ analysis, all PF objects are exploited to reconstruct events compatible with the HH signal: muons, electrons, and hadronically decaying τ leptons (τ_h) are used to identify the H → $\tau\tau$ candidate, while hadronic jets identified with b-tagging algorithms are used to select the H → bb candidate (Sect. 5.2). Finally, a careful definition of the signal region and the categorization of the events guarantees the ability to further and highly increase the sensitivity of the analysis, exploiting state-of-the-art machine learning techniques for signal extraction (Sect. 5.3). All along this process, the construction of Monte Carlo (MC) simulations that carefully reproduce the behaviour of data is performed, and the systematic uncertainties related to it are evaluated (Sects. 5.4 and 5.5). The results of the exploration of the Higgs boson pair (HH) production in the bb$\tau\tau$ decay channel are presented in this chapter alongside the statistical framework employed for the purpose.

The results of the exploration of the Higgs boson pair (HH) production in the bb$\tau\tau$ decay channel are presented in this chapter alongside the statistical framework employed for the purpose. Nevertheless, it is worth it at this point to assemble a panoramic view of all the steps necessary to produce these results as described in the previous chapters and schematically depicted in Fig. 5.3.

The bb$\tau\tau$ analysis, as any other analysis in CMS, starts from the information deposited in each subdetector of the CMS experiment, which is optimally exploited

J. Motta, *Development of Machine Learning τ Trigger Algorithms and Search for Higgs Boson Pair Production*, Springer Theses,
https://doi.org/10.1007/978-3-031-96288-2_6

and combined by Level-1 (L1) trigger and High-Level Trigger (HLT) to select the most interesting events to be recorded for storage (Sect. 2.3) based on a trigger menu targeting multiple compelling signatures. The journey continues with the offline analysis of the data, whose first step is the reconstruction of physics objects via the Particle Flow (PF) algorithm (Sect. 2.4). In the $bb\tau\tau$ analysis, all PF objects are exploited to reconstruct events compatible with the HH signal: muons, electrons, and hadronically decaying τ leptons (τ_h) are used to identify the H → $\tau\tau$ candidate, while hadronic jets identified with b-tagging algorithms are used to select the H → bb candidate (Sect. 5.2). Finally, a careful definition of the signal region and the categorization of the events guarantees the ability to further and highly increase the sensitivity of the analysis, exploiting state-of-the-art machine learning techniques for signal extraction (Sect. 5.3). All along this process, the construction of Monte Carlo (MC) simulations that carefully reproduce the behaviour of data is performed, and the systematic uncertainties related to it are evaluated (Sects. 5.4 and 5.5).

The effort summarized in the previous paragraph has but one single aim: to identify a discriminating observable that, being maximally sensible to the presence of the investigated signal, can be used as input to a statistical method. The first part of this chapter is devoted to the description of these two steps. The second part of this chapter will then cover the results obtained within the Standard Model (SM) framework as well as their interpretation in various Beyond the SM (BSM) scenarios, with special regard for Effective Field Theories (EFT).

6.1 Statistical Methods

The statistical method employed in this search is the modified frequentist approach originally developed at the time of LEP [1, 2], which was refined for the discovery of the Higgs boson (H) in 2012 and already used for the combination of the results from the ATLAS and CMS collaborations [3, 4]. This methodology involves the evaluation of the incompatibility between the observed data and two distinct hypotheses: the signal plus background hypothesis, also referred to as the *null hypothesis* (H_0), and the background-only hypothesis, also referred to as the *alternative* or *alternate hypothesis* (H_1). This approach relies on the utilization of binned or parametric distributions of selected observables within each category of the search (in this analysis, binned distributions are employed). Therefore, after finalizing the analysis strategy summarised in the introduction to this chapter, a discriminating observable is selected and strategically optimized to effectively distinguish signals from background contributions.

6.1.1 Discriminating Observable

The choice of the discriminating observable is crucial to maximize the sensitivity of the search and to provide information on the nature of the signal in case an excess is found. Various configurations were assessed to determine the optimal approach for extracting the final limits; multiple strategies were explored, such as merging different categories and testing different discriminating variables across various categories. After an exhaustive evaluation, the most successful combination emerged, proving to be highly effective for both Gluon Fusion (ggF) and Vector Boson Fusion (VBF) studies. This successful approach involves leveraging the ggF-based Deep Neural Network (DNN) score, detailed in Sect. 5.3.4 across the eight mutually exclusive categories presented in Sect. 5.3.3: the `res2b`, `res1b`, and `boosted` categories targeting the ggF production mechanisms, and the five VBF dedicated categories obtained from the multi-classification process, i.e. `classGGF`, `classVBF`, `classttH`, `classTT`, and `classDY`. The schematic representation of the categorization strategy can be found in Fig. 5.3.

After having selected a discriminating observable, the selection of its binning strategy significantly influences the analysis sensitivity. To ensure optimal results, a specialized algorithm is employed to determine the most advantageous binning scheme for each category, channel, and data-taking year. This algorithm is designed to minimize the expected limits for the SM ggF and VBF HH production in each category used in the analysis. Bayesian optimization techniques [5, 6] are leveraged to fine-tune the bin-splitting process, allowing for an efficient approach to determining the optimal binning scheme. Through this approach, the analysis sensitivity is greatly enhanced, providing a robust and precise evaluation of the limits on HH production. This strategy is described in the following.

Initially, finely binned histograms of the DNN score are prepared for each category using a precision of 0.001, ensuring a highly granular representation of the DNN score in MC simulated samples. As computational complexity can be a concern in such an optimization, an iterative approach is adopted where one category is successively added for optimization while freezing the binning of all previously considered categories that have already been optimized. By doing so, the computational burden is effectively managed, as each additional category is processed without the need to recalibrate the binning of previously included histograms. The choice of the stacking order is based on the sensitivity hierarchy of the categories as follows:

1. optimize `res2b` minimizing a loss function based on the inclusive HH cross section limit,
2. freeze previous category and optimize `res1b` minimizing a loss function based on the inclusive HH cross section limit,
3. freeze previous categories and optimize `boosted` minimizing a loss function based on the inclusive HH cross section limit,
4. freeze previous categories and optimize `classVBF` minimizing a loss function based on the VBF HH cross section limit,

5. freeze previous categories and optimize `classGGF` minimizing a loss function based on the VBF HH cross section limit,
6. freeze previous categories and optimize `classttH` minimizing a loss function based on the VBF HH cross section limit,
7. freeze previous categories and optimize `classTT` minimizing a loss function based on the VBF HH cross section limit,
8. freeze previous categories and optimize `classDY` minimizing a loss function based on the VBF HH cross section limit.

Moreover, specific care is taken to prevent the binning optimization from introducing a bias in the result by selecting particularly intricate schemes with poor modelling of the background yield. To this end, each bin of any binning scheme considered is required to satisfy the following conditions:

- presence of at least one MC simulated Drell-Yan event,
- presence of at least one MC simulated $t\bar{t}$ event,
- weighted sum of all background yields $\sum$ bkg. > 0.03; this threshold is selected as it lies in the 95% confidence interval for observation of zero events provided one event is expected,
- validity of the three previous conditions for each jet and τ energy scale uncertainty variations (cf. Sect. 5.5.2),
- relative statistical error of the expected background yield $\epsilon_{rel} < 1$, to avoid bins with a sizeable fraction of events with negative MC weights.

To accomplish the minimization process, a tailored loss function is introduced to be used by the Bayesian optimization, taking into account both the expected limit and the number of bins. Specifically, the loss function is formulated as the sum of two terms: the expected limit scaled by a factor 10^6, and the number of bins. This formulation effectively strikes a balance between achieving higher statistical significance and favouring less sophisticated binning schemes that simplify the representation of data. To reach the maximal performance and, at the same time, to account for differences in the recorded data throughout the Run-2 data-taking period, as well as the DNN dependence on the $\tau\tau$ final state, the process of minimization is performed for each individual channel and year. The pre-fit distributions of the DNN score discriminating observable are reported in Figs. 6.1, 6.2, 6.3, 6.4, 6.5, 6.6, 6.7, 6.8, and 6.9, showcasing the binning obtained from the optimization outlined here, for each of the 72 regions (i.e. 3 years $\times$ 3 channels $\times$ 8 categories) that are simultaneously explored in the maximum likelihood fit described in Sect. 6.1.2.

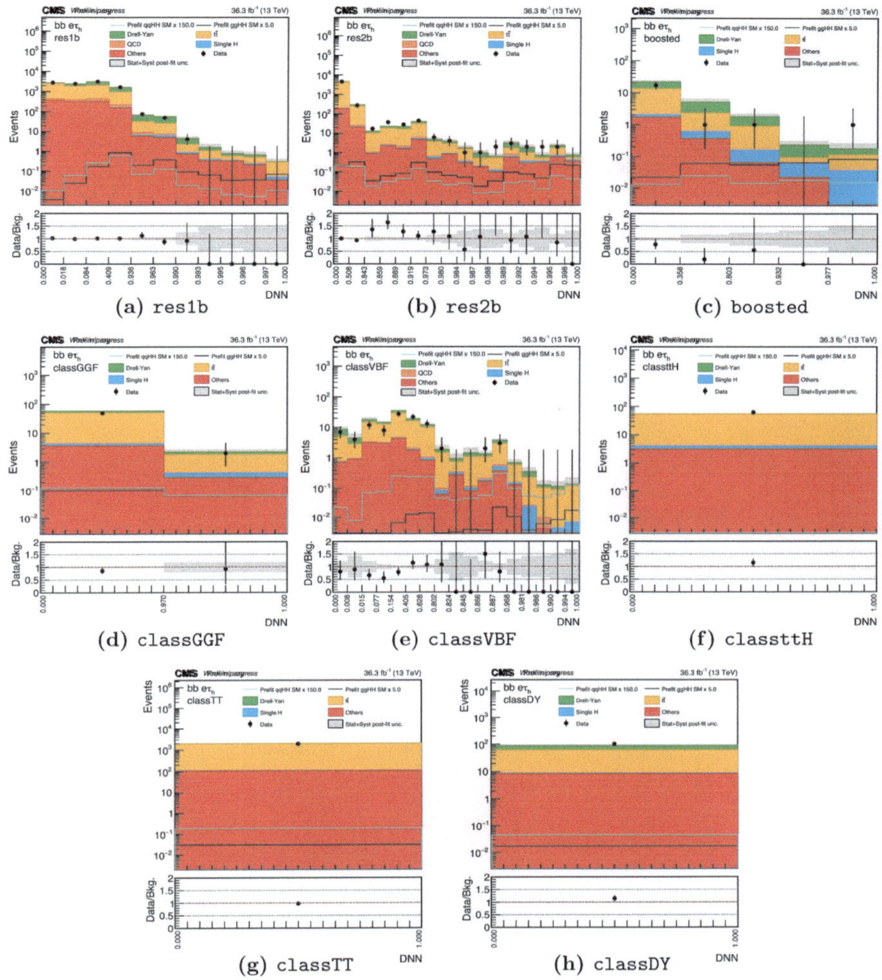

Fig. 6.1 Distributions of the events observed in the signal regions of the $\tau_e\tau_h$ final state in the 2016 data-taking period. In panels **a** through **h**, for the eight categories simultaneously fitted, the upper sub-panels report the distribution of the sore of the DNN used for the HH signal extraction, while the lower sub-panels show the data over background ratio. Data are represented by points with error bars and expected signal contributions are represented by the black (ggF HH) and cyan (VBF HH) solid lines. Expected background contributions (shaded histograms) and associated systematic uncertainties (dashed areas) are stacked while the signal histograms are not

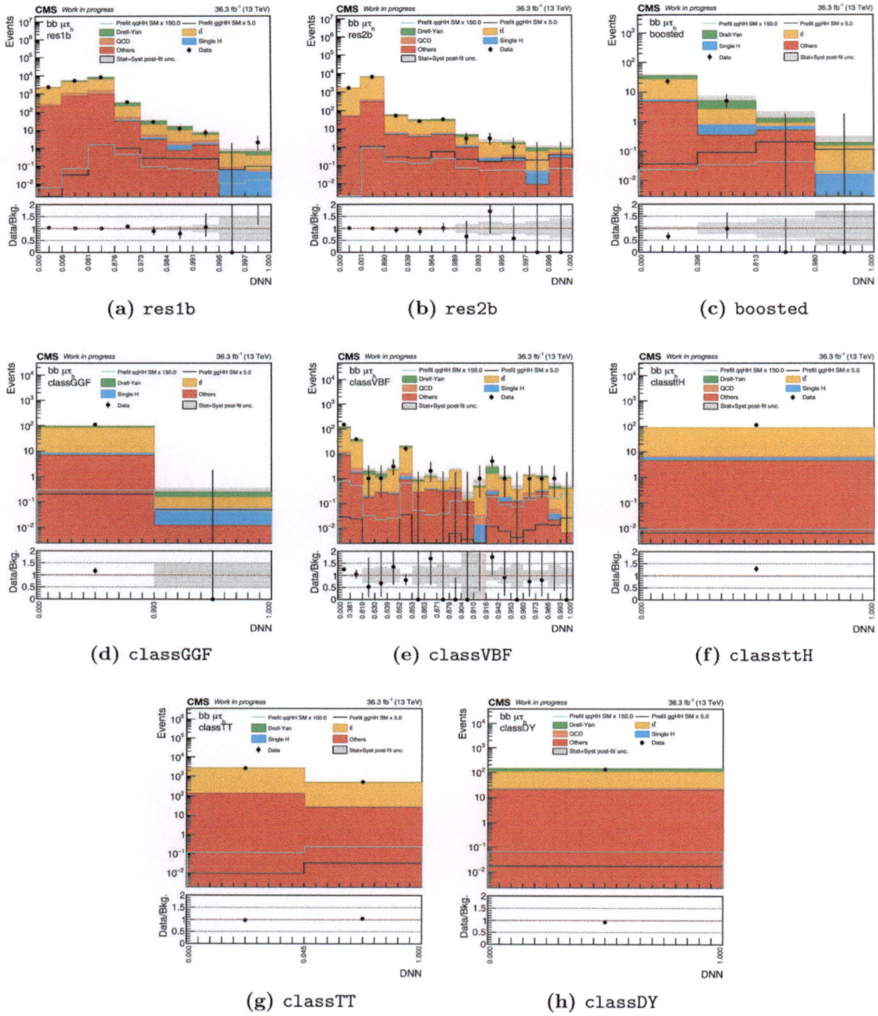

(a) res1b (b) res2b (c) boosted

(d) classGGF (e) classVBF (f) classttH

(g) classTT (h) classDY

Fig. 6.2 Distributions of the events observed in the signal regions of the $\tau_\mu\tau_h$ final state in the 2016 data-taking period. In panels **a** through **h**, for the eight categories simultaneously fitted, the upper sub-panels report the distribution of the sore of the DNN used for the HH signal extraction, while the lower sub-panels show the data over background ratio. Data are represented by points with error bars and expected signal contributions are represented by the black (ggFHH) and cyan (VBF HH) solid lines. Expected background contributions (shaded histograms) and associated systematic uncertainties (dashed areas) are stacked while the signal histograms are not

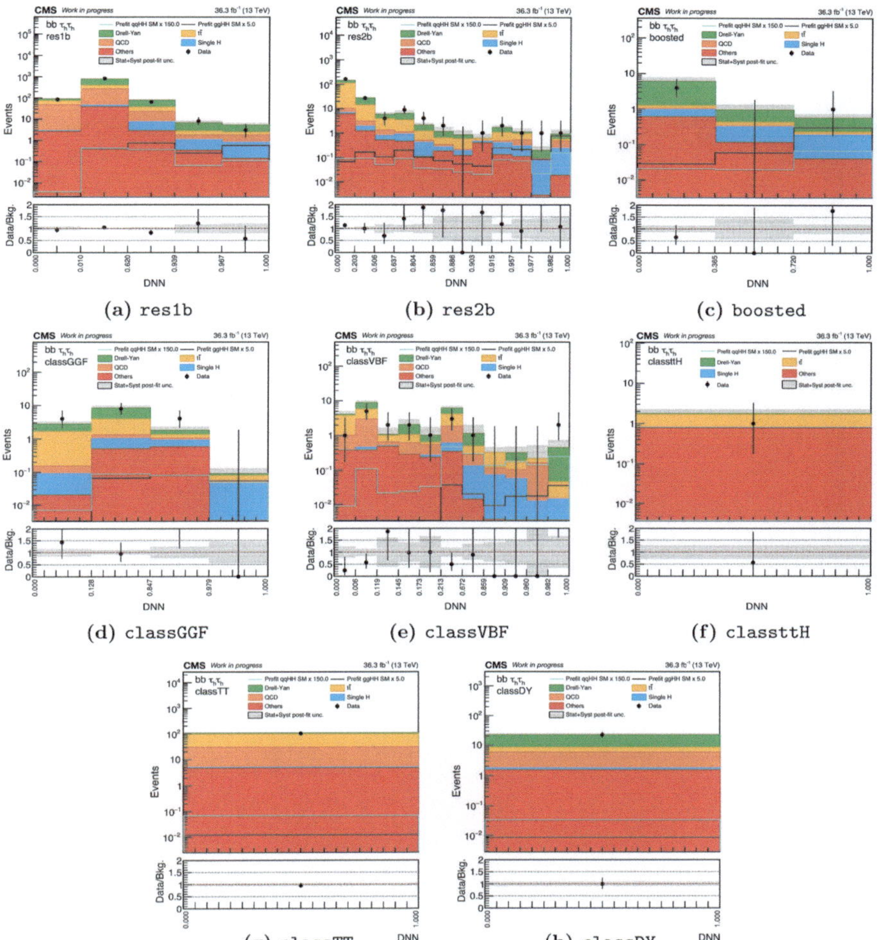

(a) res1b (b) res2b (c) boosted (d) classGGF (e) classVBF (f) classttH (g) classTT (h) classDY

Fig. 6.3 Distributions of the events observed in the signal regions of the $\tau_h\tau_h$ final state in the 2016 data-taking period. In panels **a** through **h**, for the eight categories simultaneously fitted, the upper sub-panels report the distribution of the sore of the DNN used for the HH signal extraction, while the lower sub-panels show the data over background ratio. Data are represented by points with error bars and expected signal contributions are represented by the black (ggF HH) and cyan (VBF HH) solid lines. Expected background contributions (shaded histograms) and associated systematic uncertainties (dashed areas) are stacked while the signal histograms are not

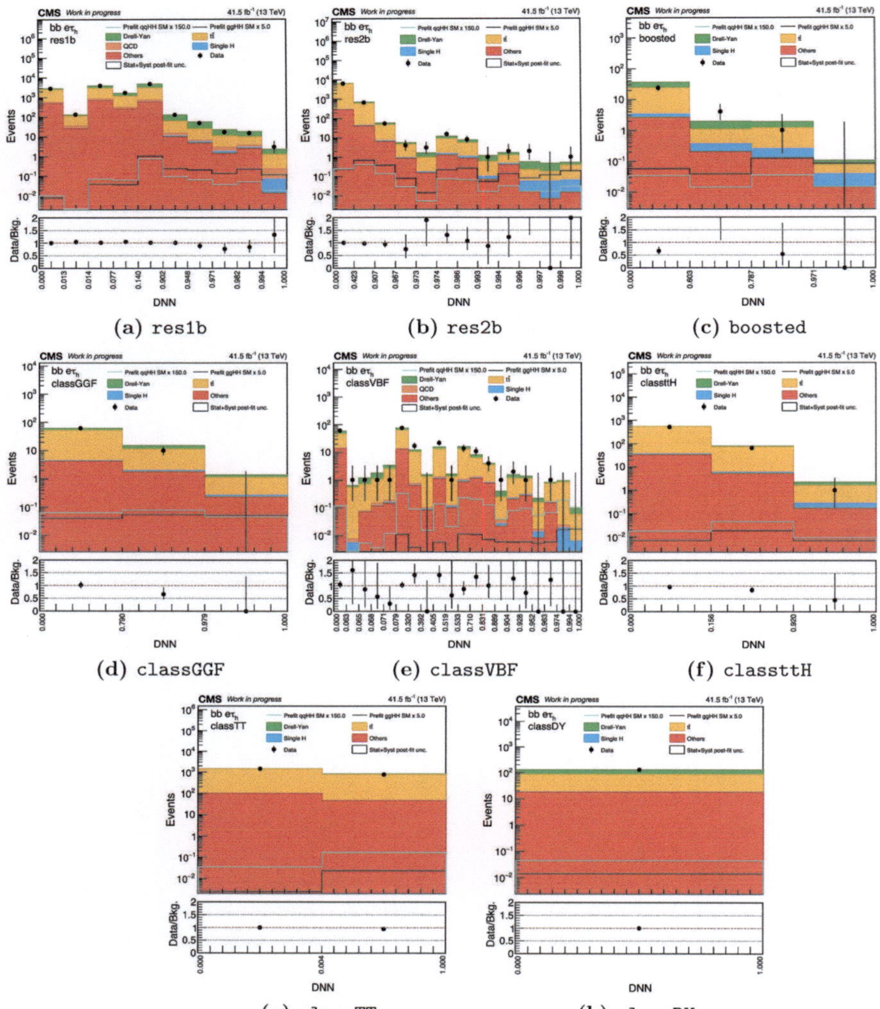

(a) res1b (b) res2b (c) boosted

(d) classGGF (e) classVBF (f) classttH

(g) classTT (h) classDY

Fig. 6.4 Distributions of the events observed in the signal regions of the $\tau_e\tau_h$ final state in the 2017 data-taking period. In panels **a** through **h**, for the eight categories simultaneously fitted, the upper sub-panels report the distribution of the sore of the DNN used for the HH signal extraction, while the lower sub-panels show the data over background ratio. Data are represented by points with error bars and expected signal contributions are represented by the black (ggF HH) and cyan (VBF HH) solid lines. Expected background contributions (shaded histograms) and associated systematic uncertainties (dashed areas) are stacked while the signal histograms are not

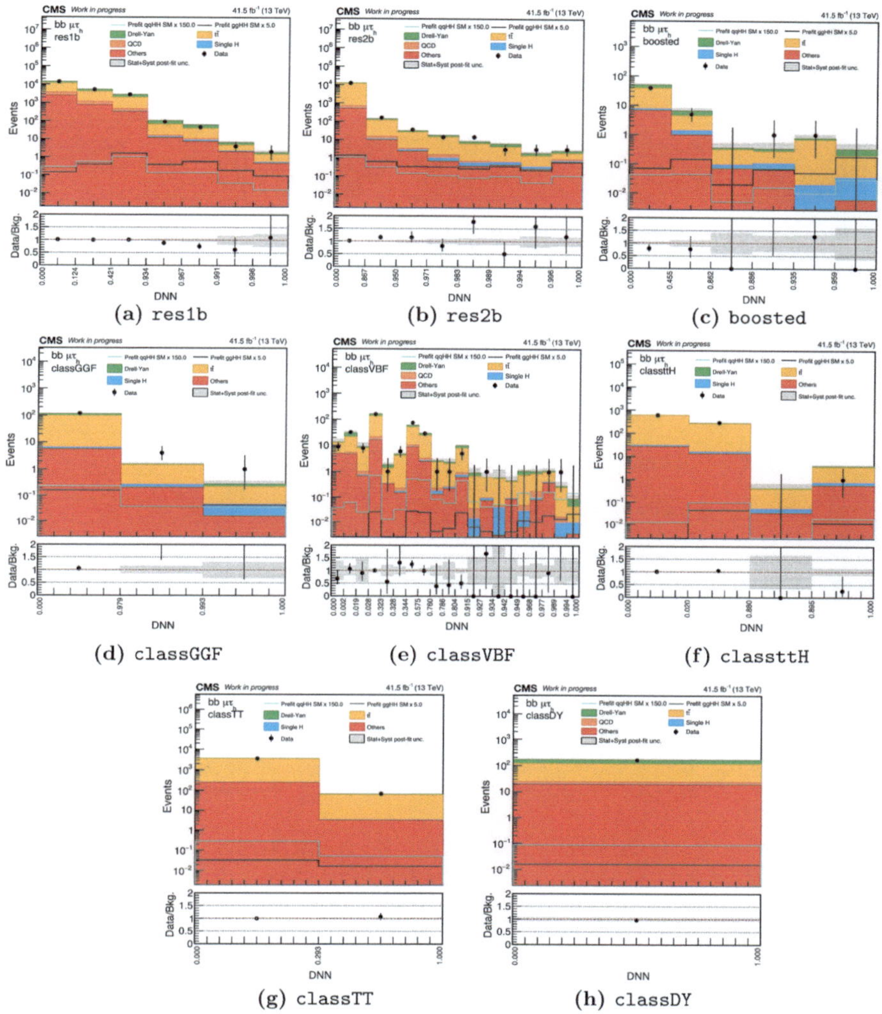

(a) res1b (b) res2b (c) boosted (d) classGGF (e) classVBF (f) classttH (g) classTT (h) classDY

Fig. 6.5 Distributions of the events observed in the signal regions of the $\tau_\mu \tau_h$ final state in the 2017 data-taking period. In panels **a** through **h**, for the eight categories simultaneously fitted, the upper sub-panels report the distribution of the sore of the DNN used for the HH signal extraction, while the lower sub-panels show the data over background ratio. Data are represented by points with error bars and expected signal contributions are represented by the black (ggFHH) and cyan (VBF HH) solid lines. Expected background contributions (shaded histograms) and associated systematic uncertainties (dashed areas) are stacked while the signal histograms are not

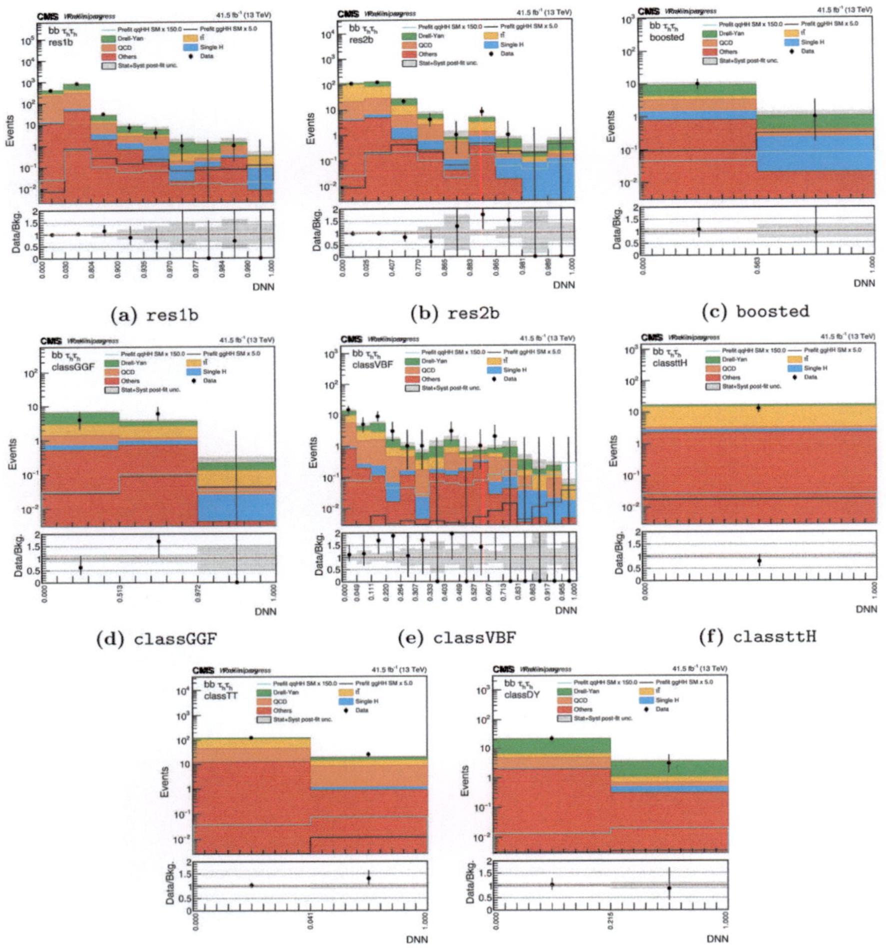

(a) res1b (b) res2b (c) boosted

(d) classGGF (e) classVBF (f) classttH

(g) classTT (h) classDY

Fig. 6.6 Distributions of the events observed in the signal regions of the $\tau_h\tau_h$ final state in the 2017 data-taking period. In panels **a** through **h**, for the eight categories simultaneously fitted, the upper sub-panels report the distribution of the sore of the DNN used for the HH signal extraction, while the lower sub-panels show the data over background ratio. Data are represented by points with error bars and expected signal contributions are represented by the black (ggFHH) and cyan (VBF HH) solid lines. Expected background contributions (shaded histograms) and associated systematic uncertainties (dashed areas) are stacked while the signal histograms are not

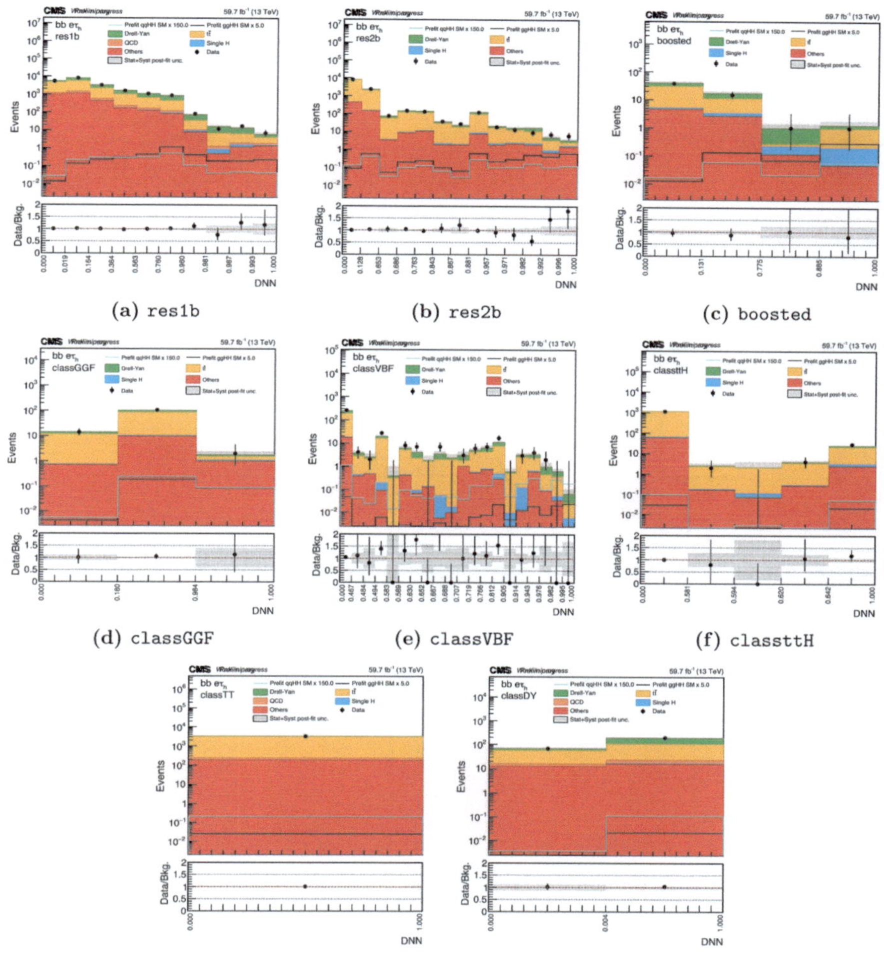

(a) res1b (b) res2b (c) boosted (d) classGGF (e) classVBF (f) classttH (g) classTT (h) classDY

Fig. 6.7 Distributions of the events observed in the signal regions of the $\tau_e\tau_h$ final state in the 2018 data-taking period. In panels **a** through **h**, for the eight categories simultaneously fitted, the upper sub-panels report the distribution of the sore of the DNN used for the HH signal extraction, while the lower sub-panels show the data over background ratio. Data are represented by points with error bars and expected signal contributions are represented by the black (ggFHH) and cyan (VBF HH) solid lines. Expected background contributions (shaded histograms) and associated systematic uncertainties (dashed areas) are stacked while the signal histograms are not

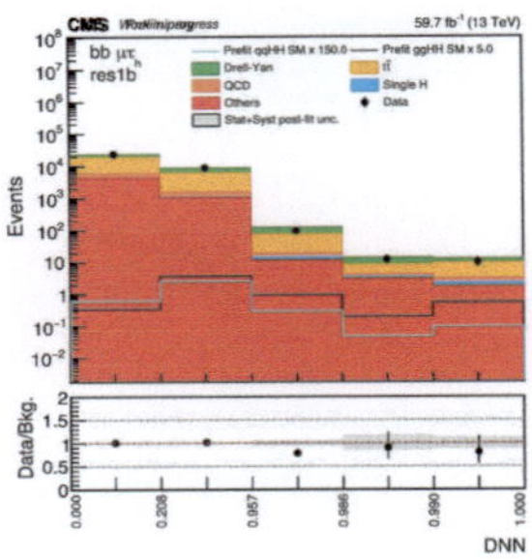

(a) res1b

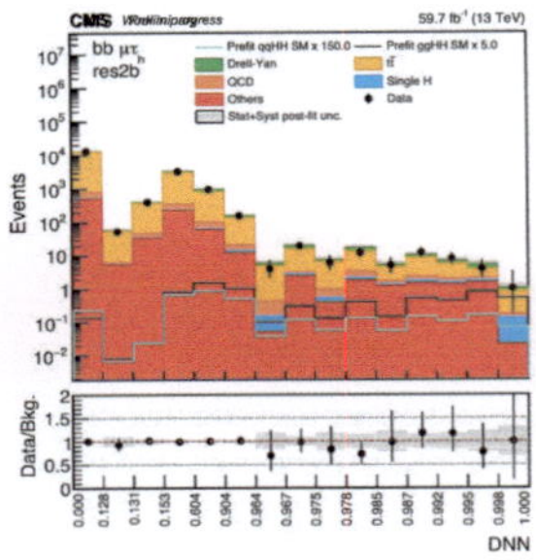

(b) res2b

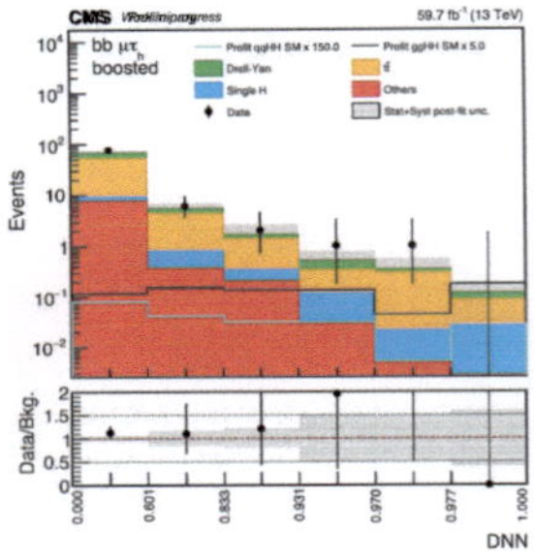

(c) boosted

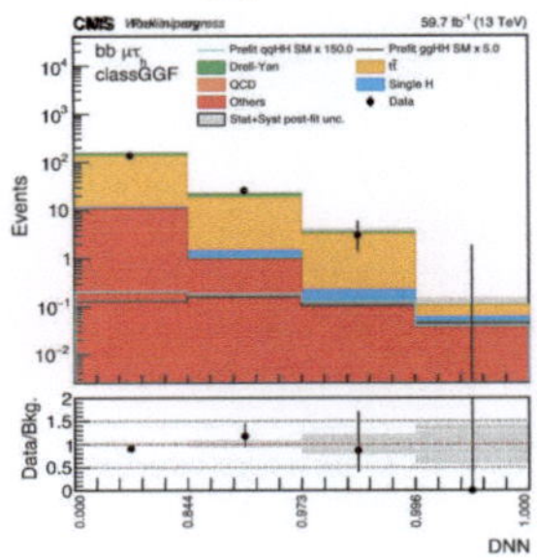

(d) classGGF

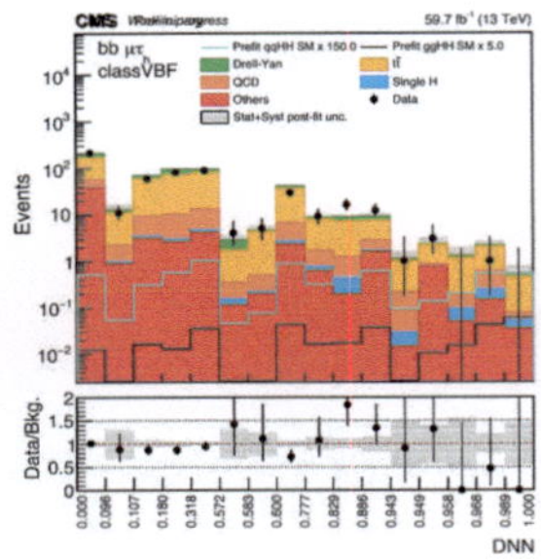

(e) classVBF

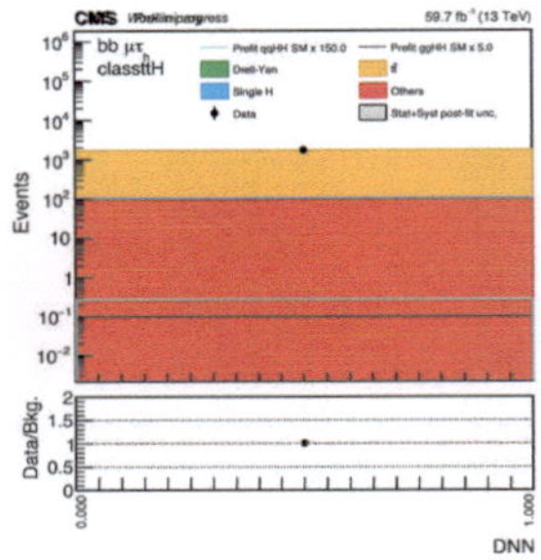

(f) classttH

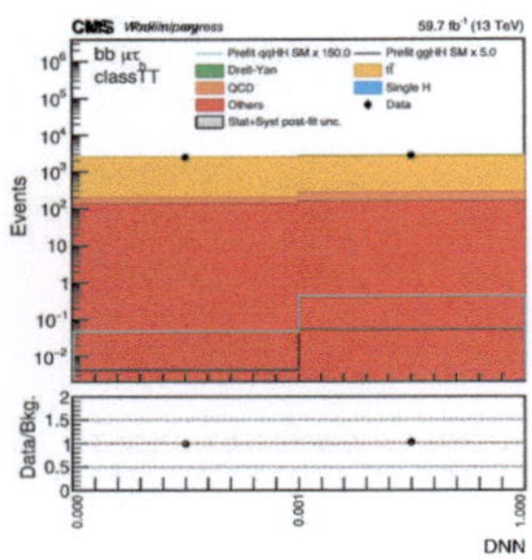

(g) classTT

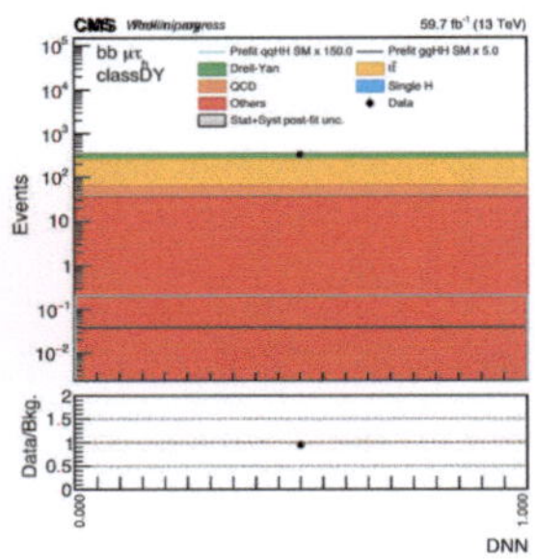

(h) classDY

Fig. 6.8 Distributions of the events observed in the signal regions of the $\tau_\mu\tau_h$ final state in the 2018 data-taking period. In panels **a** through **h**, for the eight categories simultaneously fitted, the upper sub-panels report the distribution of the sore of the DNN used for the HH signal extraction, while the lower sub-panels show the data over background ratio. Data are represented by points with error bars and expected signal contributions are represented by the black (ggF HH) and cyan (VBF HH) solid lines. Expected background contributions (shaded histograms) and associated systematic uncertainties (dashed areas) are stacked while the signal histograms are not

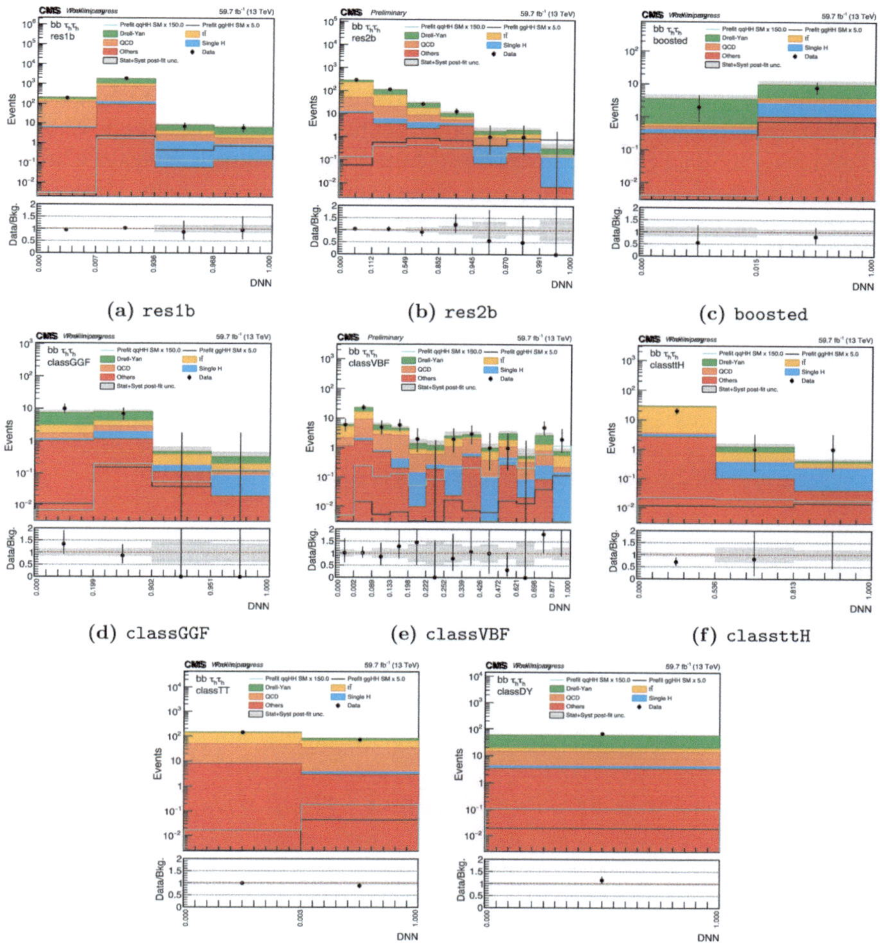

(a) res1b (b) res2b (c) boosted

(d) classGGF (e) classVBF (f) classttH

(g) classTT (h) classDY

Fig. 6.9 Distributions of the events observed in the signal regions of the $\tau_h\tau_h$ final state in the 2018 data-taking period. In panels **a** through **h**, for the eight categories simultaneously fitted, the upper sub-panels report the distribution of the sore of the DNN used for the HH signal extraction, while the lower sub-panels show the data over background ratio. Data are represented by points with error bars and expected signal contributions are represented by the black (ggFHH) and cyan (VBF HH) solid lines. Expected background contributions (shaded histograms) and associated systematic uncertainties (dashed areas) are stacked while the signal histograms are not

6.1.2 The Maximum Likelihood Method

The cornerstone of the statistical framework used in this and all other analyses at the LHC lies in the construction of the likelihood function, which captures the probability of observing the data given a particular set of model parameters. It serves as a bridge between the theoretical predictions and the experimental observations, allowing us to quantify the compatibility between the two. A brief discussion of the theoretical origin and use of the likelihood function is described in the following.

Given the dataset $\vec{x} = \{x_1, x_2, ..., x_n\}$ containing a certain number of observed events, each of these points is distributed according to a Probability Density Function (pdf) $f(x|\vec{\xi})$, where $\vec{\xi}$ is the vector of parameters of the pdf. Therefore, the probability of observing a certain set of data $\vec{x}$ given the set of parameters $\vec{\xi}$ can be written as the joint probability:

$$\mathcal{P}(\vec{x}|\vec{\xi}) = f(x_1|\vec{\xi})dx_1 \cdot f(x_2|\vec{\xi})dx_2 \cdot ... \cdot f(x_n|\vec{\xi})dx_n \tag{6.1}$$

which encodes all the information of the observed data. The vector of parameters of the pdf is generally split into two separate portions, the so-called *Parameters Of Interest* (POIs) generally identified by the symbol μ (or sometimes r) and the *nuisance parameters* usually denoted by the symbol θ; therefore $\vec{\xi} = \{\vec{\mu}, \vec{\theta}\}$. It should be noted that in the context of high energy physics, the number of occurrences of any specific process is not known due to its aleatory nature; therefore, the probability defined in Eq. 6.1 needs to be extended to:

$$\mathcal{P}(\vec{x}|\vec{\mu}, \vec{\theta}) = \Pr(n|\nu(\vec{\mu}, \vec{\theta})) \cdot \prod_{i=1}^{n} f(x_i|\vec{\mu}, \vec{\theta})dx_i \tag{6.2}$$

where $\Pr(n|\nu(\vec{\mu}, \vec{\theta}))$ is the Poisson probability of observing a total number of events equal to n given an expected total number of event ν.

In high energy physics, the number of POIs is generally reduced to be just one, the so-called *signal strength modifier* defined as the ratio between the observed cross section and the corresponding SM prediction: $\mu = \sigma_{\text{obs.}}/\sigma_{\text{SM}}$. In the $bb\tau\tau$ analysis, the expected event yield regulating the shape of the Poisson distribution is estimated from the MC simulation and process modelling of the signal (s) and backgrounds (b) described in Sect. 5.4 to be $\nu = \mu \cdot s(\vec{\theta}) + b(\vec{\theta})$, with explicit dependence on both the POI and the nuisance parameters. Moreover, the nuisance parameters in the probability of Eq. 6.2 are nothing but a representation of the systematic uncertainties detailed in Sect. 5.5: normalization uncertainties modify the total expected yield of events, whereas shape uncertainties modify their differential distribution.

The estimation of the nuisance parameters is done either from a priori considerations, e.g. theoretical uncertainty stemming from the choice of a specific set of parton distribution functions, or from auxiliary measurements performed with events independent from those entering the signal regions of the analysis, e.g. the uncertainty on the normalization of the $t\bar{t}$ background obtained in a dedicated control region as described in Sect. 5.4.4. Such estimation is denoted as $\tilde{\vec{\theta}}$ and our degree of belief on what is the real value of the nuisances given their estimation is represented with the Bayesian pdf $\rho(\vec{\theta}|\tilde{\vec{\theta}})$. By Bayes' theorem, this can be reinterpreted as a posterior frequentist probability arising from the auxiliary measurement $\tilde{\vec{\theta}}$ of $\vec{\theta}$ as:

$$\rho(\vec{\theta}|\tilde{\vec{\theta}}) \sim p(\tilde{\vec{\theta}}|\vec{\theta}) \cdot \pi_\theta(\vec{\theta}) \tag{6.3}$$

where $p(\tilde{\vec{\theta}}|\vec{\theta})$ is the probability density function for the auxiliary measurement and $\pi_\theta(\vec{\theta})$ is the prior of the measurement, which is generally chosen to be uniform to encode complete prior ignorance about it. With this ingredient established, Eq. 6.2 can be further extended to

$$\mathcal{P}(\vec{x}, \tilde{\vec{\theta}}|\mu, \vec{\theta}) = \Pr(n|\mu \cdot s(\vec{\theta}) + b(\vec{\theta})) \cdot \prod_{i=1}^{n} f(x_i|\vec{\theta})dx_i \cdot p(\tilde{\vec{\theta}}|\vec{\theta}) \tag{6.4}$$

This equation is built using all available data points in a parametric manner without loss of information; nevertheless, as in many other analyses in high energy physics, the HH $\rightarrow$ bb$\tau\tau$ search is performed employing histogramming of the data to reduce the computational needs of the statistical inference. Therefore, defining

$$\mu \cdot s_j(\vec{\theta}) + b_j(\vec{\theta}) = \int_{x_j^{\min}}^{x_j^{\max}} f(x|\vec{\theta})dx \tag{6.5}$$

the expectation value of the number of entries per each bin j evaluated from MC simulation, we can finally explicitly formulate the binned likelihood function to be:

$$\mathcal{L}(n, \tilde{\vec{\theta}}|\mu, \vec{\theta}) \equiv \prod_j \frac{(\mu \cdot s_j(\vec{\theta}) + b_j(\vec{\theta}))^{n_j}}{n_j!} e^{-(\mu \cdot s_j(\vec{\theta}) + b_j(\vec{\theta}))} \cdot p(\tilde{\vec{\theta}}|\vec{\theta}) \tag{6.6}$$

This approach to the definition of the likelihood function is useful to highlight the scalability power of the method: given a set of multiple categories, the likelihood function can be computed for all categories and the combined likelihood function will simply be the product of all categories' likelihoods. More than that, if two experiments perform the measurement of the same parameter, the two results can be easily combined in the same manner by obtaining the joint likelihood of the two experiments as the product of each experiment's likelihood. This property is the direct consequence of the fact that the likelihood function is based on the conditional probability that, given a set of parameters $\vec{\theta}$, we observe the dataset we have.

One last ingredient is now necessary to fully define $\mathcal{L}(n, \tilde{\vec{\theta}}|\mu, \vec{\theta})$: the functional form of $\rho(\vec{\theta}|\tilde{\vec{\theta}})$. In the bbττ analysis, the log-normal function is used, and it is defined as:

$$\rho(\vec{\theta}|\tilde{\vec{\theta}}; \vec{\kappa}) = \frac{1}{\sqrt{2\pi}\ln\vec{\kappa}} \cdot \exp\left\{ -\left(\frac{\ln(\vec{\theta})/\tilde{\vec{\theta}}}{2\ln\vec{\kappa}} \right)^2 \right\} \cdot \frac{1}{\vec{\theta}} \tag{6.7}$$

with $\vec{\kappa} = 1 + \vec{\epsilon}$, being $\vec{\epsilon}$ the vector of the relative scales of the uncertainties. This function is asymptotically equivalent to a Gaussian distribution of width ϵ but has the advantage of correctly describing positively defined observables by going to zero for $\theta_i = 0$. This functional form is used both for normalization and shape uncertainties; in the case of the latter, the log-normal function takes into account only the normalization effect of the shape uncertainties. The differential distribution uncertainty is modelled with a *vertical template morphing* technique that takes as an input the alternative differential distribution obtained by the $\pm 1\sigma$ variations of the uncertainty and interpolates them bin-by-bin to introduce the associated nuisances in the likelihood.

Having formulated the likelihood function, and all of its components, the next crucial step is to extract the best-fit values of the model parameters, i.e. those that maximize the likelihood. This process is known as Maximum Likelihood Estimation (MLE) and is central to our analysis. The MLE provides us with the most probable parameter values, effectively determining the best model parameters given the observed data. Therefore, in this framework, results are presented as MLE of the POI.

In the remainder of the Chapter, for simplicity of notation, all the vector symbols will be dropped.

6.1.3 Hypothesis Testing

The likelihood formalism is the foundation of the hypothesis testing procedure that is used at the LHC experiments to quantify the presence or absence of signal. To discriminate between the two hypotheses, a function of the observed data needs to be defined, the so-called *test statistics*. If built in the proper way, the test statistics should be distributed differently under the null and the alternate hypotheses. More than that, according to Neyman-Pearson lemma [7], given two simple hypotheses (i.e. where data are drawn from a fully specified pdf), the most powerful test statistics for a given significance level is the likelihood ratio test. Following this lemma, the common method used in LHC analyses is the Profile Likelihood Ratio (PLR)

$$\lambda(\mu) \equiv \frac{\mathcal{L}(n, \widetilde{\theta}|\mu, \hat{\hat{\theta}}_\mu)}{\mathcal{L}(n, \widetilde{\theta}|\hat{\mu}, \hat{\theta})} \tag{6.8}$$

where:

- μ is the specific value of the signal strength modifier being tested,
- $\hat{\hat{\theta}}_\mu$ is the MLE of the nuisance parameters once μ has been fixed; we say in this case that θ is profiled given μ,
- $\hat{\mu}$ and $\hat{\theta}$ are the MLE of μ and θ when both are left floating in the likelihood.

By construction, the PLR is bound between 0.0 and 1.0, and higher values indicate better compatibility between the data and the null hypothesis. It is worth noticing that in the most general treatment, no bound is enforced on the MLE $\hat{\mu}$, which is allowed to take both positive and negative (unphysical) values even if the search is explicitly targeting a positive signal (provided that $\mu \cdot s + b$ in the Poisson probability remains positive).

The definition of the test statistics in Eq. 6.8 differs from the test statistics used for searches at the LEP and Tevatron colliders in the profiling of the likelihood of the denominator and in the possibility of $\hat{\mu}$ to assume negative values. These two choices ensure that the PRL has asymptotic properties that make it possible to derive the $\lambda(\mu)$ distribution under the H_0 and H_1 hypotheses with analytical formulae [3] instead of using pseudo-experiments when a large number of background events is expected. This asymptotic approximation is used in deriving all the results presented in this chapter.

The PLR definition is then employed in different fashions depending on the need to set upper limits or quantify an excess in the data. This is detailed in the following.

Setting Upper Limits

The test statistics q_μ used to set exclusion limits is generally referred to as *limit test statistics* and is defined as:

$$q_\mu \equiv \begin{cases} -2\ln\lambda(\mu) & \text{if } \hat{\mu} \leq \mu \\ 0 & \text{if } \hat{\mu} > \mu \end{cases} \tag{6.9}$$

where the enforcement of the $q_\mu = 0$ for $\hat{\mu} > \mu$ ensures that upwards fluctuations of the observed data, if larger than expected for a signal of strength μ, are not regarded as evidence against the signal hypothesis itself.

The limit test statistics can then be employed to compute exclusion limits based on the modified frequentist confidence level criterion [2, 8]. In this approach, we can define the $\mathrm{CL_s}$ quantity as:

$$\mathrm{CL_s}(\mu) \equiv \frac{\mathrm{CL_{s+b}}(\mu)}{\mathrm{CL_b}(\mu)} \equiv \frac{\mathcal{P}(q_\mu \geq q_\mu^{\mathrm{obs.}}|H_0)}{\mathcal{P}(q_\mu \geq q_\mu^{\mathrm{obs.}}|H_1)} \tag{6.10}$$

where the numerator and denominator are the probabilities of q_μ being larger than any observed value $q_\mu^{\mathrm{obs.}}$ under the null or alternate hypothesis, respectively. In the $\mathrm{CL_s}$ framework, a signal of strength μ is said to be excluded at a Confidence Level (CL) α if $\mathrm{CL_s} < 1 - \alpha$. Conventionally, the confidence level $\alpha = 95\%$ is used to compute exclusion limits; this is also the case in this analysis. It should be noted that by its own construction, CL_s is not a confidence level, it is just a ratio of p-values. This ensures that the CL_s value will deviate from the standard p-value the smaller the separation power of the test. The price to pay for this desirable behaviour is that the intervals obtained with the CL_s procedure will by construction over-cover, resulting in conservative limits.

Quantifying an Excess

The test statistics q_0 used to quantify an excess of events is generally referred to as *discovery test statistics* and is defined as:

$$q_0 \equiv \begin{cases} -2\ln\lambda(0) & \text{if } \hat{\mu} \geq 0 \\ 0 & \text{if } \hat{\mu} < 0 \end{cases} \tag{6.11}$$

where the enforcement of the $q_0 = 0$ for $\hat{\mu} < 0$ ensures that downward fluctuations of the observed data, if smaller than expected for the absence of signal, are not regarded as evidence against the background hypothesis itself.

This can be used to define the significance of an excess of events based on the probability that, given an observed value $q_0^{\text{obs.}}$ of the test statistics, q_0 would be larger or equal to it under the alternate hypothesis:

$$\mathrm{p} = \mathcal{P}(q_0 \geq q_0^{\text{obs.}} | H_1) \tag{6.12}$$

which corresponds to the probability of the background having an upward fluctuation at least as large as the one observed in data, i.e. the probability of the excess being a *fluke*. By expressing this p-value as a one-sided Gaussian integral, we can define the significance Z such that:

$$\mathrm{p} = \int_Z^{+\infty} \frac{1}{\sqrt{2\pi}} e^{-x^2/2} dx \tag{6.13}$$

Conventionally, two fixed levels of significance are established as thresholds in high energy physics: a significance of 3σ ($Z = 3$, $\mathrm{p} = 1.3 \times 10^{-3}$) is used to claim an *evidence*, while a significance of 5σ ($Z = 5$, $\mathrm{p} = 2.8 \times 10^{-7}$) is used to claim an *observation* (or discovery). It is worth noticing that the definition given here is that of *local* p-value, and the more general procedure for testing multiple hypotheses at the same time would require the definition of the *global* p-value [4].

6.1.4 Validation of the Statistical Model: Expected Versus Observed

The statistical method established above provides us with an objective way of quantifying the compatibility of data with specific hypotheses in a frequentist approach that allows us to interpret p-value and CL_s as probabilities. Nevertheless, this approach is not immune from bias introduced by the choices of the analyzer. One example of such bias would be the definition of the selections applied to the events entering the signal regions based on the events in the signal regions themselves. This bad practice is counteracted in CMS with the definition of the *blinding* and *unblinding* procedure.

To avoid said bias, all analyses are required to be developed in a *blinded* manner, meaning that all analysis strategies need to be defined only by using MC simulated events complemented by data that is strictly outside of the signal regions of interest. Under these conditions, it is useful to define the concept of *expected result* for assessing the analysis' sensitivity. This would a priori require the need of generating an extensive ensemble of pseudo-data, generally referred to as *toys*, which are distributed according to the pdf of the test statistics and the median of the toys is regarded as the expected test statistics with its associated CL_s and p-value. The drawbacks of this method are its computational intensity and its time consumption. These can be overcome by a property stemming from the asymptotic formulae that hold for the PLR: the existence of a single representative dataset, the Asimov dataset [3], in which the parameter estimates are set precisely to their expected values and statistical fluctuations are suppressed. The unblinding of the analysis consists then of the process of looking at the data in the signal regions and performing the maximum likelihood fit to obtain the *observed result* of the search.

Having established the concept of expected and observed results, an important check to be performed is the control of the impact of the nuisance parameters on the result of the analysis by evaluating the so-called *pulls* and *impacts*. The pull is defined as the deviation of the value of each nuisance MLE $\hat{\theta}$ with respect to its nominal value, divided by its uncertainty. The impact is defined as the shift $\Delta\mu$ (or Δr) induced on the POI when each nuisance is fixed to its MLE $\hat{\theta} \pm 1\sigma$ value, while all other nuisances are fixed to their maximum profiled likelihood estimate $\hat{\hat{\theta}}$. The pulls and impacts of the leading 25 nuisances are shown in Fig. 6.10 where the top panel reports the expected values obtained from an Asimov dataset, while the bottom panel shows the observed values after the maximum likelihood fit of the data. The largest sources of systematic uncertainties come from the imperfect knowledge of the ggFHH production cross section, the statistical fluctuations affecting the multijet background estimation, the modelling of jet and τ leptons identification and reconstruction in simulated samples, and bin-by-bin uncertainties introduced via the Barlow-Beeston approach. The expected and observed results are mostly in agreement, with the majority of the highest-ranking impacts being the same in the two cases. The only uncertainties that are significantly overconstrained are those arising from the pileup jet identification (`PUJET_id`) scale factors. This effect is understood under the light that the a priori uncertainty due to pileup jet identification is

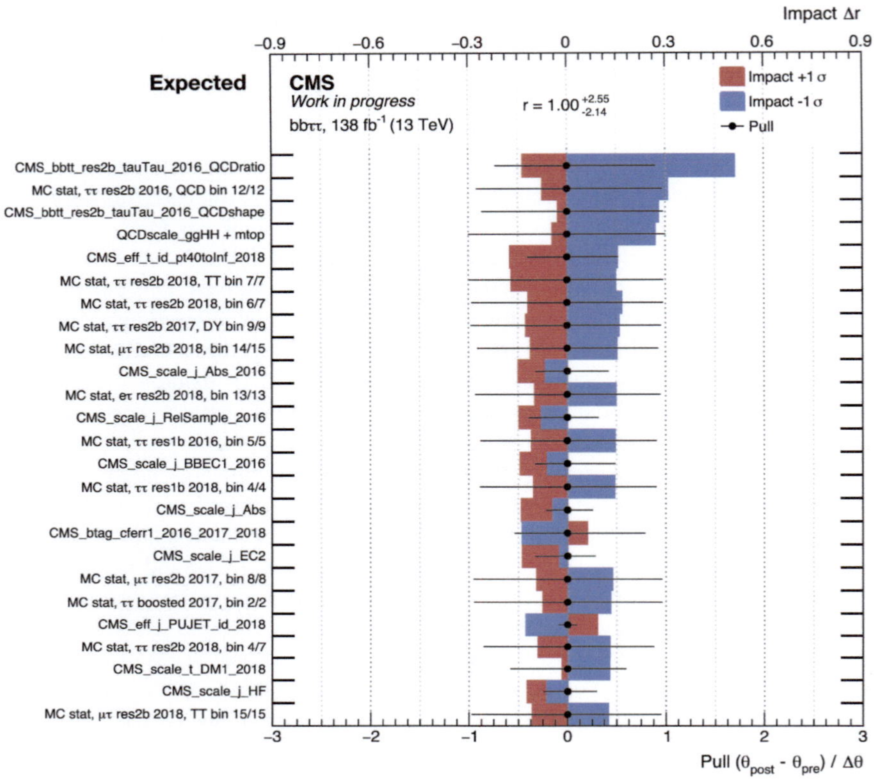

Fig. 6.10 Highest impact systematic uncertainties and relative impact on the signal strength. The black points and associated uncertainties report the deviation of the value of each nuisance θ from the maximum likelihood estimation with respect to their nominal value, divided by its uncertainty (pull). The blue and red bars report the impact of each nuisance parameter on the parameter of interest r, and they denote a positive and negative shift of the parameter by 1σ, respectively. The top panel reports the expected values obtained from an Asimov dataset, while the bottom panel shows the observed values after the maximum likelihood fit of the data. The CMS-specific name used in this figure can be grasped in Table 5.19

relatively large with shifts of up to 40% in background enriched bins, causing an average shift of $+29/-17\%$. Due to the large jet activity in the HH → bbττ signal region phase space (especially in category `res2b`) it is expected to be highly sensitive to jet-related quantities, especially to those uncertainties that are known to be derived in a very conservative approach.

Further verification of the correct modelling of the data is provided by the Goodness Of Fit (GOF) test with the saturated method [9]. This technique serves as an extension of the χ^2 test, tailored to handle non-normally distributed data, such as the number of events in the binned distributions employed in this analysis, which are distributed according to a Poissonian pdf. The test is built as a likelihood ratio,

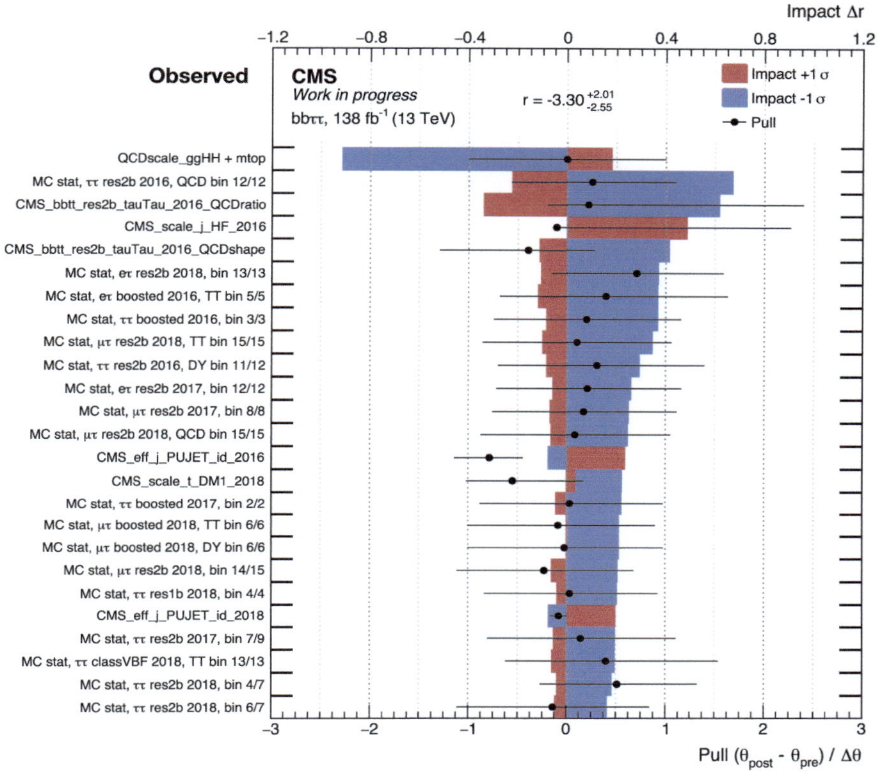

Fig. 6.10 (continued)

but the alternate hypothesis is considered to be the one that matches the data exactly. The distribution of this test statistics is obtained from pseudo-experiments generated from the modelled processes and is compared to the one computed with the observed data. The result of the GOF test is reported in Fig. 6.11, where the test is reported per each combination of ττ final state and year, as well as inclusively per each year and for the whole Run-2 dataset. In most cases, the observed value is found to lie in the bulk of the expected distribution; in the few cases in which the observed value lies in the tail of the distribution, the p-value still assumes reasonable values. This allows us to conclude that the statistical model describes the observations well.

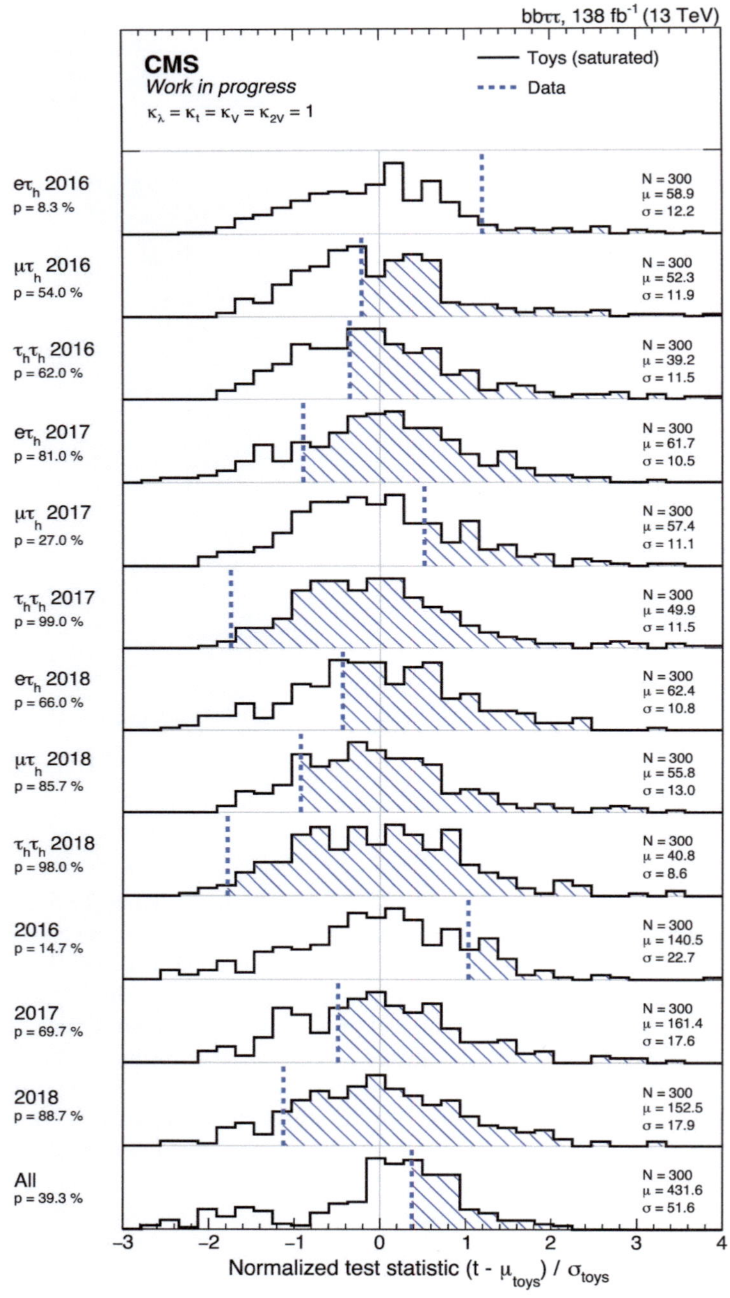

Fig. 6.11 Distribution of the normalized test statistics (χ^2 goodness-of-fit variable) computed with the saturated model (black line) and observed value (blue dashed line) from the fit to the data, reported per each combination of ττ final state and year, and inclusively per each year and for the whole Run-2 dataset. The dashed area represents the area corresponding to the calculated p-value

6.2 Results

This section presents the results obtained from the exploration of the HH production in the $bb\tau\tau$ decay channel using the statistical method presented in Sect. 6.1. The data is analyzed to set limits on the production cross section of Higgs boson pairs as predicted by the SM as well as from BSM scenarios. After the description of the dataset analyzed, this section moves to the presentation of the independent SM and EFT interpretation of the data.

6.2.1 Dataset Analyzed

The results presented in this section are obtained by analysing the whole dataset collected during the Run-2 data-taking with the CMS experiment at the CERN LHC. Systematically, a centralized procedure is executed to certify the quality of the data, ensuring that only datasets recorded during periods when all detector subsystems were operational and correctly functioning are subject to analysis, the so-called *golden certification*. The total integrated luminosity of the certified data corresponds to 138 fb^{-1} broken down as follows: 36.3 fb^{-1} in 2016, 41.5 fb^{-1} in 2017, and 59.7 fb^{-1} in 2018. All data events are divided into primary datasets according to the specific subsets of HLT paths that were used for their recording. Each of these datasets covers a separate data-taking period (Era) within each data-taking year, denoted by a letter from B through H in 2016, B through F in 2017, and A through D in 2018; the separation into Eras corresponds to notable changes in data-tacking conditions (e.g. LHC filling scheme, CMS configurations, etc.). The complete list of datasets analyzed for the HH $\rightarrow$ $bb\tau\tau$ search, organized per $\tau\tau$ pair final state and year of data-taking, is reported in Table 6.1 where the CMS-specific name of each dataset is stated alongside the corresponding run ranges and integrated luminosity.

6.2.2 Standard Model Interpretation

The set of results presented in this section is derived to test the SM prediction for the HH $\rightarrow$ $bb\tau\tau$ cross section times branching fraction value. As discussed in Sect. 1.2.1, the two main production mechanisms of Higgs boson pairs in the SM are the ggF and the VBF mechanisms. The former highly dominates, being roughly 18 times larger than the latter. Nevertheless, through the dedicated strategy for VBF events selection detailed in Sect. 5.1, this analysis is able to obtain precise results that target the exclusive VBF production alone. On the other hand, given its much larger contribution, the ggF cross section is targeted only inclusively. Therefore, in the following, the results obtained in the $bb\tau\tau$ search will be split into inclusive

Table 6.1 List of datasets analyzed arranged per $\tau\tau$ final state and year. The second column reports the internal CMS dataset name, for which the HLT triggers set, the data-taking period, and the data-reconstruction campaign are specified. The third and fourth columns report the run intervals associated with a specific dataset and the corresponding integrated luminosity

	Year	Dataset name	Run range	Int. lumi [fb^{-1}]
$\tau_\mu\tau_h$ final state	2016	/SingleMuon/Run2016B-17Jul2018_ver2-v1/MINIAOD	273150–275376	5.829
		/SingleMuon/Run2016C-17Jul2018-v1/MINIAOD	275656–276283	2.618
		/SingleMuon/Run2016D-17Jul2018-v1/MINIAOD	276315–276811	4.286
		/SingleMuon/Run2016E-17Jul2018-v1/MINIAOD	276831–277420	4.066
		/SingleMuon/Run2016F-17Jul2018-v1/MINIAOD	277932–278808	3.135
		/SingleMuon/Run2016G-17Jul2018-v1/MINIAOD	278820–280385	7.653
		/SingleMuon/Run2016H-17Jul2018-v1/MINIAOD	281613–284044	8.740
	2017	/SingleMuon/Run2017B-31Mar2018-v1/MINIAOD	297047–299329	4.794
		/SingleMuon/Run2017C-31Mar2018-v1/MINIAOD	299368–302029	9.633
		/SingleMuon/Run2017D-31Mar2018-v1/MINIAOD	302018–302663	4.248
		/SingleMuon/Run2017E-31Mar2018-v1/MINIAOD	303824–304797	9.315
		/SingleMuon/Run2017F-31Mar2018-v1/MINIAOD	305040–306462	13.54
	2018	/SingleMuon/Run2018A-17Sep2018-v2/MINIAOD	315257–316995	14.03
		/SingleMuon/Run2018B-17Sep2018-v1/MINIAOD	317080–319310	7.066
		/SingleMuon/Run2018C-17Sep2018-v1/MINIAOD	319337–320065	6.899
		/SingleMuon/Run2018D-22Jan2019-v2/MINIAOD	320413–325175	31.75
$\tau_e\tau_h$ final state	2016	/SingleElectron/Run2016B-17Jul2018_ver2-v1/MINIAOD	273150–275376	5.829
		/SingleElectron/Run2016C-17Jul2018-v1/MINIAOD	275656–276283	2.618
		/SingleElectron/Run2016D-17Jul2018-v1/MINIAOD	276315–276811	4.286
		/SingleElectron/Run2016E-17Jul2018-v1/MINIAOD	276831–277420	4.066
		/SingleElectron/Run2016F-17Jul2018-v1/MINIAOD	277932–278808	3.135
		/SingleElectron/Run2016G-17Jul2018-v1/MINIAOD	278820–280385	7.653
		/SingleElectron/Run2016H-17Jul2018-v1/MINIAOD	281613–284044	8.740

(continued)

Table 6.1 (continued)

	Year	Dataset name	Run range	Int. lumi [fb^{-1}]
	2017	/SingleElectron/Run2017B-31Mar2018-v1/MINIAOD	297047–299329	4.794
		/SingleElectron/Run2017C-31Mar2018-v1/MINIAOD	299368–302029	9.633
		/SingleElectron/Run2017D-31Mar2018-v1/MINIAOD	302018–302663	4.248
		/SingleElectron/Run2017E-31Mar2018-v1/MINIAOD	303818–304797	9.315
		/SingleElectron/Run2017F-31Mar2018-v1/MINIAOD	305040–306460	13.54
	2018	/EGamma/Run2018A-17Sep2018-v2/MINIAOD	315257–316995	14.03
		/EGamma/Run2018B-17Sep2018-v1/MINIAOD	317080–319310	7.066
		/EGamma/Run2018C-17Sep2018-v1/MINIAOD	319337–320065	6.899
		/EGamma/Run2018D-22Jan2019-v2/MINIAOD	320413–325175	31.75
$\tau_h \tau_h$ final state	2016	/Tau/Run2016B-17Jul2018_ver2-v1/MINIAOD	273150–275376	5.829
		/Tau/Run2016C-17Jul2018-v1/MINIAOD	275656–276283	2.618
		/Tau/Run2016D-17Jul2018-v1/MINIAOD	276315–276811	4.286
		/Tau/Run2016E-17Jul2018-v1/MINIAOD	276831–277420	4.066
		/Tau/Run2016F-17Jul2018-v1/MINIAOD	277932–278808	3.135
		/Tau/Run2016G-17Jul2018-v1/MINIAOD	278820–280385	7.653
		/Tau/Run2016H-17Jul2018-v1/MINIAOD	281613–284044	8.740
	2017	/Tau/Run2017B-31Mar2018-v1/MINIAOD	297047–299329	4.794
		/Tau/Run2017C-31Mar2018-v1/MINIAOD	299368–302029	9.633
		/Tau/Run2017D-31Mar2018-v1/MINIAOD	302031–302663	4.248
		/Tau/Run2017E-31Mar2018-v1/MINIAOD	303824–304797	9.315
		/Tau/Run2017F-31Mar2018-v1/MINIAOD	305040–306462	13.54
	2018	/Tau/Run2018A-17Sep2018-v1/MINIAOD	315257–316995	14.03
		/Tau/Run2018B-17Sep2018-v1/MINIAOD	317080–319310	7.066
		/Tau/Run2018C-17Sep2018-v1/MINIAOD	319337–320065	6.899
		/Tau/Run2018D-PromptReco-v2/MINIAOD	320413–325175	31.75

and exclusive VBF results. In both cases, all BSM couplings are set to zero, i.e. $c_2 = c_g = c_{2g} = 0$.

Inclusive HH Production

The results obtained on the inclusive production are largely dominated by the contribution from gluon fusion due to its cross section being considerably larger than that of the other mechanisms. For this reason, the study of inclusive production is not only a good testbench for the exploration of the κ_λ modifier but also of the κ_t coupling, whose study will only suffer from minor bias due to the residual events originating from VBF production. Both model-independent and model-dependent results are presented in this section. The former is a test of the κ_λ and κ_t coupling modifiers assuming all other couplings are SM-like, i.e. $\kappa_V = \kappa_{2V} = 1$, while the latter is a dedicated test of the SM prediction of the HH production cross section when all couplings are SM-like, i.e. $\kappa_\lambda = \kappa_t = \kappa_V = \kappa_{2V} = 1$.

Limits on the HH production cross section times $HH \to bb\tau\tau$ branching fraction as a function of the κ_λ coupling modifier are shown in Fig. 6.12 and compared to the theoretical prediction, while all other couplings are assumed to be as predicted by the SM. These limits are computed following the asymptotic CL_s prescription and the observed constraints inferred from the scan are $-1.7 < \kappa_\lambda < 8.7$, while the expected constraints are $-2.9 < \kappa_\lambda < 9.8$. The exclusion limit reveals a distinctive pattern arising from the interplay between the two dominant diagrams governing the process. Notably, in the proximity of the critical interference point with $\kappa_\lambda = 2.45$, even minor adjustments of the coupling yield significant repercussions on the signal kinematics, thereby substantially impacting the overall sensitivity of the analysis. However, as the absolute value of κ_λ increases, the triangle diagram becomes the leading contribution, and the excluded cross sections for positive and negative κ_λ converge asymptotically.

The model-independent κ_λ-scan in Fig. 6.12 can be interpreted in a fully SM-dependent result by enforcing the trilinear H self-coupling modifier to be $\kappa_\lambda = 1$. Furthermore, by including the theoretical uncertainties, a 95% CL upper limit can be set on the signal strength modifier, i.e. on the $\sigma_{ggF+VBF}(pp \to HH)/\sigma^{SM}_{ggF+VBF}$ ratio. The observed 95% CL limit on this POI is 3.3 times the SM prediction, amounting to 7.5 fb and 102 fb when including or not the decay branching fraction respectively. The expected 95% CL limit on this POI is 5.2 times the SM prediction, amounting to 11.6 fb and 159 fb when including or not the decay branching fraction, respectively. A summary of the SM-dependent upper limits is reported in Table 6.2.

The separate contributions from the $\tau\tau$ final states and categories are detailed in Fig. 6.13. For clarity of visualization, the eight categories used in this analysis are simplified to only three macro-categories, the `boosted` category is left unchanged, the `res2b` and `res1b` categories are merged into the *resolved* macro-category, and the five VBF multi-categories are merged into one. As can be appreciated from the figure, the $\tau_h\tau_h$ final state is the most sensitive for all the values of κ_λ, while the resolved categories largely outperform all other categories. The boosted final state

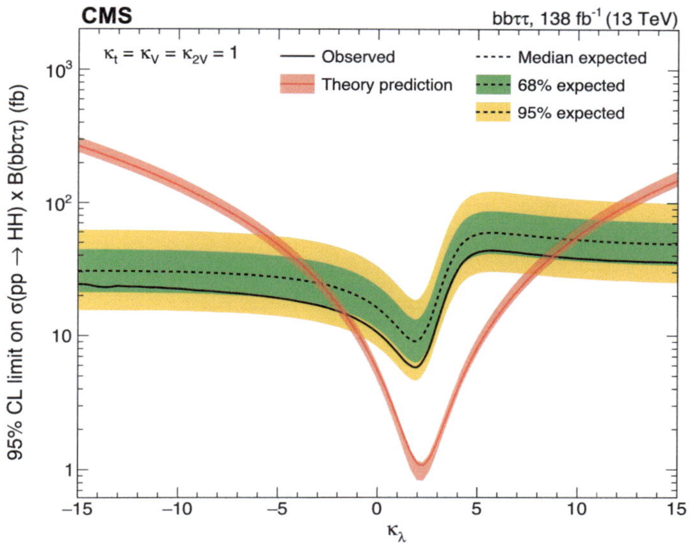

Fig. 6.12 Observed and expected upper limits at 95% CL on the inclusive HH production cross section times the HH $\rightarrow$ bb$\tau\tau$ branching ratio as a function of κ_λ when $\kappa_t = \kappa_V = \kappa_{2V} = 1$. The red solid line shows the theoretical prediction for the HH production cross section and its uncertainty (red shaded band)

Table 6.2 Expected and observed upper limits at 95% CL for the standard model point ($\kappa_\lambda = \kappa_t = \kappa_V = \kappa_{2V} = 1$), where $\sigma^{SM}_{ggF+VBF}$ is the standard model prediction of the HH cross sections

		2016	2017	2018	Combined
Expected limits	$\sigma_{ggF+VBF}(pp \rightarrow HH)/\sigma^{SM}_{ggF+VBF}$	10.6	11.7	8.2	5.2
	$\sigma_{ggF+VBF}(pp \rightarrow HH)$ [fb]	324	356	249	159
	$\sigma_{ggF+VBF}(pp \rightarrow HH \rightarrow bb\tau\tau)$ [fb]	23.6	26.0	18.2	11.6
Observed limits	$\sigma_{ggF+VBF}(pp \rightarrow HH)/\sigma^{SM}_{ggF+VBF}$	8.9	9.5	5.5	3.3
	$\sigma_{ggF+VBF}(pp \rightarrow HH)$ [fb]	272	291	169	102
	$\sigma_{ggF+VBF}(pp \rightarrow HH \rightarrow bb\tau\tau)$ [fb]	19.6	21.2	12.4	7.5

becomes very important for limit setting at the maximum interference point due to the increase of high m_{HH} events that populate this category.

Assuming that an HH signal exists with the properties predicted by the SM, constraints on the κ_λ and κ_t coupling modifiers can also be computed based on a direct scan of the likelihood. Such scans are reported in Fig. 6.14, where for computational reasons the usual choice of scanning the negative log-likelihood is taken. The observed confidence interval on κ_λ corresponds to [−1.77, +8.73] at 95% CL

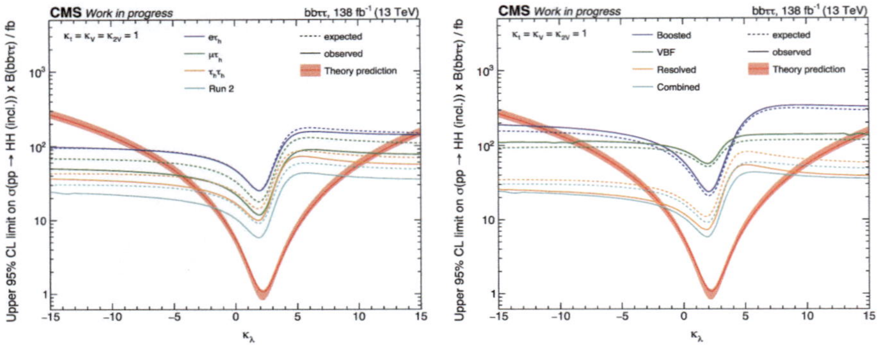

Fig. 6.13 Observed and expected upper limits at 95% CL on the inclusive HH cross section as a function of κ_λ when $\kappa_t = \kappa_V = \kappa_{2V} = 1$; the contribution from separate channels (left) and categories (right) is highlighted. The red solid line shows the theoretical prediction for the HH production cross section and its uncertainty (red shaded band)

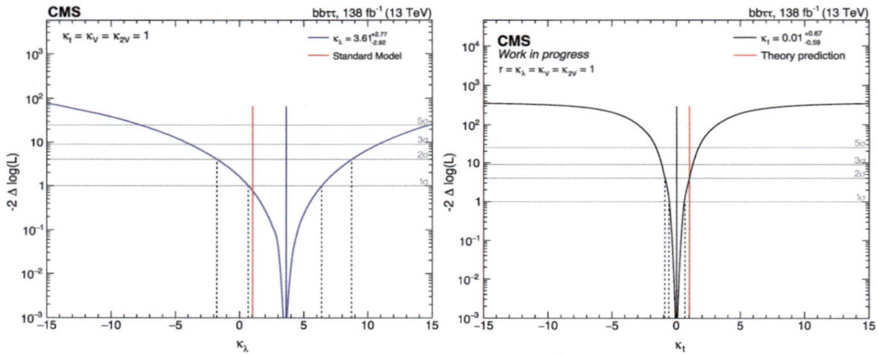

Fig. 6.14 Observed negative log-likelihood scan as a function of κ_λ (left) and κ_t (right) for the full Run-2 combination. The dashed lines show the intersection with threshold values one and four, corresponding to 68 and 95% confidence intervals, respectively

for a best fit value $\kappa_\lambda = 3.61$, while the observed confidence interval on κ_t corresponds to $[-0.90, +1.01]$ at 95% CL for a best fit value $\kappa_t = 0.01$. While the results obtained for the trilinear H self-coupling modifier are in good agreement with the ones obtained based on the CL_s approach, the best fit value of κ_t seems to be in contrast with the SM itself. Nevertheless, this surprising value can be understood by noticing that in Fig. 6.12, there is a consistent under-fluctuation of the observed data compared to the expected background. Given this under-fluctuation, the statistical model will pull the coupling of the t quark to the H boson toward zero, as in this way not only it will reduce the yield of the HH contribution but more importantly it will suppress the $t\bar{t}H$ contribution which, having a cross section of roughly 0.5 pb, will result in a better morphing of the MC toward the data.

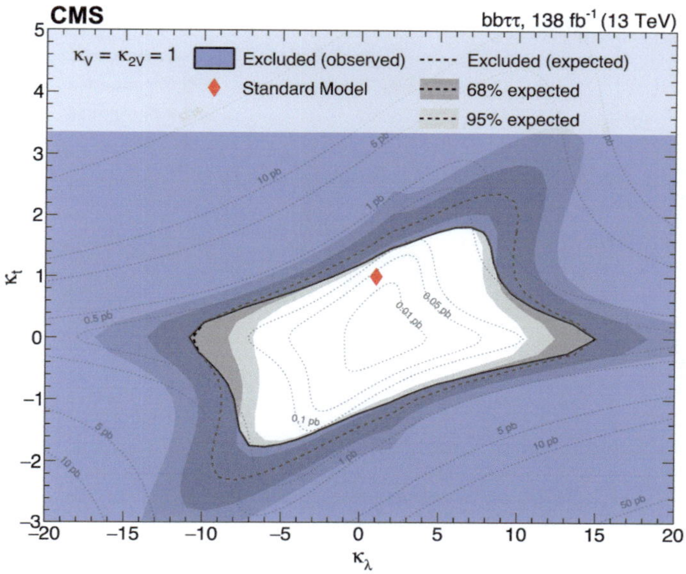

Fig. 6.15 Observed and expected two-dimensional exclusion regions as a function of the κ_λ and κ_t couplings for the full 2016–2018 combination, when both κ_{2V} and κ_V are fixed to unity (left). Observed and expected two-dimensional exclusion regions as a function of κ_{2V} and κ_V, when both κ_λ and κ_t are set to unity (right). The expected uncertainties on the exclusion boundaries are inferred from the uncertainty bands of the limit calculation and are denoted by dark and light-grey areas. The blue area marks parameter combinations that are observed to be excluded. For visual guidance, theoretical cross section values are illustrated by thin, labelled contour lines; a red diamond denotes the SM prediction

Finally, the exclusion limits are used to set two-dimensional exclusion regions as a function of κ_λ and κ_t as shown in Fig. 6.15. A point in the two-dimensional parameter space is excluded when the upper limit on the overall rate of HH production at 95% CL is measured to be below the theoretical cross section, evaluated at the corresponding parameter values. The almost symmetric exclusion region is understood by noticing that the ggF cross section in Eq. 1.54 is itself symmetric under the transformation $(\kappa_\lambda, \kappa_t) \leftrightarrow (-\kappa_\lambda, -\kappa_t)$; the residual asymmetry is to be searched in Eq. 1.55 not being symmetric under the same transformation.

Vector Boson Fusion HH Production

As discussed in Sect. 1.2.1, in the SM the production of Higgs boson pairs via vector boson fusion at LO is totally determined by three Feynman diagrams. Moreover, as it can be visually appreciated in Fig. 1.12, the VBF is particularly sensitive to the κ_{2V} coupling modifier for which these results serve a bench testing purpose. Both model-independent and model-dependent results are presented in this section. The

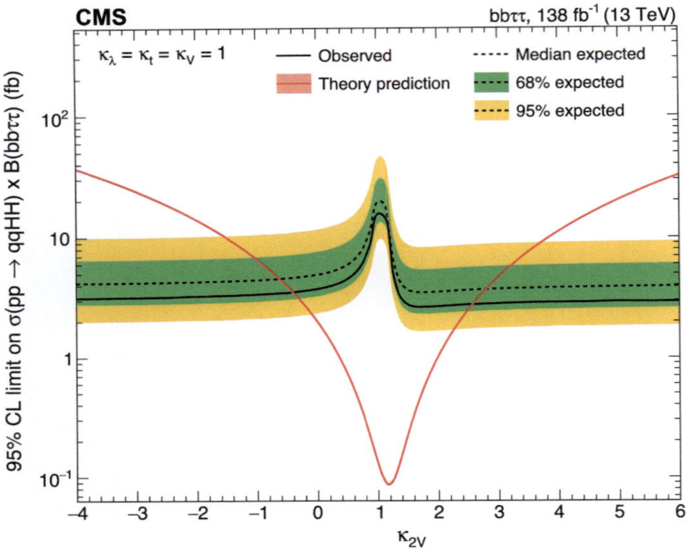

Fig. 6.16 Observed and expected upper limits at 95% CL on the vector boson fusion HH cross section times the HH $\rightarrow$ bb$\tau\tau$ branching ratio as a function of κ_{2V} when $\kappa_\lambda = \kappa_t = \kappa_V = 1$. The red solid line shows the theoretical prediction for the HH production cross section and its uncertainty (red shaded band)

former is a test of the κ_{2V} and κ_V coupling modifiers assuming all other couplings are SM-like, i.e. $\kappa_\lambda = \kappa_t = 1$, while the latter is a dedicated test of the SM prediction of the HH production cross section when all couplings are SM-like, i.e. $\kappa_\lambda = \kappa_t = \kappa_V = \kappa_{2V} = 1$.

Limits on the VBF HH production cross section times HH $\rightarrow$ bb$\tau\tau$ branching fraction as a function of the κ_{2V} coupling modifier are shown in Fig. 6.16 and compared to the theoretical prediction, while all other couplings are assumed to be as predicted by the SM. These limits are computed following the asymptotic CL_s prescription and the observed constraints inferred from the scan are $0.4 < \kappa_{2V} < 2.6$, while the expected constraints are $-0.6 < \kappa_{2V} < 2.8$. Also in this case, the exclusion limit reveals the distinctive pattern arising from the critical interference point with $\kappa_{2V} = 1$ where minimal modifications of the coupling yield significant changes in the signal kinematics and the analysis' sensitivity. However, as the absolute value of κ_{2V} increases, only one diagram becomes the leading contribution, and the excluded cross section for positive and negative κ_{2V} converges asymptotically.

As was done for the inclusive exclusion limits, the model-independent scan can be interpreted in a fully SM-dependent result requiring $\kappa_{2V} = 1$. The observed 95% CL limit on the VBF signal strength modifier is 124 times the SM prediction, amounting to 15.5 fb and 212 fb when including or not the decay branching fraction, respectively. The expected 95% CL limit on this POI is 5.2 times the SM prediction, amounting to 19.6 fb and 266 fb when including or not the decay branching fraction, respectively.

Table 6.3 Expected and observed upper limits at 95% CL for the SM point ($\kappa_\lambda = \kappa_V = \kappa_{2V} = 1$), where $\sigma_{VBF}^{SM} = 1.726$ fb is the standard model prediction of the VBF HH cross sections

		2016	2017	2018	Combined
Expected limits	$\sigma_{VBF}(pp \to qqHH)/\sigma_{VBF}^{SM}$	357	392	226	154
	$\sigma_{VBF}(pp \to qqHH)$ [fb]	616	676	391	266
	$\sigma_{VBF}(pp \to qqHH \to bb\tau\tau)$ [fb]	45.0	49.3	28.5	19.4
Observed limits	$\sigma_{VBF}(pp \to qqHH)/\sigma_{VBF}^{SM}$	283	280	241	124
	$\sigma_{VBF}(pp \to qqHH)$ [fb]	487	485	414	212
	$\sigma_{VBF}(pp \to qqHH \to bb\tau\tau)$ [fb]	35.6	35.3	30.2	15.5

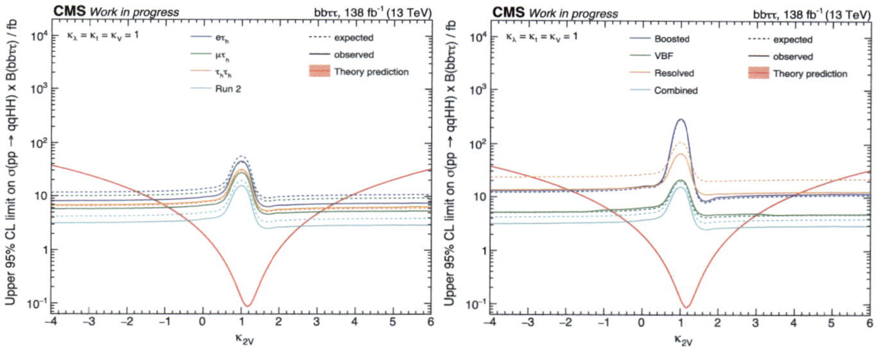

Fig. 6.17 Observed and expected upper limits at 95% CL on the VBF HH cross section as a function of κ_{2V} when $\kappa_\lambda = \kappa_t = \kappa_V = 1$; the contribution from separate channels (left) and categories (right) is highlighted. The red solid line shows the theoretical prediction for the HH production cross section and its uncertainty (red shaded band)

A summary of the SM-dependent upper limits for the VBF production mechanism is reported in Table 6.3. These results currently represent the world's most stringent limits on the VBF production mechanism.

The separate contributions from the three final states and of the categories are detailed in Fig. 6.17, where the same categorization strategy is used as for Fig. 6.13. As can be appreciated, the $\tau_h\tau_h$ final state is the most sensitive closely followed by the $\tau_\mu\tau_h$ channel, while as expected the VBF categories largely outperform all others.

In the same manner as already described for the inclusive results, under the assumption that an HH SM-like signal exists, constraints on the κ_V and κ_{2V} coupling modifiers can be computed from a direct scan of the likelihood, as reported in Fig. 6.18. The observed confidence interval on κ_V corresponds to $[-1.46, +1.55]$ at 95% CL for a best fit value $\kappa_V = 0.86$, while the observed confidence interval on κ_{2V}

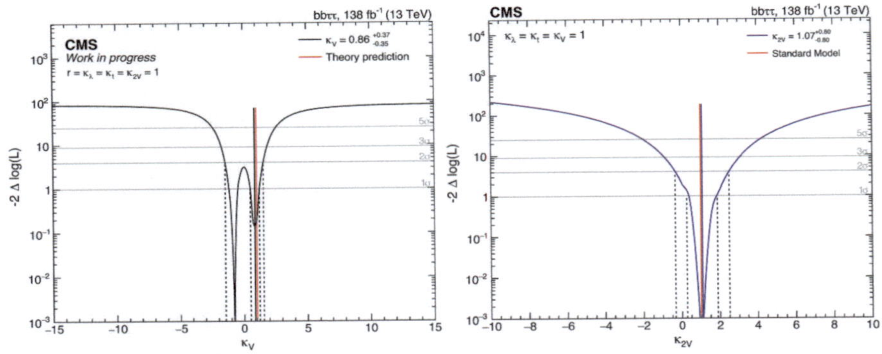

Fig. 6.18 Observed negative log-likelihood scan as a function of κ_V (left) and κ_{2V} (right) for the full Run 2 combination. The dashed lines show the intersection with threshold values one and four, corresponding to 68 and 95% confidence intervals, respectively

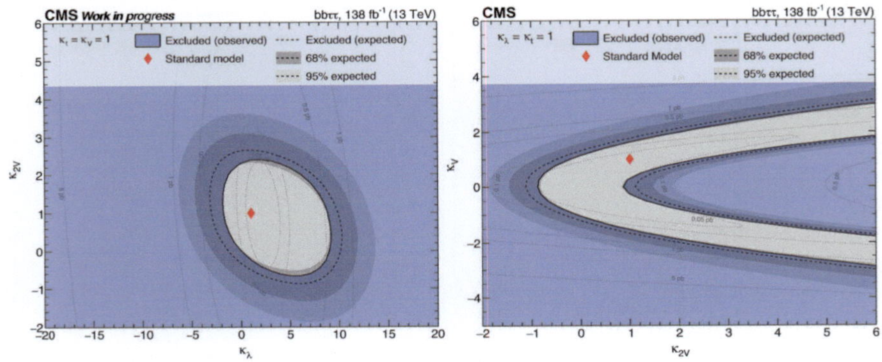

Fig. 6.19 Observed and expected two-dimensional exclusion regions as a function of the κ_λ and κ_{2V} couplings for the full 2016–2018 combination, when both κ_t and κ_V are fixed to unity (left), and as a function of κ_{2V} and κ_V, when both κ_λ and κ_t are set to unity (right). The expected uncertainties on the exclusion boundaries are inferred from the uncertainty bands of the limit calculation and are denoted by dark and light-grey areas. The blue area marks parameter combinations that are observed to be excluded. For visual guidance, theoretical cross section values are illustrated by thin, labelled contour lines; a red diamond denotes the SM prediction

corresponds to $[-0.31, +2.46]$ at 95% CL for a best fit value $\kappa_{2V} = 1.08$, which is in good agreement with the ones obtained based on the CL_s approach.

Finally, the exclusion limits are used to set two-dimensional exclusion regions as a function of κ_λ and κ_{2V}, as well as κ_V and κ_{2V}, as shown in Fig. 6.19. A point in the two-dimensional parameter space is excluded when the upper limit on the overall rate of HH production at 95% CL is measured to be below the theoretical cross section, evaluated at the corresponding parameter values.

6.2.3 EFT Interpretation

In addition to the primary results presented in the previous section, a supplementary set of findings is presented in the form of upper limits on the cross section for distinct BSM signal shape benchmarks. These benchmarks represent specific points within the five-dimensional effective Lagrangian parametrization of BSM physics in Eq. 1.61, which predict unique kinematic features for the HH system.

A total of 20 independent BSM benchmarks is explored in this analysis and defined as follows: benchmarks 1 through 12 have been designed in Reference [10] performing a clustering based on binned likelihood ratio test statistics aimed at regrouping different models that showcase a certain degree of similarity in the kinematics they predict; benchmark 8a has been defined in Reference [11] as being an outlier of the cluster of benchmark 8; benchmarks 1b through 7b have been designed in Reference [12] by visualizing how distinct classes of shapes are related to the chosen values of the underlying coupling from the five-dimensional parameter space through an unsupervised learning technique based on an autoencoder neural network. In Fig. 6.20a are reported the 95% CL upper limits on $\sigma(gg \to HH) \times \mathcal{B}(HH \to bb\tau\tau)$ for the first 13 BSM benchmarks, while in Fig. 6.20a the same quantity is show for the remaining seven BSM benchmarks. The SM observed and expected upper exclusion limit is also reported for reference in both cases. The numerical differences of the excluded cross sections among the different scenarios directly correlate with the m_{HH} distribution of the respective shape benchmarks. Variations in the shape and behaviour of the m_{HH} distribution for different benchmarks lead to diverse final state kinematics and hence diverse sensitivity levels; thus affecting the magnitude of the upper limits on the cross section. As can be expected, a better sensitivity is observed for BSM benchmarks with an enhanced cross section at high values of m_{HH} (cf. Figs. 1.14 and 1.15).

6.2.4 Comparison of the Results

This section compares the results obtained within this Thesis to other present and past findings in HH searches. First, the comparison is made with the previous CMS $bb\tau\tau$ search performed in 2016 [13] to highlight the key enhancements incorporated in the analysis approach and their influence on the sensitivity. Then, the results are put in the more comprehensive scope of HH searches performed at the LHC Run-2; the results of the analysis presented here are compared to analyses exploiting both direct and indirect measurements of the HH production.

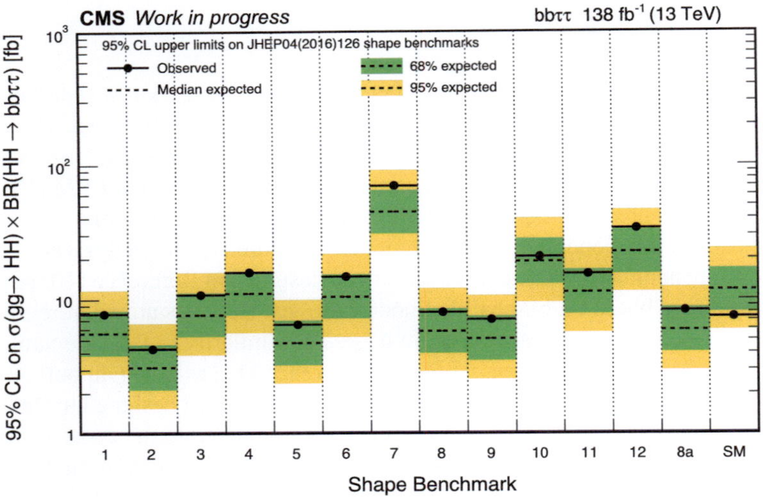

(a) Benchmark scenarios from Reference [10] and [11]

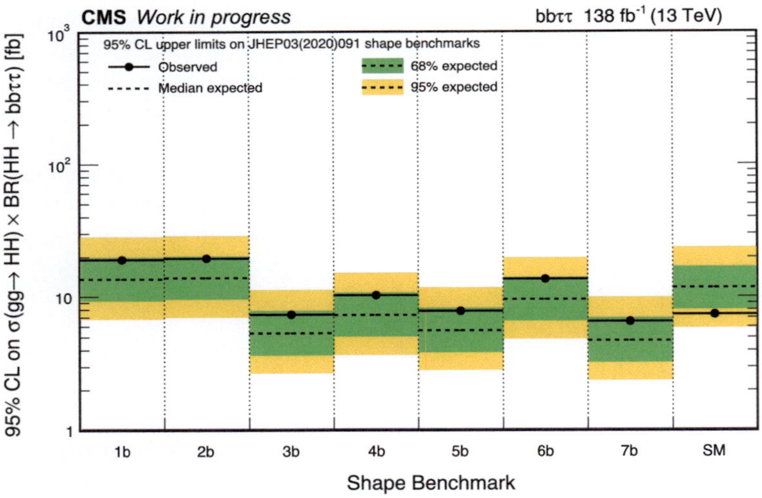

(b) Benchmark scenarios from Reference [12]

Fig. 6.20 Observed and expected upper limits at 95% CL on the HH cross section times the $bb\tau\tau$ branching fraction as a function of the twelve benchmarks defined in Reference [10] plus the benchmark 8a defined in Reference [11], Fig. 6.20a, and as a function of the seven benchmarks defined in Reference [12], Fig. 6.20b. The SM 95% CL upper limit is also reported for comparison. The differences in the upper exclusion limits are due to the dependence of the signal kinematics on the BSM scenario

Earlier bb$\tau\tau$ Results

The analysis strategy illustrated above is largely comparable to the one employed in the previous bb$\tau\tau$ search [14]; nevertheless, important changes have been implemented for the analysis presented here. These improvements are all summarised in the following to highlight the driving factors of the sensitivity enhancement.

The revised trigger strategy presented in Sect. 5.2.1 represents the first major advancement. While the original analysis employed only single-ℓ ($\ell = \mu, e$) and double-τ_h triggers, the present search extends the trigger coverage by introducing cross-triggers and VBF-oriented paths. Their implementation ensures an increase of the trigger efficiency from ~30% in the original analysis to ~50% in the current one. Moreover, the VBF trigger allows the direct investigation of the VBF production mechanism.

The second important amelioration is the enhanced identification of final state particles. Several new algorithms are exploited in this analysis compared to the 2016 search: the τ_h identification now employs the highly performant DeepTau algorithm, and the tagging of b jets is done with the improved DeepJet algorithm. Moreover, the association of the b jets from the H candidate is now optimally performed with the HHBTag approach, which ensures a selection purity enhancement of ~10% on all benchmark models compared to the DeepCSV method used in the previous publication.

The third important enhancement of the analysis is the introduction of a dedicated VBF production search. To achieve this, a new dedicated VBF trigger path was developed in 2017 and included in the analysis. Moreover, the data is optimally categorised by the VBF multi-classifier network detailed in Sect. 5.3.3. These two key upgrades ensure the possibility of setting the most stringent limits to date on the production cross section of HH via the VBF mechanism. Furthermore, among the five categories produced with this approach, one is enriched in ggF events, thus giving a further handle in the limit setting; at the same time, the background enriched categories serve for the constraining of nuisance parameters.

The most important advancement to enhance the analysis sensitivity is the development of the new DNN discriminator for signal extraction, detailed in Sect. 5.3.4. The new multivariate method is considerably more sophisticated than the BDT used in the previous analysis; it uses a number of input variables considerably larger and exploits them optimally. Comparison studies between the new method and the older one show that exploiting information through the DNN ensures an improvement of the 95% CL upper limits of ~20% in most categories.

One final important refinement is represented by the optimization strategy employed for the final discriminating observable distribution. No procedure of this kind was ever employed in the original analysis, in which the discriminators had bins of uniform dimension. The optimization procedure employed in this analysis, based on a Bayesian optimization technique, ensures the lowering of the 95% CL upper limits by an additional ~25%.

Figure 6.21 gives a visual comparison of the results from the analysis published in References [13, 15] with those obtained within this Thesis and published in Refer-

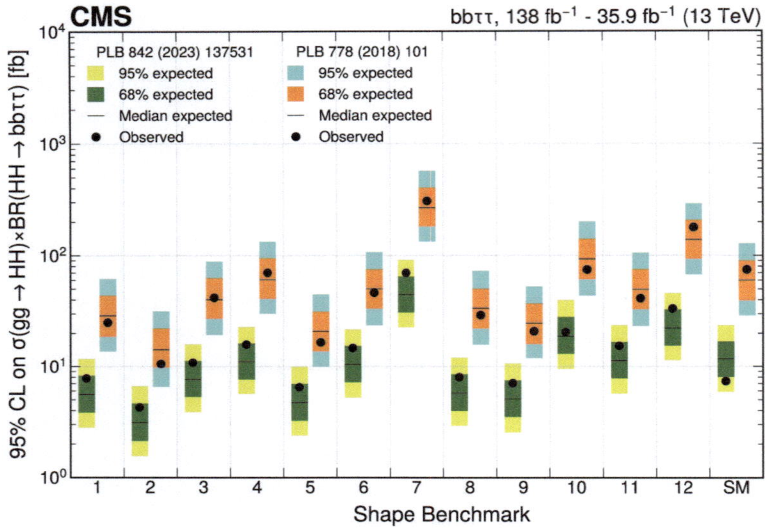

Fig. 6.21 Comparison of the 95% CL upper limits obtained in the bbττ analyses using the full Run-2 dataset (left, green and yellow) and the partial dataset of 35.9 fb^{-1} (right, orange and cyan) [15]. The results are reported for the 12 benchmarks defined in Reference [10] and for the SM

ence [16]. The 95% CL upper limit is compared for diverse signal model hypotheses: for the SM and the 12 benchmarks defined in Reference [10]. As the HH → bbττ search is intrinsically statistically limited, the evolution of the upper limits is expected to scale with inverse proportionality to the square root of the integrated luminosity $1/\sqrt{L}$; therefore, an improvement of the 95% CL upper limit by a factor two would be expected by the use of the full Run-2 dataset. Nevertheless, as can be appreciated from the Figure, the introduction of the new techniques summarised above ensures a five-fold enhancement of the result. A similar improvement is also appreciable across all tested EFT benchmarks. This proves the far-reaching performance of the analysis techniques developed during the Run-2 data-taking and beyond.

Other Run-2 HH Searches

The study of the λ_{HHH} coupling is currently one of the most active fields at colliders. To probe it, both the CMS and ATLAS Collaborations have performed extensive analyses to search for the direct production of Higgs boson pairs in diverse final states. Both experiments have analyzed the full Run-2 dataset collected at a centre-of-mass of $\sqrt{s} = 13$ TeV, amounting to 139 fb^{-1} and 138 fb^{-1} for ATLAS and CMS, respectively. The studies at both experiments mainly focus on the three most sensitive direct channels: HH → bbbb, HH → bbττ, and HH → bbγγ; nevertheless, multiple investigations of additional final states have been performed within both Collaborations.

The bbbb channel has become particularly competitive during Run-2 owing to the large improvement in jet taggers' performance at CMS and ATLAS. This final state benefits from its large branching ratio $\mathcal{B}(\text{bbbb}) = 34\%$, but was originally considered as an almost unfeasible path due to the large background contamination. Nonetheless, the great improvement in software tools that was seen over Run-2 brought this final state to yield results comparable to the other cleaner channels both at CMS [17] and ATLAS [18]. Both Collaborations have also used this channel to set ever-better limits on the VBF production mechanism and to determine 95% CL intervals on the value of the κ_{2V} modifier.

The bb$\tau\tau$ final state and its exploration in CMS were extensively discussed in this Thesis; its main strength is the balance between a sizeable branching ratio $\mathcal{B}(\text{bb}\tau\tau) = 7.3\%$ and the selection purity of the $\tau\tau$ pair. The ATLAS Collaboration has also explored this final state yielding very competitive results on the inclusive HH production cross section, as well as 95% CL intervals on the value of the κ_λ modifier [19]. In contrast to the CMS results, ATLAS did not employ this channel to set limits on κ_{2V}. The CMS bb$\tau\tau$ analysis currently sets the world's most stringent limit on the VBF HH production cross section.

The bb$\gamma\gamma$ channel has historically been the best final state due to its clean photon signature while suffering from a small branching ratio $\mathcal{B}(\text{bb}\gamma\gamma) = 0.3\%$. Nevertheless, the technical improvements of the other analysis now make this the second/third most sensitive channel. Both Collaborations employ the bb$\gamma\gamma$ search to set limits on the inclusive cross section of HH production and set 95% CL intervals on κ_λ [20, 21]. In addition, within CMS, this channel is also employed to directly tackle the VBF production mechanism and set limits on κ_{2V} [21].

Further final states explored in the two Collaborations are: bbWW*, WW*WW*, $\tau^+\tau^-\tau^+\tau^-$, bbZZ^*, and WW$^*\gamma\gamma$. These channels, while suffering from very limited statistics due to their small branching fractions, are an important additional handle to set limits on the production cross section of HH.

Tables 6.4 and 6.5 report the most recent results from the ATLAS and CMS Collaborations. In the former, the 95% CL upper limits on the inclusive HH production cross section and the 95% CL interval on κ_λ are reported; in the latter, the 95% CL upper limits on the VBF HH production cross section and the 95% CL interval on κ_{2V} are reported. The observed and expected results are shown in all cases, alongside the main Reference they were taken from. These results only include those final states used in the respective CMS and ATLAS combinations [22, 23] reported in the last row of each table.

The 95% CL upper limits on the inclusive HH production cross section set by the CMS Collaboration are also reported in a visual display in Fig. 6.22. From this, it can be appreciated how the bb$\tau\tau$ analysis is the one with the second lowest expected upper limit and the lowest observed limit.

It should be noted that these results constitute an outstanding achievement by both Collaborations. Only a few years ago, the direct searches for Higgs boson pair production were considered to be feasible only with the statistical power foreseen for the High-Luminosity LHC (HL-LHC) dataset. Nevertheless, the technical improvements attained throughout Run-2 now give hope for outstanding results already in Run-3, the most exciting being the possible exclusion of $\kappa_\lambda = 0$. Nonetheless, a possible first observation would still require the HL-LHC dataset.

One additional important avenue for the study of λ_{HHH} has been opened during Run-2: its probe through single Higgs production and decay [24, 25]. While HH production is sensitive to λ_{HHH} at LO, single H production is sensitive to it at NLO via loop corrections. Namely, electroweak quantum corrections featuring an anomalous trilinear H self-coupling could highly impact the H production cross section. Therefore, λ_{HHH}-dependent prediction of the H production and decay can be computed under the assumption that new physics couples to the SM via the Brout-Englert-Higgs potential and dominantly affects only the H self-couplings. These corrections have been derived for different production and decay modes [26]. The constraints on κ_λ extracted by the CMS Collaboration from single H production are compared to the HH combination in Fig. 6.23. This result showcases how the bounds on λ_{HHH} are already competitive between the two approaches.

The ATLAS Collaboration has also performed this measurement and has combined the single- and double-Higgs results in a comprehensive result on the κ_λ modifier [23]. In this effort, two different hypotheses are made: the first assumes all κ-modifiers to be SM-like and lets only κ_λ be the free parameter of the fit; the second assumes that only κ_{2V} is SM-like and lets all other modifiers be the floating parameters (κ_λ, κ_t, κ_b, κ_τ, and κ_V). The former hypothesis yields an observed (expected)

Table 6.4 Summary of the sensitivities to non-resonant SM HH production in the ATLAS and CMS Collaborations searches using the data collected at $\sqrt{s} = 13$ TeV. All quoted results are obtained with the full Run-2 dataset, corresponding to 139 and 138 fb^{-1} for the ATLAS and CMS Collaborations, respectively. The third and fourth columns present the observed and expected 95% CL upper limits on SM HH production normalized to the SM prediction; the fifth and sixth show the observed and expected 95% CL allowed interval for κ_λ when all other H couplings are set to their SM prediction. The last column reports the most up-to-date Reference for each search

Channel	Collab.	95% CL upper limit		Allowed κ_λ interval at 95% CL		References
		Observed	Expected	Observed	Expected	
bbbb	ATLAS	5.4	8.1	[−3.9, 11.1]	[−4.6, 10.8]	[18]
	CMS	6.4	4.0	(−2.3, 9.4)	(−5.0, 12.0)	[17, 22]
$bb\tau\tau$	ATLAS	4.7	3.9	–	–	[19]
	CMS	3.3	5.2	(−1.7, 8.7)	(−2.9, 9.8)	[16]
$bb\gamma\gamma$	ATLAS	4.2	5.7	(−1.5, 6.7)	(−2.4, 7.7)	[20]
	CMS	8.4	5.5	[−3.3, 8.5]	[−2.5, 8.2]	[21]
Multilepton	ATLAS	40	29	–	–	[27]
	CMS	21	19	(−6.9, 11.1)	(−6.9, 11.7)	[28]
$bbZZ$	ATLAS	–	–	–	–	
	CMS	32	40	(−8.8, 13.4)	(−9.8, 15.0)	[29]
Combined	ATLAS†	2.4	2.9	(−0.6, 6.6)	(−2.1, 7.8)	[23]
	CMS	2.5	3.4	(−1.24, 6.49)	(−1.23, 7.2)	[22]

† This combined result does not include the multilepton final state reported above

Table 6.5 Summary of the sensitivities to non-resonant SM VBF HH production in the ATLAS and CMS Collaborations searches using the data collected at $\sqrt{s} = 13$ TeV. All quoted results are obtained with the full Run-2 dataset, corresponding to 139 and 138 fb^{-1} for the ATLAS and CMS Collaborations, respectively. The third and fourth columns present the observed and expected 95% CL upper limits on SM HH production normalized to the SM prediction; the fifth and sixth show the observed and expected 95% CL allowed interval for κ_{2V} when all other H couplings are set to their SM prediction. The last column reports the most up-to-date Reference for each search

Channel	Collab.	95% CL upper limit		Allowed κ_{2V} interval at 95% CL		References
		Observed	Expected	Observed	Expected	
$bb\tau\tau$	ATLAS	–	–	–	–	
	CMS	124	154	(−0.4, 2.6)	(−0.6, 2.8)	[16]
$bb\gamma\gamma$	ATLAS	–	–	–	–	
	CMS	225	208	[−1.3, 3.5]	[−0.9, 3.1]	[21]
bbbb	ATLAS	130	130	[−0.03, 2.11]	[−0.05, 2.12]	[18]
	CMS	226	412	[0.62, 1.41])	[0.66, 1.37)	[17, 30]
Combined	ATLAS	–	–	–	–	
	CMS	–	–	(0.67, 1.38)	(−1.23, 7.2)	[22]

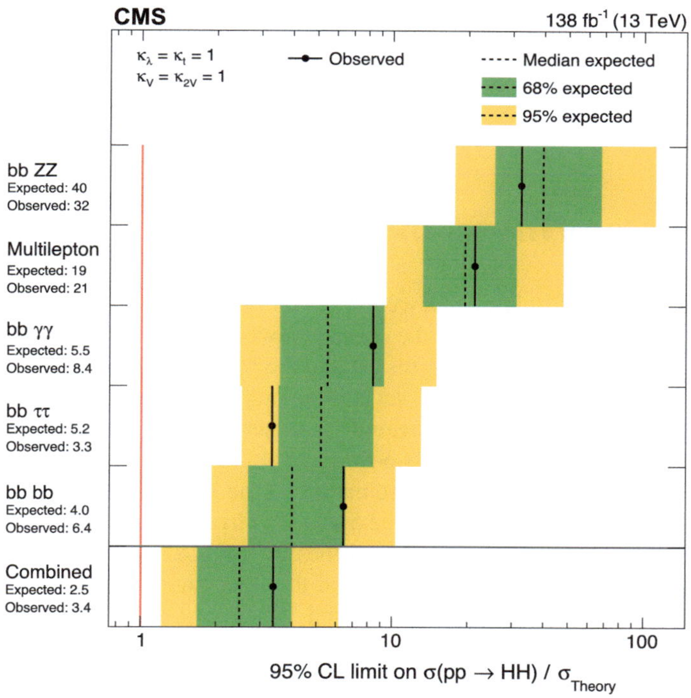

Fig. 6.22 The expected and observed limits on the ratio of experimentally estimated production cross section and the expectation from the SM (σ_{Theory}) in searches using different final states and their combination. The search modes are ordered, from upper to lower, by their expected sensitivities, from the least to the most sensitive. The overall combination of all searches is shown by the lowest entry. The red line represents the SM expectation of $\kappa_\lambda = 1$ [22]

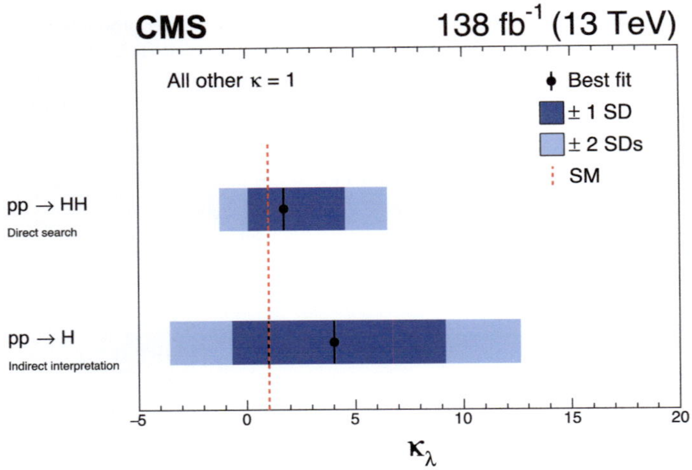

Fig. 6.23 Constraint on the Higgs boson self-coupling modifier κ_λ from single H production and HH production. The red dashed line represents the SM expectation of $\kappa_\lambda = 1$ [22]

$-0.4 < \kappa_\lambda < 6.3$ ($-1.9 < \kappa_\lambda < 7.6$) interval at 95% CL, the latter $-1.4 < \kappa_\lambda < 6.1$ ($-2.2 < \kappa_\lambda < 7.7$).

6.3 Prospects of HH Searches

The results presented in this chapter were derived by exploiting the full Run-2 dataset, amounting to 138 fb^{-1}. However, the context in which this Thesis has been carried out is that of the Run-3 operations restart and the preparation for the HL-LHC era. Both in Run-3 and at the HL-LHC, several upgrades of the hardware and software capabilities of the CMS detector currently represent and will represent paramount opportunities in the search for Higgs boson pairs.

The latest CMS Collaboration combination set the 95% CL expected upper limit on the HH production cross section to $2.5 \times \sigma_{\mathrm{HH}}^{\mathrm{SM}}$. Given the current statistical limitation of all HH searches, the evolution of the upper limits is expected to scale with inverse proportionality to the square root of the integrated luminosity $1/\sqrt{L}$. The Run-3 data-taking is currently underway, and a dataset of ~250 fb^{-1} is foreseen to be delivered by 2025. This means that combining the Run-2 and Run-3 datasets, an expected 95% CL upper limit of $\sim 1.5 \times \sigma_{\mathrm{HH}}^{\mathrm{SM}}$ can be foreseen, without accounting for any improvement in the analysis strategies. This result would already represent an important milestone as it could set stringent limits on some theoretical models like those discussed in Sect. 1.1.6 in the context of the electroweak phase transition and electroweak baryogenesis. This limit can be further improved via the amelioration of the analysis strategies, as discussed in the following, with particular attention to the HH → bbττ search.

The trigger strategy represents the first important improvement for Run-3 analyses. To this end, many advancements can be brought forward by exploiting additional HLT paths that are sensible to the HH signal and can increase its acceptance efficiency. In the case of the bbττ analysis, two examples are the use of single-τ and Missing Energy Transverse (MET) paths. Both these path types were already available during Run-2, but their use was avoided to simplify the derivation of the trigger scale factors. Nevertheless, within the analysis team, it has been recently shown that their inclusion can be achieved with dedicated studies of their overlap. Moreover, during Run-3, two new crucial improvements have been introduced. The first is a new HLT path targeting double-τ+jet, which is highly relevant owing to its topology. The second are the so-called *HH* and *VBF parking* [31], which consists of writing events directly to tape during the latest stages of an LHC fill when the rate is substantially decreased due to the lower instantaneous luminosity. This strategy also profits from the recent implementation at HLT of the new and highly performant PARTICLENET algorithm [32]. The joint use of all these improvements could boost the efficiency of the bbττ trigger strategy above 60%, and that of bbbb above 80%.

The second central advancement for the HH → bbττ search is represented by the possibility of targeting topologies of the final state currently not probed. The Run-2 HH → bbbb search [17, 30] has shown how considerable improvements can be achieved by exploiting boosted topologies. In this context, the first area of development is the exploration of the boosted ττ topology, which is now unexplored. Moreover, the `boosted` category of the analysis presented here only focuses on semi-boosted b jets, which are reconstructed both as small- and large-radius jets. Nevertheless, an important portion of the phase space is represented by jets which fully overlap and are reconstructed as a single small-radius jet. In both cases, the PARTICLENET algorithm represents our golden spear toward the exploration of fully boosted topologies.

These improvements, presented focusing on the HH → bbττ analysis, are just a subset of the extensive work that is being done within the CMS Collaboration to strive for ever-better limits on the HH production cross section. Assuming that a boost in analysis capabilities similar to that obtained during Run-2 will also be attained in Run-3, the target of a $\sim 1.5 \times \sigma_{\text{HH}}^{\text{SM}}$ limit could already be a pessimistic one. Owing to these crucial refinements, one substantial goal is now in sight by the end of Run-3: the exclusion of $\lambda_{\text{HHH}} = 0$. The attainment of this result at a 3σ significance level under the hypothesis of the existence of SM HH production would effectively amount to the first evidence of HH production at the LHC.

The Run-3 dataset will also be instrumental in probing possible BSM physics. Of particular interest will be to continue the exploration of the five-dimensional EFT phase space, already discussed in the pages of this Thesis, which gives indirect access to BSM physics currently outside the reach of the LHC.

Notwithstanding the crucial importance of striving for the results highlighted above, the ultimate goal for HH searches is set by the statistical power granted by the HL-LHC dataset. With a foreseen $\sim$4000 fb^{-1} integrated luminosity, the HL-LHC represents our first chance at observing the direct production of Higgs boson pairs. The most recent CMS Collaboration projection of the sensitivity to HH production showcased that a potential discovery could arrive at the end of the HL-LHC even in the worst scenario of adiabatic improvement of the analyses only based on the integrated luminosity increase [22]. In this context, to enhance further our chances of reaching such a goal and possibly attain it before the end of the HL-LHC, it is crucial to devise data-taking strategies that can guarantee the current selection efficiency and possibly extend it. In this context, of central importance for the $bb\tau\tau$ analysis will be the ability of the L1 trigger and HLT to reconstruct τ_h candidates efficiently. One important advancement in this direction will be the L1 trigger TAUMINATOR algorithm designed as part of this Thesis work.

The prospect of reaching the important HH observation goal in the 2040s could seem a *far and gloomy* possibility to many. Nevertheless, in my opinion, it only stresses the exciting times of technical development that await the high energy physics community. The already planned data-taking periods represent fertile ground for new hardware, software, and analysis strategy development. Moreover, it should not be forgotten how surprises could await us on the road, as BSM physics could be hiding in the large portions of phase space we have not probed yet.

In conclusion, the exploration of Higgs boson pair production represents both the present and the future of high energy physics. It constitutes a collective effort which requires the exploitation of several decay channels by both the CMS and ATLAS Collaborations. In this context, the $bb\tau\tau$ final state will play a central role owing to its many qualities and high sensitivity. In my opinion, the upcoming years of HH search will be very exciting and rewarding, and I hope to continue to be part of this effort.

References

1. Read AL (2000) Modified frequentist analysis of search results (the CL_s method). In: (2000). https://doi.org/10.5170/CERN-2000-005.81, https://cds.cern.ch/record/451614
2. Read AL (2002) Presentation of search results: the CL_s technique. J Phys G: Nucl Part Phys 28(10):2693–2704. https://doi.org/10.1088/0954-3899/28/10/313
3. Cowan G et al (2011) Asymptotic formulae for likelihood-based tests of new physics. Eur Phys J C 71(2). https://doi.org/10.1140/epjc/s10052-011-1554-0
4. Procedure for the LHC Higgs boson search combination in Summer 2011. Tech. rep. Geneva: CERN (2011). https://cds.cern.ch/record/1379837
5. Brochu E, Cora VM, de Freitas N (2010) A tutorial on Bayesian optimization of expensive cost functions, with application to active user modeling and hierarchical reinforcement learning. arXiv:1012.2599 [cs.LG]
6. Frazier PI (2018) A tutorial on Bayesian optimization. arXiv: 1807.02811 [stat.ML]
7. Neyman J, Pearson ES, Pearson K (1933) IX. On the problem of the most efficient tests of statistical hypotheses. Philos Trans R Soc London Ser A, Contain Pap Math Phys Character 231(694-706):289–337. https://doi.org/10.1098/rsta.1933.0009, https://royalsocietypublishing.org/doi/abs/10.1098/rsta.1933.0009

8. Junk T (1999) Confidence level computation for combining searches with small statistics. Nucl Instrum Methods Phys Res Sect A: Accel, Spectrometers, Detect Assoc Equip 434(2-3):435–443. https://doi.org/10.1016/s0168-9002(99)00498-2
9. Cousins RD (2013) Generalization of chisquare goodness-of-fit test for binned data using saturated models, with application to histograms. In: 2013
10. Carvalho A et al (2016) Higgs pair production: choosing benchmarks with cluster analysis. J High Energy Phys 2016(4):1–28. https://doi.org/10.1007/jhep04(2016)126
11. Buchalla G et al (2018) Higgs boson pair production in non-linear Effective Field Theory with full mt-dependence at NLO QCD. JHEP 09:57. https://doi.org/10.1007/JHEP09(2018)057, arXiv: 1806.05162
12. Capozi M, Heinrich G (2020) Exploring anomalous couplings in Higgs boson pair production through shape analysis. JHEP 03:91. https://doi.org/10.1007/JHEP03(2020)091, arXiv: 1908.08923
13. CMS Collaboration (2018) Search for Higgs boson pair production in events with two bottom quarks and two tau leptons in proton-proton collisions at $\sqrt{s} = 13$TeV. Phys Lett B 778:101–127. https://doi.org/10.1016/j.physletb.2018.01.001
14. CMS Collaboration (2018) Observation of the Higgs boson decay to a pair of τ leptons with the CMS detector. Phys Lett B 779:283–316. https://doi.org/10.1016/j.physletb.2018.02.004
15. Cadamuro L (2018) Search for Higgs boson pair production in the $BBTauTausps$ decay channel with the CMS detector at the LHC. Springer. ISBN: 978-3-030-04055-0. https://doi.org/10.1007/978-3-030-04055-0
16. CMS Collaboration (2023) Search for nonresonant Higgs boson pair production in final state with two bottom quarks and two tau leptons in proton-proton collisions at $\sqrt{s} = 13$TeV. Phys Lett B 842:137531. ISSN: 0370-2693, https://doi.org/10.1016/j.physletb.2022.137531, https://www.sciencedirect.com/science/article/pii/S0370269322006657
17. CMS Collaboration (2022) Search for Higgs Boson pair production in the four b quark final state in proton-proton collisions at $\sqrt{s} = 13$TeV. Phys Rev Lett 129:081802. https://doi.org/10.1103/PhysRevLett.129.081802
18. ATLAS Collaboration (2023) Search for nonresonant pair production of Higgs bosons in the $b\bar{b}b\bar{b}$ final state in pp collisions at $\sqrt{s} = 13$ TeV with the ATLAS detector. arXiv: 2301.03212 [hep-ex]
19. ATLAS Collaboration (2023) Search for resonant and non-resonant Higgs boson pair production in the decay channel using 13 TeV pp collision data from the ATLAS detector. J High Energy Phys 2023(7). https://doi.org/10.1007/jhep07(2023)040
20. ATLAS Collaboration (2022) Search for Higgs boson pair production in the two bottom quarks plus two photons final state in pp collisions at $\sqrt{s} = 13$ TeV with the ATLAS detector. Phys Rev D 106:052001. https://doi.org/10.1103/PhysRevD.106.052001
21. CMS Collaboration (2021) Search for nonresonant Higgs boson pair production in final states with two bottom quarks and two photons in proton-proton collisions at $\sqrt{s} = 13$TeV. J High Energy Phys 2021(3). https://doi.org/10.1007/jhep03(2021)257
22. CMS Collaboration (2022) A portrait of the Higgs boson by the CMS experiment ten years after the discovery. Nature 607(7917):60–68. https://doi.org/10.1038/s41586-022-04892-x
23. ATLAS Collaboration (2023) Constraints on the Higgs boson self-coupling from single- and double-Higgs production with the ATLAS detector using pp collisions at s=13 TeV. Phys Lett B 843:137745. ISSN: 0370-2693. https://doi.org/10.1016/j.physletb.2023.137745, https://www.sciencedirect.com/science/article/pii/S0370269323000795
24. Degrassi G et al (2016) Probing the Higgs self coupling via single Higgs production at the LHC. J High Energy Phys 2016(12). https://doi.org/10.1007/jhep12(2016)080
25. Maltoni F et al (2017) Trilinear Higgs coupling determination via single-Higgs differential measurements at the LHC. Eur Phys J C 77(12). https://doi.org/10.1140/epjc/s10052-017-5410-8
26. Monti F et al (2022) Modelling of the single-Higgs simplified template cross-sections (STXS 1.2) for the determination of the Higgs boson trilinear self-coupling. Tech. rep. Geneva: CERN, 2022. https://cds.cern.ch/record/2803606

27. ATLAS Collaboration (2020) Search for non-resonant Higgs boson pair production in the bb$\ell\nu\ell\nu$ final state with the ATLAS detector in pp collisions at $sqrts = 13$TeV. Phys Lett B 801:135145. ISSN: 0370-2693. https://doi.org/10.1016/j.physletb.2019.135145, https://www.sciencedirect.com/science/article/pii/S0370269319308676
28. CMS Collaboration (2023) Search for Higgs boson pairs decaying to $\mathbf{WW^*WW^*}$,$\mathbf{WW^*}\tau\tau$ and $\tau\tau\tau\tau$ in proton-proton collisions at $\sqrt{s} = 13$TeV. J High Energy Phys 2023(7). https://doi.org/10.1007/jhep07(2023)095
29. CMS Collaboration (2023) Search for nonresonant Higgs boson pair production in the four leptons plus twob jets final state in proton-proton collisions at $\sqrt{s} = 13$ TeV. J High Energy Phys 2023(6). https://doi.org/10.1007/jhep06(2023)130
30. CMS Collaboration (2023) Search for nonresonant pair production of highly energetic Higgs bosons decaying to bottom quarks. Phys Rev Lett 131:041803. https://doi.org/10.1103/PhysRevLett.131.041803
31. CMS Collaboration (2023) Novel strategy targeting HH and HHH production at high level trigger in Run 3. In: CMS Detector Performance Summaries. https://cds.cern.ch/record/2868787
32. Qu H, Gouskos L (2020) Jet tagging via particle clouds. Phys Rev D 101(5). https://doi.org/10.1103/physrevd.101.056019

Chapter 7
Conclusions

On July 4th 2012, the ATLAS and CMS Collaborations at CERN announced the discovery of a scalar particle compatible with the Higgs boson (H) predicted by the Standard Model (SM) of particle physics. With the value of the H mass experimentally measured to be 125.08 ± 0.12 GeV, the structure of the Brout-Englert-Higgs (BEH) scalar field potential and the strength of the H self-coupling (λ_{HHH}) are precisely predicted. While the measured properties are so far consistent with the SM, the determination of λ_{HHH} provides an independent test of the SM and is fundamental to probe experimentally the shape of the scalar potential that is at the base of the electroweak symmetry breaking mechanism. The precise measurement of the H self-coupling will also serve to shed light on the current open question of the SM, raging from the (meta)stability problem of the Universe to the observed matter-antimatter asymmetry.

On December 3rd 2018, the Run-2 operation of the LHC came to an end after three years of outstanding operation. The dataset collected in this period represents the largest ever accumulated at a particle accelerator at the unprecedented centre-of-mass energy of $\sqrt{s} = 13$ TeV. Exploiting this dataset to target the precision measurement of the properties of the H and detecting possible signs of physics Beyond the SM (BSM) has been the central effort of the particle physics community in the past years. Among these measurements, the search for Higgs boson pair (HH) production is one of the most sought-after goals, which allows for the direct investigation of λ_{HHH} and the shape of the BEH potential.

This Thesis presented the search for HH production in the final state with two bottom quarks and two τ leptons (bb$\tau\tau$) exploiting the full Run-2 statistics for a total of 138 fb^{-1} analyzed. The predictions of the SM have been tested for inclusive HH production as well as for exclusive VBF HH production. The observed (expected) 95% CL upper limit on inclusive HH production has been set to 3.3 (5.2) times the SM prediction, while the 95% CL upper limit on exclusive VBF HH production has been set to 124 (154) times the SM prediction; the former is the current second most stringent limit while the latter is the tightest limit ever set. The analysis also sets

J. Motta, *Development of Machine Learning τ Trigger Algorithms and Search for Higgs Boson Pair Production*, Springer Theses,
https://doi.org/10.1007/978-3-031-96288-2_7

observed (expected) 95% CL confidence intervals on the H self-coupling κ-modifier $-1.7 < \kappa_\lambda < 8.7$ ($-2.9 < \kappa_\lambda < 9.8$), and the κ-modifier of the coupling between two H and two vector bosons $0.4 < \kappa_{2V} < 2.6$ ($-0.6 < \kappa_{2V} < 2.8$), both derived from limits on the HH production cross section times the bb$\tau\tau$ branching fraction. Moreover, one-dimensional maximum likelihood scans are employed to set observed 95% CL confidence intervals on the κ-modifier of the Yukawa coupling between the t quark and the H $-0.90 < \kappa_t < +1.01$ for a best fit value $\kappa_t = 0.01$, and on the κ-modifier of the coupling between two H and one vector boson $-1.46 < \kappa_V < +1.55$ for a best fit value $\kappa_V = 0.86$. The results are also interpreted in the EFT Lagrangian framework to set limits on the production cross section of 20 different BSM benchmarks.

On July 5th 2022, the Run-3 of the LHC officially started, and it is expected to last until the end of 2025. This third operational run represents a very important data-taking period for the HH production searches; the dataset accumulated during Run-3 is crucial for two main reasons: it can allow for the possible first evidence of HH production in the form of an exclusion of $\lambda_{HHH} = 0$ at a 3σ significance level, and it can be exploited to test innovative techniques that will prove fundamental for the success of the ensuing High-Luminosity LHC (HL-LHC) era. In this context, of particular importance is the performance of the Level-1 (L1) trigger, on which relies the sensitivity of all analyses. This Thesis presented two major contributions to the commissioning of the L1 trigger for Run-3. The first contribution has been the development of a simple and yet more informative approach to optimizing the parameters of the L1 τ_h and e/γ algorithms, backed by a new estimation of the isolation of the L1 candidates more resilient to the limited statistical power of the dataset used in the process. The importance of this work has been shown, highlighting the excellent performance of both L1 algorithms during the 2022 data-taking. The second contribution has been the development of the CALIBRATON algorithm, an innovative technique based on a neural network architecture aimed at calibrating single detector objects by exploiting offline reconstructed electrons and jets from data. The expected performance of the CALIBRATON in Run-3 has been presented, and its possible application in future contexts has been highlighted.

In 2029, the HL-LHC era will start, and the CMS experiment will collect a total of 4000 fb^{-1} over the ensuing ten years, allowing for the investigation of many open questions on the SM and its possible BSM extensions. Concurrently, the L1 trigger will undergo its ultimate upgrade, implementing powerful state-of-the-art FPGAs and exploiting an increased latency. This Thesis presented the design and firmware implementation of the innovative TAUMINATOR L1 trigger algorithm for the reconstruction, calibration, and identification of τ_h candidates, which is currently considered one of the baselines for the Phase-2 L1 trigger. The algorithm is based on a convolutional neural network architecture and addresses the τ_h identification problem with an image recognition approach. The TAUMINATOR design is strictly intertwined with the need for its firmware implementation, and the technical solutions implemented mirror the need for contained FPGA resources usage. The algorithm has been implemented in firmware within the scope of this Thesis, and the bit-wise agreement with simulation is attained. The results show a significant increase in the

CMS triggering capabilities on τ_h candidates with respect to traditional triggering methods, with a large positive impact on the ambitious CMS physics program.

This Thesis was opened by the following question: *What are the basic constituents that compose the Universe, and what are the fundamental laws that regulate them?*, which can be identified as the *red thread* underlying the history of physics. As this Thesis comes to an end, the question remains as valid as it was 2500 years ago. The SM has proven to be a remarkable theory, capable of predicting and explaining the collider physics of the past 60 years, and constitutes our current best description of the subatomic scales. Nevertheless, at the time of writing this Thesis, the exploration of the TeV scale has just started, and the surprises to be found are endless. HH searches constitute the keystone of the CMS physics program to further test the SM, and the scalar sector represents our golden spear in the search for BSM physics. We do not know what the Run-3 and HL-LHC data will reveal, nor do we know if the SM will stand tall as our best theory; all I know is that it will be fascinating to follow the best attempts of the CMS Collaboration to prove the SM wrong and open the door to a new era of ever more exciting physics.

[...] all laws, all theories, remain essentially tentative, or conjectural, or hypothetical, even when we feel unable to doubt them any longer.
Karl Raimund Popper

Jona Motta | Curriculum Vitae

jona.motta@cern.ch
0000-0003-0985-913X
github.com/jonamotta
gitlab.cern.ch/jmotta

J. Motta, *Development of Machine Learning τ Trigger Algorithms and Search for Higgs Boson Pair Production*, Springer Theses,
https://doi.org/10.1007/978-3-031-96288-2

Experience

University of Zürich CERN, Geneva, Switzerland
Postdoctoral Researcher 2024–Ongoing
CMS group of Prof. Dr. Florencia Canelli.
Search for Higgs boson pair production, search for heavy Higgs boson candidates, development of electronics modules for the Phase-2 upgrade of the Inner Tracker, and development statistical methods in data analysis.
Supervision of Ph.D. and M.Sc. students.

CMS Collaboration CERN, Geneva, Switzerland
HLepRare group convener 2025–Ongoing
Level-3 conevenrship of the HLepRare group within the Higgs PAG (Physics Analysis Group) supervising, coordinating, and reviewing all CMS analyses targeting Higgs bosons decays into leptons or rare decay channels

CMS Collaboration CERN, Geneva, Switzerland
HIG PAG Trigger Contact 2025–Ongoing
Contact person in the Higgs PAG (Physics Analysis Group) supervising the development and performance L1+HLT trigger algorithms/strategies useful for analyses targeting the Higgs bosons and related BSM particles

University of Zürich Zürich, Switzerland
Students Supervision 2024–Ongoing
Supervisor of three Ph.D. students and one Master student working on the search for Higgs boson pair production in several final states and the search for heavy Higgs boson candidates

University of Zürich Zürich, Switzerland
Teaching Assistant 2024–Ongoing
Frontal lectures on the fundamentals of modern physics, in particular: optics, electromagnetism, waves, relativity, and quantum mechanics
Laboratory supervision of experiments for the study of the Zeeman effect, the Compton scattering, the Stern-Gerlach effect, electron diffraction, the Mach-Zehnder interferometer, and the laser resonator.
Referents: Prof. Christof Aegerter and Prof. Matthias Hengsberger

Laboratoire Leprince-Ringuet (École Polytechnique) Palaiseau, France

Ph.D. in High Energy Physics 2020–2023

Search for Higgs boson pair production in the bb$\tau\tau$ decay channel, development of machine learning based τ trigger algorithms, development of deep-learning detector calibration models, firmware implementation of fast-machine learning techniques.
Supervisors: Dr. Alexandre Zabi, Dr. Roberto Salerno

CMS Collaboration CERN, Geneva, Switzerland

Level-1 Trigger Phase-1 e/γ *and* τ *DPG Contact* 2022–2024

Level-3 DPG (Detector Performance Group) contact person for the development and performance monitoring of the Level-1 trigger e/g and τ objects during the Run 3 data taking.
Supervising and coordinating a team of 8 collaborators from 5 institutes.

École Polytechnique Palaiseau, France

Teaching Assistant 2020–2023

Frontal and practical lectures on experimental aspects and analysis techniques of high-energy physics, with particular attention to LHC experiments.
Laboratory supervision of experiments for the detection of cosmic muons, the study of matter-antimatter annihilation, the detection of β-radiation, and the study of the Compton effect.
Referents: Prof. Alexandre Zabi

École Polytechnique Palaiseau, France

MSc Thesis Supervision 2021

During my Ph.D. I co-supervised the M.Sc. thesis project of Mathis Gueneau, titled "Développement d'un algorithme de déclenchement sur les leptons τ utilisant les informations du calorimètre à haute granularité de CMS pour la Phase-2 du LHC". The thesis has been awarded the "Felicitations du Jury" for the best internship work in the academic year.

Research Awards

CMS Ph.D. Thesis Award 2023
Awarded by the CMS Collaboration *"recognizing exceptional doctoral research within the CMS Collaboration that pushes the boundaries of high-energy physics"* and *"rewarding the most outstanding theses for the CMS experiment, providing inspiration and motivation for the new generations of students in CMS and elsewhere"*

CMS Award 2022
Awarded by the CMS Collaboration with the motivation: *"For the outstanding contribution to the development and optimisation of the Level-1 τ lepton trigger algorithm for the Run 3 of the LHC and the HL-LHC"*

Best Parallel Talk Award
Award at the Higgs conference, Stony Brook 2021. Title: "Searches for Resonant and Non-resonant HH/YH Production in Final States with Tau Leptons and W Bosons at CMS".

Education

ETH & École Polytechnique Zürich, Switzerland & Palaiseau, France
Joint MSc in High Energy Physics 2018–2020
ECTS: 120—GPA: 5.96/6—Final Grade: 5.96/6 with Distinction
Dissertation: Study of the Higgs boson self-coupling in the $bb\tau\tau$ decay channel. Supervisor: Dr. Roberto Salerno
Project: Testing Lepton Flavour Universality in semi-leptonic decays of the B_c^+ meson: a feasibility study in CMS. Supervisors: Prof. Dr. Günther Dissertori, Dr. Riccardo Manzoni

Università di Milano Bicocca Milano, Italy
BSc in Physics 2015–2018
ECTS: 180—GPA: 29/30—Final Grade: 110/110 cum Laude
Dissertation: Performance studies for Higgs pair searches at LHC with the CMS detector. Supervisor: Dr. Pietro Govoni, Dr. Chiara Amendola

Selected List of Publications

As a member of the CMS Collaboration, I am an author of all the papers published by the Collaboration itself.
Here is a list of public results of which I am either a single author or a major contributor:

1. Jona Motta, "Overview of the HL-LHC Upgrade for the CMS Level-1 Trigger". PoS EPS-HEP2023 (2024), p. 534
2. Jona Motta, "Development and firmware implementation of a Machine Learning based hadronic Tau lepton Level-1 Trigger algorithm in CMS for the HL-LHC". PoS EPS-HEP2023 (2024), p. 590

3. CMS Collaboration, “Search for nonresonant Higgs boson pair production in final state with two bottom quarks and two tau leptons in proton-proton collisions at $\sqrt{s} = 13\,\text{TeV}$”.
   Physics Letters B 842 (2023), p. 137531
4. CMS Collaboration, “Hadronic Tau Reconstruction in the CMS Phase-2 Level-1 Trigger using NNs with Calorimetric Information”.
   CMS-DP-2023-062
5. CMS Collaboration, “Performance of Level-1 Trigger e/gamma and tau in Run 3 (2023)”.
   CMS-DP-2023-055
6. CMS Collaboration, “Performance of Level-1 trigger e/gamma and tau in Run 3 (2022)”.
   CMS-DP-2023-008
7. Jona Motta, “Development of a L1 tau lepton trigger algorithm for the HL-LHC using the CMS high-granularity calorimeter information”.
   PoS LHCP2021 (2021), p. 228

International Conferences

COMETA workshop on EFT in HH and VBS at LHC Paris, France
Invited talk March 2025
Title: Experimental review of the EFT MC modelling in HH searches

Higgs Conference Uppsala, Sweden
Parallel talk November 2024
Title: Di-Higgs Searches at CMS

Higgs Hunting Paris, France
Plenary talk September 2023
Title: Higgs Self-coupling

EPS-HEP2023 Hamburg, Germany
Parallel talk August 2023
Title: Overview of the HL-LHC Upgrade for the CMS Level-1 Trigger

EPS-HEP2023 Hamburg, Germany
Poster August 2023
Title: Development and Firmware Implementation of a Machine Learning Based Hadronic τ Lepton Level-1 Trigger Algorithm in CMS for the HL-LHC

IRN Terascale Bonn, Germany
Plenary talk March 2022
Title: Search for Non-resonant Higgs Boson Pairs Production in the bb$\tau\tau$ Final State in CMS

Lake Louise Winter Institute Lake Louise, Canada
Plenary talk February 2022
Title: Performing Precision Measurements and New Physics Searches at the HL-LHC with the Upgraded CMS Level-1 Trigger

Higgs Conference Virtual World
Parallel talk October 2021
Title: Searches for Resonant and Non-resonant HH/YH Production in Final States with τ Leptons and W Bosons at CMS
Awarded the Best Parallel Talk prize

LHCP Conference Virtual World
Poster July 2021
Title: Development of a L1 τ Lepton Trigger Algorithm for the HL-LHC Using the CMS High-Granularity Calorimeter Information